Free Energy Computations

A Mathematical Perspective

Free Energy Computations

A Mathematical Perspective

Tony Lelièvre
Ecole des Ponts ParisTech & INRIA Rocquencourt, France

Mathias Rousset
INRIA Lille - Nord Europe, France

Gabriel Stoltz
Ecole des Ponts ParisTech & INRIA Rocquencourt, France

ICP

Imperial College Press

Published by

Imperial College Press
57 Shelton Street
Covent Garden
London WC2H 9HE

Distributed by

World Scientific Publishing Co. Pte. Ltd.
5 Toh Tuck Link, Singapore 596224
USA office: 27 Warren Street, Suite 401-402, Hackensack, NJ 07601
UK office: 57 Shelton Street, Covent Garden, London WC2H 9HE

Library of Congress Cataloging-in-Publication Data
Lelièvre, Tony.
Free energy computations : a mathematical perspective / by Tony Lelièvre, Gabriel Stoltz & Mathias Rousset.
p. cm.
Includes bibliographical references and index.
ISBN-13: 978-1-84816-247-1 (hardcover : alk. paper)
ISBN-10: 1-84816-247-2 (hardcover : alk. paper)
1. Gibbs' free energy. 2. Statistical physics--Mathematical models. I. Stoltz, Gabriel. II. Rousset, Mathias. III. Title.
QC318.E57L45 2010
536'.7--dc22

2010005252

British Library Cataloguing-in-Publication Data
A catalogue record for this book is available from the British Library.

Printed in Singapore.

Preface

The computation of free-energy differences is a very important and active research field in computational statistical physics. Being applied mathematicians, we were unable to find a textbook that suited our needs when we first approached numerical problems in computational statistical physics. The references we found were either theoretical statistical physics books, focusing on model problems treated analytically and perturbations thereof; or textbooks for practitioners, with recipes and comments on the physical models, but almost no mathematical analysis of the numerical techniques. We hope that the present book contributes to filling this gap and usefully complements some of the recent references concerning the long-time integration of Hamiltonian dynamics (such as [Hairer *et al.* (2006)]).

The audience we have in mind while writing these lines is composed both of mathematicians and scientists from the applied communities (physics, chemistry, biology, etc.), who use free-energy techniques as one tool among many to study the complex systems they are interested in. We conceive these notes as a self-contained presentation of what we believe are currently the most standard computational methods for free-energy computations. We hope that this presentation will be of interest to researchers in the field, while still being accessible for graduate students.

Free-energy computation is an opportunity for mathematicians to study many theoretical concepts and numerical strategies, such as techniques to sample multimodal probability measures, constrained stochastic dynamics to sample measures on submanifolds, adaptive importance sampling strategies, etc. In this book, we insist on the numerical analysis of the methods at hand, giving error estimates or rates of convergence.

Besides, for those interested in the applications, this book is an opportunity to learn more about the mathematical underpinning of the numerical techniques used on a daily basis for the computation of free-energy differences. We want to highlight the similarities between techniques presented as very different in nature in the current literature. We hope that a more abstract viewpoint on the problems at hand will be inspiring for practitioners. To this audience, this book may also be seen as a companion book, more biased towards mathematical analysis, of the recent review book on free energy methods [Chipot and Pohorille (2007b)].

Let us also insist at this point on the many possible application fields of the techniques presented in this book. Besides the obvious application domains where a free-energy difference is an important quantity *per se* since it has an experimental meaning (biology, physics, etc.), there also exist scientific fields where ratios of partition functions are required for computational purposes. An example is computational statistics (in particular Bayesian statistics). Moreover, it is often the case that some adequate importance sampling function is needed to enhance the sampling of multimodal probability distributions, and it is in general difficult to think of good candidates for high-dimensional problems. The free energy can be used as an efficient and automatic importance sampling function, once some "slowly evolving", "frustrated", or "metastable" degree of freedom has been identified. In particular, adaptive methods are very interesting since they provide an adaptive importance sampling strategy.

The book is organized linearly, but sometimes we need to anticipate on notions presented later on to motivate some techniques or concepts (especially in the introductory chapter). The notation is homogeneous throughout the book, and, for the reader's convenience, it is summarized in the Appendix. This will hopefully help the reader to keep track of the objects manipulated, in particular in the sections on constrained processes.

We now briefly describe the structure of the book, which is based on the mathematical classification explained at the end of Chapter 1. The introductory chapter presents the notions of statistical physics which will be constantly used in this book. It also gives the definition of *free energy*. This presentation is deliberately very different from that of standard physics textbooks, since our aim is primarily to describe how to compute average properties predicted by statistical physics, rather than motivating the physical relevance of these expressions. Chapter 2 presents in its first two sections standard techniques to compute averages in computational statistical physics, before describing a first set of techniques to compute free-energy differences using only these standard methods (*free-energy perturbation, histogram methods*). Chapter 3 is more technical, and deals mainly with *constrained processes*, which are a fundamental tool for the so-called *thermodynamic integration* method. *Nonequilibrium processes* are considered in Chapter 4, with variations and extensions around the famous *Jarzynski equality*. *Adaptive methods* are at the heart of Chapter 5. Our understanding of these recently proposed methods is not complete to date. We hope that our general mathematical presentation will motivate further research. The last chapter is a short presentation of *selection strategies*, which can be used as a complement to other methods. Chapter 5 is the part of the book where most of the open problems are listed.

Software deserves some comment. Most research groups in computational statistical physics are developing an in-house code, or are organized around the use of a well-known simulation package, dedicated to a family of applications. This code is often efficient, flexible, and allows one to treat challenging systems of current interest. These characteristics mean that the source files are not easily understood and modified, and that it is difficult to precisely know which numerical methods are used, and how they are actually implemented. For some users, the code is a black-box, whose results are too quickly trusted and too rarely questioned. We believe that running a code as a black-box and blindly trusting the outcome is risky. Questioning the validity of the results is a necessity. Besides, the reproducibility of the numerical simulations is as fundamental a rule as the reproducibility of any experiment in experimental sciences. For these reasons, we propose a series of codes for the two running examples considered throughout the book. They are freely available on Gabriel Stoltz's webpage[1] so that the numerical computations presented throughout this book can be double-checked. Our programs are not implemented in a computationally optimal way. On the other hand, the simplicity of the code and the documentation should make it simple enough for the reader to fully understand the details of one possible implementation of the methods and to test different numerical strategies.

We conclude this preface by emphasizing that computational statistical physics in general, and free-energy based techniques in particular, are a relatively recent research domain in applied mathematics. Many questions therefore remain open. Any comments about the scientific content, the pedagogical or non-pedagogical approaches used, typos, etc., are welcome! We

[1] Visit the webpage http://cermics.enpc.fr/~stoltz/ or download directly the file from http://cermics.enpc.fr/~stoltz/FreeEnergyCodes.tar.gz.

are very much indebted to many of our coworkers, for a careful re-reading of parts of this book: Manuel Athènes, Chris Chipot, John Chodera, Claude Le Bris, Frédéric Legoll, Kimiya Minoukadeh, and Raphaël Roux. We finally acknowledge the hospitality of the Hausdorff Institute of Mathematics, in Bonn (Germany), where we three authors attended a research program in spring 2008 and where the writing of this book was initiated. We are also grateful to the Banff International Research Station (Canada), where we were invited to a timely and stimulating workshop on "Mathematical and Numerical Methods for Free Energy Calculations in Molecular Systems" in June 2008.

Tony Lelièvre, Mathias Rousset and Gabriel Stoltz
Paris, January 2010

Contents

Chapter 1

Introduction

This chapter presents the physical and mathematical framework to understand the basics of molecular simulation and computational statistical physics techniques. Section 1.1 recalls the aims of computational statistical physics, gives some historical landmarks, and provides the orders of magnitude of the quantities to be computed. Section 1.2 is a short summary of the most important concepts of statistical physics which will be of constant use throughout this book. It is decomposed into three parts: the static description of microscopic systems (unknowns, boundary conditions, interaction potentials), the dynamics of isolated systems (the Hamiltonian dynamics), and some elements on thermodynamic ensembles. We are then in a position to define free energies in Section 1.3, discuss their relationships with metastability issues, and finally classify the most common methods currently used to compute free energy differences in terms of the mathematical objects at hand. This classification is the basis of the construction of the book, see Section 1.4 for more details.

1.1 Computational statistical physics: some landmarks

Before giving a detailed mathematical framework of computational statistical physics, we first describe the scientific context, by recalling in Section 1.1.1 some order of magnitudes for the quantities under investigation, and by expliciting in Section 1.1.2 what we understand to be the current aims of molecular simulation.

Table 1.1 Some important physical constants or quantities in quantum and statistical physics.

Physical constant	Usual notation	Value
Avogadro number	$\mathcal{N}_A$	6.02×10^{23}
Boltzmann constant	k_B	1.381×10^{-23} J/K
Reduced Planck constant	$\hbar$	1.054×10^{-34} Js
Elementary charge	e	1.602×10^{-19} C
Electron mass	m_e	9.11×10^{-31} kg
Proton mass	m_p	1.67×10^{-27} kg
Electron-Volt	eV	1.602×10^{-19} J

1.1.1 *Some orders of magnitude*

In the framework of statistical physics, matter is most often described at the atomic level, either in a quantum or classical framework. Some of the concepts developed in this introduction may however be used in other physical frameworks than molecular simulation (for instance, the Hamiltonian dynamics presented in Section 1.2.2 is the fundamental evolution equation in celestial mechanics).

In this book, only classical systems are considered. Some important physical constants are recalled in Table 1.1. From those constants, the orders of magnitudes of the classical description of matter at the microscopic level can be inferred. The typical distances are expressed in Å (10^{-10} m), the energies are of the order of $k_B T \simeq 4 \times 10^{-21}$ J at room temperature, and the typical times are of the order of 10^{-15} s when the proton mass is the reference mass.

The orders of magnitude used in the microscopic description of matter are far from the orders of magnitude of the macroscopic quantities we are used to. For instance, the number of particles under consideration in a macroscopic sample of material is of the order of the Avogadro number $\mathcal{N}_A \sim 10^{23}$. For practical numerical computations of matter at the microscopic level, following the dynamics of every atom would require simulating $\mathcal{N}_A$ atoms and performing $O(10^{15})$ time integration steps, which is of course impossible! These numbers should be compared with the current orders of magnitude of the problems that can be tackled with classical molecular simulation, such as the simulation of the complete satellite tobacco mosaic virus [Freddolino *et al.* (2006)], which involved 1 million atoms over 50 ns, or the folding simulations of the Villin headpiece,[1] where

[1]See the website of the Folding@Home project: `http://folding.stanford.edu/`

a trajectory of 500 μs was integrated for 2×10^4 atoms.

To give some insight into such large numbers, it is helpful to compute the number of moles of water on earth. Recall that one mole of water corresponds to 18 mL, so that a standard glass of water contains roughly 10 moles, and a typical bathtub contains 10^5 mol. On the other hand, there are approximately 1.3×10^{18} m^3 of water in the oceans, *i.e.* 7.2×10^{22} mol, a number comparable to the Avogadro number. This means that inferring the macroscopic behavior of physical systems described at the microscopic level by the dynamics of several millions of particles only is like inferring the ocean's dynamics from hydrodynamics in a bathtub...

Describing the macroscopic behavior of matter knowing its microscopic description therefore seems out of reach. Statistical physics allows us to bridge the gap between microscopic and macroscopic descriptions of matter, at least on a conceptual level. The question is whether the estimated quantities for a system of N particles correctly approximate the macroscopic property, formally obtained in the thermodynamic limit $N \to +\infty$ (the density being kept fixed). In some cases, in particular for simple homogeneous systems, the macroscopic behavior is well approximated from small-scale simulations, see Section 1.1.2.1. However, the convergence of the estimated quantities as a function of the number of particles involved in the simulation should be checked in all cases.

1.1.2 *Aims of molecular simulation*

Despite its intrinsic limitations on spatial and timescales, molecular simulation, has been used and developed over the past 50 years, and its number of users keeps increasing. As we understand it, it has two major aims nowadays.

First, it can be used as a *numerical microscope*, which allows us to perform "computer" experiments. This was the initial motivation for simulations at the microscopic level: physical theories were tested on computers. This use of molecular simulation is particularly clear in its historic development, which was triggered and sustained by the physics of simple liquids. Indeed, there was no good analytical theory for these systems, and the observation of computer trajectories was very helpful to guide the physicists' intuition about what was happening in the system, for instance the mechanisms leading to molecular diffusion. In particular, the pioneering works on Monte-Carlo methods [Metropolis *et al.* (1953)], and the first molecular dynamics simulation [Alder and Wainwright (1956)] were performed because of such motivations. Today, understanding the behavior of matter at the microscopic level can still be difficult from an experimental viewpoint (because of the high resolution required, both in time and in space), or because we simply do not know what to look for! Numerical simulations are then a valuable tool to test some ideas or obtain some data to process and analyze in order to help assessing experimental setups. This is particularly true for current nanoscale systems.

Another major aim of molecular simulation, maybe even more important than the previous one, is to compute macroscopic quantities or thermodynamic properties, typically through averages of some functionals of the system. In this case, molecular simulation is a way to obtain *quantitative* information on a system, instead of resorting to approximate theories, constructed for simplified models, and giving only qualitative answers. Sometimes, these properties are accessible through experiments, but in some cases only numerical computations are possible since experiments may be unfeasible or too costly (for instance, when high pressure or large temperature regimes are considered, or when studying materials not yet synthesized). More generally, molecular simulation is a tool to explore the links between the microscopic and macroscopic properties of a material, allowing to address modelling questions such as "Which microscopic ingredients are necessary (and which are not) to observe a given macroscopic behavior?"

1.1.2.1 *An example: the equation of state of Argon*

Let us detail to some extent the second approach, and illustrate it with a simple but realistic example. We consider microscopic systems composed of N particles (typically atoms, *i.e.* nuclei together with their electronic clouds), described by the positions of the particles $q = (q_1, \cdots, q_N) \in \mathcal{D}$ and the associated momenta $p = (p_1, \cdots, p_N) \in \mathbb{R}^{3N}$. For physical and biological systems currently studied, N is typically between 10^3 and 10^9. The vector (q, p) is called the *microscopic state* or the *configuration* of the system.

In the framework of statistical physics, macroscopic quantities of interest are written as averages over thermodynamic ensembles, which are probability measures on all the admissible microscopic configurations:

$$\mathbb{E}_\mu(A) = \int_{T^*\mathcal{D}} A(q,p)\, \mu(dq\, dp). \tag{1.1}$$

In this expression, the function A is called an *observable*. The position variable $q = (q_1, \ldots, q_N)$ belongs to $\mathcal{D}$, which is called the configuration space.

The set $\mathcal{D}$ is an open subset (possibly the whole) of $\mathbb{R}^n$ with $n = 3N$, or $\mathcal{D} = \mathbb{T}^n$ (where $\mathbb{T} = \mathbb{R}/\mathbb{Z}$ denotes the one-dimensional torus). The choice of $\mathcal{D}$ depends on the boundary conditions at hand, see Section 1.2.1.1. For the two choices mentioned above, the momentum variable $p = (p_1, \ldots, p_N)$ belongs to $\mathbb{R}^n$, so that the cotangent space $T^*\mathcal{D}$ used in (1.1) can be identified with $\mathcal{D} \times \mathbb{R}^n$. The set of all possible microscopic configurations (q, p) is called the *phase space*. The probability measure μ has support on the phase space and depends on the thermodynamic ensemble used, see Section 1.2 for further precision on the most common choices.

Remark 1.1 (Generalization to other configuration spaces). *All the results presented in this book may be generalized to the case when the configuration space $\mathcal{D}$ is not $\mathbb{R}^n$, but some open subset of $\mathbb{R}^n$, with a potential energy function which goes sufficiently fast to ∞ on $\partial\mathcal{D}$ to prevent the dynamics from leaving the domain $\mathcal{D}$. For the case of molecular constraints, we refer to Section* 3.3.6.2.

A statistical description through a probability measure μ is a convenient description since the whole microscopic information is both unimportant (what matters are average quantities, and not the positions of all particles composing the system) and too large to be processed.

An example of an observable is the bulk pressure P in a Lennard-Jones liquid. For particles of masses m_i, described by their positions q_i and their momenta p_i, it is given by $P = \mathbb{E}_\mu(A)$ with

$$A(q,p) = \frac{1}{3|\mathcal{D}|} \sum_{i=1}^{N} \left(\frac{|p_i|^2}{m_i} - q_i \cdot \frac{\partial V}{\partial q_i}(q) \right),$$

where $|\mathcal{D}|$ is the physical volume of the box occupied by the fluid, and the potential energy function V is made precise below, see (1.4)-(1.5).

In practice, such averages may yield results that are very close to experimental measurements, even for systems small in comparison to the actual sizes of macroscopic systems (provided the interaction potentials are short-ranged). For example, the equation of state of Figure 1.1 has been computed with a system of a few thousand particles only, a number which is 20 orders of magnitude lower than the Avogadro number. The computed results are compared with experimental measurements.[2] The agreement is very good in the case of Argon. Notice also that high-pressure results, not easily obtained with experimental setups, can be computed.

[2]See for instance the NIST webpage `http://webbook.nist.gov/chemistry/fluid/`

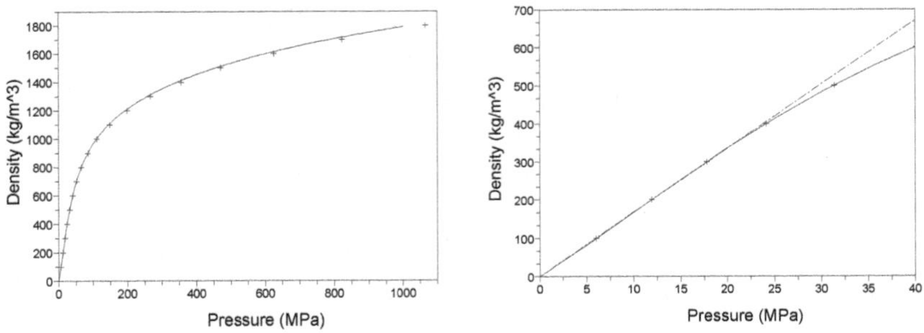

Fig. 1.1 Numerical equation of state of argon at $T = 300$ K ("+") and experimental reference curve (solid line). The picture on the right is a zoom on the low density/low pressure part of the curve, where the ideal gas regime is plotted in dash-dotted line.

We will restrict ourselves in this book to static equilibrium properties of the form of (1.1), and will not consider dynamical properties depending on the actual time evolution of the system (autocorrelation functions, transport coefficients such as thermal conductivity, exit times out of some region in phase space, ...).

1.2 Microscopic description of physical systems

The description of systems in statistical physics requires several ingredients: microscopic interaction laws between the constituents of matter and possibly the environment (see Section 1.2.1), time evolution equations for isolated systems (see Section 1.2.2), and the notion of thermodynamic ensembles, which are probability measures on the set of all possible microscopic configurations, consistent with the macroscopic state of the system (see Section 1.2.3).

1.2.1 *Interactions*

The interactions between the particles are taken into account through a potential function V, depending on the positions q only. The total energy of the system is given by the Hamiltonian

$$H(q,p) = E_{\rm kin}(p) + V(q), \tag{1.2}$$

where the kinetic energy is

$$E_{\mathrm{kin}}(p) = \frac{1}{2} p^T M^{-1} p, \qquad M = \begin{pmatrix} m_1 \,\mathrm{Id}_3 & & 0 \\ & \ddots & \\ 0 & & m_N \,\mathrm{Id}_3 \end{pmatrix}.$$

The matrix M is called the mass matrix. A Hamiltonian such as (1.2) is said to be *separable* since the energetic contributions of the momentum and position variables can be added independently. An instance of a non-separable Hamiltonian is the case when the mass-matrix depends on the configuration q of the system.

Most Hamiltonians encountered in applications are separable, and we will in any case restrict ourselves to separable Hamiltonians in this book. Non-separable Hamiltonians may be considered for modelling purposes (when working with internal coordinates, for rigid body dynamics for instance), or for mathematical convenience (such as the modified Hamiltonians used in the backward analysis of Hamiltonian dynamics, see the references at the end of Section 1.2.2.4).

In order to describe more precisely the interactions between the elementary constituents of the system, several points have to be made precise. First, the boundary conditions of the system must be specified (see Section 1.2.1.1). Then, we give more detail on the interaction potential V in Section 1.2.1.2. This function is very important since it incorporates almost all the physics of the problem. It is therefore no surprise that obtaining reliable potential functions is still a very active research field.

1.2.1.1 *Boundary conditions*

Several boundary conditions can be imposed onto the system:

(i) Many current simulations are performed using periodic boundary conditions, so that surface effects can be avoided and configurations typically encountered in the bulk of the system can be obtained. In this case, a particle interacts not only with all the particles in the systems, but also with their periodic images (see Figure 1.2). In practice, interactions are set to 0 when the distance between two or several particles exceeds a given cut-off radii r_{cut}. When cubic domains of length L are considered as in Figure 1.2, the domain length should be chosen so that $r_{\mathrm{cut}} < L/2$. This ensures that a particle interacts either with the primitive particle, or at most one of its periodic images;

(ii) In some simulations, the system is allowed to visit the entire physical space ($\mathcal{D} = \mathbb{R}^{3N}$). This is the case for isolated systems, such as molecules *in vacuo*. It may be convenient however to quotient out rigid body motions in this case since the potential energy is usually invariant under translations and rotations of the system;

(iii) It is sometimes necessary to consider confined systems. In this case, the positions of the particles are restricted to some predefined region of space, and some rules have to be set for reflections on the boundaries of the system (such as specular reflection of the momenta).

Let us finally mention that open systems with inflows or outflows of energy, particles etc., are sometimes considered. In this case, there may be some exchanges or forcing at the boundaries. Such situations are not considered in this book.

1.2.1.2 *Potential functions*

***Ab initio* interaction potentials.** Ideally, the interaction potentials between the particles should be obtained in a non-empirical approach by resorting to *ab initio* computations. Relying on the standard Born-Oppenheimer assumption, the positions q_i of the nuclei of charges Z_i are kept fixed, and the energy of the system is obtained by adding the Coulomb interaction energies between the nuclei, and the electronic ground-state energy:

$$V(q_1, \ldots, q_N) = \sum_{1 \le i < j \le N} \frac{Z_i Z_j}{|q_i - q_j|} + V_{\mathrm{elec}}(q_1, \ldots, q_N). \tag{1.3}$$

Denote by $M = Z_1 + \cdots + Z_N$ the number of electrons. The system is assumed to be neutral. The electronic ground-state energy is obtained by minimizing the electronic problem over the Hilbert space $\mathcal{H}$ of admissible wavefunctions, which is a subset of the space $\bigwedge_{m=1}^{M} \mathrm{L}^2(\mathbb{R}^3, \mathbb{C})$ of antisymmetric functions. We omit the spin variable for notational simplicity although this variable is very important for quantitative computations. The electronic problem then reads

$$V_{\mathrm{elec}}(q_1, \ldots, q_N) = \inf \left\{ \left\langle \psi, \widehat{H}_{q_1,\ldots,q_N} \psi \right\rangle_{\mathcal{H}} \;\middle|\; \psi \in \mathcal{H},\ \|\psi\|_{L^2} = 1 \right\},$$

where the electronic Hamiltonian operator reads

$$\widehat{H}_{q_1,\ldots,q_N} = -\sum_{m=1}^{M} \frac{1}{2} \Delta_{x_m} - \sum_{m=1}^{M} \sum_{i=1}^{N} \frac{Z_i}{|x_m - q_i|} + \sum_{1 \le n < m \le M} \frac{1}{|x_n - x_m|}.$$

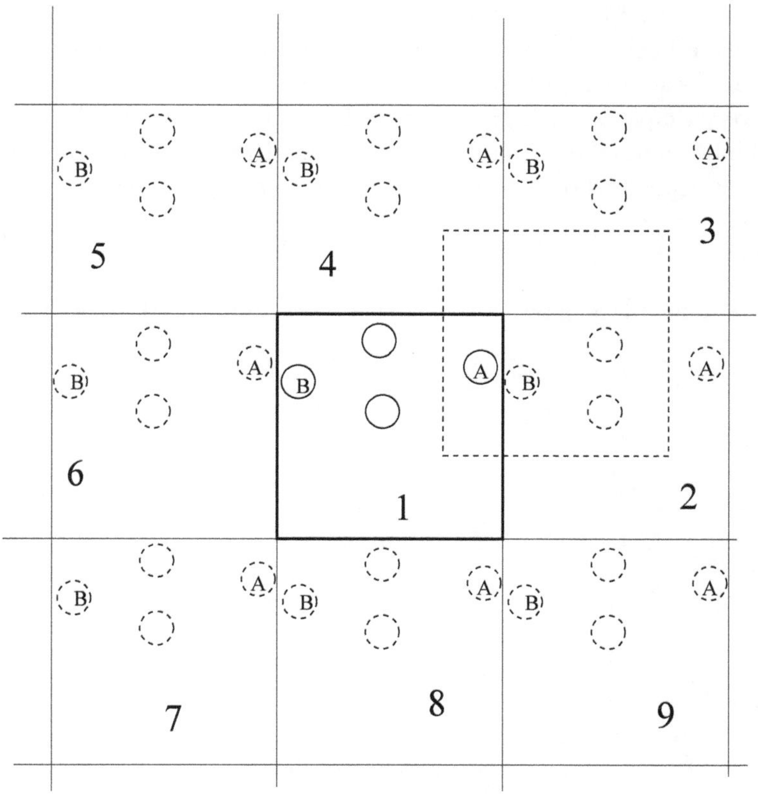

Fig. 1.2 System with periodic boundary conditions. The simulation cell is numbered "1", and the other cells are obtained by translation. The particles inside the primitive cell have interactions with particles in all the other cells.

We refer for instance to [Cancès *et al.* (2003)] for further precision on the computation of *ab initio* interaction potentials. Such computations are however very time-consuming, so that only small systems can be simulated this way (using Born-Oppenheimer molecular dynamics [Niklasson *et al.* (2006)] or the Car-Parrinello approach [Car and Parrinello (1985)]).

Empirical potentials. In practice, empirical formulas for the potential energy function are used to study larger systems. These empirical formulae are obtained by assuming a functional form for the interaction potential, which depends on a set of parameters. These parameters may be chosen so that the potential energy function is as close as possible to the function (1.3) obtained from small *ab initio* computations. Alternatively, the parameters may be such that average properties computed from molecular simulations match experimental thermodynamic properties such as the equation of state of the material, its heat capacity, etc.

A very simple example of an empirical potential is the potential function of a fluid composed of N particles, interacting through a pairwise additive potential depending only on the distance between the particles:

$$V(q_1, \ldots, q_N) = \sum_{1 \le i < j \le N} \mathcal{V}(|q_i - q_j|). \tag{1.4}$$

For example, noble gases are well described using (1.4) when $\mathcal{V}$ is the

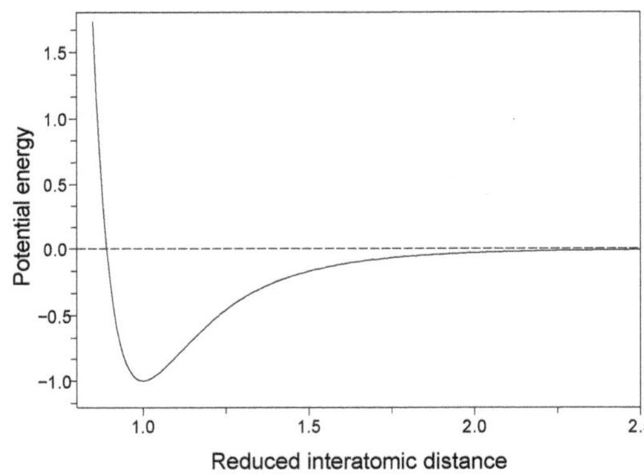

Fig. 1.3 Lennard-Jones potential (1.5) where the distance and the energy are expressed in terms of the equilibrium distance $2^{1/6}\sigma$ and the reference energy ε.

Lennard-Jones potential (depicted in Figure 1.3)

$$\mathcal{V}(r) = 4\,\varepsilon \left(\left(\frac{\sigma}{r}\right)^{12} - \left(\frac{\sigma}{r}\right)^{6} \right). \tag{1.5}$$

This potential depends on two parameters: an energy ε and a distance σ. For argon for instance, $\varepsilon = 1.66 \times 10^{-21}$ J, and $\sigma = 3.405$ Å. The model (1.4)-(1.5) is suitable for noble gases since these systems are monatomic and the corresponding atoms have closed electronic shells. Therefore, the dominant physical interaction is the weakly attractive long-range van der Waals contribution, which scales as r^{-6}.

Potential functions for molecules. Many molecular systems contain molecules. Therefore, interaction potentials describing the existence of bonds between atoms are required. This is modelled through interactions involving several atoms. To describe these potentials, it is convenient to introduce the vector $r_{i,j} = q_j - q_i$.

(1) The interactions of two atoms involved in a covalent bond can be modelled *via* a harmonic potential energy

$$\mathcal{V}_2(q_i, q_{i+1}) = \frac{k_0}{2}\left(|r_{i,i+1}| - l_{\text{eq}}\right)^2,$$

where l_{eq} is the equilibrium length;

(2) Three atoms can interact *via* the three-body interaction potential energy

$$\mathcal{V}_3(q_i, q_{i+1}, q_{i+2}) = \frac{k_\theta}{2}(\theta_i - \theta_{\text{eq}})^2,$$

where the bond angle θ_i is

$$\theta_i = \arccos\left(\frac{r_{i,i+1}}{|r_{i,i+1}|} \cdot \frac{r_{i+1,i+2}}{|r_{i+1,i+2}|}\right),$$

while θ_{eq} is the equilibrium bond angle;

(3) Four atoms may experience the four-body interaction potential energy

$$\mathcal{V}_4(q_i, q_{i+1}, q_{i+2}, q_{i+3}) = u_{\text{tors}}(\cos\phi_i), \tag{1.6}$$

where the dihedral angle ϕ_i is obtained from the relation

$$\cos\phi_i = -\frac{r_{i,i+1} \times r_{i+1,i+2}}{|r_{i,i+1} \times r_{i+1,i+2}|} \cdot \frac{r_{i+1,i+2} \times r_{i+2,i+3}}{|r_{i+1,i+2} \times r_{i+2,i+3}|}.$$

Local interactions have to be complemented by non-bonded interactions: van der Waals forces modelled by Lennard-Jones potentials, and Coulomb interactions, see [Schlick (2002)] for further precision.

A typical example of a simple molecular system is depicted in Figure 1.4 (left), which corresponds to the pentane molecule in the so-called united-atom representation (see [Ryckaert and Bellemans (1978)]). In this representation, the hydrogen atoms are not explicitly represented. We label by $q_1, \ldots, q_5$ the positions of the carbon atoms in the pentane molecule, while $q_6, \ldots, q_N$ are the positions of the solvent molecules. The solvent molecules are assumed to interact with all the other atoms through a pairwise potential $\mathcal{V}_{\text{sol}}$ depending only on the relative distance. The total interaction energy then reads

$$V(q) = V_{\text{pentane}}(q_1, \ldots, q_5) + V_{\text{solvent}}(q_6, \ldots, q_N) + V_{\text{interaction}}(q),$$

with

$$V_{\text{solvent}}(q_6,\ldots,q_N) = \sum_{6\le i<j\le N} \mathcal{V}_{\text{sol}}(|q_i - q_j|),$$

and

$$V_{\text{interaction}}(q) = \sum_{i=1}^{5} \sum_{6\le j\le N} \mathcal{V}_{\text{sol}}(|q_i - q_j|).$$

The interactions within the molecule are

$$V_{\text{pentane}}(q_1,\ldots,q_5) = \sum_{i=1}^{4} \mathcal{V}_2(q_i,q_{i+1}) + \sum_{i=1}^{3} \mathcal{V}_3(q_i,q_{i+1},q_{i+2}) + \sum_{i=1}^{2} \mathcal{V}_4(q_i,q_{i+1},q_{i+2},q_{i+3}),$$

where the dihedral potential function u_{tors} in (1.6) is given by an expression of the form

$$u_{\text{tors}}(x) = c_1(1-x) + 2c_2(1-x^2) + c_3(1+3x-4x^3).$$

The parameters c_i $(i = 1,2,3)$ used in the united-atom model of [Ryckaert and Bellemans (1978)] are such that there are three stable dihedral angles, the one at $\phi = 0$ being energetically more favorable than the others (see Figure 1.4, right).

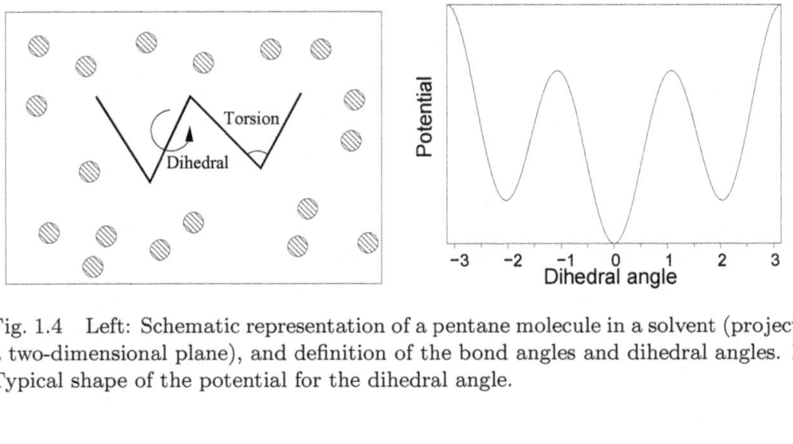

Fig. 1.4 Left: Schematic representation of a pentane molecule in a solvent (projected on a two-dimensional plane), and definition of the bond angles and dihedral angles. Right: Typical shape of the potential for the dihedral angle.

More realistic force fields. Pairwise additive potentials such as (1.4), and two-, three- or four-body bonded interactions may however not be a good approximation of the many-body *ab initio* potential function (1.3). Many studies aim at proposing better empirical potential functions (or force fields). Recent instances of such potentials are the (Modified) Embedded-Atom Model potentials [Baskes (1992)], or bond-order potentials of REBO [Tersoff (1989)] or ReaxFF [van Duin *et al.* (2001)] types, which contain term depending on the local coordination of the atoms. The latter potentials can even account for chemical reactions (*i.e.* bond breakings and bond formations).

Non-dimensional units. In practice, it is more convenient (and numerically more stable) to work with non-dimensional quantities. In this case, the manipulated numbers are all of order 1. In general, reduced units require the following reference quantities:

- a reference mass m_0, for instance the mass of the heaviest or the lightest atom in the system;
- a reference energy ε_0, given by the magnitude of a typical interaction energy, or alternatively by $k_B T$. This energy is therefore of the order of 10^{-21} J;
- a reference length l_0, given by the typical interaction distance, for instance a covalent bond length when molecules are present in the system. Usually, l_0 is of the order of several angströms.

Moreover, other reference quantities can be derived from the above fundamental reference quantities. For instance, a reference time t_0 is obtained by requiring that the typical kinetic energy is of the order of magnitude of the reference energy:

$$t_0 = \frac{m_0^{1/2} l_0}{\varepsilon_0^{1/2}}. \tag{1.7}$$

This time is typically of the order of the pico-second.

1.2.2 *Dynamics of isolated systems*

We consider in this section the time evolution of isolated systems described at the microscopic level. After a general presentation of the Hamiltonian dynamics in its usual form in Section 1.2.2.1, some equivalent reformulations are proposed in Section 1.2.2.2. We then recall some important properties of the dynamics in Section 1.2.2.3, and close the section with some elements on the numerical analysis of time-discretization schemes (Section 1.2.2.4).

1.2.2.1 *The Hamiltonian dynamics*

For separable Hamiltonians, the evolution of isolated systems is governed by the Hamiltonian dynamics

$$\begin{cases} \dfrac{dq(t)}{dt} = \nabla_p H(q(t), p(t)) = M^{-1} p(t), \\ \dfrac{dp(t)}{dt} = -\nabla_q H(q(t), p(t)) = -\nabla V(q(t)). \end{cases} \tag{1.8}$$

Initial conditions

$$(q(0), p(0)) = (q^0, p^0) \tag{1.9}$$

should be provided. Introducing the matrix

$$J = \begin{pmatrix} 0 & \mathrm{Id}_{3N} \\ -\mathrm{Id}_{3N} & 0 \end{pmatrix}, \tag{1.10}$$

and denoting $x = (q, p) \in T^*\mathcal{D}$, the Hamiltonian dynamics can be seen as the first-order ordinary differential equation:

$$\frac{dx}{dt} = J\nabla H(x) = J \begin{pmatrix} \nabla_q H(q, p) \\ \nabla_p H(q, p) \end{pmatrix}. \tag{1.11}$$

The existence and uniqueness of trajectories typically follows from the Cauchy-Lipschitz theorem. A sufficient condition is that ∇H is locally Lipschitz continuous and H is bounded below. We will always assume in the sequel that the Cauchy problem (1.8)-(1.9) is well-posed.

We denote by ϕ_t the flow of the Hamiltonian dynamics, *i.e.* the application which associates to some initial condition (q^0, p^0) the solution $(q(t), p(t)) = \phi_t(q^0, p^0)$ to (1.8) at time t. Let us emphasize that (1.8) is an autonomous equation since the system is assumed to be isolated, so that the flow only depends on the duration time t of the trajectory, and not on the initial and final times separately.

The flow is a semi-group: $\phi_{t+u} = \phi_t \circ \phi_u$ for all $t, u \geq 0$. Actually, it is possible to define the backward evolution ϕ_{-t} for $t \geq 0$, using for instance the reversibility of the dynamics (see (1.19) below), so that $\phi_{t+u} = \phi_t \circ \phi_u$ for all $t, u \in \mathbb{R}$.

1.2.2.2 *Equivalent reformulations*

When the Hamiltonian equation (1.8) is reformulated in terms of the positions only, it reads

$$M \frac{d^2q(t)}{dt^2} = -\nabla V(q),$$

which is Newton's second law.

There are also more abstract reformulations of (1.8), which will be useful below. The Poisson bracket for two smooth observables A_1, A_2 (*i.e.* functions of (q,p)) is defined as

$$\{A_1, A_2\} = (\nabla_q A_1)^T \nabla_p A_2 - (\nabla_p A_1)^T \nabla_q A_2, \qquad (1.12)$$

where C^T denotes the transpose of a matrix C. Notice that the Poisson bracket can be rewritten as

$$\{A_1, A_2\} = (\nabla A_1)^T J \nabla A_2.$$

Hamilton's equations of motion, (1.8), are then equivalent to the following transport equation: for any smooth observable A,

$$\frac{d}{dt}\left[A(q(t),p(t))\right] = \{A, H\}(q(t),p(t)). \qquad (1.13)$$

The concept of the Poisson bracket will be particularly useful to study generalized Hamiltonian dynamics for system with mechanical constraints (see Section 3.3). Some important properties of the Poisson bracket (1.12), which can be checked by simple computations, are the following:

- *Non-degeneracy*: if for any compactly-supported smooth test function φ_1, $\{\varphi_1, \varphi_2\} = 0$, then φ_2 is a constant function.

For any compactly supported smooth test functions $\varphi_1, \varphi_2, \varphi_3$,

- *Skew-symmetry*:
$$\{\varphi_1, \varphi_2\} = -\{\varphi_2, \varphi_1\}.$$
- *Jacobi identity*:
$$\{\varphi_1, \{\varphi_2, \varphi_3\}\} + \{\varphi_2, \{\varphi_3, \varphi_1\}\} + \{\varphi_3, \{\varphi_1, \varphi_2\}\} = 0. \qquad (1.14)$$
- *Leibniz' rule*:
$$\{\varphi_1\varphi_2, \varphi_3\} = \varphi_1\{\varphi_2, \varphi_3\} + \varphi_2\{\varphi_1, \varphi_3\}. \qquad (1.15)$$
- *Divergence formula*:
$$\int_{T^*\mathcal{D}} \{\varphi_1, \varphi_2\}\, dq\, dp = 0. \qquad (1.16)$$

- *Integration by parts*:

$$\int_{T^*\mathcal{D}} \{\varphi_1, \varphi_2\}\, \varphi_3 \, dq \, dp = \int_{T^*\mathcal{D}} \{\varphi_2, \varphi_3\}\, \varphi_1 \, dq \, dp. \qquad (1.17)$$

The transport equation (1.13) (or equivalently the Hamiltonian equation (1.8)) may also be restated as an evolution equation for the phase space density of the particles. Assume that the initial conditions (q^0, p^0) are distributed according to some measure with density $\psi^0(q,p)$ with respect to the phase space Lebesgue measure, and that each initial phase space configuration is evolved according to the dynamics (1.8). Then the configurations $\phi_t(q^0, p^0)$ at time t are distributed according to a measure with density $\psi(t,q,p)$, whose evolution is governed by the following partial differential equation (called the Liouville equation):

$$\partial_t \psi = \nabla_q H \cdot \nabla_p \psi - \nabla_p H \cdot \nabla_q \psi = \{H, \psi\}, \qquad \psi(0,q,p) = \psi^0(q,p).$$

1.2.2.3 *Properties of the Hamiltonian dynamics*

The Hamiltonian dynamics has several interesting mathematical and structural properties:

(1) *Symmetry.* Since $\phi_t \circ \phi_{-t} = \mathrm{Id}$,

$$\phi_{-t} = \phi_t^{-1}. \qquad (1.18)$$

(2) *Reversibility.* Consider the momentum reversal function

$$S(q,p) = (q, -p).$$

Then, the time-reversed evolution ϕ_{-t} for $t \geq 0$, defined by (1.18), is easily seen to be equal to a forward evolution with momenta reversed (the so-called backward flow):

$$\phi_{-t} = S \circ \phi_t \circ S. \qquad (1.19)$$

(3) *Energy conservation.* The choice $A = H$ in the Poisson bracket reformulation of the Hamiltonian dynamics (1.13) leads to $\dfrac{dH(q(t),p(t))}{dt} = 0$, which means that

$$H(q(t),p(t)) = H(q^0,p^0).$$

(4) *Volume preservation.* For all measurable sets $B \subset T^*\mathcal{D}$, and for all $t \in \mathbb{R}$,

$$\int_{\phi_t(B)} dq \, dp = \int_B dq \, dp. \qquad (1.20)$$

This identity, often referred to as Liouville's theorem, is a consequence of the equality

$$|\text{jac}\,\phi_t(q,p)| = 1,$$

where $\text{jac}\,\phi_t(q,p) = \det(\nabla\phi_t(q,p))$. The proof of the latter assertion relies on the fact that the Hamiltonian vector field is divergence-free. Lemma VII.3.1 in [Hairer *et al.* (2006)] then allows us to conclude.

(5) *Symplecticity.* The matrix J defined by (1.10) is antisymmetric and orthogonal ($J^T = -J = J^{-1}$). For an open set $U \in T^*\mathcal{D}$, a mapping $g : U \to \mathbb{R}^{2dN}$ is symplectic if $\nabla g(q,p)$ satisfies

$$(\nabla g)^T J \nabla g = J \tag{1.21}$$

for all $(q,p) \in U$. It is easily shown that the flow ϕ_t is symplectic for all $t \in \mathbb{R}$ (this is a result due to Poincaré). Actually, any symplectic map is locally Hamiltonian (see Section VI.2 in [Hairer *et al.* (2006)] for further precision), which shows that the symplecticity of the flow is indeed a characteristic feature of Hamiltonian systems. Note that the volume preservation property (1.20) is recovered as a consequence of the symplecticity property since (1.21) with $g = \phi_t$ implies that $(\det\nabla\phi_t)^2 = 1$. The symplecticity property is stronger, and can be understood as the conservation of oriented elementary parallelograms, see Section 3.5 in [Leimkuhler and Reich (2005)] for a pedagogical presentation.

1.2.2.4 *Numerical integration*

We discuss in this section numerical schemes to integrate (1.8). Let us first mention a few reasons why it is both hopeless and useless to integrate precisely the Hamiltonian dynamics in the context of molecular simulation:

(i) The Hamiltonian dynamics is known to be strongly sensitive to the initial conditions, or to numerical errors such as round-off errors: Small differences between two initially close configurations are exponentially magnified as time passes. Since the initial conditions can never be known exactly for physical reasons in molecular systems (in particular because there are too many atoms whose positions and momenta are required) and very long integration times are needed, this is a first reason not to try to integrate too precisely the Hamiltonian dynamics. The situation may of course be different in other application fields where Hamiltonian dynamics are used for systems with less degrees of freedom, such as celestial mechanics.

(ii) Moreover, given the large number of particles in molecular simulations (hence the numerical cost of evaluating forces) and the very small time-steps that would be needed to integrate precisely the trajectory are prohibitive.
(iii) Finally, the aim of many current computations in computational statistical physics is the evaluation of average properties along a long trajectory (see the ergodicity assumption (1.30) below). Therefore, it is sufficient to ensure a correct sampling rather than integrating precisely the trajectory. In particular, a basic requirement is the preservation of the energy over long trajectories.

The above arguments led to the development of numerical techniques devoted to Hamiltonian systems, fully taking into account the energy preservation as a basic first requirement. This requirement is more important than the scheme's order (*i.e.* the integer p such that the error between the exact solution over a time interval Δt and the numerical solution after one step of the numerical scheme is of order Δt^{p+1}), which determines the convergence rate of the numerical approximation only on finite-time intervals.

A very convenient algorithm to approximately preserve the energy was proposed in [Verlet (1967)] (actually rediscovered by Verlet, since it was already known by Störmer in the context of celestial mechanics at the beginning of the 20th century, and even by Newton; see Section 1.3 in [Hairer *et al.* (2003)] for historical precisions). The Verlet algorithm is nowadays the standard integration scheme for Hamiltonian dynamics. Denoting by (q^n, p^n) an approximation of $(q(t_n), p(t_n))$ at time $t_n = n\Delta t$ (where Δt is the time step), it reads

$$\begin{cases} p^{n+1/2} = p^n - \dfrac{\Delta t}{2} \nabla V(q^n), \\ q^{n+1} \quad = q^n + \Delta t \, M^{-1} p^{n+1/2}, \\ p^{n+1} \quad = p^{n+1/2} - \dfrac{\Delta t}{2} \nabla V(q^{n+1}). \end{cases} \tag{1.22}$$

The numerical flow associated with this scheme is denoted by $\Phi_{\Delta t}^{\mathrm{Verlet}}$ in the sequel: $(q^{n+1}, p^{n+1}) = \Phi_{\Delta t}^{\mathrm{Verlet}}(q^n, p^n)$. It is easy to check that the scheme is of order 2.

Stability requirements limit the time-step Δt which can be used in practice. We now detail the study of the *linear stability*, for the one-dimensional harmonic potential $V(q) = \omega^2 q^2/2$ of mass $m = 1$. In this case, the Verlet scheme reads

$$\begin{pmatrix} q^{n+1} \\ p^{n+1} \end{pmatrix} = A_{\Delta t} \begin{pmatrix} q^n \\ p^n \end{pmatrix}, \quad A_{\Delta t} = \begin{pmatrix} 1 - \dfrac{(\omega\Delta t)^2}{2} & \Delta t \\ -\omega^2 \Delta t \left(1 - \dfrac{(\omega\Delta t)^2}{4}\right) & 1 - \dfrac{(\omega\Delta t)^2}{2} \end{pmatrix}.$$

The eigenvalues of the matrix $A_{\Delta t}$ have modulus 1 if and only if $\omega\Delta t < 2$. In this case, the trajectory $(q^n, p^n)_{n\geq 0}$ is bounded. Otherwise, one eigenvalue has a modulus strictly larger than 1, so that the trajectory $(q^n, p^n)_{n\geq 0}$ is not bounded in general. Besides, it is easily shown that the modified energy

$$H_{\Delta t}(q,p) = H(q,p) - \frac{(\omega\Delta t)^2}{4} q^2$$

is preserved exactly: $H_{\Delta t}(q^n, p^n) = H_{\Delta t}(q^0, p^0)$ for all $n \geq 0$. Therefore, when $\omega\Delta t < 2$, the boundedness of the trajectory implies

$$\sup_{n\in\mathbb{N}} \left| H(q^n, p^n) - H(q^0, p^0) \right| \leq C\,\Delta t^2.$$

For more general potentials, there is no simple rule to place an upper bound on the time-step. However, the linear stability requirement suggests that an admissible time-step should be a fraction of the fastest vibration period in the system.

Actually, the positions q^n obtained by the Verlet scheme (1.22) satisfy

$$M \frac{q^{n+1} - 2q^n + q^{n-1}}{\Delta t^2} = -\nabla V(q^n),$$

which is the simple centered finite-difference discretization for the equation $M \dfrac{d^2 q}{dt^2}(t) = -\nabla V(q(t))$. However, the very good properties of the numerical method cannot be understood from this equation. It is important to keep both variables q and p, and study the numerical flow $\Phi_{\Delta t}^{\text{Verlet}}$ of (1.22). Indeed, this application shares many qualitative properties with the exact flow ϕ_t of (1.8); in particular, it is time reversible:

$$S \circ \Phi_{\Delta t}^{\text{Verlet}} \circ S = \Phi_{-\Delta t}^{\text{Verlet}},$$

where $S(q,p) = (q, -p)$ is the momentum reversal operator; symmetric:

$$\left(\Phi_{\Delta t}^{\text{Verlet}}\right)^{-1} = \Phi_{-\Delta t}^{\text{Verlet}};$$

and symplectic: $(\nabla \Phi_{\Delta t}^{\text{Verlet}})^T J \nabla \Phi_{\Delta t}^{\text{Verlet}} = J$. The latter property is of paramount importance for the longtime integration of the Hamiltonian dynamics. A well-established result, recalled in the reference book [Hairer *et al.* (2006)] on geometric numerical integration (see in particular Chapters VIII and IX), is that, when Δt is small enough, the energy of the system is conserved up to $O(\Delta t^p)$ error terms over very long times when symplectic numerical schemes of order p are used (under some technical assumptions).

More generally, the longtime stability properties of symplectic numerical methods applied to symplectic flows can be studied with the help of the so-called backward analysis. Contrarily to standard error analysis where the numerical trajectory is considered as an approximation of the true trajectory of the exact problem, the backward analysis consists in interpreting the numerical trajectory as the *exact trajectory of some modified ordinary differential equation*, and then to study the properties of the modified problem. For symplectic methods approaching symplectic flows, the modified equation is still Hamiltonian. Therefore, *some modified energy is preserved exactly*. This property is finally used to show that the *exact energy is preserved approximately*. In fact, some rather involved analysis has to be used since the modified Hamiltonian is defined as a formal series, which does not converge in general. Some optimal truncations should then be considered, and the modified energy is therefore not strictly preserved, but the error terms are very small.

1.2.3 *Thermodynamic ensembles*

The macroscopic state of a system is described, within the framework of statistical physics, by a probability measure μ on the phase space $T^*\mathcal{D} = \mathcal{D} \times \mathbb{R}^{3N}$. Macroscopic features of the system are then computed as averages of an observable A with respect to this measure, as given by (1.1):

$$\mathbb{E}_\mu(A) = \int_{T^*\mathcal{D}} A(q,p)\,\mu(dq\,dp).$$

We therefore call the measure μ the *macroscopic state* of the system.

The practical computation of the ensemble average requires numerical techniques to sample configurations (q^n, p^n) according to the probability measure μ (or possibly according to a measure $\tilde{\mu}$ very close to μ, the difference between μ and $\tilde{\mu}$ originating from errors in the numerical integration of a continuous dynamics for instance, see Section 2.3.1.1 for further precision). The ensemble average (1.1) is then approximated by

$$\lim_{N\to+\infty} \frac{1}{N} \sum_{n=1}^{N} A(q^n, p^n). \tag{1.23}$$

The numerical techniques of course depend on the thermodynamic ensemble at hand. Most methods generate a sequence of microscopic configurations

$(q^n, p^n)_{n\geq 1}$ from a time-discrete dynamics, so that the successive configurations are not independent.

We present more thoroughly in this section two very commonly used thermodynamic ensembles, namely the microcanonical ensemble (Section 1.2.3.1) and the canonical ensemble (Section 1.2.3.2). These ensembles describe respectively isolated systems, and systems at a fixed temperature (in contact with a so-called thermostat or energy reservoir). We also mention some other thermodynamic ensembles in Section 1.2.3.3, for the sake of completeness.

1.2.3.1 *The microcanonical ensemble*

The thermodynamic ensemble naturally associated with the Hamiltonian dynamics (1.8) is the *microcanonical ensemble*, which describes isolated systems at constant energy. This ensemble is also often termed *NVE ensemble*, the capital letters referring to the invariants of the system, namely the number of particles N, the volume of the simulation box V, and the energy E.

The corresponding probability measure is the normalized uniform probability measure on the set $\mathcal{S}(E)$ of configurations at the given energy level E:

$$\mathcal{S}(E) = \left\{ (q,p) \in T^*\mathcal{D} \;\middle|\; H(q,p) = E \right\}.$$

We present three ways to understand this idea.

An explicit construction. The building block for the construction of the microcanonical measure is the measure $\delta_{H(q,p)-E}(dq\,dp)$, where the conditioning relies on level sets of constant total energy. This measure can be obtained by an explicit construction, using a limiting procedure. Consider a given energy level E, some small energy variation $\Delta E > 0$, and define

$$\mathcal{N}_{\Delta E}(E) = \left\{ (q,p) \in T^*\mathcal{D} \;\middle|\; E \leq H(q,p) \leq E + \Delta E \right\}.$$

Then, the following integral of a given test function A expresses the fact that the set $\mathcal{N}_{\Delta E}(E)$ is endowed with a uniform measure:

$$\Pi_{E,\Delta E}(A) = \frac{1}{\Delta E} \int_{\mathcal{N}_{\Delta E}(E)} A(q,p)\, dq\, dp.$$

In the limit $\Delta E \to 0$, a measure supported on the submanifold $\mathcal{S}(E)$ is recovered. Notice that this measure is not normalized to 1 *a priori*. The measure $\delta_{H(q,p)-E}(dq\,dp)$ is defined through the expectations of any observable A as

$$\int_{\mathcal{S}(E)} A(q,p)\, \delta_{H(q,p)-E}(dq\,dp) = \lim_{\Delta E \to 0} \frac{1}{\Delta E} \int_{\mathcal{N}_{\Delta E}(E)} A(q,p)\, dq\, dp. \quad (1.24)$$

The construction highlights the fact that the regions where $|\nabla H|$ is large have a lower weight in the average since the volume of the infinitesimal domain included in $\mathcal{N}_{\Delta E}(E)$ and centered at $(q,p) \in \mathcal{S}(E)$ is proportional to $|\nabla H(q,p)|^{-1}$, see Figure 1.5. This observation is consistent with the result (1.26) below, obtained with the co-area formula, and motivates the factor $|\nabla H(q,p)|^{-1}$ on the right-hand side of (1.26).

Once the measure $\delta_{H(q,p)-E}(dq\,dp)$ is defined, the microcanonical measure is obtained by a suitable normalization:

$$\mu_{\mathrm{mc},E}(dq\,dp) = Z_E^{-1}\,\delta_{H(q,p)-E}(dq\,dp),$$

where the partition function used in the normalization

$$Z_E = \int_{\mathcal{S}(E)} \delta_{H(q,p)-E}(dq\,dp)$$

is assumed to be finite. See the discussion after (1.28) for some sufficient conditions to this end.

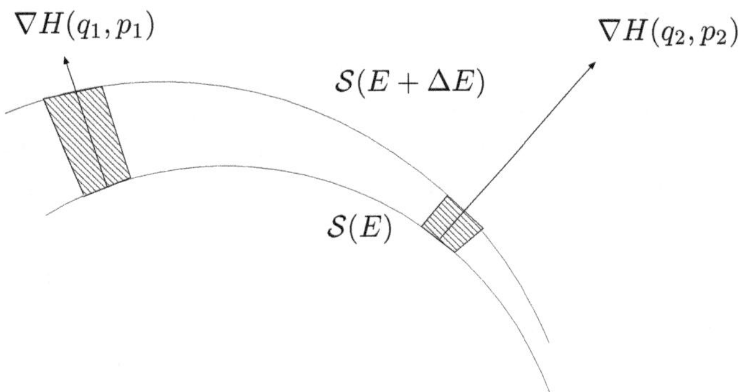

Fig. 1.5 Limiting procedure used to construct the microcanonical measure. The volume of the infinitesimal domain between $\mathcal{S}(E)$ and $\mathcal{S}(E+\Delta E)$ centered at a given point $(q,p) \in T^*\mathcal{D}$ is proportional to $|\nabla H|^{-1}$.

An alternative definition of the microcanonical measure. The measure $\delta_{H(q,p)-E}(dq\,dp)$ for a given energy level E has support in $\mathcal{S}(E)$, and is defined by the following relation: for all test functions $f : T^*\mathcal{D} \to \mathbb{R}$

and $g : \mathbb{R} \to \mathbb{R}$,

$$\int_{T^*\mathcal{D}} g(H(q,p))\, f(q,p)\, dq\, dp = \int_{\mathbb{R}} g(E) \int_{\mathcal{S}(E)} f(q,p)\, \delta_{H(q,p)-E}(dq\, dp)\, dE. \tag{1.25}$$

By the co-area formula (see (3.12)), an alternative expression of the measure $\delta_{H(q,p)-E}(dq\, dp)$ is

$$\delta_{H(q,p)-E}(dq\, dp) = \frac{\sigma_{\mathcal{S}(E)}(dq\, dp)}{|\nabla H(q,p)|}, \tag{1.26}$$

where $\sigma_{\mathcal{S}(E)}(dq\, dp)$ is the area measure induced by the Lebesgue measure on the manifold $\mathcal{S}(E)$ when the phase space is endowed with the standard Euclidean scalar product (see Remark 3.4 and Section 3.3.2.1 for further precision on the definition of surface measures).

The microcanonical measure can then be rewritten as

$$\mu_{\mathrm{mc},E}(dq\, dp) = Z_E^{-1}\, \delta_{H(q,p)-E}(dq\, dp) = Z_E^{-1} \frac{\sigma_{\mathcal{S}(E)}(dq\, dp)}{|\nabla H(q,p)|}, \tag{1.27}$$

with

$$Z_E = \int_{\mathcal{S}(E)} \delta_{H(q,p)-E}(dq\, dp) = \int_{\mathcal{S}(E)} \frac{\sigma_{\mathcal{S}(E)}(dq\, dp)}{|\nabla H(q,p)|}. \tag{1.28}$$

The partition function Z_E is finite for instance when $\mathcal{S}(E)$ is bounded and $|\nabla H| \neq 0$ on this set. Since we consider only separable Hamiltonians, the condition $|\nabla H(q,p)| = 0$ is equivalent to $p = 0$ and $\nabla V(q) = 0$. Therefore, $|\nabla H| \neq 0$ is ensured as soon as $\nabla V(q) \neq 0$ for all configurations $(q,0) \in \mathcal{S}(E)$.

The microcanonical measure as an ergodic limit. Practitioners often see microcanonical averages as ergodic limits over Hamiltonian trajectories. Notice first that $\mu_{\mathrm{mc},E}(dq\, dp)$ is invariant by the Hamiltonian dynamics flow ϕ_t for all energy levels E. Indeed, by the conditioning formula (1.25),

$$\begin{aligned}
&\int_{\mathbb{R}} g(E) \int_{\mathcal{S}(E)} f(\phi_t(q,p))\, \delta_{H(q,p)-E}(dq\, dp)\, dE \\
&\quad = \int_{T^*\mathcal{D}} g(H(q,p))\, f(\phi_t(q,p))\, dq\, dp \\
&\quad = \int_{T^*\mathcal{D}} g(H \circ \phi_{-t}(Q,P))\, f(Q,P)\, dQ\, dP \\
&\quad = \int_{T^*\mathcal{D}} g(H(Q,P))\, f(Q,P)\, dQ\, dP \\
&\quad = \int_{\mathbb{R}} g(E) \int_{\mathcal{S}(E)} f(q,p)\, \delta_{H(q,p)-E}(dq\, dp)\, dE,
\end{aligned}$$

where we have used the change of variables $(Q,P) = \phi_t(q,p)$ and the invariance of the Hamiltonian by the flow ϕ_t. Therefore,

$$\int_{\mathcal{S}(E)} f(q,p)\,\delta_{H(q,p)-E}(dq\,dp) = \int_{\mathcal{S}(E)} f\circ\phi_t(q,p)\,\delta_{H(q,p)-E}(dq\,dp) \quad (1.29)$$

for all times $t \in \mathbb{R}$, which shows the claimed invariance. A more intuitive way to understand this equality is to realize that

$$\frac{1}{\Delta E}\int_{\mathcal{N}_{\Delta E}(E)} f(Q,P)\,dQ\,dP = \frac{1}{\Delta E}\int_{\mathcal{N}_{\Delta E}(E)} f\circ\phi_t(q,p)\,dq\,dp$$

by the same change of variables as above, and then to use (1.24) to obtain (1.29) in the limit $\Delta E \to 0$.

In view of the preservation of the microcanonical measure by the Hamiltonian flow, the following ergodicity assumption can therefore be considered: Thermodynamic integrals of the form (1.1) are computed as trajectorial averages

$$\int_{\mathcal{S}(E)} A(q,p)\,\mu_{\mathrm{mc},E}(dq\,dp) = \lim_{T\to+\infty}\frac{1}{T}\int_0^T A(\phi_t(q,p))\,dt, \quad (1.30)$$

where ϕ_t is the flow of the Hamiltonian dynamics (1.8), and the initial condition (q^0,p^0) is such that $H(q^0,p^0) = E$.

Ergodicity can be rigorously shown for completely integrable systems and their perturbations (see for instance [Arnol'd (1989)]). In general however, no convergence result can be stated, and examples of non-ergodicity can be found. A simple instance of non-ergodicity is the following. Consider the one-dimensional double-well potential

$$V(q) = \left(q^2-1\right)^2. \quad (1.31)$$

The submanifolds $\mathcal{S}(E)$ for $E < 1$ are composed of two simply connected subdomains, and ergodicity can only be expected in a given connected component (see Figure 1.6). Other instances of non-ergodicity cases are situations when there are other invariants than the energy (such as the total momentum of the system, for instance). In those cases, ergodicity is possible only with respect to the Lebesgue measure conditioned to the set of all the invariants of the dynamics.

From a numerical viewpoint, the computation of averages according to the right-hand side of (1.30) requires very stable algorithms allowing a

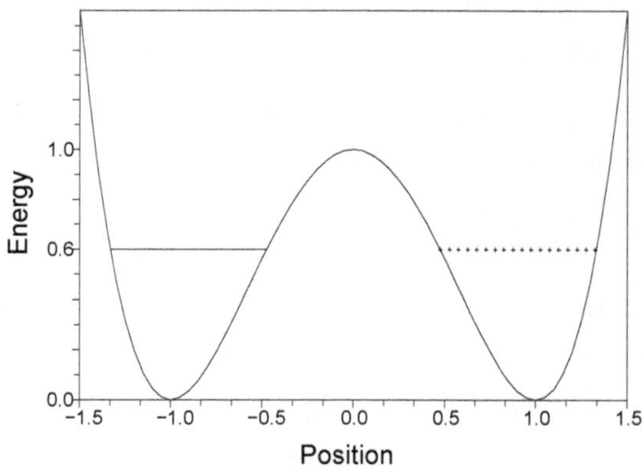

Fig. 1.6 Accessible positions in the energy surface $H(q, p) = 0.6$ for the double-well potential (1.31). If the dynamics starts in one of the connected components, it remains there.

longtime integration of the Hamiltonian dynamics with a very good preservation of the energy, such as the Verlet algorithm (1.22). The numerical analysis of microcanonical sampling methods based on these properties (in the very particular case of completely integrable systems) can be read in [Cancès *et al.* (2004, 2005)]. There exist also stochastic methods based on constrained diffusion processes to sample the microcanonical measure, see [Faou (2006); Faou and Lelièvre (2009)]. The aim of these methods is to destroy all invariants of the dynamics, except the energy.

1.2.3.2 *The canonical ensemble*

In many physical situations, systems in contact with some energy thermostat are considered, rather than isolated systems with a fixed energy. In this case, the energy of the system fluctuates. It however has a fixed temperature. In this situation, the microscopic configurations are distributed according to the so-called *canonical measure*. The canonical ensemble is also often termed *NVT ensemble*, since the number of particles N, the volume V and the temperature T are fixed.

We first define the canonical measure, then give some elements on its derivation from a principle of entropy maximization under constraints, and close this section with a brief presentation of some techniques to sample the canonical measure.

Definition of the canonical measure. We assume in the sequel that $e^{-\beta V} \in L^1(\mathcal{D})$. The canonical probability measure μ on $T^*\mathcal{D}$ reads

$$\mu(dq\,dp) = Z_\mu^{-1} \exp(-\beta H(q,p))\,dq\,dp, \tag{1.32}$$

where $\beta = 1/(k_B T)$ (T denotes the temperature and k_B the Boltzmann constant). The normalization constant

$$Z_\mu = \int_{T^*\mathcal{D}} \exp(-\beta H(q,p))\,dq\,dp$$

in (1.32) is called the *partition function*. When the Hamiltonian H is separable, the canonical measure is of the tensorized form

$$\mu(dq\,dp) = \nu(dq)\,\kappa(dp),$$

where ν and κ are the two following probability measures:

$$\nu(dq) = Z_\nu^{-1} e^{-\beta V(q)}\,dq, \qquad Z_\nu = \int_{\mathcal{D}} e^{-\beta V(q)}\,dq, \tag{1.33}$$

and

$$\kappa(dp) = \left(\frac{\beta}{2\pi}\right)^{3N/2} \prod_{i=1}^{N} m_i^{-3/2} \exp\left(-\frac{\beta}{2} p^T M^{-1} p\right)\,dp. \tag{1.34}$$

Under μ, the position a and the momentum p are independent random variables. Therefore, sampling configurations (q,p) according to the canonical measure $\mu(dq\,dp)$ can be performed by independently sampling positions according to $\nu(dq)$ and momenta according to $\kappa(dp)$.

It is straightforward to sample from κ since the momenta are Gaussian random variables. The actual issue is therefore to sample from ν. Appropriate methods are presented in Sections 2.1 and 2.2.

Some elements on the derivation of the canonical measure. The expression (1.32) of the canonical probability measure can be obtained by maximizing the entropy under the constraint that the energy is fixed *in average*. Such a derivation is performed in [Balian (2007)] for instance. The constraint that the average energy of the system is fixed formalizes the idea that the system under study exchanges energy with the thermostat or energy reservoir to which it is coupled. The energy is therefore not fixed, but it has nonetheless a well-defined average value.

Consider a measure which has a density $\rho(q,p)$ with respect to the Lebesgue measure. The constraints on the admissible functions $\rho(q,p)$ are

$$\rho \geq 0, \qquad \int_{T^*\mathcal{D}} \rho(q,p)\,dq\,dp = 1, \qquad \int_{T^*\mathcal{D}} H\rho(q,p)\,dq\,dp = E \tag{1.35}$$

for some energy level E. The first two conditions ensure that ρ is the density of a probability measure, while the last one expresses the conservation of the energy in average.

The statistical entropy is defined as

$$\mathfrak{S}(\rho) = -\int_{T^*\mathcal{D}} \rho(q,p) \ln \rho(q,p) \, dq \, dp. \tag{1.36}$$

It quantifies the amount of information missing, or the "degree of disorder" as is sometimes stated in a more physical language. The entropy is non-positive since $x \ln x - x + 1 \geq 0$ for all $x > 0$. We refer to Chapter 3 in [Balian (2007)] for further precision on the properties of $\mathfrak{S}$. The statistical entropy allows us to give a rigorous meaning to the idea that a thermodynamic measure is (quoting [Balian (2007)], Section 4.1.3) "*the most disordered macrostate possible compatible with the data,*" or, equivalently, the measure which "*contains no more information than is strictly necessary to take the data into account.*" The amount of information or disorder is quantified by the entropy.

The canonical measure is recovered as the solution to the following optimization problem

$$\sup \left\{ \mathfrak{S}(\rho), \ \rho \in L^1(T^*\mathcal{D}), \ \rho \geq 0, \ \int_{T^*\mathcal{D}} \rho = 1, \ \int_{T^*\mathcal{D}} H\rho = E \right\}. \tag{1.37}$$

Formally, the Euler-Lagrange equation satisfied by an extremum reads

$$\mathfrak{S}'(\rho) + \lambda + \gamma H = 0,$$

where λ, γ are the Lagrange multipliers associated with the two constraints in (1.37) (normalization and average energy fixed). Since $\mathfrak{S}'(\rho) = 1 + \ln \rho$, a candidate maximizer in (1.37) is the measure with density

$$\exp\left(-1 - \lambda - \gamma H(q,p)\right).$$

Usually, the Lagrange multiplier γ associated with the energy constraint is denoted by β, and $\exp(1+\lambda) = Z$ is a normalization constant. The Lagrange multiplier β exists and is unique since

$$\beta \mapsto \mathcal{E}(\beta) = \frac{\displaystyle\int_{T^*\mathcal{D}} H \, \mathrm{e}^{-\beta H}}{\displaystyle\int_{T^*\mathcal{D}} \mathrm{e}^{-\beta H}}$$

is an increasing function. This is a consequence of the positivity of the derivative of the average energy

$$\mathcal{E}'(\beta) = \frac{\displaystyle\int_{T^*\mathcal{D}} (H - \mathcal{E}(\beta))^2 \, \mathrm{e}^{-\beta H}}{\displaystyle\int_{T^*\mathcal{D}} \mathrm{e}^{-\beta H}}$$

when H is not constant.

It is easy to verify that the canonical measure is indeed the unique maximizer of (1.37), as shown in Section 4.2 of [Balian (2007)]. For the sake of completeness, we sketch the proof of this statement. Consider an arbitrary function satisfying (1.35). Using the inequality $\ln x \leq x - 1$ (with equality if and only if $x = 1$):

$$\int_{T^*\mathcal{D}} \rho_1 \ln \rho_2 - \int_{T^*\mathcal{D}} \rho_1 \ln \rho_1 = \int_{T^*\mathcal{D}} \rho_1 \ln \left(\frac{\rho_2}{\rho_1} \right) \leq \int_{T^*\mathcal{D}} \rho_1 - \rho_2 = 0$$

when ρ_1 and ρ_2 satisfy the constraints (1.35). Equality holds if and only if $\rho_1(q,p) = \rho_2(q,p)$ almost everywhere. Then, choosing $\rho_2(q,p) = Z^{-1} \exp(-\beta H(q,p))$, it holds, for any ρ satisfying the constraints (1.35):

$$- \int_{T^*\mathcal{D}} \rho \ln \rho \leq - \int_{T^*\mathcal{D}} \rho \ln \left(Z^{-1} \mathrm{e}^{-\beta H} \right) \leq \ln Z + \beta \int_{T^*\mathcal{D}} H \, \rho.$$

In view of the energy constraint (last condition in (1.35)),

$$\mathfrak{S}(\rho) \leq \ln Z + \beta E = \mathfrak{S} \left(Z^{-1} \mathrm{e}^{-\beta H} \right),$$

with equality if and only if $\rho(q,p) = Z^{-1} \exp(-\beta H(q,p))$. This shows that the canonical measure is indeed the unique maximizer of the entropy under the constraints (1.35).

Sampling the canonical measure. Let us now describe briefly some techniques to sample the canonical measure (1.32). We will rely on these methods in Section 1.3.4 when we present methods to compute free energy differences. The techniques we consider here are stochastic dynamics $t \mapsto (q_t, p_t)$ which are ergodic for the canonical measure, in the sense that the expectation of a given observable

$$\mathbb{E}_\mu(A) = \int_{T^*\mathcal{D}} A(q,p) \, \mu(dq \, dp),$$

where μ is the canonical measure (1.32), can be obtained as an ergodic limit

$$\mathbb{E}_\mu(A) = \lim_{T \to +\infty} \frac{1}{T} \int_0^T A(q_t, p_t) \, dt \tag{1.38}$$

over one realization of the stochastic dynamics. The dynamics we use are motivated solely by the ergodicity property (1.38), and should therefore be seen as a sampling mean. We do not care whether the evolution is physically relevant since we are only interested in time-independent equilibrium properties.

As will be seen in more detail in Section 2.2.3, (1.38) holds under mild assumptions on the potential for the Langevin dynamics

$$\begin{cases} dq_t = M^{-1} p_t \, dt, \\ dp_t = -\nabla V(q_t) \, dt - \gamma \, M^{-1} p_t \, dt + \sigma \, dW_t, \end{cases} \tag{1.39}$$

where W_t is a standard dN-dimensional Brownian motion, and $\gamma, \sigma > 0$ verify

$$\sigma^2 = \frac{2\gamma}{\beta}. \tag{1.40}$$

The relation (1.40) is called *fluctuation-dissipation relation* since it relates the magnitude of the dissipative term $-\gamma \, M^{-1} p_t \, dt$ and the magnitude of the random term $\sigma \, dW_t$. The Langevin dynamics may be seen as the superposition of a Hamiltonian dynamics, which preserves the energy, and a stochastic process on the momenta which ensures that the energy levels are visited according to their weight in the canonical ensemble. The latter condition fixes the magnitude of the random term.

Actually, since the difficult task is the sampling of the configurational part ν of the canonical measure, we could consider a dynamics on the configurational space only, such as the *overdamped Langevin dynamics* (see Section 2.2.2 for further precision, and Section 2.2.4 for a motivation of the terminology):

$$dq_t = -\nabla V(q_t) \, dt + \sqrt{\frac{2}{\beta}} \, dW_t, \tag{1.41}$$

where W_t is again a standard dN-dimensional Brownian motion. Under reasonable assumptions on the potential, this dynamics satisfies

$$\mathbb{E}_\nu(A) = \int_{\mathcal{D}} A(q) \, \nu(dq) = \lim_{T \to +\infty} \frac{1}{T} \int_0^T A(q_t) \, dt,$$

where ν is given in (1.33). Intuitively, each term in (1.41) can be motivated as follows: the gradient force $-\nabla V(q_t)$ ensures that the energy decreases (which in turn, ensures that the visited configurations are likely enough), while the random noise term supplies some energy so that the temperature is correct. The precise balance between the drift term which removes energy in average and the stochastic term is determined by the condition that the canonical measure is preserved by the dynamics (1.41).

1.2.3.3 *Other thermodynamic ensembles*

We saw in Section 1.2.3.2 that the Boltzmann-Gibbs probability measure (1.32) can be seen as the phase space probability measure maximizing the statistical entropy among the set of phase space probability measures compatible with the observed macroscopic data (in this case, average energy given). The derivation performed for an average energy fixed may be performed for any average thermodynamic quantity, leading to other thermodynamic ensembles. The choice of the ensemble amounts to choosing which quantities are fixed exactly or in average.

We present in this section a general derivation of thermodynamic ensembles associated with a given set of constraints, and next focus on two examples, the isobaric-isothermal ensemble (NPT) where the number of particles, the pressure and the temperature are fixed, and the grand-canonical ensemble (μVT) where the chemical potential, the volume and the temperature are fixed. Many other cases could be treated in a similar fashion (fixed temperature and magnetization for a spin system, fixed temperature and average velocity for a fluid, etc.). This section is not necessary for understanding the remainder of the book, and can be omitted in a first reading.

General derivation. Assume that the microscopic state of the system is described by (q,p,x), where (q,p) denotes as above a phase space configuration, and where $x \in \mathcal{X}$ is some additional degree of freedom. We denote by $\mathcal{D}_x$ and $T^*\mathcal{D}_x$ the set of admissible positions q and configurations (q,p) for a given value of x, so that the set of admissible configurations (q,p,x) is the space

$$\mathcal{E} = \bigcup_{x \in \mathcal{X}} T^*\mathcal{D}_x \times \{x\}.$$

Denote by $\lambda(dq\,dp\,dx)$ some reference measure on $\mathcal{E}$. This measure expresses the *a priori* information available on the system. Here, we will consider a reference measure of the form

$$\lambda(dq\,dp\,dx) = 1_{(q,p) \in T^*\mathcal{D}_x}\, dq\,dp\,\pi(dx).$$

The conditional measure with respect to the parameter x (*i.e.* the measure obtained in the (q,p) variables when the parameter x is kept fixed) is the usual Lebesgue measure on the set of admissible configurations. The reference measure π on the variable x depends on the problem at hand.

Consider then a measure $\rho(dq\,dp\,dx)$ describing the macroscopic state of the system, and several observables $A_1, \dots, A_M$, functions of (q,p,x), whose averages are fixed. We assume that the measure $\rho(dq\,dp\,dx)$ is absolutely continuous with respect to the reference measure $\lambda(dq\,dp\,dx)$, and denote, with an abuse of notation, by $\rho(q,p,x)$ the corresponding density. In this setting, the entropy is defined as

$$\mathfrak{S}_\lambda(\rho) = -\int_{\mathcal{E}} \rho(q,p,x) \ln \rho(q,p,x)\, \lambda(dq\,dp\,dx),$$

and the probability measure describing the system is obtained as the solution of the following maximization problem:

$$\sup_{\rho \in \mathcal{S}(A_1^0,\dots,A_M^0)} \left\{ \mathfrak{S}_\lambda(\rho) \right\}, \tag{1.42}$$

with

$$\mathcal{S}(A_1^0,\dots,A_M^0) \\ = \left\{ \rho \in L^1(\lambda) \;\middle|\; \rho \geq 0,\ \int_{\mathcal{E}} \rho\, d\lambda = 1,\ \int_{\mathcal{E}} A_i \rho\, d\lambda = A_i^0,\quad \forall i \in \{1,\dots,M\} \right\}.$$

The necessary condition to be satisfied by an extremum of (1.42) reads

$$\mathfrak{S}'_\lambda(\rho) + \alpha_0 + \sum_{i=1}^M \alpha_i A_i = 0.$$

Therefore,

$$\rho(q,p,x) = Z^{-1} \exp\left(\sum_{i=1}^M \alpha_i A_i(q,p,x) \right). \tag{1.43}$$

Remark 1.2 (Nonequilibrium steady states). *Let us stress that the above derivation is performed under the assumption that the system is at equilibrium. In particular,* no notion of dynamics is required. *For nonequilibrium systems in a steady state, the dynamics has to be made precise. It is not always clear whether a stationary probability measure exists, and, when it exists, whether it is unique and whether the distribution of the microscopic configurations converges to it. There are some positive results, see [Rey-Bellet (2006)] in the case of heat transport in one-dimensional atom chains. In general however, no explicit expression of the invariant measure is available, in contrast to formulas such as* (1.43).

Isobaric-isothermal ensemble (NPT). Let us now present a first application of the above general derivation. Isobaric-isothermal ensembles are characterized by the fact that the energy and the volume of the system are fixed in average only. Consider for example a periodic system for which the size of the unit cell can vary in one direction, and denote by $x > 0$ the length of unit cell in this direction (while it is fixed to L in the two remaining directions). Then,

$$\mathcal{D}_x = \left[x\mathbb{T} \times (L\mathbb{T})^2\right]^N, \qquad T^*\mathcal{D}_x = \left[x\mathbb{T} \times (L\mathbb{T})^2\right]^N \times \mathbb{R}^{3N}.$$

We choose a uniform measure on all possible volumes:

$$\mathcal{X} = (0, +\infty), \qquad \lambda(dq\,dp\,dx) = 1_{(q,p)\in T^*\mathcal{D}_x} 1_{x>0}\, dq\,dp\,dx.$$

The constraints to be taken into account are $A_1 = H$ (average energy fixed), and $A_2(x, q, p) = xL^2$ (average volume fixed).

Applying the results of the general derivation to the NPT case, it is easily seen that the probability measure describing the equilibrium is

$$\rho_{\mathrm{NPT}}(dq\,dp\,dx) = Z_{\mathrm{NPT}}^{-1}\, \mathrm{e}^{-\beta P L^2 x}\, \mathrm{e}^{-\beta H(q,p)}\, 1_{\{q\in[x\mathbb{T}\times(L\mathbb{T})^2]^N\}}\, dq\,dp\,dx,$$

where the Lagrange multiplier associated with the volume constraint is written as βP. The quantity P can be identified with the pressure.

Grand canonical ensemble (μVT). We now describe a second application of the above general derivation. Consider a fluid of N indistinguishable particles. The additional variable describing the microscopic state of the system is the number $N \in \mathbb{N}^*$ of particles contained in a periodic cubic box of volume L^3. For a given number N of particles, the set of admissible configurations is

$$T^*\mathcal{D}_N = (L\mathbb{T})^{3N} \times \mathbb{R}^{3N}.$$

The reference measure for the number N of particles

$$\sum_{n=1}^{+\infty} \frac{1}{n!}\,\delta_n(dN)$$

is the uniform measure on the set of positive integers, up to factors $n!$ which are related to the indistinguishability of the particles. (See for instance Chapter 3 in [Minlos (2000)] for further precision on the construction of the reference measure for the grand-canonical ensemble.) Therefore,

$$\lambda(dq\,dp\,dN) = \sum_{n=1}^{+\infty} \frac{1}{n!}\, 1_{(q,p)\in T^*\mathcal{D}_n}\, dq\,dp\,\delta_n(dN).$$

We denote by H_n the Hamiltonian function on each space $T^*\mathcal{D}_n$, which is a function of the variables $(q_1, \ldots, q_n, p_1, \ldots, p_n)$. The Hamiltonian H is then defined as $H(q,p,n) = H_n(q,p)$ for $(q,p) \in T^*\mathcal{D}_n$.

The constraints to be taken into account are $A_1 = H$ (average energy fixed) and $A_2(x,q,N) = N$ (average number of particles fixed). Applying the results of the general derivation, the grand-canonical equilibrium measure reads:

$$\rho_{\mu\mathrm{VT}}(dq\,dp\,dN) = Z_{\mu\mathrm{VT}}^{-1} \sum_{n=1}^{+\infty} \frac{\mathrm{e}^{\beta\mu n}}{n!}\,\mathrm{e}^{-\beta H_n(q,p)}\,1_{(q,p)\in T^*\mathcal{D}_n}\,dq\,dp\,\delta_n(dN), \tag{1.44}$$

where $\beta\mu$ is the Lagrange multiplier associated with the average number constraint.[3] The parameter μ can be identified with the chemical potential.

1.3 Free energy and its numerical computation

Free energy is a central concept in thermodynamics and in modern studies on biochemical and physical systems. The statistical physics definition of this quantity as the logarithm of the partition function

$$F = -\frac{1}{\beta}\ln \int_{T^*\mathcal{D}} \mathrm{e}^{-\beta H(q,p)}\,dq\,dp$$

is motivated in Section 1.3.1.

In many applications, the important quantity is actually the *free energy difference* between various macroscopic states of the system, rather than the free energy itself. Free energy differences allow to quantify the relative likelihood of different states. A state should be understood here as either

(i) the collection of all possible microscopic configurations, distributed according to the canonical measure (1.32), and satisfying a given macroscopic constraint $\xi(q) = z$, where $\xi : \mathcal{D} \to \mathbb{R}^m$ with m small. In this case, z is the index of the state; or
(ii) the collection of all possible microscopic configurations distributed according to the canonical measure associated with a Hamiltonian depending on some parameter λ. The parameter λ is then the index of the state.

[3]The notation μ for the chemical potential, standard in the physics and chemistry literature, should not be confused with the notation used for the canonical measure throughout this book.

This is explained in more detail in Section 1.3.2, where two examples are also provided.

Beside these practical motivations to compute free energy differences, a more numerical motivation is to use the free energy to devise algorithms which overcome sampling barriers. Indeed, it is often the case in practice that approximations (1.23) to ensemble averages exhibit a slow convergence. The trajectory generated by the numerical method typically remains trapped for a long time in some region of the phase space, and hops only occasionally to another region, where it also remains trapped for a long time. This occurs as soon as there exist several regions of phase space separated by very low probability areas. Such regions are called *metastable*. The concept of metastability may be formalized in various ways, see Section 2.3.2.2. Chemical and physical intuitions may guide the practitioners of the field toward the identification of some slowly evolving degree of freedom responsible for the metastable behavior of the system. This quantity is a function $\xi(q)$ of the configuration of the system, where $\xi : \mathcal{D} \to \mathbb{R}^m$ with m small. The framework to consider is therefore the case of transitions indexed by a reaction coordinate. If the function ξ is well chosen (*i.e.* if the dynamics in the direction orthogonal to ξ is not too metastable), the free energy can be used as a biasing potential to accelerate the sampling (1.23), see Section 1.3.3.

It is thus important to accurately compute free energy differences in order to assess the relative likelihood of physical states or to build efficient algorithms to overcome sampling barriers. The most important techniques to this end are briefly presented in Section 1.3.4, and then detailed in the following chapters.

1.3.1 *Absolute free energy*

We first define the free energy in Section 1.3.1.1, and then motivate this definition from a macroscopic thermodynamics perspective (Section 1.3.1.2).

1.3.1.1 *Definition*

We restrict ourselves throughout the book to the canonical ensemble, though most of the concepts and numerical methods considered can be extended to other thermodynamic ensembles (see Section 1.2.3.3). The *absolute free energy* of a system is defined as

$$F = -\frac{1}{\beta} \ln Z_\mu, \tag{1.45}$$

where Z_μ is the partition function

$$Z_\mu = \int_{T^*\mathcal{D}} \mathrm{e}^{-\beta H(q,p)}\, dq\, dp. \tag{1.46}$$

Since the potential energy function V (hence the Hamiltonian H) is defined only up to an additive constant when empirical potential functions are used, so is the absolute free energy. However, this has no consequence on free energy differences, see Section 1.3.2 below.

The free energy (1.45) is called the Helmholtz free energy. Similar free energies can be considered for other thermodynamic ensembles. They are also logarithms of the partition functions multiplied by a factor $-\beta^{-1}$. When the isobaric-isothermal ensemble (NPT) is considered, the associated free energy is called the Gibbs free energy.

For separable Hamiltonians (1.2), the partition function can be rewritten as

$$Z_\mu = Z_\nu \left(\frac{2\pi}{\beta}\right)^{3N/2} \prod_{i=1}^{N} m_i^{3/2}, \qquad Z_\nu = \int_{\mathcal{D}} \mathrm{e}^{-\beta V(q)}\, dq,$$

and the only difficulty is the computation of the configurational partition function Z_ν. This partition function cannot be computed as such in general. It however has a simple expression for some specific systems, such as the ideal gas, or solids at low temperature (resorting to the phonon spectrum, *i.e.* assuming that the interactions can be approximated by a sum of harmonic interactions), see [Frenkel and Smit (2002); Rickman and LeSar (2002)].

1.3.1.2 *Relationship with macroscopic thermodynamics*

We now motivate the definition (1.45) of the free energy, and also comment on the limits of the theory.

Analogy with the definition in thermodynamics. A first motivation for the definition (1.45) relies on an analogy with macroscopic thermodynamics, where the Helmholtz free energy of a system at constant temperature T is defined as

$$F = U - TS, \tag{1.47}$$

U being the internal energy of the system, and S its entropy. The microscopic definition of the internal energy is the average energy as given by the laws of statistical physics:

$$\mathbb{E}_\mu(H) = Z_\mu^{-1} \int_{T^*\mathcal{D}} H(q,p)\, \mathrm{e}^{-\beta H(q,p)}\, dq\, dp, \tag{1.48}$$

where the normalization constant Z_μ is given by (1.46). Besides, the microscopic quantity, proportional to the statistical entropy (1.36) encountered in Section 1.2.3.2,

$$\Sigma = -k_\mathrm{B} \int_{T^*\mathcal{D}} \ln\left(\frac{d\mu}{dq\, dp}\right) d\mu \tag{1.49}$$

is proportional to the mathematical relative entropy of the canonical measure (1.32) with respect to the Lebesgue measure $dq\, dp$. It has many similarities with the (macroscopic) thermodynamic entropy S, as shown in [Gibbs (1902)]. Replacing U and S in (1.47) by their microscopic counterparts defined in (1.48) and (1.49) respectively, we obtain a quantity $\mathcal{F}$ which should be *similar* to some free energy:

$$\mathcal{F} = \mathbb{E}_\mu(H) - T\Sigma = -\frac{1}{\beta} \ln Z_\mu. \tag{1.50}$$

Work and heat exchanges. A second motivation for the definition (1.45) relies on a decomposition of energy exchanges into work and heat for isothermal transformations. This requires however the notion of free energy differences, so that the corresponding discussion is postponed to Section 1.3.2.1.

Validity and relevance of these motivations. In spite of the formal analogies highlighted above, the relationships between the microscopic definition of the free energy or entropy, and their counterparts in classical macroscopic thermodynamics are still not completely clear. Quoting [Balian (2007)] (see Section 3.4.6):

> Notwithstanding the many interrelations which have been established between the different kinds of entropy, the identification of the thermodynamic entropy and the statistical entropy has not yet been accepted universally. While the former can be measured more or less directly for systems in thermodynamic equilibrium and thus appears to be a property of the system itself, the latter refers to the knowledge of the system by an observer and does have a nature which is partially subjective, or at least anthropocentric and relative. It certainly may appear paradoxical that these two quantities would be equal to one another.

1.3.2 *Relative free energies*

As mentioned above, the quantity of interest in many applications and in this book is not the absolute free energy, but the *free energy differences* between various states. Typical examples studied by computer simulations include the solvation free energy (which is the free energy difference between a molecule *in vacuo* and its counterpart surrounded by solvent molecules), and the binding free energy of two molecules (this free energy difference determines whether a new drug can have an efficient action on a given protein for example). See [Chipot and Pohorille (2007b)] for other relevant examples in chemistry and biophysics.

In this section, we describe more precisely what we mean by *states*, and how a transition between two states can be defined. As already hinted at in the introduction to this section, two cases should be considered: alchemical transitions (Section 1.3.2.1) and transitions indexed by a reaction coordinate (Section 1.3.2.2). In order to fix the ideas, we illustrate each type of transition with a typical example: computation of chemical potential through Widom insertion in the alchemical case (see Section 1.3.2.3), and dimer molecule in a solvent in the reaction coordinate case (see Section 1.3.2.4). These examples will also be our running examples for the numerical illustrations throughout the book.

1.3.2.1 *Alchemical transitions*

The so-called *alchemical case* considers transitions indexed by an external parameter λ, independent of the microscopic phase space configuration (q,p). Typical examples are the intensity of an applied magnetic field for a spin system, or the constants used in the empirical force fields (such as the energy ε or the length σ in the Lennard-Jones potential (1.5)). See Section 1.3.2.3 below and Section 2.8 in [Chipot and Pohorille (2007a)] for more examples. The name "alchemical" refers to the fact that the nature of the particles at hand can be modified in the computer simulation by changing the parameters of the potential describing the molecular interactions.

For a given value of λ, the system is described by a Hamiltonian H_λ. A state is then the collection of all possible microscopic configurations $T^*\mathcal{D}$, distributed according to the canonical measure

$$\mu_\lambda(dq\,dp) = \frac{1}{Z_\lambda}\mathrm{e}^{-\beta H_\lambda(q,p)}\,dq\,dp, \qquad Z_\lambda = \int_{T^*\mathcal{D}} \mathrm{e}^{-\beta H_\lambda(q,p)}\,dq\,dp. \quad (1.51)$$

An alchemical transition transforms the state $\lambda = 0$ into the state $\lambda = 1$.

The corresponding free energy difference is

$$F(1) - F(0) = -\beta^{-1} \ln \left(\frac{\displaystyle\int_{T^*\mathcal{D}} \mathrm{e}^{-\beta H_1(q,p)}\, dq\, dp}{\displaystyle\int_{T^*\mathcal{D}} \mathrm{e}^{-\beta H_0(q,p)}\, dq\, dp} \right). \tag{1.52}$$

It is often the case that H_λ depends on λ only through the potential energy V_λ. In this case, the free energy difference simplifies as

$$F(1) - F(0) = -\beta^{-1} \ln \left(\frac{\displaystyle\int_{\mathcal{D}} \mathrm{e}^{-\beta V_1(q)}\, dq}{\displaystyle\int_{\mathcal{D}} \mathrm{e}^{-\beta V_0(q)}\, dq} \right). \tag{1.53}$$

Work and heat exchanges in a reversible isothermal transformation. Now that alchemical transitions have been defined, we can come back to the motivation of the definition of free energy in terms of work and heat exchanges, see Section 1.3.1.2. An alchemical transformation can be considered as isothermal since there is a common thermodynamic temperature in the family of measures (1.51), defined through the factor β.

In accordance with (1.48), the energy of the state described by H_λ is

$$U(\lambda) = \mathbb{E}_{\mu_\lambda}(H_\lambda) = Z_\lambda^{-1} \int_{T^*\mathcal{D}} H_\lambda(q,p)\, \mathrm{e}^{-\beta H_\lambda(q,p)}\, dq\, dp,$$

while, in view of (1.49), the corresponding microscopic entropy is

$$\mathfrak{S}(\lambda) = -k_{\mathrm{B}} \int_{T^*\mathcal{D}} \ln \left(\frac{d\mu_\lambda}{dq\, dp} \right) \mu_\lambda(dq\, dp).$$

A simple computation shows that

$$\frac{d}{d\lambda}\mathbb{E}_{\mu_\lambda}(H_\lambda) = \mathbb{E}_{\mu_\lambda}\left(\frac{\partial H_\lambda}{\partial \lambda}\right) - \beta \left[\mathbb{E}_{\mu_\lambda}\left(H_\lambda \frac{\partial H_\lambda}{\partial \lambda}\right) - \mathbb{E}_{\mu_\lambda}(H_\lambda)\, \mathbb{E}_{\mu_\lambda}\left(\frac{\partial H_\lambda}{\partial \lambda}\right) \right].$$

This relation can be rewritten by decomposing the energy variation of the system (as λ changes) as a work contribution, supplemented with some heat exchange:

$$\frac{dU}{d\lambda} = \frac{dW}{d\lambda} + \frac{dQ}{d\lambda}, \qquad \frac{dQ}{d\lambda} = T\, \frac{d\mathfrak{S}}{d\lambda}. \tag{1.54}$$

Here, W is the so-called reversible work, by definition equal to the variation of the free-energy $F(\lambda) = -\beta^{-1} \ln Z_\lambda$:

$$\frac{dW}{d\lambda} = F'(\lambda) = \mathbb{E}_{\mu_\lambda}\left(\frac{\partial H_\lambda}{\partial \lambda}\right).$$

The equality (1.54) has a well-known analogue in standard thermodynamics: During a reversible isothermal transformation $d\lambda$, the heat exchanged with the thermostat, defined as the energy variation minus the exerted work $dU - \delta W = \delta Q$, is an exact differential $\delta Q = T d\mathfrak{S}$ involving the entropy $\mathfrak{S}$ of the system. This interpretation of (1.54) in terms of standard thermodynamics is due to Boltzmann in its original research on the microscopic interpretation of macroscopic thermodynamics (see Section 1.5 in [Gallavotti (1999)] for further precision). Note that in this microscopic framework the heat exchange is related to the variation of the weight $Z_\lambda^{-1}\,\mathrm{e}^{-\beta H_\lambda(q,p)}$ of the configurations during a reversible transformation.

1.3.2.2 *Transitions indexed by a reaction coordinate*

In the *reaction coordinate case*, the Hamiltonian of the system is kept fixed. A state is a measure on a submanifold of the phase space. These submanifolds are the level sets of some function, the so-called *reaction coordinate*,

$$\xi : \mathcal{D} \to \mathbb{R}^m,$$

with $m \leq 3N$. Examples of such functions are dihedral angles, or distances between two molecular subgroups, as in the example presented in Section 1.3.2.4 below. To ξ is associated a foliation of the phase space into submanifolds $\Sigma(z) = \{q \in \mathcal{D} \mid \xi(q) = z\}$, so that

$$\mathcal{D} = \bigcup_{z \in \mathbb{R}^m} \Sigma(z).$$

We assume in the sequel that the submanifolds $\Sigma(z)$ are simply connected,[4] and that $\Sigma(z_1) \neq \Sigma(z_2)$ when $z_1 \neq z_2$.

The free energy difference is related to the relative likelihoods of marginal distributions with respect to ξ. For the canonical measure (1.32), the marginal distribution is by definition

$$\mu^\xi(dz) = \left(\frac{1}{Z_\mu} \int_{\Sigma(z)\times\mathbb{R}^{3N}} \mathrm{e}^{-\beta H(q,p)}\, \delta_{\xi(q)-z}(dq)\, dp \right) dz.$$

It is the image of the measure μ by ξ. The measure $\delta_{\xi(q)-z}(dq)$ is defined as in (1.25) and (1.26) through the relation $\delta_{\xi(q)-z}(dq)\, dz = dq$, see also Sections 3.2.1 and 3.3.2 for further precision. In particular, it can be written as

$$\delta_{\xi(q)-z}(dq) = \frac{\sigma_{\Sigma(z)}(dq)}{|\nabla \xi(q)|}, \tag{1.55}$$

[4]This will be important to state ergodicity results for dynamics constrained to remain on $\Sigma(z)$.

where $\sigma_{\Sigma(z)}(dq)$ is the area measure induced by the Lebesgue measure on the manifold $\Sigma(z)$ when $\mathcal{D}$ is equipped with the standard Euclidean scalar product. The free energy is then defined as the log-density of the marginal distribution:

$$\mathrm{e}^{-\beta F(z)}\, dz = \mu^{\xi}(dz).$$

Thus, $\exp(-\beta[F(1) - F(0)])$ can be interpreted as the relative likelihood of states in $\Sigma(1)$ compared to states in $\Sigma(0)$. More explicitly,

$$F(z) = -\beta^{-1} \ln \left(\frac{1}{Z_\mu} \int_{\Sigma(z)\times\mathbb{R}^{3N}} \mathrm{e}^{-\beta H(q,p)}\, \delta_{\xi(q)-z}(dq)\, dp \right). \qquad (1.56)$$

The free energy can therefore also be seen as some effective potential associated with ξ. The function $z \mapsto F(z)$ is called *potential of mean force.* This terminology is motivated by the fact that $F'(z)$, called the *mean force,* is some average force exerted on the system when the reaction coordinate is kept constant, see Chapter 3 for further precision.

The free energy difference between the state $\Sigma(0)$ and the state $\Sigma(1)$ is finally defined as

$$F(1) - F(0) = -\beta^{-1} \ln \left(\frac{\displaystyle\int_{\Sigma(1)\times\mathbb{R}^{3N}} \mathrm{e}^{-\beta H(q,p)}\, \delta_{\xi(q)-1}(dq)\, dp}{\displaystyle\int_{\Sigma(0)\times\mathbb{R}^{3N}} \mathrm{e}^{-\beta H(q,p)}\, \delta_{\xi(q)}(dq)\, dp} \right). \qquad (1.57)$$

For separable Hamiltonians, (1.2), the free energy difference can be rewritten as

$$F(1) - F(0) = -\beta^{-1} \ln \left(\frac{\displaystyle\int_{\Sigma(1)} \mathrm{e}^{-\beta V(q)}\, \delta_{\xi(q)-1}(dq)}{\displaystyle\int_{\Sigma(0)} \mathrm{e}^{-\beta V(q)}\, \delta_{\xi(q)}(dq)} \right). \qquad (1.58)$$

Notice that when $|\nabla \xi|$ is constant, the free energy difference $F(1) - F(0)$ only depends upon ξ through $\Sigma(1)$ and $\Sigma(0)$ in view of (1.55).

Remark 1.3 (Choice of the reaction coordinate). *For a given foliation of the configurational space, the free energy difference depends in general on the choice of the reaction coordinate indexing this foliation. Indeed, consider another reaction coordinate $\widetilde{\xi}$, defining the same level sets, with in particular*

$$\widetilde{\Sigma}(0) = \left\{ q \;\middle|\; \widetilde{\xi}(q) = 0 \right\} = \left\{ q \mid \xi(q) = 0 \right\} = \Sigma(0), \qquad (1.59)$$

and a similar relation for $z = 1$. For instance, $\widetilde{\xi} = f(\xi)$ with any one-to-one increasing function $f : [0,1] \to [0,1]$ has the same level sets as ξ, and satisfies (1.59).

In general, the associated free energy differences $F(1) - F(0)$ and $\widetilde{F}(1) - \widetilde{F}(0)$ are different. *Indeed, the surface measure $\sigma_{\Sigma(0)}(dq)$ is somehow intrinsic (it depends only on the submanifold $\Sigma(0) = \widetilde{\Sigma}(0)$ and on the ambient scalar product), while the measure $\delta_{\xi(q)-z}(dq)$ depends on the gradients of the reaction coordinates through the factors $|\nabla\xi(q)|^{-1}$, see the right-hand side of* (1.55). *It is therefore a* modelling choice *to decide which reaction coordinate to use, in particular when comparing results of numerical simulations to experimental measurements. Of course, there are no such issues in the alchemical case.*

Remark 1.4 (Relation with the alchemical setting).
Alchemical transitions can be considered as a special case of transitions indexed by a reaction coordinate, upon introducing the extended variable $Q = (\lambda, q)$ and the reaction coordinate $\xi(Q) = \lambda$. In this case, the geometry of the submanifolds is very simple since $|\nabla\xi(Q)| = 1$. The level sets are $\Sigma(\lambda) = \{\lambda\} \times \mathcal{D}$, and the measure $\delta_{\xi(Q)-\lambda}(dQ)$ in the extended space is the Lebesgue measure dq on $\mathcal{D}$.

Besides, the reaction coordinate case is sometimes considered as a limiting case of the alchemical case, using the family of Hamiltonians

$$H_\lambda^\eta(q) = V(q) + \frac{1}{2\eta}\Big(\xi(q) - \lambda\Big)^2 + \frac{1}{2}p^T M^{-1} p,$$

and letting $\eta \to +\infty$. See Section 5.1.2 *for further precision.*

1.3.2.3 *A typical alchemical transition: Widom insertion*

We describe here the running example used to illustrate simulation results for alchemical transitions. We consider a fluid composed of N particles and enclosed in a domain $\mathcal{D} = (L\mathbb{T})^d$, where the physical dimension is $d = 2$. The chemical potential is defined as

$$\mu = F(N+1) - F(N),$$

where $F(N)$ is the free energy of the system composed of N particles in the canonical ensemble for the domain $(L\mathbb{T})^d$ at a given inverse temperature β. In fact, the chemical potential is equal, in the thermodynamic limit, to the Lagrange multiplier μ used to define the grand-canonical measure (1.44) (see Section 5.6.3 in [Balian (2007)]).

The chemical potential can be rewritten as

$$\mu = \mu_{\mathrm{id}} + \mu_{\mathrm{ex}},$$

where the so-called "ideal gas contribution" μ_{id} comes from the kinetic part of the partition function, and has an analytic expression (see [Frenkel and Smit (2002)]). The challenge is the computation of the "excess chemical potential" μ_{ex}, which arises from interactions between the fluid particles. Denoting by $q^N \in \mathcal{D}_N = (L\mathbb{T})^{dN}$ the positions of N particles,

$$\mu_{\mathrm{ex}} = -\beta^{-1} \ln \left(\frac{\displaystyle\int_{\mathcal{D}_{N+1}} \mathrm{e}^{-\beta V_{N+1}(q^{N+1})}\, dq^{N+1}}{\displaystyle L^d \int_{\mathcal{D}_N} \mathrm{e}^{-\beta V_N(q^N)}\, dq^N} \right), \tag{1.60}$$

where $V_N(q^N)$ is the potential energy function for a fluid composed of N particles, and $\mathcal{D}_N = (L\mathbb{T})^{dN}$ is the associated configuration space. Notice that $\mu_{\mathrm{ex}} = 0$ when the potential functions are $V_N = V_{N+1} = 0$ thanks to the factor L^d in (1.60).

Denoting the canonical measure for a fluid of N particles by

$$\mu_N(dq^N) = Z_N^{-1}\, \mathrm{e}^{-\beta V_N(q^N)}\, dq^N, \qquad Z_N = \int_{\mathcal{D}_N} \mathrm{e}^{-\beta V(q^N)}\, dq^N,$$

and defining the energy difference between a fluid of N and $N+1$ particles as

$$\Delta_N V(q^N, q) = V_{N+1}(q^{N+1}) - V_N(q^N),$$

when $q^{N+1} = (q^N, q)$, the chemical potential (1.60) can be rewritten in the form (1.53) upon defining a potential function on $\mathcal{D}_{N+1}$:

$$V_\lambda(q^N, q) = V_N(q^N) + \lambda \Delta_N V(q^N, q). \tag{1.61}$$

Indeed,

$$\mu_{\mathrm{ex}} = -\frac{1}{\beta} \ln \left(\frac{\displaystyle\int_{\mathcal{D}_{N+1}} \mathrm{e}^{-\beta V_1(q^{N+1})}\, dq^{N+1}}{\displaystyle\int_{\mathcal{D}_{N+1}} \mathrm{e}^{-\beta V_0(q^{N+1})}\, dq^{N+1}} \right).$$

Another expression of the excess chemical potential, which will be useful for later purposes, is:

$$\mu_{\mathrm{ex}} = -\beta^{-1} \ln \left(\frac{1}{L^d} \int_{\mathcal{D}_N \times \mathcal{D}_1} \mathrm{e}^{-\beta \Delta_N V(q^N, q^1)}\, \mu_N(dq^N)\, dq^1 \right). \tag{1.62}$$

The idea of the alchemical transition is therefore to go from a system of $N+1$ particles where one of the particles does not have any interaction with the others, to a system of $N+1$ fully interacting particles. When there are sufficiently many particles in the simulation box, adding an extra one requires some energy since some space must be created. The alchemical transition consists in progressively switching on the interactions with the $(N+1)$-th particle.

The computational results presented in this book have been obtained for a system with pairwise interactions, so that

$$V_N(q^N) = \sum_{1 \le i < j \le N} \mathcal{V}(|q_i - q_j|).$$

As in [Hendrix and Jarzynski (2001); Oberhofer *et al.* (2005)], we use a smoothed Lennard-Jones potential (in order to avoid the singularities at the origin). This potential reads

$$\mathcal{V}(r) = \begin{cases} a - br^2, & 0 \le r \le 0.8\,\sigma, \\ \Phi_{\rm LJ}(r) + c(r - r_{\rm cut}) - d, & 0.8\,\sigma \le r \le r_{\rm cut}, \\ 0, & r \ge r_{\rm cut}, \end{cases} \tag{1.63}$$

where

$$\Phi_{\rm LJ}(r) = 2\varepsilon \left(\frac{1}{2} \left(\frac{\sigma}{r}\right)^{12} - \left(\frac{\sigma}{r}\right)^{6} \right),$$

is the Lennard-Jones potential expressed in length units such that the equilibrium position corresponds to $r = \sigma$: $\Phi'_{\rm LJ}(\sigma) = 0$. The value $r_{\rm cut} = 2.5\,\sigma$ is a prescribed cut-off radius. The numbers a, b, c, d ensure that the potential is C^1.

1.3.2.4 *A typical transition indexed by a reaction coordinate: Dimer in a solvent*

We now describe the running example used to illustrate simulation results for transitions indexed by a reaction coordinate. We consider a system composed of N particles in a two-dimensional periodic box of side length L. Among these particles, two particles (numbered 1 and 2 in the following) are designated to form a dimer while the others are solvent particles.

All particles, except the two particles forming the dimer, interact through the purely repulsive WCA pair potential, which is a truncated Lennard-Jones potential [Dellago *et al.* (1999); Straub *et al.* (1988)]:

$$V_{\rm WCA}(r) = \begin{cases} 4\varepsilon \left[\left(\frac{\sigma}{r}\right)^{12} - \left(\frac{\sigma}{r}\right)^{6} \right] + \varepsilon & \text{if } r \le r_0, \\ 0 & \text{if } r > r_0, \end{cases}$$

where r denotes the distance between two particles, ε and σ are two positive parameters and $r_0 = 2^{1/6}\sigma$. The interaction potential between the two particles of the dimer is a double-well potential

$$V_S(r) = h\left[1 - \frac{(r - r_0 - w)^2}{w^2}\right]^2, \tag{1.64}$$

where h and w are two positive parameters. The total energy of the system is therefore, for $q \in (L\mathbb{T})^{dN}$ with $d = 2$,

$$V(q) = V_S(|q_1 - q_2|) + \sum_{3 \le i < j \le N} V_{WCA}(|q_i - q_j|) + \sum_{i=1,2} \sum_{3 \le j \le N} V_{WCA}(|q_i - q_j|),$$

where q_1 and q_2 are the positions of the two particles forming the dimer.

The potential V_S has two energy minima. The first one, at $r = r_0$, corresponds to the compact state. The second one, at $r = r_0 + 2w$, corresponds to the stretched state. The height of the energy barrier separating the two states is h. Figure 1.7 presents a schematic view of the system.

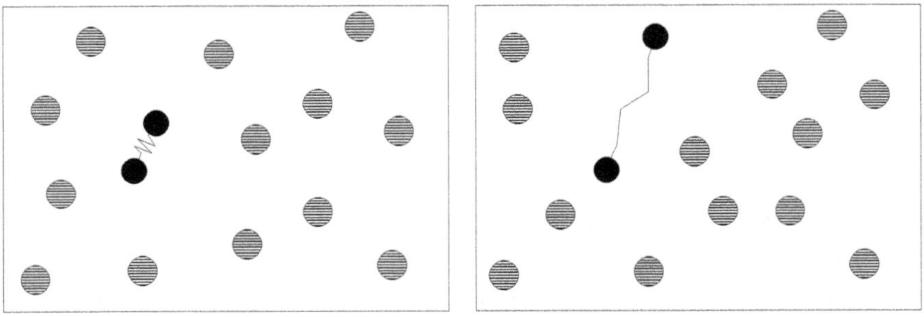

Fig. 1.7 Schematic views of the system, when the dimer is in the compact state (Left), and in the stretched state (Right). The interaction of the particles forming the dimer is described by a double-well potential. All the other interactions are of WCA form.

The reaction coordinate used to describe the transition from the compact to the stretched state is the normalized bond length

$$\xi(q) = \frac{|q_1 - q_2| - r_0}{2w}, \tag{1.65}$$

of the dimer molecule. The compact state (resp. the stretched state) corresponds to the value $z = 0$ (resp. $z = 1$) of the reaction coordinate.

1.3.3 *Free energy and metastability*

Standard molecular simulation techniques such as those that will be presented in Chapter 2 often experience difficulties in sampling *metastable* potentials. Potentials are called metastable when the corresponding canonical measure has several regions of high probability separated by low-probability regions. Typical numerical methods spend a lot of time in one given metastable basin, and only rarely escape it to visit another basin. These escapes are rare but fast events. The notion of metastability may be formalized and quantified in several ways, see Section 2.3.2.2 for a more detailed discussion. Some examples of metastable potentials are described in Sections 1.3.3.1 and 1.3.3.2 below.

We motivate in this section the interest of free energy methods for the sampling of metastable potentials. Such methods can be used provided the low- and high-probability regions of the systems are the level sets of some function $\xi(q)$, which is still called a reaction coordinate. Alternatively, $\xi(q)$ can be seen as some slowly evolving degrees of freedom encoding some coarse-grained information on the system. The free energy associated with ξ may then be used as a biasing potential enforcing transitions from one metastable basin to another. We show an instance of this strategy in Section 1.3.3.3 for the potentials considered in Sections 1.3.3.1 and 1.3.3.2. Of course the reliability of the method crucially depends on the choice of the reaction coordinate. This is a very important problem in practice, unfortunately rather ill-posed.

1.3.3.1 *A simple example of metastable dynamics*

Consider the potential energy

$$V(x,y) = \frac{1}{6}\Big[4(1-x^2-y^2)^2 + 2(x^2-2)^2 + ((x+y)^2-1)^2 + ((x-y)^2-1)^2\Big], \tag{1.66}$$

and a single particle $q = (x,y)$ evolving according to the overdamped Langevin dynamics:

$$dq_t = -\nabla V(q_t)\,dt + \sqrt{\frac{2}{\beta}}\,dW_t.$$

Figure 1.8 presents the level sets of the potential (1.66) and a typical trajectory.

The overdamped Langevin dynamics can be shown to be ergodic for the canonical probability measure $\nu(dq) = Z^{-1}\exp(-\beta V(q))\,dq$ (see Section 2.2.2 for more detail on the overdamped Langevin dynamics and its numerical implementation). The dynamics projected in the y variable is irrelevant, whereas the time evolution of the x variable shows that it is a "slow" variable. If the average position $\mathbb{E}_\nu(x)$ is computed as a time-average along a trajectory, the convergence is very slow (compared to the convergence of the average $\mathbb{E}_\nu(y)$ for instance). This suggests to choose $\xi(x,y) = x$.

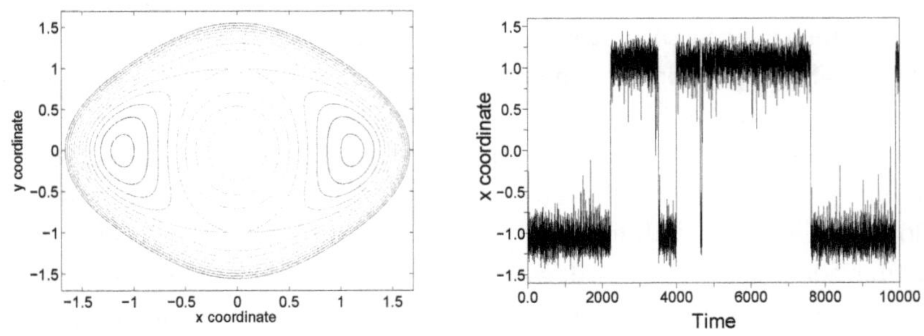

Fig. 1.8 Left: Level sets of the potential (1.66). Right: Projected trajectory in the x variable for $\Delta t = 0.01$, $\beta = 6$.

For later purposes, we compute the free energy profile for the reaction coordinate $\xi(x,y) = x$:

$$F(x_2) - F(x_1) = -\beta^{-1} \ln \left(\frac{\psi^\xi(x_2)}{\psi^\xi(x_1)} \right), \tag{1.67}$$

where the marginals ψ^ξ of the equilibrium canonical distribution are

$$\psi^\xi(x) = \int_{\mathbb{R}} \mathrm{e}^{-\beta V(x,y)} \, dy.$$

This profile is illustrated in Figure 1.9, together with

$$F'(x) = \frac{\displaystyle\int_{\mathbb{R}} \partial_x V(x,y) \, \mathrm{e}^{-\beta V(x,y)} \, dy}{\displaystyle\int_{\mathbb{R}} \mathrm{e}^{-\beta V(x,y)} \, dy}.$$

Notice that F' is the opposite of the averaged force experienced in the direction of the reaction coordinate (the so-called mean force). There is a high free energy barrier at $x = 0$, which corresponds to a small value of $\psi^\xi(x)$. This barrier is at the origin of the metastable behavior since it separates two regions of high probability.

1.3.3.2 *Entropic and energetic barriers*

Free energy barriers can have two origins, related to either energetic or entropic bottlenecks. We give below two toy examples of purely energetic and purely entropic barriers. Of course, in general, both components are

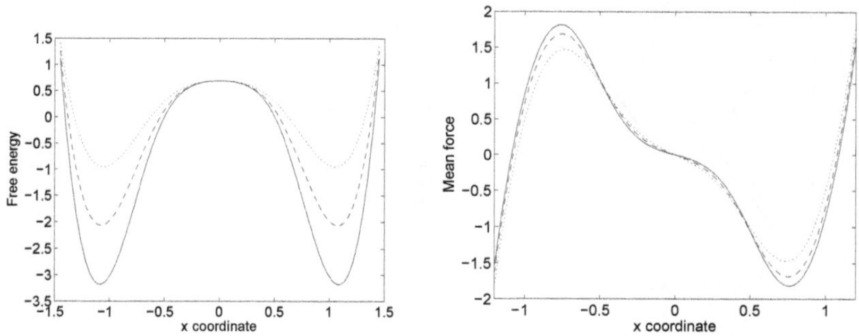

Fig. 1.9 Left: Potential of mean force for the potential plotted in Figure 1.8, using the x coordinate as reaction coordinate. From top to bottom: $\beta = 2$ (dotted line), $\beta = 3$ (dashed line), $\beta = 4$ (solid line). Right: Associated mean forces.

mixed, and it is not so obvious to decide whether the metastability of the dynamics rather has an energetic or an entropic origin (except in some limiting temperature regime, see the discussion at the end of this section).

Purely energetic barrier. Consider $q = (q_1, \ldots, q_N) \in \mathbb{R}^N$, $p \in \mathbb{R}^N$, and

$$H(q,p) = W(q_1) + V(q_2, \ldots, q_N) + \frac{1}{2} p^T M^{-1} p, \tag{1.68}$$

where W is a one-dimensional double-well potential $W(q_1) = h(q_1^2 - 1)^2$ with h large enough. Then, choosing the first coordinate q_1 as a reaction coordinate: $\xi(q) = q_1$, it holds (up to a multiplicative constant which does not depend on z):

$$\mathrm{e}^{-\beta F(z)} = \int_{\mathbb{R}^{2N-1}} \mathrm{e}^{-\beta H(z, q_2, \ldots, q_N, p_1, \ldots, p_N)} \, dq_2 \ldots dq_N \, dp_1 \ldots dp_N,$$

so that

$$F(z_2) - F(z_1) = W(z_2) - W(z_1).$$

In this case, it is clear that free energy barriers are purely of energetic origin.

Purely entropic barrier. Entropic barriers are often encountered in complex systems with many degrees of freedom. In this case, the system typically has enough energy to overcome the energetic barriers it can encounter, but has not, somehow, got its energy concentrated in the right modes or directions. It is expected that entropic barriers increase with the

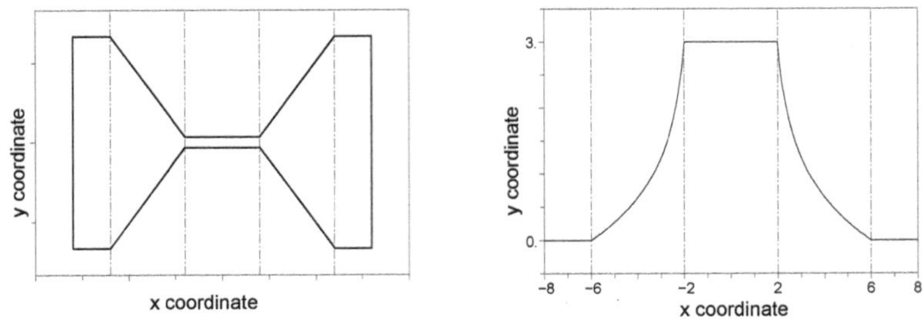

Fig. 1.10 Left: Potential for which entropic barriers have to be overcome, in the case $L_1 = 2$, $L_2 = 4$ and $L_3 = 2$. The potential is 0 in the region enclosed by the curve, and $+\infty$ outside. Right: Associated free energy profile when the x coordinate is the reaction coordinate ($\beta = 1$).

dimensionality of the system (think of a random walk in a high-dimensional space).

A toy model of an entropic barrier is the potential presented in Figure 1.10. The potential is zero inside the curve, and $+\infty$ outside, so that the system is confined in the bone-shaped region. Here, $q = (x, y) \in \mathcal{D} = \{q \in \mathbb{R}^2 \mid V(q) = 0\}$. Denote by d the width of the tunnel between the two metastable regions, by $2L_1$ its length, by L_2 the length of the transition region, and by L_3 the length of the initial and final rectangular domains, which are of heights Δ. We choose $\xi(q) = x$ as the reaction coordinate. Then,

$$F(x) = \begin{cases} -\beta^{-1} \ln d & \text{when } |x| \le L_1, \\ -\beta^{-1} \ln \left(d + \dfrac{\Delta - \delta}{L_2}(|x| - L_1) \right) & \text{when } L_1 \le |x| \le L_1 + L_2, \\ -\beta^{-1} \ln \Delta & \text{when } L_1 + L_2 \le |x| \le L_1 + L_2 + L_3. \end{cases} \tag{1.69}$$

There is a free energy barrier in the tunnel region, arising from the contraction of the phase space volume: Less configurations are accessible, although the energy has not changed. This barrier has no energy component in it since the average energy for a fixed value of the reaction coordinate is zero.

Figure 1.11 presents a typical trajectory in the case $L_1 = L_3 = 2$, $L_2 = \Delta = 4$, $\delta = 0.2$, for a Metropolis random walk with isotropic Gaussian moves of variance $2\tau/\beta$ (see Section 2.1.2 for further precision on the Metropolis algorithm). Here, this amounts to proposing a new position $\tilde{q}^{n+1}$ as

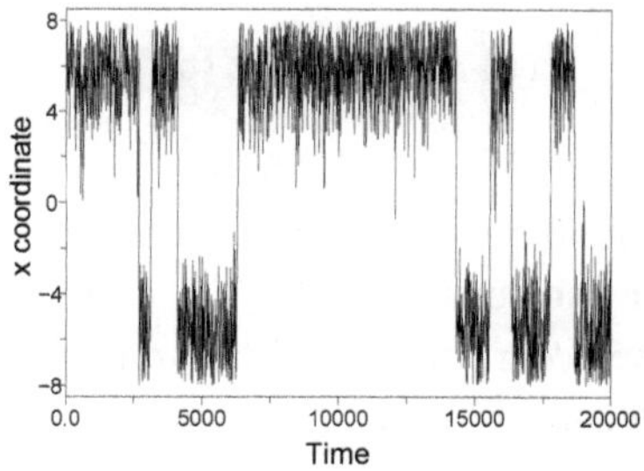

Fig. 1.11 Typical trajectory of the variable x for the potential presented in Figure 1.10, when a Metropolis dynamics is used, for the parameters $\tau = 0.1$ and $\beta = 1$. The time variable is defined as the number of iterations times the typical time τ.

$$\tilde{q}^{n+1} = q^n + \sqrt{\frac{2\tau}{\beta}}\, G^n,$$

where $(G^n)_{n \geq 0}$ are independent and identically distributed centered Gaussian random variables of identity covariance; and setting $q^{n+1} = \tilde{q}^{n+1}$ when $\tilde{q}^{n+1} \in \mathcal{D}$, and $q^{n+1} = q^n$ otherwise. The simulation results show that the x coordinate only significantly varies on long timescales, which is a typical signature of metastability.

Temperature dependence of the free energy barrier. The temperature dependence of the free energy barrier is a good indicator of the nature of the bottleneck. Indeed, in the case of a purely energetic barrier (1.68), the ratio of the marginal distributions

$$\mathrm{e}^{-\beta(F(z_1)-F(z_0))} = \mathrm{e}^{-\beta(W(z_1)-W(z_0))}$$

varies exponentially as a function of β, whereas, for the example (1.69) of purely entropic barrier, this ratio does not depend on β. In general, it is expected that free energy barriers at low temperatures (*i.e.* in the limit $\beta \to +\infty$) are mostly of energetic nature, in accordance with large deviation principles [Freidlin and Wentzell (1998)]. On the other hand, at high temperatures (in the limit $\beta \to 0$),

$$\begin{aligned} F(z_1) - F(z_0) &= -\frac{1}{\beta} \ln \left(\frac{\displaystyle\int_{\mathcal{D}} \mathrm{e}^{-\beta V(q)}\, \delta_{\xi(q)-z_1}(dq)}{\displaystyle\int_{\mathcal{D}} \mathrm{e}^{-\beta V(q)}\, \delta_{\xi(q)-z_0}(dq)} \right) \\ &\simeq -\frac{1}{\beta} \ln \left(\frac{\displaystyle\int_{\mathcal{D}} \delta_{\xi(q)-z_1}(dq)}{\displaystyle\int_{\mathcal{D}} \delta_{\xi(q)-z_0}(dq)} \right), \end{aligned}$$

provided the integrals

$$I(z) = \int_{\mathcal{D}} \delta_{\xi(q)-z}(dq)$$

are finite for z_0 and z_1. In this case the free energy difference is controlled at first order by the entropic contribution. Indeed, $I(z)$ measures the accessible phase space for the constraint $\xi(q) = z$, and some entropy can be defined from this volume according to Boltzmann's definition of the entropy as the logarithm of a density of states.

1.3.3.3 *Free energy biased sampling*

In the simple examples considered in Sections 1.3.3.1 and 1.3.3.2, the slowly evolving variable is known. There is a clear free energy barrier when using $\xi(x, y) = x$ as a reaction coordinate for (1.67) and (1.69). It is then possible to bias the dynamics in the x variable in order to remove the free energy barrier. More precisely, we now sample the modified potential

$$V(q) - F(\xi(q)).$$

Notice that the free energy associated to the reaction coordinate ξ for this modified potential is constant:

$$-\beta^{-1} \ln \int_{\Sigma(z)} \mathrm{e}^{-\beta(V-F\circ\xi)(q)}\, \delta_{\xi(q)-z}(dq) = \beta^{-1} \ln Z_\mu.$$

The above formula is a consequence of the definition (1.56) of the free energy $F(z)$, using also the equality $F(\xi(q)) = F(z)$ on $\Sigma(z)$. The marginal law of $Z^{-1}\,\mathrm{e}^{-\beta(V-F\circ\xi)(q)}\,dq$ along ξ is therefore the uniform law.

If ξ completely describes the metastability of the potential V as in the previous examples, the modified potential $V - F\circ\xi$ is no longer metastable. An efficient importance sampling method can then be obtained, especially when F does not vary too much (see Section 2.4.1.4 for further precision on importance sampling). We now numerically illustrate this strategy.

Application to the two-dimensional double-well potential. Consider the system described by the potential (1.66). Figure 1.12 presents the new potential $V - F\circ\xi$ (where the free energy bias, computed with standard quadrature rules, has been applied for $|x| \leq 1.7$) and a typical trajectory of the overdamped Langevin dynamics for the potential $V - F \circ \xi$, projected on the x coordinate. The comparison with Figure 1.8 shows that the transitions from the region $x < 0$ to the region $x > 0$ are now sufficiently frequent in order to attain good sampling accuracies.

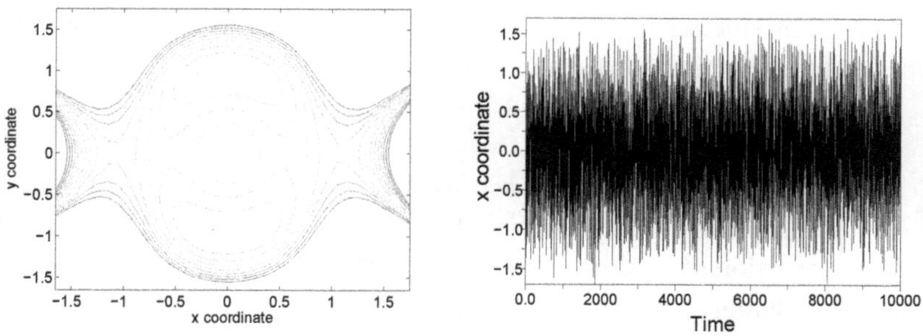

Fig. 1.12 Left: Modified potential $V - F \circ \xi$. Right: Projected trajectory in the x variable for $\Delta t = 0.01$, $\beta = 6$ for the dynamics associated with the modified potential.

Application to the entropic barrier problem. Figure 1.13 presents the results for a Metropolis random-walk dynamics biased by the free energy (1.69) in the case of the potential presented in Figure 1.10 (see Section 1.3.3.2 for a brief description of the dynamics). As in the previous case, the metastability is removed, and many transitions are observed from one well to the other (compare with Figure 1.11). The effect of the free energy bias is to increase the likelihood of regions close to the transition zone, so that many more crossings are attempted.

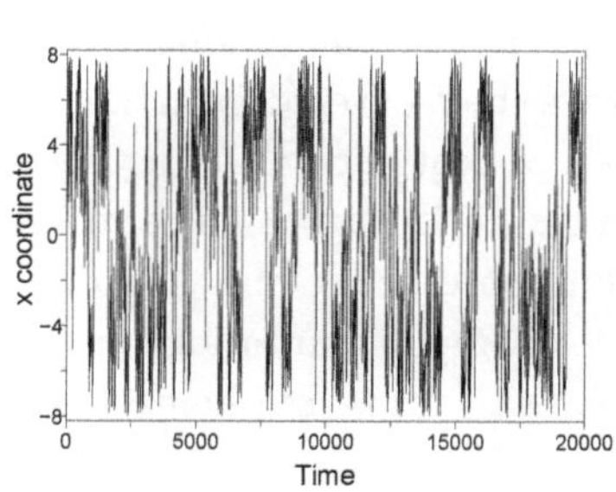

Fig. 1.13 Typical trajectory for the potential exhibiting an entropic barrier when the dynamics is biased by the analytically-known free energy. The numerical parameters are the same as for Figure 1.11.

1.3.4 *Computational techniques for free energy differences*

We present in this section the key ideas behind the methods currently available to compute free energy differences. Some of these techniques are suited both for alchemical transitions and transitions indexed by a reaction coordinate, but not all of them. In our opinion, the currently available techniques fall within the following four classes:

(i) The first technique, dating back to [Kirkwood (1935)], is *thermodynamic integration*, which mimics the quasi-static evolution of a system as a succession of equilibrium samplings (this amounts to an infinitely slow switching between the initial and final states). In practice, it allows to compute free energy differences by integrating the derivative of the free energy, which happens to be a canonical average for a fixed value of the reaction coordinate or alchemical parameter. This technique can be used both for alchemical transitions and transitions indexed by reaction coordinates, see Chapter 3;

(ii) The second one is based on straightforward sampling methods. In the alchemical case, the *free energy perturbation method*, introduced in [Zwanzig (1954)], recasts free energy differences as usual canonical averages (see Section 2.4.1). In the reaction coordinate case, usual sampling methods can also be employed, relying on *histogram methods* (see Section 2.5);

(iii) A more recent class of methods relies on dynamics with an imposed schedule for the reaction coordinate or the alchemical parameter. These techniques therefore use *nonequilibrium dynamics*. Equilibrium properties can however be recovered from the nonequilibrium trajectories with a suitable exponential reweighting, see [Jarzynski (1997b, a)]. This technique can handle both alchemical transitions and transitions indexed by reaction coordinates, see Chapter 4. It also has many similarities with free-energy perturbation since the corresponding free-energy estimators have the same mathematical structure (exponential averages);

(iv) Finally, *adaptive biasing dynamics* may be used in the reaction coordinate case. The switching schedule is not imposed *a priori*, but a biasing term in the dynamics forces the transition by penalizing the regions which have already been visited. This biasing term can be a biasing force as for the *Adaptive Biasing Force* technique of [Darve and Porohille (2001)], or a biasing potential as for the Wang-Landau method [Wang and Landau (2001b, a)], *nonequilibrium metadynamics* [Iannuzzi *et al.* (2003)] or Self-Healing Umbrella Sampling [Marsili *et al.* (2006)].

We refer to Figure 1.14 for a schematic comparison of the computational methods in the reaction coordinate case.

We now give a flavor of these approaches, preferentially in the alchemical setting for simplicity. The remainder of the book is devoted to a thorough presentation of these techniques. Recall that the free energy is defined, up to an additive constant (unimportant as long as free energy differences are concerned), by (1.52) or (1.53) in the alchemical case, and by (1.57) or (1.58) in the reaction coordinate case.

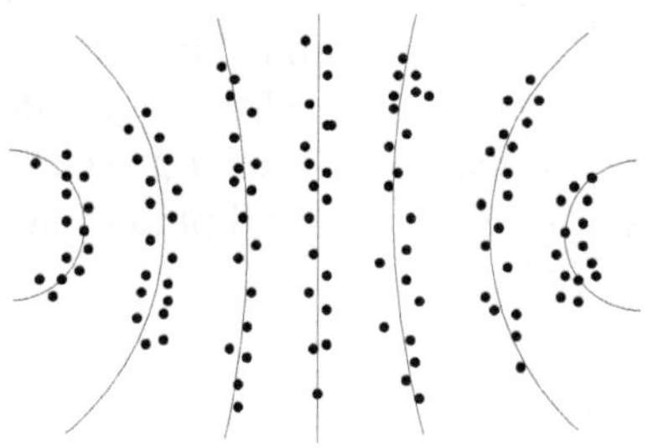

(a) Histogram method: sample points around the level sets are generated.

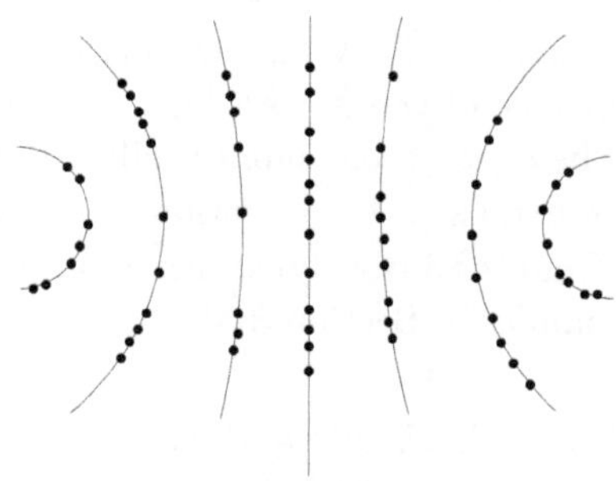

(b) Thermodynamic integration: a projected dynamics is used to sample each "slice" of the phase space.

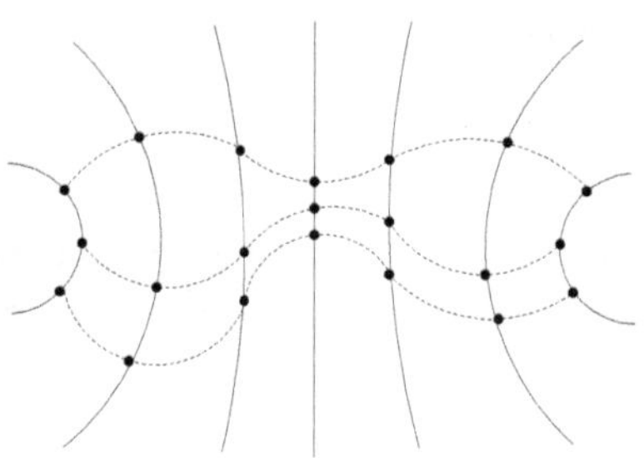

(c) Nonequilibrium dynamics: the switching is imposed *a priori* and is the same for all trajectories.

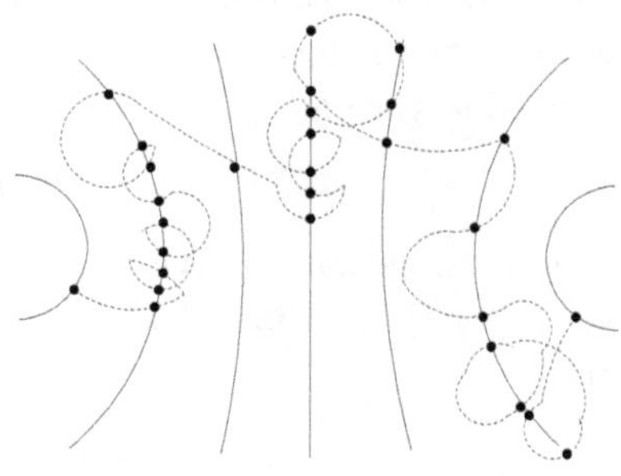

(d) Adaptive dynamics: the system is forced to leave regions where the sampling is sufficient.

Fig. 1.14 Cartoon comparison of the different techniques to compute free energy differences in the reaction coordinate case.

1.3.4.1 *Thermodynamic integration*

Thermodynamic integration consists in remarking that

$$F(\lambda) - F(0) = \int_0^\lambda F'(s)\,ds, \tag{1.70}$$

and that the derivative

$$F'(\lambda) = \frac{\displaystyle\int_{T^*\mathcal{D}} \frac{\partial H_\lambda}{\partial \lambda}(q,p)\, \mathrm{e}^{-\beta H_\lambda(q,p)}\, dq\, dp}{\displaystyle\int_{T^*\mathcal{D}} \mathrm{e}^{-\beta H_\lambda(q,p)}\, dq\, dp}$$

is the canonical average of $\partial_\lambda H_\lambda$ with respect to the canonical measure

$$\mu_\lambda(dq\, dp) = Z_\lambda^{-1}\, \mathrm{e}^{-\beta H_\lambda(q,p)}\, dq\, dp.$$

In practice, $F'(\lambda_i)$ is computed using classical sampling techniques for a sequence of values $\lambda_i \in [0,1]$. The integral on the right-hand side of (1.70) is then integrated numerically to obtain the free energy difference profile. The extension to transitions indexed by a reaction coordinate is presented in Chapter 3 (for dynamics in position space in Section 3.2 and phase space dynamics in Section 3.3).

1.3.4.2 *Methods based on straightforward sampling*

Free energy perturbation. Free energy perturbation is a technique which is restricted to the computation of free energy differences in the alchemical case (see however Remark 1.4 for an extension of the alchemical setting to the reaction coordinate case). It consists in rewriting the free energy difference as

$$\Delta F = -\beta^{-1} \ln \int_{T^*\mathcal{D}} \mathrm{e}^{-\beta(H_1 - H_0)}\, d\mu_0.$$

An approximation of ΔF is then obtained by generating configurations (q^n, p^n) distributed according to μ_0 and computing the empirical average

$$\frac{1}{N} \sum_{n=1}^{N} \mathrm{e}^{-\beta(H_1 - H_0)(q^n, p^n)}.$$

However, the initial and the final distributions μ_0 and μ_1 often hardly overlap. Intermediate steps should then be considered, or some importance sampling strategy should be used to improve the numerical accuracy, see Section 2.4.1.

It is also possible to resort to bridge sampling. In this case, the free energy difference ΔF is estimated using sample points from μ_0 and μ_1, see Section 2.4.2.

Histogram methods. In the reaction coordinate case, a naive algorithm to compute approximate free energy differences would be to sample configurations using a simple dynamics ergodic with respect to the canonical measure (see Chapter 2 for further precision), and to compute approximations of the marginal law in the reaction coordinate. More precisely, this can be done in practice by discretizing the values of the reaction coordinate into small intervals, and approximating the free energy by computing the canonical average of the indicator function of these intervals in the limit when the interval width Δz goes to 0. Defining

$$\chi_{z,\Delta z}(q) = \frac{1}{\Delta z} 1_{|\xi(q)-z| \leq \Delta z/2},$$

it holds

$$-\frac{1}{\beta} \ln \mathbb{E}_\mu(\chi_{z,\Delta z}) = -\frac{1}{\beta} \ln \left(\frac{1}{Z_\mu} \int_{T^*\mathcal{D}} \frac{1_{|\xi(q)-z| \leq \Delta z/2}}{\Delta z} \mathrm{e}^{-\beta H(q,p)} \, dq \, dp \right)$$
$$\longrightarrow F(z) = -\frac{1}{\beta} \ln \left(\frac{1}{Z_\mu} \int_{T^*\mathcal{D}} \mathrm{e}^{-\beta H(q,p)} \, \delta_{\xi(q)-z}(dq) \, dp \right) \qquad (1.71)$$

when $\Delta z \to 0$. However, the metastable features of the dynamics used for sampling usually prevent such a simple strategy from being efficient, see Section 1.3.3. The idea of histogram methods is to sample configurations centered on some level set $\Sigma(z)$, typically by sampling canonical measures associated with modified potentials

$$V(q) + \frac{1}{2\eta} \Big(\xi(q) - z \Big)^2,$$

where $\eta > 0$ is a small parameter, and to construct a global sample for the canonical measure $\mu(dq\,dp)$ by concatenating the sample points (with some appropriate weighting factor), see Section 2.5 for further precision. Once this global sample is obtained, an approximation of the free energy is obtained with (1.71) (for Δz small enough).

1.3.4.3 *Nonequilibrium dynamics*

Free energy differences can be expressed as a nonlinear average over nonequilibrium trajectories, using the so-called Jarzynski equality, see (1.74) below. This equality can easily be obtained for a system governed by Hamiltonian dynamics, with *initial conditions at equilibrium*, canonically distributed according to μ_0, and subjected to a switching schedule $\Lambda : [0,T] \to \mathbb{R}$ with $\Lambda(0) = 0$ and $\Lambda(T) = 1$. More precisely, we consider initial conditions $(q(0),p(0)) \sim \mu_0$, which are evolved according to the following non-autonomous ordinary differential equation for $0 \leq t \leq T$ (compare with (1.8)):

$$\begin{cases} \dfrac{dq}{dt}(t) = \nabla_p H_{\Lambda(t)}(q(t), p(t)), \\ \dfrac{dp}{dt}(t) = -\nabla_q H_{\Lambda(t)}(q(t), p(t)). \end{cases} \tag{1.72}$$

Defining by ϕ^Λ the associated flow, the work performed on the system starting from some initial conditions (q, p) is

$$\mathcal{W}(q,p) = \int_0^T \frac{\partial H_{\Lambda(t)}}{\partial \lambda}(\phi_t^\Lambda(q,p))\, \Lambda'(t)\, dt = H_1(\phi_T^\Lambda(q,p)) - H_0(q,p). \tag{1.73}$$

The last equality is obtained by noticing that

$$\frac{d}{dt}\Big(H_{\Lambda(t)}(\phi_t^\Lambda(q,p))\Big) = \\ \frac{\partial H_{\Lambda(t)}}{\partial \lambda}(\phi_t^\Lambda(q,p))\, \Lambda'(t) + \begin{pmatrix} \nabla_q H_{\Lambda(t)}(\phi_t^\Lambda(q,p)) \\ \nabla_p H_{\Lambda(t)}(\phi_t^\Lambda(q,p)) \end{pmatrix} \cdot \partial_t \phi_t^\Lambda(q,p)),$$

and the second term on the right-hand side vanishes in view of (1.72). Then,

$$\int_{T^*\mathcal{D}} \mathrm{e}^{-\beta \mathcal{W}(q,p)}\, d\mu_0(q,p) = Z_0^{-1} \int_{T^*\mathcal{D}} \mathrm{e}^{-\beta H_1(\phi_T^\Lambda(q,p))}\, dq\, dp.$$

Since ϕ_T^Λ defines a change of variables of Jacobian 1, the above equality can be restated as

$$\mathbb{E}_{\mu_0}(\mathrm{e}^{-\beta \mathcal{W}}) = \frac{Z_1}{Z_0} = \mathrm{e}^{-\beta(F(1)-F(0))}, \tag{1.74}$$

where the expectation is taken with respect to initial conditions distributed according to μ_0. The extension to stochastic dynamics, for transitions indexed by a reaction coordinate or an alchemical parameter, is presented in Chapter 4.

In view of the equality (1.74), it is already clear that the lowest values of the work dominate the nonlinear average (1.74), and the distribution of weights $\mathrm{e}^{-\beta \mathcal{W}(q,p)}$ is often degenerate in practice. This prevents in general an accurate numerical computation of the (1.74), and raises issues very similar to the ones encountered with free-energy perturbation. Refined strategies are therefore needed to use nonequilibrium methods in practice (see Chapters 4 and 6).

1.3.4.4 *Adaptive dynamics*

Adaptive dynamics may be seen as some adaptive importance sampling strategy, with a biasing potential at time t function of the reaction coordinate. The biasing potential converges in the longtime limit to the free energy by construction of the dynamics.

To illustrate this strategy, we consider the case of the Adaptive Biasing Force (ABF) method [Darve and Porohille (2001); Hénin and Chipot (2004)] in the simple example when the reaction coordinate $\xi(q) = q_1$ has values in $\mathbb{T}$, while the remaining coordinates $q_{2\dots N}$ belong to $\mathbb{R}^{N-1}$. Recall that, when ξ adequately describes the metastabilities of the system, the dynamics biased by the free energy is less metastable than the original dynamics (see Section 1.3.3 for two typical examples).

Let us assume that we know the free energy F. Denoting by $q_t = (q_{1,t}, q_{2\dots N,t})$ the current configuration of the system, the overdamped Langevin dynamics associated with the modified potential $V - F \circ \xi$ reads

$$\left\{ \begin{aligned} dq_t &= -\Big(\nabla V(q_t) - F'(q_{1,t})\, e_1\Big) dt + \sqrt{\frac{2}{\beta}}\, dW_t, \\ F'(z) &= \mathbb{E}_\nu\Big(\partial_{q_1} V(q) \,\Big|\, \xi(q) = z\Big) = \frac{\displaystyle\int_{\mathbb{R}^{N-1}} \partial_{q_1} V(z, q_{2\dots N})\, \mathrm{e}^{-\beta V(z,q_{2\dots N})}\, dq_{2\dots N}}{\displaystyle\int_{\mathbb{R}^{N-1}} \mathrm{e}^{-\beta V(z,q_{2\dots N})}\, dq_{2\dots N}}, \end{aligned} \right. \tag{1.75}$$

where $e_1 = (1, 0, \dots, 0)^T$ is the unit vector in the q_1 direction. Denote by

$$\widetilde{\nu}(dq) = \widetilde{Z}^{-1} \exp\Big(- \beta(V(q) - F(q_1))\Big)\, dq$$

the stationary measure of the process (1.75). The equilibrium mean force $F'(z)$ can actually be rewritten as a canonical average with respect to $\widetilde{\nu}$, conditionally on $q_1 = z$:

$$F'(z) = \mathbb{E}_\nu\Big(\partial_{q_1} V(q) \,\Big|\, \xi(q) = z\Big) = \mathbb{E}_{\widetilde{\nu}}\Big(\partial_{q_1} V(q) \,\Big|\, \xi(q) = z\Big). \tag{1.76}$$

Indeed, the bias $F(\xi(q))$ is constant when $\xi(q)$ is kept constant. Therefore, conditional averages with respect to $\widetilde{\nu}$ for $\xi(q) = z$ fixed are equal to conditional averages with respect to the canonical measure (1.33) since the factor $\mathrm{e}^{-\beta F(\xi(q))}$ cancels out in the numerator and denominator of the conditional average.

Now, of course, F is not known in practice. In view of (1.75)-(1.76), it seems natural to replace, in the dynamics (1.75), the conditional expectation with respect to the stationary measure in the expression of the equilibrium mean force, by the conditional expectation with respect to the current law of q_t:

$$\begin{cases} dq_t = -\Big(\nabla V(q_t) - F_t'(q_{1,t})\, e_1\Big)dt + \sqrt{\frac{2}{\beta}}\, dW_t, \\ F_t'(z) = \mathbb{E}\Big(\partial_{q_1} V(q_t) \,\Big|\, \xi(q_t) = z\Big). \end{cases} \tag{1.77}$$

Notice that the biasing potential F_t now explicitly depends on the time variable. Denoting by $\psi_t(q)\, dq$ the law of q_t at time t (intuitively, the distribution of configurations obtained by simulating an infinite number of replicas interacting only through the common bias they are constructing), the biasing force $F'(z)$ can be rewritten in a form closer to the expression in (1.75):

$$F_t'(z) = \frac{\displaystyle\int_{\mathbb{R}^{N-1}} \partial_{q_1} V(z, q_{2\dots N})\, \psi_t(z, q_{2\dots N})\, dq_{2\dots N}}{\displaystyle\int_{\mathbb{R}^{N-1}} \psi_t(z, q_{2\dots N})\, dq_{2\dots N}}.$$

We now motivate why the adaptive dynamics (1.77) may be relevant. The distribution of the variable $\xi(q_t) = q_{1,t}$ is given by the marginal law with density

$$\psi_t^\xi(z) = \int_{\mathbb{R}^{N-1}} \psi_t(z, q_{2\dots N})\, dq_{2\dots N}.$$

A simple computation (see Section 5.2.3.1) shows that

$$\partial_t \psi_t^\xi(z) = \frac{1}{\beta} \partial_z^2 \psi_t^\xi(z).$$

The above diffusion equation implies that ψ_t^ξ converges (exponentially fast) to the uniform distribution on $\mathbb{T}$. Therefore, the metastable features associated with ξ are suppressed. Heuristically, the simple diffusion equation in the direction q_1 is not too surprising since the biasing force F_t' aims precisely at counteracting in average the force experienced by the system in the direction q_1.

Besides, the dynamics (1.77) in the $q_{2\dots N}$ variable (at fixed z) is an overdamped Langevin dynamics associated with the potential $V(z, q_{2\dots N})$. Assuming that the dynamics is at equilibrium conditionnally on the z variable, the distribution of the variable $q_{2\dots N}$ at fixed z is equal to the canonical conditional distribution:

$$\frac{\psi_t(z, q_{2\dots N})}{\psi_t^\xi(z)}\, dq_{2\dots N} = Z_z^{-1}\, \mathrm{e}^{-\beta V(z, q_{2\dots N})}\, dq_{2\dots N}.$$

Recall also that the marginal law ψ_t^ξ converges to the uniform law. On the other hand, $\widetilde{\nu}(dq)$ is the unique probability measure whose marginal distribution in the ξ variable is the uniform law, while the conditional distributions at fixed values of ξ are equal to the canonical conditional distributions. This motivates the convergence of $\psi_t(q)\,dq$ towards $\widetilde{\nu}(dq)$, and therefore the convergence of F_t towards F.

The above presentation naturally suggests a parallel implementation of the dynamics through many replicas constructing a shared biasing potential. This plain parallel implementation can be enhanced through some selection process on the replicas (see Section 6.2). There exist also adaptive dynamics where the biasing potential F_t is updated, in contrast to the method presented here where the derivative of the biasing potential is updated. See Section 5.1 for further precision.

1.4 Summary of the mathematical tools and structure of the book

Table 1.2 presents in a synthetic manner the techniques used from a mathematical viewpoint for each of the methods presented in Section 1.3.4. This explains the construction of the book: we present the methods in what we consider to be the increasing order of mathematical complexity.

Table 1.2 Mathematical theories used for each free energy technique (MCs = Markov chains, SDEs = Stochastic differential equations).

Free energy perturbation	Time homogeneous MCs and SDEs	Chapter 2
Histogram methods	Time homogeneous MCs and SDEs	Chapter 2
Thermodynamic integration	Projected SDEs and MCs	Chapter 3
Nonequilibrium dynamics	Nonhomogenous MCs and SDEs	Chapter 4
Adaptive dynamics	Nonlinear SDEs and MCs	Chapter 5
Selection procedures	Particle systems and jump processes	Chapter 6

For the reader's convenience, see the dependency diagram in Figure 1.15, which highlights the prerequisites for each chapter. In particular, Sections 2.1 and 2.2 cover some material which will be of constant use for the remainder of the book.

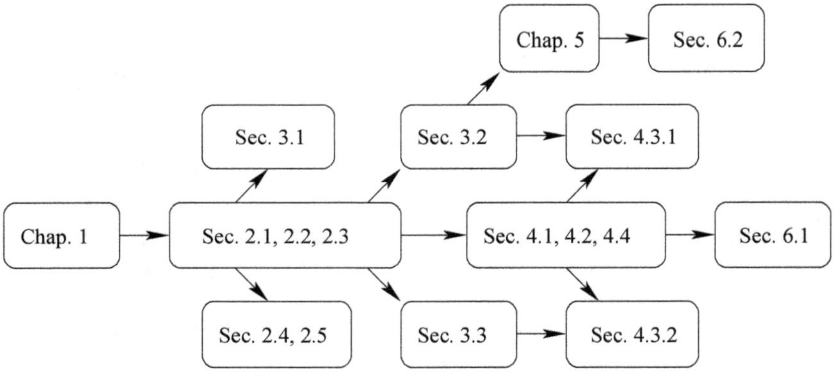

Fig. 1.15 Interdependence of the chapters and sections.

Chapter 2

Sampling methods

This chapter presents some standard methods used in practice to sample the canonical Boltzmann-Gibbs distribution

$$\mu(dq\,dp) = Z_\mu^{-1} \mathrm{e}^{-\beta H(q,p)}\, dq\, dp, \qquad Z_\mu = \int_{T^*\mathcal{D}} \mathrm{e}^{-\beta H(q,p)}\, dq\, dp. \tag{2.1}$$

The goal is to generate, by an appropriate numerical method, a sequence of microscopic configurations $(q^i, p^i)_{i \geq 0}$ such that

$$\lim_{n \to +\infty} \frac{1}{n} \sum_{i=0}^{n-1} A(q^i, p^i) = \int_{T^*\mathcal{D}} A(q,p)\, \mu(dq\, dp). \tag{2.2}$$

If the Hamiltonian is separable (see (1.2)), as is usually the case when cartesian coordinates are used, the measure (2.1) has a tensorized form, and the components of the momenta are distributed according to independent Gaussian distributions. It is therefore straightforward to sample the kinetic part of the canonical measure. The real difficulty consists in sampling positions distributed according to the canonical measure

$$\nu(dq) = Z_\nu^{-1} \mathrm{e}^{-\beta V(q)}\, dq, \qquad Z_\nu = \int_{\mathcal{D}} \mathrm{e}^{-\beta V(q)}\, dq, \tag{2.3}$$

which is typically a high dimensional distribution, with many local concentrated modes (the distribution is multimodal). This is the reason why many sampling methods focus on sampling the configurational part ν of the canonical measure.

Since most concepts needed for sampling purposes can be used either in the configuration space or in the phase space, the following notation will be used throughout this chapter: The state of the system is denoted by

$$x \in \mathcal{S} \subset \mathbb{R}^d,$$

which can be the position space $q \in \mathcal{D}$ (and then $d = 3N$), or the full phase space $(q, p) \in T^*\mathcal{D}$ with $d = 6N$. The measure $\pi(dx)$ is the canonical distribution to be sampled (ν in configuration space, μ in phase space).

From a mathematical point of view, most sampling methods may be classified as (see [Cancès *et al.* (2007)]):

(1) "Direct" probabilistic methods, such as the standard rejection method, which generate identically and independently distributed (i.i.d) configurations;
(2) Markov chain techniques, which are presented in Section 2.1;
(3) Markovian stochastic dynamics, which are presented in Section 2.2;
(4) Purely deterministic methods, such as the Nosé-Hoover method [Nosé (1984); Hoover (1985)] and its variants (see for instance the Hamiltonian reformulation presented in [Bond *et al.* (1999)]).

Direct probabilistic methods are typically based on some prior probability measure used to sample configurations, which are then accepted or rejected according to some criterion (as for the rejection method, for instance). Usually, a prior probability measure which is easy to sample should be used. However, due to the high dimensionality of the problem, it is extremely difficult to design a prior sufficiently close to the canonical distribution to achieve a reasonable acceptance rate. Direct probabilistic methods are therefore rarely used in practice. Their convergence follows from a simple application of the Law of Large Numbers or Central Limit Theorem.

In Markov chain methods, the prior is replaced by a *proposal move* which generates a new configuration from a former one. The latter is then accepted or rejected according to a so-called Metropolis rule. Here again, designing a relevant proposal move is the cornerstone of the method, and this proposal depends crucially on the model at hand.

A particular class of proposition moves is obtained by considering time discretizations of a stochastic dynamics (such as a Langevin or overdamped Langevin dynamics), which are then accepted or rejected according to a Metropolis-type criterion. This is therefore a systematic approach to construct a generic Metropolis algorithm, and/or to remove time-step errors when computing average quantities by ergodic limits of time-discretized dynamics. It is in general not difficult to prove the ergodicity of the dynamics (and thus the convergence of the estimator) but the rate of convergence is often difficult to estimate and depends on the relevance of the proposal move with respect to the averaged quantity.

Deterministic methods *à la* Nosé-Hoover have been motivated as some perturbation of the physical Hamiltonian dynamics, but their convergence properties are unclear. Indeed, it can be shown that although these methods preserve the right measure, they may be non-ergodic. This has been observed numerically, and recently proved rigorously in the case of a harmonic oscillator [Legoll *et al.* (2007, 2009)]. We therefore discard those methods in this book, and concentrate on stochastic methods for which ergodicity can be proved. However, many (if not all) techniques presented in the following chapters to compute free energy differences could be used with Nosé-Hoover like sampling methods as a basis.

This chapter is organized as follows. Sections 2.1 and 2.2 present some basic techniques to sample configurations according to the canonical probability measures (2.1) or (2.3). Markov chain techniques, in particular the Metropolis-Hastings algorithms and its variant, are presented in Section 2.1. Methods based on discretizations of stochastic differential equations, such as the Langevin equations, are described in Section 2.2. These sections cover some material which will be of constant use in the remainder of the book. Section 2.3 describes error estimates for the computed quantities, which is an important issue in practice to assess the quality of the estimation. Since stochastic techniques are used, the error can be decomposed as some intrinsic error (bias), and some random fluctuations related to the finiteness of the number of sampled configurations. Finally, some applications of the simple sampling techniques presented in Sections 2.1 and 2.2 to the computation of free energy differences are considered at the end of the chapter: free energy perturbation and bridge sampling in the alchemical case (Section 2.4), and histogram methods in the reaction coordinate case (Section 2.5).

2.1 Markov chain methods

After recalling some background material on Markov chains in Section 2.1.1, we present one of the most popular methods in computational statistical physics, the Metropolis-Hastings algorithm (Section 2.1.2). The remaining two subsections introduce methods which will be useful to remove time-step errors in the discretization of continuous dynamics: The so-called Hybrid Monte-Carlo algorithm is described in Section 2.1.3, and a generalization of the Metropolis-Hastings algorithms is proposed in Section 2.1.4.

2.1.1 *Some background material on the theory of Markov chains*

A time-homogeneous Markov chain $(x^n)_{n\geq 0}$ is a sequence of random variables sampled from a *probability transition kernel* $P(x, dx')$: At each iteration n, the new state x^{n+1} is sampled knowing only x^n (and not the previous iterations), according to the probability distribution $P(x^n, dx')$. Notice that, since $P(x^n, dx')$ is a probability distribution, the following normalization condition is satisfied:

$$\forall x \in \mathcal{S}, \qquad \int_{\mathcal{S}} P(x, dx') = 1.$$

In case $P(x, dx')$ has a density with respect to the Lebesgue measure, with a slight abuse of notation, we still denote P the *probability transition density*, so that the transition kernel is in this case $P(x, x')\, dx'$.

A Markov chain can generically be written as follows:

$$x^{n+1} = F(x^n, \Theta_n),$$

where $(\Theta_n)_{n\geq 0}$ is a sequence of independent identically distributed random variables. In this case, the transition kernel is characterized by the following equality: for any state x and for any observable ϕ (*i.e.* a bounded and continuous function),

$$\int_{\mathcal{S}} \phi(x')\, P(x, dx') = \mathbb{E}\Big[\phi\big(F(x, \Theta_1)\big)\Big].$$

When the transition kernel depends on the time index n, the chain is called time-inhomogeneous.

To study the longtime properties of a time-homogeneous chain, three features are of interest:

- *Stationarity.* A probability distribution π is a stationary probability distribution of P (or is said invariant for P) as soon as:

  $$\int_{x\in\mathcal{S}} P(x, dx')\, \pi(dx) = \pi(dx'),$$

  which may be equivalently restated as the following equality of averages of any bounded continuous test function ϕ:

  $$\int_{\mathcal{S}} \int_{\mathcal{S}} \phi(x')\, P(x, dx')\, \pi(dx) = \int_{\mathcal{S}} \phi(x)\, \pi(dx). \qquad (2.4)$$

  This condition means that, if the random variable x^0 is distributed according to π, then so is x^1, and, by induction, x^n as well.

- *Reversibility.* The chain P is said to be reversible with respect to π as soon as the following *detailed balance condition* is satisfied:
$$P(x, dx')\, \pi(dx) = P(x', dx)\, \pi(dx'). \tag{2.5}$$
The reversibility condition implies the stationarity of π. Indeed, a simple computation shows that, for a bounded continuous function ϕ and using (2.5),
$$\begin{aligned} \int_{\mathcal{S}} \int_{\mathcal{S}} \phi(x')\, P(x, dx')\, \pi(dx) &= \int_{\mathcal{S}} \int_{\mathcal{S}} \phi(x')\, P(x', dx)\, \pi(dx') \\ &= \int_{\mathcal{S}} \left(\int_{\mathcal{S}} P(x', dx) \right) \phi(x')\, \pi(dx') \\ &= \int_{\mathcal{S}} \phi(x')\, \pi(dx'), \end{aligned}$$
which is the stationarity condition (2.4). The detailed balance condition (2.5) is equivalent to the following statement: If x^0 is distributed according to π, then for any n, the sequence $(x^0, \dots, x^n)$ has the same probability distribution as the time-reversed sequence $(x^n, \dots, x^0)$.
- *Irreducibility.* Let us define by induction the nth step transition probability as
$$P^n(x, dx') = \int_{y \in \mathcal{S}} P(x, dy)\, P^{n-1}(y, dx'),$$
and $P^1(x, dx') := P(x, dx')$. The Markov chain P is said to be (aperiodically) irreducible with respect to π if for any measurable set A such that $\pi(A) > 0$, and π-almost all initial conditions x_0, there exists $n_0 > 0$ such that for any $n \geq n_0$,
$$P^n(x_0, A) > 0\,. \tag{2.6}$$
This means that the set A can be reached in n steps with positive probability starting from x_0.

There is a variant of the reversibility condition, which will be useful when dealing with phase space models such as Langevin processes: reversibility up to a one-to-one transformation S of the state space. In the models considered in this book,
$$S : \mathcal{S} \to \mathcal{S},$$
is a one-to-one transformation of $\mathcal{S}$. In practical applications, $S = \mathrm{Id}$ in configuration space, and
$$S : (q, p) \mapsto (q, -p) \tag{2.7}$$

is the momentum reversal operator in phase space. If x is distributed according to $\pi(dx)$, then $\pi(S^{-1}(dx'))$ denotes the distribution of $x' = S(x)$, so that

$$\int_{\mathcal{S}} \phi(S(x))\, \pi(dx) = \int_{\mathcal{S}} \phi(x')\, \pi(S^{-1}(dx')).$$

To state it differently, $\pi(S^{-1}(dx'))$ is the measure π transported by S. Finally, the transformation S is said to leave the distribution π invariant as soon as $\pi(dx') = \pi(S^{-1}(dx'))$. Here, (2.7) leaves the canonical Boltzmann-Gibbs distribution μ invariant. However, the presentation proposed below is generic, and more general transformations S may be considered.

- *Reversibility up to a one-to-one transformation.* Suppose the one-to-one transformation S leaves the probability distribution π invariant. The probability kernel $P(x, dx')$ is said to be reversible up to S with respect to the probability measure π as soon as the following modified, detailed balance condition is satisfied:

$$P(x, dx')\, \pi(dx) = P\big(S^{-1}(x'), S^{-1}(dx)\big)\, \pi(dx')\,. \tag{2.8}$$

  Reversibility up to S implies the stationarity of π. Reversibility up to S is equivalent to the following statement: If x^0 is distributed according to π, then for any n the sequence $(x^0, \ldots, x^n)$ has the same distribution as the sequence $(S(x^n), \ldots, S(x^0))$.

Since π is assumed to be invariant by S, the condition (2.8) may be restated as the following equality for any test function ϕ:

$$\int_{\mathcal{S}}\int_{\mathcal{S}} \phi(x, x')\, P(x, dx')\, \pi(dx) = \int_{\mathcal{S}}\int_{\mathcal{S}} \phi(S(x), S(x')) P(x', dx)\, \pi(dx').$$

Notice that the composition of two transition kernels π-reversible up to S is also π-reversible up to S.

Stationarity and aperiodic irreducibility imply ergodicity (see Theorem 17.1.7 in [Meyn and Tweedie (1993)]):

Proposition 2.1. *Let $(x^n)_{n\geq 0}$ be a Markov chain in $\mathcal{S}$ with a stationary probability measure π. If $(x^n)_{n\geq 0}$ is aperiodically irreducible, then it is* pathwise ergodic, *meaning that for any bounded measurable function A and π-almost all initial conditions x^0:*

$$\lim_{N\to+\infty} \frac{1}{N} \sum_{n=1}^{N} A(x^n) = \int_{\mathcal{S}} A(x)\, \pi(dx) \quad \text{a.s.}$$

2.1.2 *The Metropolis-Hastings algorithm*

The Metropolis-Hastings algorithm is a widely used method in molecular simulation. It allows to sample a canonical probability distribution $\pi(dx)$ known up to its normalization constant (which is indeed the case in almost all situations of interest).

2.1.2.1 *Presentation of the method*

The Metropolis-Hastings algorithm generates a Markov chain of the system configurations $(x^n)_{n\geq 0}$ having the distribution of interest $\pi(dx)$ as a stationary distribution. It consists in a two-step procedure. First, a move is generated, according to some given proposition transition kernel $T(x, dx')$. Then, the latter proposal move is either accepted or rejected, according to a rule such that the probability distribution $\pi(dx)$ is an invariant measure of the corresponding Markov chain. The original Metropolis algorithm was proposed in [Metropolis *et al.* (1953)], and relied on symmetric proposals in the configuration space, meaning that

$$T(x, dx')\,dx = T(x', dx)\,dx'.$$

It was later refined in [Hastings (1970)] in order to allow for non-symmetric propositions which can bias proposals towards higher probability regions with respect to the target distribution π.

Given a target probability distribution π and a proposition transition kernel T, the Metropolis-Hastings algorithm constructs in a systematic way a Markov chain reversible with respect to π. The detailed balance condition (2.5) is usually not verified for T and π, and a correction has therefore to be considered. For this correction to be possible, the proposition kernel $T(x, dx')$ must have some reversibility property, in the sense that the measures $T(x, dx')\,\pi(dx)$ and $T(x', dx)\,\pi(dx')$ have to be mutually absolutely continuous (or equivalent) for all x, x', in order for the Metropolis-Hastings ratio to be defined and positive (see below). Under this assumption, the algorithm is the following:

Algorithm 2.2 (Metropolis-Hastings algorithm). *Assume that the two measures $T(x', dx)\,\pi(dx')$ and $T(x, dx')\,\pi(dx)$ are equivalent and define the Metropolis-Hastings ratio:*

$$r(x, x') = \frac{T(x', dx)\,\pi(dx')}{T(x, dx')\,\pi(dx)}.$$

The ratio $r(x,x')$ is defined and positive for almost any couple of states (x,x') with respect to the measure $T(x,dx')\,\pi(dx)$. Consider an initial configuration x^0 and iterate on $n \geq 0$,

(1) Propose a new state $\tilde{x}^{n+1}$ from x^n according to the proposition kernel $T(x^n,\cdot)$;
(2) Accept the proposition with probability

$$\min\left(1, r(x^n, \tilde{x}^{n+1})\right),$$

and set in this case $x^{n+1} = \tilde{x}^{n+1}$; otherwise, set $x^{n+1} = x^n$.

In practice, the second step consists in drawing (independently) a random variable U^n with uniform law on $(0,1)$, and to accept (resp. reject) the move if $U^n \leq \min\left(1, r(x^n, \tilde{x}^{n+1})\right)$ (resp. if $U^n > \min\left(1, r(x^n, \tilde{x}^{n+1})\right)$). Notice that, as mentioned above, the distribution π has to be known only up to a multiplicative constant to perform this algorithm.

It may be worth emphasizing that, as for "direct" probabilistic methods such as the rejection method, rejected moves are discarded, but, in contrast to the "direct" probabilistic methods, a new configuration x^{n+1}, equal to the previous one x^n, is obtained in any case. This is important to estimate correctly canonical averages: Configurations where many propositions are rejected are counted several times (and possibly many times) in the average.

As an example, let us consider the case of the canonical sampling of positions using a symmetric proposition kernel. In this case, the invariant measure is $\nu(dq) = Z_\nu^{-1} \exp(-\beta V(q))\, dq$, and the transition kernel satisfies:

$$T(q, dq')\, dq = T(q', dq)\, dq'.$$

It is therefore reversible with respect to the Lebesgue measure, whereas we would like the Markov chain to be reversible with respect to the canonical measure. This is why an acceptance/rejection step is required, the corresponding Metropolis-Hastings ratio being simply the Metropolis ratio

$$r(q, q') = \exp\left[-\beta(V(q') - V(q))\right].$$

In this case, the interpretation of the algorithm is particularly simple. If the proposed move has a lower energy, it is always accepted, which allows to visit more frequently the states of higher probability. On the other hand, transitions to less likely states of higher energies are not forbidden (but accepted less often), which is important to observe transitions from one metastable region to another when these regions are separated by some energy barrier.

2.1.2.2 *Mathematical properties*

The probability transition kernel of the Metropolis-Hastings chain is:

$$P(x, dx') = \min(1, r(x, x'))\, T(x, dx') + (1 - \alpha(x))\, \delta_x(dx'), \tag{2.9}$$

where $\alpha(x) \in [0, 1]$ is the probability to accept a move starting from x (considering all possible propositions):

$$\alpha(x) = \int_S \min(1, r(x, y))\, T(x, dy).$$

The first part of the transition kernel corresponds to the accepted transitions from x to x', which occur with probability $\min(1, r(x, x'))$; while the term $(1 - \alpha(x))\delta_x(dx')$ encodes all the rejected steps.

A simple computation shows that the Metropolis-Hastings transition kernel P given in (2.9) is reversible with respect to π. Indeed, consider

$$\begin{aligned} P(x, dx')\pi(dx) = \min(1, r(x, x'))\, T(x, dx')\pi(dx) \\ + (1 - \alpha(x))\, \delta_x(dx')\pi(dx). \end{aligned} \tag{2.10}$$

Using the identity $r(x, x') = 1/r(x', x)$ and the algebraic equality for $r > 0$:

$$\min(1, r) = r \min\left(1, \frac{1}{r}\right),$$

the first term on the right-hand side of (2.10) can be rewritten as

$$\begin{aligned} \min(1, r(x, x'))\, T(x, dx')\pi(dx) &= \min(1, r(x', x))\, r(x, x') T(x, dx')\pi(dx) \\ &= \min(1, r(x', x))\, T(x', dx)\pi(dx'). \end{aligned}$$

On the other hand, for a given test function ϕ,

$$\begin{aligned} \int_S \int_S \phi(x, x')\, (1 - \alpha(x))\, \delta_x(dx')\pi(dx) &= \int_S \phi(x, x)(1 - \alpha(x))\pi(dx) \\ = \int_S \int_S \phi(x, x')(1 - \alpha(x'))\delta_{x'}(dx)\pi(dx'), \end{aligned}$$

so that $(1 - \alpha(x))\, \delta_x(dx')\pi(dx) = (1 - \alpha(x'))\delta_{x'}(dx)\pi(dx')$. This shows finally the reversibility property $P(x, dx')\pi(dx) = P(x', dx)\pi(dx')$.

The Metropolis-Hastings algorithm relies on the acceptance ratio $R(r(x, x'))$ with $R(r) = \min(1, r)$ but other functions R satisfy $R(r) = rR(1/r)$ and thus lead to a dynamics reversible with respect to π. For instance, the Barker rule corresponds to $R(r) = r/(r+1)$. However, it can be shown that the Metropolis rule is optimal in terms of asymptotic variance (see [Peskun (1973)]).

To conclude to the pathwise ergodicity of the algorithm using Proposition 2.1, it remains to check whether the chain is (aperiodically) irreducible. This property depends on the proposal kernel T, and should be checked for the model at hand. Note that as soon as the Metropolis-Hastings ratio $r(x, y) > 0$ for all $(x, y) \in \mathcal{S}$ for instance, the aperiodic irreducibility of the proposal transition T alone (with respect to the reference measure π) is equivalent to the aperiodic irreducibility of the Markov chain induced by the Metropolis Algorithm 2.2.

Besides determining the theoretical convergence of the algorithm, the proposed kernel is also a key element in devising efficient algorithms. It is observed in practice that the optimal acceptance/rejection rate, in terms of the variance of the estimator (a mean of some functional over a trajectory) for example, is often around 0.5, ensuring some balance between

- large moves that decorrelate the iterates when they are accepted (hence reducing the correlations in the chain, which is interesting for the convergence to happen faster, see Section 2.3.1), but lead to high rejection rates (and thus, degenerate samples since the same position may be counted several times)
- and small moves that are less rejected but do not decorrelate the iterates much.

This trade-off between small and large proposal moves has been investigated rigorously in some simple cases in [Roberts *et al.* (1997); Roberts and Rosenthal (1998)], where some optimal acceptance rates are obtained in a limiting regime. We warn the reader however that these computations do not take into account specific important features of actual models (like metastability or phase space dynamics such as the Langevin dynamics). In practice, it is always a good idea to run some preliminary small simulations to roughly determine the optimal acceptance rate.

2.1.2.3 *Some examples of proposition transition kernels*

Symmetric moves. The most simple transition kernels are based on random walks. For instance, it is possible to modify the current configuration by a random perturbation, such as

$$q' = q + G, \quad G \sim \mathcal{N}(0, \sigma^2 \,\mathrm{Id}),$$

in which case the proposal probability kernel is

$$T(q, dq') = \left(\sigma\sqrt{2\pi}\right)^{-d} \exp\left(-\frac{|q'-q|^2}{2\sigma^2}\right) dq',$$

where d is the dimension of the ambient space. Of course, uniformly distributed displacements could be considered as well

$$q' = q + U, \quad U \sim \mathcal{U}\left((-\sigma,\sigma)^d\right)$$

in which case

$$T(q,dq') = (2\sigma)^{-d}\, 1_{(q'-q)\in(-\sigma,\sigma)^d}\, dq'.$$

Both these proposals are symmetric. The problem with such proposals is that they may be not well suited to the target probability measure (creating very correlated successive configurations for small σ, or very unlikely moves for large σ).

Another symmetric proposal which may be used in the case of many particles $q = (q_1, \dots, q_N)$ consists in applying a random walk displacement to only one particle chosen at random. This may help to propose moves which are likely to be accepted (think for example of a relatively dense fluid). For instance, for a uniformly distributed perturbation with a typical magnitude $\sigma > 0$, the transition kernel is

$$T(q,dq') = \frac{1}{N}\sum_{i=1}^{N}\left(\prod_{j\neq i}\delta_{q_j}(dq'_j)\right)\left(\prod_{\alpha=x,y,z} 1_{|q'_{i,\alpha}-q_{i,\alpha}|\leq\sigma}\right)\frac{dq'_i}{8\sigma^3},$$

where $q_{i,\alpha}$ is the α component of q_i.

For random walk proposals, under weak assumptions on the potential energy function, it is possible to transform a given configuration into another one in a finite number of steps by moving individually one particle after another. This shows the irreducibility of the chain.

Non-symmetric move. An instance of a non-symmetric proposal which may be useful to sample the canonical measure (2.3) is the one used in the so-called Metropolis-Adjusted Langevin Algorithm (MALA) [Roberts and Rosenthal (1998)]:

$$q' = q - \alpha\nabla V(q) + \sqrt{\frac{2\alpha}{\beta}}\, G, \quad G \sim \mathcal{N}(0, \mathrm{Id}).$$

The proposal is built on the time discretization (with time step $\alpha > 0$) of the overdamped Langevin dynamics which is ergodic with respect to the canonical measure (see Section 2.2.2 below). The acceptance-rejection step corrects the bias introduced by the time discretization. Roughly speaking, the added drift term proportional to $-\nabla V(q)$ brings back the system to regions of higher probability, while the random term adds some stochastic fluctuations. It holds

$$T(q,dq') = \left(\frac{\beta}{4\pi\alpha}\right)^{d/2} \exp\left(-\beta\frac{|q'-q+\alpha\nabla V(q)|^2}{4\alpha}\right) dq'.$$

Notice that in this case $T(q,dq')\,dq \neq T(q',dq)\,dq'$.

Refinements and extensions. Since the proposal kernel T lies at the heart of the method, it is no surprise that a large fraction of the literature from the applied community deals with new and creative proposal kernels, using for instance "non-physical moves" where molecules are broken and other ones are linked together. Another important example is parallel tempering strategies [Marinari and Parisi (1992)], where several replicas of the system are simulated in parallel at different temperatures, and sometimes exchanges between two replicas at different temperatures are attempted, the probability of such an exchange being given by a Metropolis-Hastings ratio.

2.1.3 *Hybrid Monte-Carlo*

A specific case of the general Metropolis-Hastings algorithm presented in Section 2.1.2 is when the proposal move is computed using the (deterministic) Hamiltonian dynamics. Since such moves do not change the energy of the system, the resulting Metropolis-type strategy should be complemented with another mechanism which allows to change energy levels, such as resampling the momenta according to the canonical measure. This leads to the so-called Hybrid Monte-Carlo algorithm, first introduced in [Duane *et al.* (1987)] and analyzed from a mathematical viewpoint in [Schütte (1999); Cancès *et al.* (2007)]. The interest of a proposal based on the Hamiltonian dynamics is that it is very likely to be accepted.

The standard HMC algorithm defines a Markov chain on the positions q of the system, as follows:

Algorithm 2.3 (Hybrid Monte-Carlo). *Consider an initial configuration* $q^0 \in \mathcal{D}$ *and* $\tau > 0$. *Iterate on* $n \geq 0$,

(1) generate momenta p^n *according to the marginal of the canonical distribution* (2.1) *on momentum (which is gaussian) and compute the energy* $H(q^n, p^n)$ *of the configuration* (q^n, p^n)*;*
(2) compute $\Phi_\tau(q^n, p^n) = (q^{n,\tau}, p^{n,\tau})$*, that is, integrate the Hamiltonian equations of motion* (1.8)*, on the time interval* $[0, \tau]$ *starting from the initial data* (q^n, p^n)*;*
(3) compute the energy $H(q^{n,\tau}, p^{n,\tau})$ *of the new phase space configuration. Accept the proposal* $q^{n,\tau}$ *with probability*

$$\min\left(1, \exp\left[-\beta\big(H(q^{n,\tau}, p^{n,\tau}) - H(q^n, p^n)\big)\right]\right),$$

and set $q^{n+1} = q^{n,\tau}$ *in this case; otherwise, set* $q^{n+1} = q^n$.

The integration of the Hamiltonian dynamics in step (2) is done in practice by applying several steps of a numerical integrator $\Phi_{\Delta t}$, and $\Phi_\tau = (\Phi_{\Delta t})^p$ with $\tau = p\Delta t$. Let us emphasize that the proposal $q^{n,\tau}$ would always be accepted at step (3) if the Hamiltonian equations of motion, that are energy conserving, were exactly integrated. In practice, for the method to sample without bias the canonical distribution, the chosen integrator Φ_τ (usually based on the Verlet scheme (1.22)) should preserve the phase space measure $dq\,dp$ and verify

$$S \circ \Phi_\tau \circ S = \Phi_\tau^{-1}. \tag{2.11}$$

The latter property is ensured when the numerical scheme is time-reversible and symmetric (see Sections 1.2.2.3 and 1.2.2.4). The proof of the invariance of the canonical measure for the HMC algorithm follows from a result in the more general context of Section 2.1.4 (see in particular Section 2.1.4.3). In a nutshell, this is a consequence of the fact that the HMC Markov chain may be rewritten as a Markov chain on (q,p), and step (1) and steps (2)-(3) preserve the phase space canonical measure. Indeed, step (1) corresponds to resampling momenta, while steps (2)-(3) is an integration of the Hamiltonian dynamics which is then "metropolized".

However, the irreducibility of the chain requires additional assumptions on the potential, for instance energy functions bounded from above [Cancès *et al.* (2007)]. The most simple example of a system for which the HMC algorithm is not ergodic is the harmonic oscillator (see [Mackenzie (1989)]). Indeed, in the case when τ is an integer multiple of the period, and the Hamiltonian equations are integrated exactly, $q^0 = q^n$ for all $n \geq 0$. A remedy to such failures is to consider random integration times [Mackenzie (1989)]. There exist of course many other refinements of the standard HMC scheme.

The time-step Δt used for the numerical integrator can be chosen larger than in standard applications since the dynamics of the system used to generate proposals needs not accurately reproduce the physical dynamics of the system. On the other hand, it should not be too large, otherwise the energy is not well preserved, the rejection rate is large, and the efficiency of the method decreases. The second parameter to fix in practice is the integration time τ. Heuristically, iterates are less correlated when τ is larger, but since the computational cost of the algorithm is proportional to $\tau/\Delta t$, some trade-off has to be found.

Finally, as already pointed out, the correctness of the HMC method relies solely on (2.11) and on the measure-preserving property of the numerical scheme. Thus the Hamiltonian used in the Verlet scheme may be modified (*e.g.* for computational purposes only a simplified Hamiltonian may be used), and the HMC method will still be reversible with respect to the canonical distribution associated with the true Hamiltonian H. See [Plechac and Rousset (2010)] for an illustration of this idea.

2.1.4 *Generalized Metropolis-Hastings variants*

We now present a more general version of the Metropolis-Hastings algorithm, which includes Hybrid Monte-Carlo type methods as a special case. The goal of this section is mainly to prove the reversibility (up to some transformation S) of any Markov chain constructed from a Metropolis-Hastings type algorithm. This general formalism will be of particular interest for numerical approximations of stochastic dynamics based on Hamiltonian dynamics (as for the HMC scheme presented in Section 2.1.3). A typical example is the Langevin dynamics (see Section 2.2.3).

2.1.4.1 *Presentation of the method*

The key point of the generalization is that, in practical cases, the proposition kernel T at hand has a reversibility property up to an involutive transformation S:

$$S = S^{-1},$$

which leaves the canonical distribution π invariant:

$$\pi(S^{-1}(dx)) = \pi(S(dx)) = \pi(dx).$$

The operator S is for instance the momentum flip operator for standard separable Hamiltonians with quadratic kinetic energies.

In this context, the generalized Hybrid Metropolis-Hastings algorithm reads as follows:

Algorithm 2.4 (Generalized Metropolis-Hastings algorithm). *Assume that the two measures $T(S(x'), S(dx))\,\pi(dx')$ and $T(x, dx')\,\pi(dx)$ are equivalent, so that the Metropolis-Hastings ratio*

$$r(x, x') = \frac{T(S(x'), S(dx))\,\pi(dx')}{T(x, dx')\,\pi(dx)}$$

is defined and positive for almost any couple of states $(x, x') \in \mathcal{S}^2$ with respect to the measure $T(x, dx')\,\pi(dx)$. Consider an initial configuration x^0 and iterate on $n \geq 0$,

(1) *propose a new state* $\tilde{x}^{n+1}$ *from* x^n *according to the proposition kernel* $T(x^n, \cdot)$;
(2) *accept the move with probability*

$$\min\left(1, r(x^n, \tilde{x}^{n+1})\right),$$

and set in this case $x^{n+1} = \tilde{x}^{n+1}$; *otherwise, set* $x^{n+1} = S(x^n)$.

In words, the transition kernel $T(S(x), S(dy))$, required to compute the acceptance rate, corresponds to the following Markov chain: from an initial configuration x, change it into $S(x)$, apply the transition kernel $T(S(x), dy)$ to obtain a new configuration y, and change it into $S(y)$. This construction may for instance be motivated by the fact that, for a given transition from (q,p) to (q',p'), the probability of the reverse move (obtaining (q,p) starting from (q',p')) is much smaller than the probability to obtain $(q,-p)$ starting from $(q',-p')$.

2.1.4.2 *Mathematical properties*

Let us now analyze the generalized Metropolis-Hastings algorithm. The probability transition kernel of the Markov chain is:

$$P(x, dx') = \min\left(1, r(x,x')\right) T(x, dx') + (1 - \alpha(x))\, \delta_{S(x)}(dx'), \qquad (2.12)$$

with

$$\alpha(x) = \int_S \min\left(1, r(x,y)\right) T(x, dy).$$

The transition kernel P satisfies the detailed balance condition up to the transformation S:

$$P(x, dx')\, \pi(dx) = P(S(x'), S(dx))\, \pi(dx'). \qquad (2.13)$$

This implies that π is an invariant probability for P. The proof is similar to the classical Metropolis-Hastings case (see Section 2.1.2), and consists in treating separately each term on the right-hand side of the above equality. The algebraic identity

$$\min(1, r) = r \min\left(1, \frac{1}{r}\right)$$

and the new symmetry property (based on the fact that S is an involution which leaves π invariant):

$$r(x, x') = \frac{1}{r(S(x'), S(x))},$$

yield (using again the invariance of π under S):

$$\min(1, r(x, x'))\, T(x, dx')\pi(dx) = \min\Big(1, r(S(x'), S(x))\Big) T(S(x'), S(dx))\pi(dx').$$

Besides,

$$(1 - \alpha(x))\delta_{S(x)}(dx')\pi(dx) = (1 - \alpha(S(x')))\delta_{x'}(S(dx))\pi(dx'),$$

since, for any test function ϕ:

$$\begin{aligned}
\int_S \int_S \phi(x, x')(1 - \alpha(S(x')))\delta_{x'}(S(dx))\pi(dx') \\
&= \int_S \int_S \phi(S(x), x')(1 - \alpha(S(x')))\delta_{x'}(dx)\pi(dx') \\
&= \int_S \phi(S(x'), x')(1 - \alpha(S(x')))\pi(dx') \\
&= \int_S \phi(y, S(y))(1 - \alpha(y))\pi(dy) \\
&= \int_S \int_S \phi(y, y')(1 - \alpha(y))\delta_{S(y)}(dy')\pi(dy).
\end{aligned}$$

Therefore, $P(x, dx')\pi(dx) = P(S(x'), S(dx))\pi(dx')$.

Remark 2.5 (About the rejection step). *In the rejection step, it is necessary to change x^n to $S(x^n)$ in order for π to be an invariant measure for P. In the particular case when the proposal is a step of the Verlet scheme (see for instance the HMC-based discretization of the Langevin dynamics in Algorithm* 2.11 *below), this means that the dynamical properties of the trajectory are lost as soon as a rejection step is performed. There is therefore a choice to be made between correcting the error on the sampled measure due to time discretization and obtaining correct statistics on the dynamics. See however the recent results on pathwise convergence for Metropolized dynamics [Bou-Rabee and Vanden-Eijnden (2009)].*

2.1.4.3 *Relationship with the Hybrid Monte-Carlo algorithm*

Steps (2)-(3) of the Hybrid Monte-Carlo algorithm (see Algorithm 2.3) can be seen as a particular case of the generalized Metropolis-Hastings algorithm with:

- $T(x, dy) = \delta_{\Phi_\tau(x)}(dy)$ where $x = (q, p)$ is a point in phase space, dy the phase space Lebesgue measure, and $\Phi_\tau(x) = \Phi_\tau(q, p)$ the flow of the Hamiltonian dynamics or the numerical integrator at hand;

- $S(q,p) = (q,-p)$ is the momentum flip.

Let us indeed show that, when the condition (2.11) is satisfied and Φ_τ preserves the phase space measure, the Metropolis-Hastings ratio is

$$r(x,x') = \frac{T(S(x'),S(dx))\,\pi(dx')}{T(x,dx')\,\pi(dx)} = \exp\Big(-\beta\big[H(x')-H(x)\big]\Big). \quad (2.14)$$

In this context, $\pi(dx) = \mathrm{e}^{-\beta H(x)}\,dx$. For a given test function ϕ, it holds (using $S^{-1} = S$)

$$\begin{aligned}
&\int_{\mathcal{S}}\int_{\mathcal{S}} \phi(x,x')T(S(x'),S(dx))\,\pi(dx') \\
&\qquad = \int_{\mathcal{S}}\int_{\mathcal{S}} \phi(x,x')\delta_{(\Phi_\tau\circ S)(x')}(S(dx))\pi(dx') \\
&\qquad = \int_{\mathcal{S}}\int_{\mathcal{S}} \phi(S(x),x')\delta_{(\Phi_\tau\circ S)(x')}(dx)\pi(dx') \\
&\qquad = \int_{\mathcal{S}} \phi((S\circ\Phi_\tau\circ S)(x'),x')\pi(dx') \\
&\qquad = \int_{\mathcal{S}} \phi(\Phi_\tau^{-1}(x'),x')\,\mathrm{e}^{-\beta H(x')}\,dx' \\
&\qquad = \int_{\mathcal{S}} \phi(x,\Phi_\tau(x))\,\mathrm{e}^{-\beta(H\circ\Phi_\tau)(x)}\,dx, \qquad (2.15)
\end{aligned}$$

where we have used in the last line the change of variable $x = \Phi_\tau(x')$ (of Jacobian equal to 1 since Φ_τ preserves the phase space measure). Besides,

$$\begin{aligned}
\int_{\mathcal{S}}\int_{\mathcal{S}} \phi(x,x')T(x,dx')\,\pi(dx) &= \int_{\mathcal{S}}\int_{\mathcal{S}} \phi(x,x')\delta_{\Phi_\tau(x)}(dx')\pi(dx) \\
&= \int_{\mathcal{S}} \phi(x,\Phi_\tau(x))\,\mathrm{e}^{-\beta H(x)}\,dx.
\end{aligned}$$

The comparison with (2.15) shows that (2.14) holds.

Therefore, the HMC algorithm may be seen as the composition of two operations: (i) a resampling of the momenta according to the momenta marginal of the canonical measure; (ii) a step of the generalized Metropolis-Hastings algorithm with the proposal function $T(x,dy) = \delta_{\Phi_\tau(x)}(dy)$. Both operations leave the canonical measure invariant, so that the phase space measure is indeed invariant for this interpretation of the HMC scheme as a Markov chain in the (q,p) variables.

2.2 Continuous stochastic dynamics

After recalling some background material on Markov processes in Section 2.2.1, we present two important dynamics which are ergodic for the canonical measure: the overdamped Langevin dynamics in Section 2.2.2, and the Langevin dynamics in Section 2.2.3. The terminology is motivated in Section 2.2.4, where we show how the overdamped Langevin dynamics is obtained from the Langevin dynamics when the magnitude of the damping term goes to infinity.

2.2.1 *Mathematical background on Markovian continuous processes*

In this section, some elements of the theory of Markov processes are presented for a generic process

$$t \mapsto x_t \in \mathcal{S} \subset \mathbb{R}^d.$$

As in the previous section, the state space $\mathcal{S}$ is either the configuration space: $x_t = q_t \in \mathcal{D}$ (and $d = 3N$) or the phase space: $x_t = (q_t, p_t) \in T^*\mathcal{D}$ (and $d = 6N$). The mathematical formalism associated with Markovian continuous stochastic dynamics shares many features with the theory of Markov chains. Indeed, Markov diffusion processes can be obtained as continuous limit of random walks, and when numerical discretizations are considered, Markov chains are recovered.

The aim of this section is to provide a rough introduction to what will be needed in the forthcoming chapters on stochastic processes. Further precision and more rigorous presentations can be found in textbooks, see for instance [Oksendal (1992)] for a nice introduction to the topic, while [Karatzas and Shreve (1988); Ikeda and Watanabe (1989); Stroock and Varadhan (1979); Ethier and Kurtz (1986)] are reference mathematical textbooks for Markov processes and diffusions.

2.2.1.1 *Infinitesimal generator of diffusion processes*

Most of the Markov stochastic processes which are considered in this book are defined as solutions of stochastic differential equations (SDEs) of the form:

$$dx_t = b_t(x_t)\, dt + \sigma_t(x_t)\, dW_t, \tag{2.16}$$

where the drift coefficient $b_t(x)$ is a d-dimensional vector, the diffusion coefficient $\sigma_t(x)$ is a $d \times n$ matrix, and $t \mapsto W_t$ is a standard n-dimensional Brownian motion. Such processes are called diffusion processes.

We recall that a one-dimensional Brownian motion W_t is a stochastic process such that: (i) $W_0 = 0$, (ii) for all $0 \leq s < t$, $(W_t - W_s)$ is independent of the past $(W_r)_{0\le r\le s}$, and (iii) for all $0 \le s < t$, $(W_t - W_s)$ has the same distribution as $(W_{t-s} - W_0)$. A standard n-dimensional Brownian motion W_t has components which are n independent one-dimensional Brownian motions. Another characterization of Brownian motion is the following: It is a Gaussian process (namely for all $k \in \mathbb{N}$ and all $0 \le t_0 < t_1 < \ldots < t_k$, $(W_{t_0}, W_{t_1}, \ldots, W_{t_k})$ is a Gaussian random vector) with mean $\mathbb{E}(W_t) = 0$ and covariance $\mathbb{E}(W_s \cdot W_t) = \min(s,t)$. This implies that a discretization of a standard n-dimensional Brownian motion $(W_0 = 0, W_{t_1}, \ldots, W_{t_i}, \ldots)$ of W_t on a grid $t_0 = 0 \le t_1 \le \cdots \le t_i \le \ldots$ is characterized by the fact that the random variables

$$W_{t_{i+1}} - W_{t_i} \in \mathbb{R}^n,$$

are independent Gaussian vectors with zero mean and $n \times n$ covariance matrices $(t_{i+1} - t_i)\mathrm{Id}$.

A stochastic process x_t is a solution to (2.16) if it satisfies for all $t \ge 0$

$$x_t = x_0 + \int_0^t b_s(x_s)\,ds + \int_0^t \sigma_s(x_s)\,dW_s.$$

The first integral $\int_0^t b_s(x_s)\,ds$ is a standard integral for a process $s \mapsto b_s(x_s)$ with finite variations (similarly to the deterministic setting). The second integral is a so-called Itô integral which can be understood as the limit (in probability) of the time discretization:

$$\lim_{\Delta t \to 0} \sum_{k=0}^{T/\Delta t} \sigma_{k\Delta t}(x_{k\Delta t})(W_{(k+1)\Delta t} - W_{k\Delta t}) = \int_0^t \sigma_s(x_s)\,dW_s. \tag{2.17}$$

Loosely speaking,

$$\sigma_t(x_t)\,dW_t \simeq \sigma_t(x_t)(W_{t+dt} - W_t).$$

Over a time interval of size Δt, the term with finite variation $\int_0^t b_s(x_s)\,ds$ has a variation of order Δt, while the Itô integral term $\int_0^t \sigma_s(x_s)\,dW_s$ (also called the martingale term, or the term with non-zero quadratic variation) has a variation of order $\sqrt{\Delta t}$. This implies that if two processes are such that $x_0 + \int_0^t H_s\,ds + \int_0^t K_s\,dW_s = \tilde{x_0} + \int_0^t \tilde{H}_s\,ds + \int_0^t \tilde{K}_s\,dW_s$, (where H_s and K_s are stochastic processes such that $\int_0^t |H_s|\,ds < \infty$ a.s. and $\int_0^t |K_s|^2\,ds < \infty$ a.s., and similar assumptions on $\tilde{H}_s$ and $\tilde{K}_s$) then, a.s., $x_0 = \tilde{x_0}$, $H_s = \tilde{H}_s$ and $K_s = \tilde{K}_s$.

Two important properties of Itô integrals hold under the condition

$$\mathbb{E}\left(\int_0^t |\sigma_s(x_s)|^2\,ds\right) < \infty.$$

First, the expectation is

$$\mathbb{E}\left(\int_0^t \sigma_s(x_s)\, dW_s\right) = 0,$$

while the variance is

$$\mathbb{E}\left(\left|\int_0^t \sigma_s(x_s)\, dW_s\right|^2\right) = \int_0^t \mathbb{E}\left(|\sigma_s(x_s)|^2\right) ds.$$

Let us insist on the fact that, in the discretization (2.17), the limit process depends on the time $t \in [k\Delta t, (k+1)\Delta t]$ used to evaluate $\sigma_t(x_t)$. The (non-anticipating) Itô integral is characterized by the choice $t = k\Delta t$. The choice $t = (k+1/2)\Delta t$ leads to the so-called Stratonovitch integral (denoted by $\circ$ in the following):

$$\sigma_t(x_t) \circ dW_t \simeq \frac{1}{2}(\sigma_t(x_t) + \sigma_{t+dt}(x_{t+dt}))(W_{t+dt} - W_t). \tag{2.18}$$

If x_t is solution to (2.16), it can be checked that

$$dx_t = \left(b_t(x_t) - \frac{1}{2}(\sigma_t^T(x_t)\cdot\nabla)\sigma_t^T(x_t)\right) dt + \sigma_t(x_t) \circ dW_t. \tag{2.19}$$

A sufficient set of conditions for the general stochastic differential equation (2.16) to have a unique (strong) continuous solution on a given time-interval is for instance the following (see [Ikeda and Watanabe (1989)] and Section III.4 in [Has'minskii (1980)]):

- b_t and σ_t are continuous functions of time and space, which are locally Lipschitz continuous with respect to the space variables, with a Lipschitz constant independent of time;
- $b_t(x)\cdot x \le C\left(1 + \|x\|^2\right)$ for some time independent constant C;
- $\mathrm{Tr}\left(\sigma(x)\sigma^T(x)\right) \le C\left(1+\|x\|^2\right)$ for some time independent constant C.

The first condition is a local condition ensuring uniqueness and existence until exit time from a bounded set, and the two last conditions are global conditions ensuring non-explosion, *i.e.* the exit time from any bounded set goes to infinity (with probability one) as the size of the set increases. To ensure this non-explosive behavior, a general approach is to prove the existence of Lyapunov functions associated with the stochastic differential equation (see Section III.4 in [Has'minskii (1980)]). In this monograph, we will always assume that the stochastic differential equations at hand are well posed.

Diffusion processes are Markov processes. Markov processes are characterized by their transition functions. For a Markov process $t \mapsto x_t$, the transition function $P_{s,t}$ is defined as: For any $0 \le s < t$ and for any test function φ,

$$P_{s,t}(\varphi)(x) = \mathbb{E}\Big(\varphi(x_t)\,\Big|\,x_s = x\Big).$$

The Markovian assumption ensures that the family $(P_{s,t})_{s,t\geq 0}$ has a semigroup structure: for $0 \le r < s < t$,

$$P_{r,s} \circ P_{s,t} = P_{r,t}.$$

In an appropriate functional setting, the infinitesimal generator associated with the semigroup $P_{s,t}$ can be defined as the following strong limit: For $t \geq 0$,

$$\lim_{s \to 0} \frac{P_{t,t+s} - \mathrm{Id}}{s} = \mathcal{L}_t.$$

Then, $P_{s,t}$ satisfies the so-called backward Kolmogorov equation: for any test function φ, for a fixed $t > 0$, and $s \in [0, t]$,

$$\partial_s P_{s,t}(\varphi) = -\mathcal{L}_s P_{s,t}(\varphi)\,. \tag{2.20}$$

The transition function $P_{s,t}$ also satisfies the so-called forward Kolmogorov equation: For any test function φ, for a fixed $s > 0$, and $t \geq s$,

$$\partial_t P_{s,t}(\varphi) = P_{s,t}(\mathcal{L}_t(\varphi))\,. \tag{2.21}$$

Let us consider the case when the random variable x_t has a density $\psi(t, x)$ for all times $t \geq 0$. By rewriting (2.21) with $s = 0$, and integrating with respect to $\psi(0, x)\,dx$, it follows that, for any test function φ,

$$\frac{d}{dt}\mathbb{E}(\varphi(x_t)) = \mathbb{E}\Big(\mathcal{L}_t(\varphi)(x_t)\Big). \tag{2.22}$$

This is a weak formulation of the so-called Fokker-Planck equation satisfied by the density ψ:

$$\partial_t \psi = \mathcal{L}_t^* \psi, \tag{2.23}$$

where $\mathcal{L}_t^*$ denotes the dual operator of $\mathcal{L}_t$ (for the $L^2(dx)$-scalar product).

In the case of a time-homogeneous Markov process (when b and σ do not depend on time in (2.16)), $P_{s,t} = P_{0,t-s}$, and the transition operator is then simply denoted

$$P_t = P_{0,t}.$$

The infinitesimal generator $\mathcal{L}$ does not depend on time so that $P_t = \exp(t\mathcal{L})$ (the Hille-Yosida theory can be used to give a precise meaning to this notation, see for example [Reed and Simon (1975); Ethier and Kurtz (1986)]). The backward and forward Kolmogorov equations are identical in this case:

$$\partial_t P_t(\varphi) = P_t(\mathcal{L}(\varphi)) = \mathcal{L}\Big(P_t(\varphi)\Big). \tag{2.24}$$

In the case of a diffusion process such as (2.16), the infinitesimal generator associated with the Markov process is given by the following second order linear differential operator:

$$\mathcal{L}_t = \sum_{i=1}^{d} b_t^i(x)\,\partial_i + \frac{1}{2}\sum_{i,j=1}^{d} a_t^{i,j}(x)\,\partial_i\partial_j, \tag{2.25}$$

with

$$a_t = \sigma_t\sigma_t^T.$$

In this setting, the Fokker-Planck equation (2.23) writes:

$$\partial_t\psi(t,x) = -\sum_{i=1}^{d} \partial_i\left(b_t^i(x)\psi(t,x)\right) + \frac{1}{2}\sum_{i,j=1}^{d} \partial_i\partial_j\left(a_t^{i,j}(x)\psi(t,x)\right).$$

Equation (2.25) is derived using the Itô calculus (see [Oksendal (1992)]), which is the chain-rule derivation formula for diffusion processes. For a smooth function $\varphi : \mathbb{R}^n \to \mathbb{R}$, and x_t solution to (2.16),

$$\begin{aligned} d\varphi(x_t) &= \nabla\varphi^T(x_t)dx_t + \frac{1}{2}a_t(x_t) : \nabla^2\varphi(x_t)\,dt \qquad (2.26)\\ &= \left(\nabla\varphi^T(x_t)b_t(x_t) + \frac{1}{2}a_t(x_t) : \nabla^2\varphi(x_t)\right)dt + \nabla\varphi^T(x_t)\sigma_t(x_t)\,dW_t, \end{aligned}$$

where the so-called Itô term

$$\frac{1}{2}a_t(x_t) : \nabla^2\varphi(x_t)\,dt = \frac{1}{2}\sum_{i,j=1}^{d}\sum_{k=1}^{n}\sigma_t^{i,k}(x_t)\,\sigma_t^{j,k}(x_t)\,\partial_i\partial_j\varphi(x_t)\,dt$$

comes from the fact that, on a time increment of size Δt, the Brownian motion typically varies as $\sqrt{\Delta t}$, so that a Taylor expansion up to order 2 is necessary to compute the limit of $(\varphi(x_{t+\Delta t}) - \varphi(x_t))/\Delta t$ as Δt goes to zero. The formal computational rules of Itô stochastic calculus are given by the typical size of Brownian motion increments

$$dW_t = \mathrm{O}\left(\sqrt{dt}\right),$$

and the rule to obtain almost sure infinitesimal square co-variations

$$dW_t \otimes dW_t = \mathrm{Id}\,dt.$$

The formula (2.25) of the generator is obtained by taking the expectation of (2.26) and identifying $\mathcal{L}$ such that (2.22) holds.

Remark 2.6 (Chain rule for Stratonovitch integration). *Notice that the chain-rule derivation formula is different for Stratonovitch integral:*

$$d\varphi(x_t) = \nabla\varphi^T(x_t) \circ dx_t, \tag{2.27}$$

where the generalization of the Stratonovitch product (2.18) *to solutions of stochastic differential equations is*

$$\begin{aligned}\nabla\varphi^T(x_t) \circ dx_t &\simeq \frac{1}{2}\Big(\nabla\varphi(x_{t+dt}) + \nabla\varphi(x_t)\Big)^T dx_t \\ &\simeq \frac{1}{2}\Big(\nabla\varphi(x_{t+dt}) + \nabla\varphi(x_t)\Big)^T \Big(b_t(x_t)\,dt + \sigma_t(x_t)(W_{t+dt} - W_t)\Big).\end{aligned}$$

To show (2.27)*, we use the formal Taylor expansion (keeping only the leading order in powers of dt):*

$$\frac{1}{2}\Big(\nabla\varphi(x_{t+dt}) - \nabla\varphi(x_t)\Big)^T \Big(\sigma_t(x_t)\,dW_t\Big) \simeq \sum_{i,j=1}^{d} \frac{1}{2} a_t^{i,j}(x_t)\partial_i\partial_j\varphi(x_t)dt,$$

and insert it into (2.26)*, which leads to*

$$\begin{aligned}d\varphi(x_t) &= \nabla\varphi^T(x_t)\,dx_t + \frac{1}{2}a_t(x_t) : \nabla^2\varphi(x_t)\,dt \\ &\simeq \nabla\varphi^T(x_t)\Big(b_t(x_t)\,dt + \sigma_t(x_t)(W_{t+dt} - W_t)\Big) \\ &\quad + \frac{1}{2}\Big(\nabla\varphi(x_{t+dt}) - \nabla\varphi(x_t)\Big)^T (\sigma_t(x_t)dW_t) \\ &\simeq \nabla\varphi^T(x_t)\,b_t(x_t)\,dt + \frac{1}{2}\Big(\nabla\varphi(x_{t+dt}) + \nabla\varphi(x_t)\Big)^T \Big(\sigma_t(x_t)\,dW_t\Big) \\ &\simeq \nabla\varphi^T(x_t) \circ dx_t.\end{aligned}$$

2.2.1.2 *Properties of time-homogeneous diffusion processes*

For time-homogeneous diffusion processes, b_t and σ_t do not depend on the time variable t. Similarly to the Markov chain case, the long time behavior of time-homogeneous Markov processes can be studied using the notion of stationary probability distribution, reversibility, and irreducibility, which can all be characterized using the (time-independent) infinitesimal generator:

$$\mathcal{L} = \sum_{i=1}^{d} b^i(x)\,\partial_i + \frac{1}{2}\sum_{i,j=1}^{d} a^{i,j}(x)\,\partial_i\partial_j. \tag{2.28}$$

The following notions will be useful.

- *Stationarity.* The measure π is a stationary probability distribution for the diffusion with generator $\mathcal{L}$ as soon as, for any observable φ,
$$\int_{\mathcal{S}} \mathcal{L}(\varphi)\, d\pi = 0\,. \tag{2.29}$$
This is equivalent to the following statement: If x_0 is a random variable distributed according to π, so is x_t for any time $t \geq 0$.
- *Reversibility.* For any observables φ_1 and φ_2, a diffusion is said to be reversible with respect to the probability distribution π, when the following detailed balance condition is satisfied:
$$\int_{\mathcal{S}} \varphi_1\, \mathcal{L}(\varphi_2)\, d\pi = \int_{\mathcal{S}} \varphi_2\, \mathcal{L}(\varphi_1)\, d\pi\,. \tag{2.30}$$
This means that the generator $\mathcal{L}$ is a self-adjoint operator on $L^2(d\pi)$. Reversibility implies stationarity (consider $\varphi_1 = 1$ so that $\mathcal{L}\varphi_1 = 0$ and compare with the stationarity requirement (2.29)) and is equivalent to the following statement: For any $T > 0$, if x_0 is distributed according to π, then the path $(x_t)_{t\in[0,T]}$ and the time-reversed path $(x_{T-t})_{t\in[0,T]}$ have the same probability distribution on the space of continuous trajectories.
- *Irreducibility.* The process is said to be Lebesgue irreducible, or simply irreducible, when its transition functions verify the following property: For any Borel set A with positive Lebesgue measure, and Lebesgue-almost all initial conditions $x_0 \in \mathbb{R}^d$, for all $t > 0$,
$$P_t(1_A)(x_0) > 0. \tag{2.31}$$
This means that, starting from x_0, the process can reach at any positive time any subset of the state space with positive probability.

As for Markov chains, it is possible to define a variant of reversibility up to a one-to-one symmetry S of the measure π (typically the momentum reversal operator):

- *Reversibility up to a one-to-one transformation.* Let S be a one-to-one transformation leaving π invariant. The generator $\mathcal{L}$ is said to be reversible up to S with respect to π, as soon as for any observables φ_1 and φ_2, the modified detailed balance condition holds:
$$\int_{\mathcal{S}} \varphi_1\, \mathcal{L}\varphi_2\, d\pi = \int_{\mathcal{S}} (\varphi_2 \circ S)\, \mathcal{L}(\varphi_1 \circ S)\, d\pi\,. \tag{2.32}$$
Reversibility up to S implies the stationarity of π, and is equivalent to the following statement: For any $T > 0$, if x_0 is distributed

according to π, then $(x_t)_{t\in[0,T]}$ and the time-reversed process $t \mapsto (S(x_{T-t}))_{t\in[0,T]}$ have the same probability distribution on the space of continuous paths.

The long time behavior of Markov diffusion processes is well understood when the associated generator (2.28) has enough smoothing properties, namely when it is hypoelliptic in the sense of Hörmander (see [Hörmander (1967)] and Theorem 2.7 below). In this case, existence of a stationary probability distribution and irreducibility imply uniqueness of the stationary distribution, and ergodicity of the process. Irreducibility is also a direct consequence of the existence of a stationary invariant probability distribution having a positive density with respect to the Lebesgue measure. We refer to [Kliemann (1987)].

The precise statement of the result requires the definition of the Lie algebra $L(A_0, \dots, A_m)$ of a family of operators $(A_0, \dots, A_m)$, which is the vectorial space of operators containing $\mathrm{Span}(A_0, \dots, A_m)$ and satisfying the stability property:

$$B \in L(A_0, \dots, A_m) \ \Rightarrow \ [B, A_i] \in L(A_0, \dots, A_m), \qquad i = 0, \dots, m,$$

where the Lie bracket between two operators U and V is

$$[V, U] := VU - UV.$$

Theorem 2.7 (Sufficient conditions for ergodicity). *Consider the time-homogeneous stochastic differential equation*

$$dx_t = b(x_t)\, dt + \sigma(x_t)\, dW_t,$$

with smooth functions b, σ, and assume that there exists a (strong) solution for all times $t \geq 0$. Define

$$A_k := \sum_{i=1}^{d} \sigma^{i,k} \partial_i \qquad k = 1, \dots, n,$$

and

$$A_0 := \sum_{i=1}^{d} b^i \partial_i - \frac{1}{2} \sum_{i,j=1}^{d} \sum_{k=1}^{n} \sigma^{i,k} \partial_i(\sigma^{j,k}) \partial_j,$$

so that the generator may be rewritten as

$$\mathcal{L} = \frac{1}{2} \sum_{k=1}^{n} A_k^2 + A_0.$$

Assume that the Hörmander condition (hypoellipticity) is verified:

$$L(A_0, \ldots, A_n) = \mathrm{Span}(\partial_1, \ldots, \partial_d), \tag{2.33}$$

where $L(A_0, \ldots, A_n)$ *is the Lie algebra generated by* $(A_0, \ldots, A_n)$. *If a stationary probability distribution* π *having a positive density with respect to the Lebesgue measure exists, then the irreducibility property is verified, the stationary probability distribution is unique, and pathwise ergodicity holds for any initial condition* x_0*:*

$$\lim_{T \to +\infty} \frac{1}{T} \int_0^T \varphi(x_t)\, dt = \int_S \varphi\, d\pi \quad \text{a.s.},$$

for any bounded measurable observable φ.

For a proof of this result, we refer to [Kliemann (1987)].

The essence of the above result is that, although the noise terms do not act in all directions, there is sufficient coupling between the degrees of freedom not directly affected by the noise and the degrees of freedom experiencing some random forcing so that actually all degrees of freedom are affected by the noise term.

2.2.2 *Overdamped Langevin process*

Overdamped processes are stochastic dynamics on the system positions $q \in \mathcal{D}$ only. They are described through the simplest stochastic differential equation reversible with respect to the canonical distribution in position:

$$\nu(dq) = Z_\nu^{-1} \exp(-\beta V(q))\, dq\,.$$

The evolution equation is given by:

$$dq_t = -\nabla V(q_t)\, dt + \sqrt{\frac{2}{\beta}} dW_t, \tag{2.34}$$

where $t \mapsto W_t$ is a standard $3N$-dimensional Wiener process.

2.2.2.1 *Detailed balance and ergodicity*

The generator associated with (2.34) acts on test functions of the variable q. For a smooth function $\phi : q \mapsto \phi(q)$,

$$\mathcal{L}\phi = \frac{1}{\beta}\Delta\phi - \nabla V \cdot \nabla\phi = \frac{\mathrm{e}^{\beta V}}{\beta} \mathrm{div}\left(\mathrm{e}^{-\beta V} \nabla\phi\right). \tag{2.35}$$

The process (2.34) is reversible with respect to the canonical probability distribution $\nu(dq)$, as can be seen from the following computation using the above expression of the generator, as well as an integration by parts: for any test functions φ_1 and φ_2,

$$\begin{aligned}\int_{\mathcal{S}} \varphi_1 \, \mathcal{L}(\varphi_2) \, \mathrm{e}^{-\beta V} &= \frac{1}{\beta} \int_{\mathcal{S}} \varphi_1 \, \mathrm{div} \left(\mathrm{e}^{-\beta V} \nabla \varphi_2 \right) \\ &= -\frac{1}{\beta} \int_{\mathcal{S}} \nabla \varphi_1(q) \cdot \nabla \varphi_2(q) \, \mathrm{e}^{-\beta V(q)} \, dq \\ &= \int_{\mathcal{S}} \varphi_2 \, \mathcal{L}(\varphi_1) \, \mathrm{e}^{-\beta V}.\end{aligned}$$

A consequence of reversibility is that the canonical measure is invariant. Besides, the Hörmander condition (2.33) of Theorem 2.7 is readily verified, which ensures the ergodicity of the overdamped process (2.34).

2.2.2.2 *Time discretization and numerical implementation*

Implicit methods are cumbersome in molecular dynamics due to the high dimensionality of the problem. The typical numerical scheme for the overdamped process (2.34) is therefore the explicit Euler-Maruyama scheme:

$$q^{n+1} = q^n - \Delta t \nabla V(q^n) + \sqrt{\frac{2\Delta t}{\beta}} \, G^n, \tag{2.36}$$

where $(G^n)_{n \geq 0}$ are i.i.d. centered Gaussian random vectors in $\mathbb{R}^{3N}$ with identity co-variance matrix:

$$\mathbb{E} \left(G^n \otimes G^n \right) = \mathrm{Id}_{3N}.$$

Remark 2.8 (Weak approximation of the Brownian increment). *In the case when only the law of q_t for a given time t is of interest, it is possible to replace G^n by any random numbers with appropriate statistical properties (correct momenta up to order two for example). This could be useful to ensure that the random fluctuations remain almost surely bounded for example. We refer to [Talay and Tubaro (1990)] for precise results concerning the so-called weak convergence for stochastic differential equations.*

The time discretization implies that the Markov chain $(q^n)_{n \geq 0}$ does not, in general, sample the canonical distribution ν. Under suitable assumptions (see [Talay and Tubaro (1990); Mattingly *et al.* (2002)]), it can be shown that the numerical scheme is still ergodic, with an invariant probability measure $\nu_{\Delta t}$ close to the canonical measure ν. The distance between $\nu_{\Delta t}$ and ν (measured in a suitable norm) is typically of order Δt.

It is however possible to correct this bias of order Δt using a Metropolis rule. This leads to the so-called Metropolis Adjusted Langevin Algorithm [Roberts and Rosenthal (1998)] (MALA), which is a particular case of the Metropolis-Hastings Algorithm 2.2.

Algorithm 2.9 (Metropolis Adjusted Langevin algorithm). *Consider the density proposal :*

$$T(q,q') := \left(\frac{\beta}{4\pi\Delta t}\right)^{d/2} \exp\left(-\frac{\beta}{4\Delta t}\left|q' - q + \Delta t \nabla V(q)\right|^2\right), \tag{2.37}$$

and the associated Metropolis-Hastings ratio

$$r(q,q') := \mathrm{e}^{-\beta(V(q')-V(q))}\,\frac{T(q',q)}{T(q,q')}. \tag{2.38}$$

Consider an initial configuration $q^0 \in \mathcal{D}$ and iterate on $n \geq 0$,

(1) *Propose a new state $\tilde{q}^{n+1}$ from q^n according to $T(q_n,\cdot)$, which corresponds to one step of the explicit Euler-Maruyama scheme* (2.36)*:*

$$\tilde{q}^{n+1} = q^n - \Delta t \nabla V(q^n) + \sqrt{\frac{2\Delta t}{\beta}}\, G^n.$$

(2) *Accept the proposition with probability*

$$\min\left(1, r(q^n, \tilde{q}^{n+1})\right),$$

and set in this case $q^{n+1} = \tilde{q}^{n+1}$; if not set $q^{n+1} = q^n$.

The Markov chain generated by the MALA is reversible with respect to the canonical distribution ν (see Sections 2.1.2 and 2.1.4), and since the aperiodic irreducibility assumption (2.6) is verified for $n_0 = 1$, the ergodicity of the Metropolis Markov chain follows.

2.2.3 *Langevin process*

Hamiltonian dynamics preserve the energy, while a sampling of the canonical measure requires visiting all the energy levels. The Langevin dynamics is a phenomenological model of a Hamiltonian system coupled with a thermostat, which is an infinite reservoir of energy. In some case studies, this phenomenological model can be derived in some limiting regime [Kupferman *et al.* (2002)], relying on the Mori-Zwanzig formalism [Zwanzig (1973)]. Historically, the model was introduced by the botanist R. Brown to describe the movement of particles in a fluid, which were undergoing many collisions.

In this book, the Langevin equation will however be considered as a sampling device only, and we will not discuss the physical relevance of this dynamics. From a numerical viewpoint, several studies advocate the use of Langevin dynamics rather than overdamped Langevin dynamics [Scemama *et al.* (2006); Cancès *et al.* (2007)].

The corresponding stochastic equations of motion read, in a very general form,

$$\begin{cases} dq_t = \nabla_p H(q_t, p_t)\, dt, \\ dp_t = -\nabla_q H(q_t, p_t)\, dt - \gamma(q_t)\nabla_p H(q_t, p_t)\, dt + \sigma(q_t)\, dW_t, \end{cases} \tag{2.39}$$

where $t \mapsto W_t$ is a $3N$-dimensional standard Brownian motion, and σ and γ are (possibly position dependent) $3N \times 3N$ real matrices. For separable Hamiltonians, namely when

$$H(q,p) = E_{\text{kin}}(p) + V(q), \qquad E_{\text{kin}}(p) = \frac{1}{2} p^T M^{-1} p, \tag{2.40}$$

the evolution equations (2.39) may be simplified as

$$\begin{cases} dq_t = M^{-1} p_t\, dt, \\ dp_t = -\nabla V(q_t)\, dt - \gamma(q_t)\, M^{-1} p_t\, dt + \sigma(q_t)\, dW_t. \end{cases} \tag{2.41}$$

The term $\sigma(q_t)\, dW_t$ is a fluctuation term bringing energy into the system, this energy being dissipated through the viscous friction term $-\gamma(q_t)\, M^{-1} p_t\, dt$. These two terms are related through the following *fluctuation-dissipation* relation, which ensures that the canonical measure at the correct temperature is sampled:

$$\sigma\sigma^T = \frac{2\gamma}{\beta}. \tag{2.42}$$

Notice that γ is therefore a symmetric matrix. Often, γ and σ are proportional to the identity matrix, or γ is proportional to the mass matrix M. It may be interesting to choose position-dependent matrices σ, γ to restrict the action of the thermostat to the boundaries only, therefore sticking to the physical Hamiltonian dynamics in the core regions of the system, but in most applications, σ and γ are constant.

The fluctuation-dissipation relation (2.42) ensures that the canonical measure is a stationary measure of the Langevin process, and that the detailed balance condition (up to momenta reversal) with respect to the canonical distribution holds (see Section 2.2.3.1 below). Although time-reversibility properties depend on the Hamiltonian at hand, the stationarity of the canonical distribution still holds for (2.39) when considering a general Hamiltonian function (*e.g.* no longer separable).

Since the process $t \mapsto q_t$ is of finite variations, and σ depends only on q and not on p, it is indifferent to write in (2.39)-(2.41) the Itô or the Stratonovitch stochastic integration:

$$\sigma(q_t)\,dW_t = \sigma(q_t) \circ dW_t.$$

As is often the case with stochastic differential equations, the law of the solution of (2.39) depends on σ only through $\sigma\sigma^T$.

2.2.3.1 *Detailed balance and ergodicity*

The generator $\mathcal{L}$ associated with (2.39) can be decomposed as the sum of two contributions:

$$\mathcal{L} = \mathcal{L}^{\mathrm{ham}} + \mathcal{L}^{\mathrm{thm}}.$$

The first contribution $\mathcal{L}^{\mathrm{ham}}$ is associated with the Hamiltonian dynamics (which is obtained by taking $\gamma = \sigma = 0$) and easily expressed using the Poisson bracket formulation (1.12): for a test function φ,

$$\mathcal{L}^{\mathrm{ham}}\varphi = \nabla_p H \cdot \nabla_q \varphi - \nabla_q H \cdot \nabla_p \varphi = \{\varphi, H\}.$$

The second contribution $\mathcal{L}^{\mathrm{thm}}$ is associated with the thermostat: for a test function φ,

$$\begin{aligned}\mathcal{L}^{\mathrm{thm}}\varphi &= -\nabla_p H \cdot (\gamma \nabla_p \varphi) + \frac{1}{\beta}\,\mathrm{div}_p(\gamma \nabla_p \varphi) \\ &= \frac{\mathrm{e}^{\beta H}}{\beta}\mathrm{div}_p\left(\gamma \mathrm{e}^{-\beta H} \nabla_p \varphi\right),\end{aligned}$$

which reduces to

$$\mathcal{L}^{\mathrm{thm}}\varphi = \frac{\mathrm{e}^{\beta E_{\mathrm{kin}}}}{\beta}\mathrm{div}_p\left(\gamma \mathrm{e}^{-\beta E_{\mathrm{kin}}} \nabla_p \varphi\right)$$

in the case of a separable Hamiltonian (see (2.40)).

The complete generator of the Langevin process is therefore

$$\mathcal{L} = \{\cdot, H\} + \frac{\mathrm{e}^{\beta H}}{\beta}\mathrm{div}_p\left(\gamma \mathrm{e}^{-\beta H} \nabla_p \cdot\right). \tag{2.43}$$

It is then easily verified that the canonical measure

$$\mu(dq\,dp) = Z_\mu^{-1} \exp(-\beta H(q,p))\,dq\,dp$$

is an invariant measure since for any smooth test function φ (using (1.16))

$$\begin{aligned}\int_{T^*\mathcal{D}} \mathcal{L}(\varphi)\,d\mu &= \int_{T^*\mathcal{D}} \left(\{\varphi, H\} + \frac{\mathrm{e}^{\beta H}}{\beta}\mathrm{div}_p\left(\gamma \mathrm{e}^{-\beta H}\nabla_p \varphi\right)\right) \mathrm{e}^{-\beta H}\,dq\,dp \\ &= -\frac{1}{\beta}\int_{T^*\mathcal{D}} \left\{\varphi, \mathrm{e}^{-\beta H}\right\}\,dq\,dp + \frac{1}{\beta}\int_{T^*\mathcal{D}} \mathrm{div}_p\left(\gamma \mathrm{e}^{-\beta H}\nabla_p \varphi\right)\,dq\,dp \\ &= 0.\end{aligned}$$

When the Hamiltonian is separable (see (2.40)), the Langevin process satisfies time-reversibility up to momentum reversal (2.32): for any smooth test functions φ_1 and φ_2,

$$\int_{\mathcal{S}} \varphi_1 \, \mathcal{L}\varphi_2 \, d\pi = \int_{\mathcal{S}} \varphi_2 \circ S \, \mathcal{L}(\varphi_1 \circ S) \, d\pi,$$

$S(q,p) := (q,-p)$ denoting the momentum reversal operator. The proof of this equality follows from the following computations, which show also that the property $H \circ S = H$ is sufficient to ensure the result. First, for the thermostat part, (for a fixed position q),

$$\begin{aligned}
\frac{1}{\beta}\int_{\mathbb{R}^{3N}} \varphi_1 \, \mathrm{div}_p\left(\mathrm{e}^{-\beta E_{\mathrm{kin}}}\gamma\nabla_p\varphi_2\right) dp &= \frac{1}{\beta}\int_{\mathbb{R}^{3N}} \varphi_2 \, \mathrm{div}_p\left(\mathrm{e}^{-\beta E_{\mathrm{kin}}}\gamma\nabla_p\varphi_1\right) dp \\
&= \frac{1}{\beta}\int_{\mathbb{R}^{3N}} \varphi_2 \circ S \, \mathrm{div}_p\left(\mathrm{e}^{-\beta E_{\mathrm{kin}}}\gamma\nabla_p(\varphi_1 \circ S)\right) dp,
\end{aligned}$$

where the symmetry of γ has been used, and where, in the last integral, the change of variable $p \mapsto -p$ has been performed. Second, for the Hamiltonian part (using (1.17)):

$$\begin{aligned}
\int_{T^*\mathcal{D}} \varphi_1 \, \{\varphi_2, H\} \, \mathrm{e}^{-\beta H} \, dq \, dp &= -\frac{1}{\beta}\int_{T^*\mathcal{D}} \varphi_1 \, \left\{\varphi_2, \mathrm{e}^{-\beta H}\right\} dq \, dp \\
&= \frac{1}{\beta}\int_{T^*\mathcal{D}} \varphi_2 \, \left\{\varphi_1, \mathrm{e}^{-\beta H}\right\} dq \, dp \\
&= -\int_{T^*\mathcal{D}} \varphi_2 \, \{\varphi_1, H\} \, \mathrm{e}^{-\beta H} \, dq \, dp \\
&= \int_{T^*\mathcal{D}} \varphi_2 \circ S \, \{\varphi_1 \circ S, H\} \, \mathrm{e}^{-\beta H} \, dq \, dp,
\end{aligned}$$

where in the last integral, the change of variable $p \mapsto -p$ has been done, noticing that

$$\begin{aligned}
\{\varphi, H\}(q,-p) &= -M^{-1}p \cdot (\nabla_q\varphi)(q,-p) - \nabla V \cdot (\nabla_p\varphi)(q,-p) \\
&= -M^{-1}p \cdot \nabla_q(\varphi(q,-p)) + \nabla V \cdot \nabla_p(\varphi(q,-p)) \\
&= -\{\varphi \circ S, H\}(q,p).
\end{aligned}$$

In conclusion, adding the contributions from the Hamiltonian and the thermostat parts, the detailed balance condition (2.32) follows.

Next, the conditions for ergodicity of Theorem 2.7 can be readily verified when $\sigma(q)$ has full rank (*i.e.* a rank equal to $3N$) for all q in position space. Indeed, using the notation of Theorem 2.7, the first order part in the Langevin generator is

$$A_0 := M^{-1}p \cdot \nabla_q - \nabla V \cdot \nabla_p - p^T M^{-1}\gamma\nabla_p,$$

and the second order terms are decomposed with respect to the rows of σ as follows:

$$A_k := (\sigma^T)_{k,\cdot}(q)\nabla_p, \qquad k = 1, \ldots, n. \tag{2.44}$$

Thus the Lie bracket between the two yields:

$$[A_0, A_k] = -(\sigma^T)_{k,\cdot}(M^{-1}\nabla_q) + \left((\sigma^T)_{k,\cdot}M^{-1}\gamma + (M^{-1}p \cdot \nabla_q)((\sigma^T)_{k,\cdot})\right)\nabla_p. \tag{2.45}$$

Since $\sigma(q)$ has full rank in $\mathbb{R}^{3N}$, the families $\{(\sigma^T)_{k,\cdot}\nabla_p\}_{k=1,\ldots,n}$ (obtained from (2.44)) and $\{(\sigma^T)_{k,\cdot}M^{-1}\nabla_q\}_{k=1,\ldots,n}$ (obtained by subtracting the ∇_p contributions from (2.45)) span the full vector space, and so,

$$\mathrm{Span}(A_0, \ldots, A_n, [A_0, A_1], \ldots, [A_0, A_n]) = \mathrm{Span}(\partial_{q_1}, \ldots, \partial_{q_N}, \partial_{p_1}, \ldots, \partial_{p_N}).$$

This shows that $\mathcal{L}$ is hypoelliptic, and ergodicity holds for the Langevin process since the stationary canonical distribution has positive density with respect to the Lebesgue phase space measure.

2.2.3.2 *Time discretization and numerical implementation*

Let us now discuss the time discretization of the Langevin process (2.41) for a separable Hamiltonian. Having in mind a splitting strategy between the Hamiltonian part and the thermostat part, let us first consider the thermostat part, namely the stochastic differential equation:

$$\begin{cases} dq_t = 0, \\ dp_t = -\gamma(q_t)\, M^{-1}p_t\, dt + \sigma(q_t)\, dW_t. \end{cases} \tag{2.46}$$

The solution to this stochastic differential equation can be obtained analytically since p_t is an *Ornstein-Uhlenbeck process.* However, the extension to geometrically constrained processes (see Section 3.3.4), needed to compute free energy differences indexed by a general reaction coordinate, or for systems with molecular constraints, may be cumbersome.

We instead consider a simple midpoint Euler scheme for the thermostat part of the dynamics:

$$\begin{cases} q^{n+1} = q^n, \\ p^{n+1} = p^n - \dfrac{\Delta t}{2}\gamma(q^n)M^{-1}(p^n + p^{n+1}) + \sigma(q^n)\sqrt{\Delta t}\, G^n, \end{cases} \tag{2.47}$$

starting from some initial condition (q^0, p^0), and where $(G^n)_{n\geq 0}$ are i.i.d. centered and normalized (with co-variance given by the identity matrix) Gaussian random vectors. By linearity of the implicit term, (2.47) can be written explicitly as follows:

$$p^{n+1} = Ap^n + BG^n,$$

where

$$A = \left(\mathrm{Id} + \frac{\Delta t}{2}\gamma M^{-1}\right)^{-1} \left(\mathrm{Id} - \frac{\Delta t}{2}\gamma M^{-1}\right),$$

and

$$B = \sqrt{\Delta t}\left(\mathrm{Id} + \frac{\Delta t}{2}\gamma M^{-1}\right)^{-1} \sigma,$$

where we omit the dependence of γ and σ (hence of A and B) on $q^n = q^0$. Then,

$$p^n = A^n p^0 + \sum_{k=0}^{n-1} A^{n-1-k} B G^k$$

and thus, stability of (2.47) requires that, in the sense of non-negative symmetric matrices,

$$\frac{\Delta t}{2}\gamma \leq M. \tag{2.48}$$

The Markov chain induced by (2.47) is reversible both in the plain sense, and up to momenta reversal. Since the position q^n is fixed, it is sufficient to consider the Markov chain on momenta. The probability transition density of (2.47) reads (up to a normalizing constant)

$$T(p,p') = \exp\left(-\frac{1}{2}(p' - Ap)^T \left(BB^T\right)^{-1} (p' - Ap)\right).$$

Multiplying by the canonical distribution density, and expanding the above expression leads to

$$\exp\left(-\beta\frac{p^T M^{-1} p}{2}\right) T(p,p') = \exp\left(-\frac{\beta}{2}\left(p^T C p + p'^T C p' - 2p^T D p'\right)\right), \tag{2.49}$$

where

$$C = \frac{1}{2\Delta t}\left(\gamma^{-1} + \Delta t\, M^{-1} + \frac{\Delta t^2}{4} M^{-1}\gamma M^{-1}\right),$$

and

$$D = \frac{1}{2\Delta t}\left(\gamma^{-1} - \frac{\Delta t^2}{4} M^{-1}\gamma M^{-1}\right).$$

Since (2.49) is invariant by permutation of p and p' and by momentum reversal $(p,p') \mapsto (-p,-p')$, the detailed balance condition (2.5) is satisfied both in the plain sense, and up to momenta reversal.

Now, a typical numerical scheme for Langevin processes is obtained by a splitting procedure, using within each time-step:

- the Verlet propagator (1.22) for the Hamiltonian part,
- and the midpoint Euler propagator (2.47) for the thermostat part.

This splitting will be referred to as the (midpoint Euler-Verlet-Midpoint Euler) splitting in this book, and reads as follows:

$$\left\{\begin{aligned} p^{n+1/4} &= p^n - \frac{\Delta t}{4}\gamma(q^n)M^{-1}(p^n + p^{n+1/4}) + \sqrt{\frac{\Delta t}{2}}\,\sigma(q^n)\,G^n, \\ p^{n+1/2} &= p^{n+1/4} - \frac{\Delta t}{2}\nabla V(q^n), \\ q^{n+1} &= q^n + \Delta t\; M^{-1}p^{n+1/2}, \\ p^{n+3/4} &= p^{n+1/2} - \frac{\Delta t}{2}\nabla V(q^{n+1}), \\ p^{n+1} &= p^{n+3/4} - \frac{\Delta t}{4}\gamma(q^{n+1})M^{-1}(p^{n+1} + p^{n+3/4}) \\ &\quad + \sqrt{\frac{\Delta t}{2}}\,\sigma(q^{n+1})\,G^{n+1/2}. \end{aligned}\right. \tag{2.50}$$

A well-known variant of the latter is the Brünger-Brooks-Karplus (BBK) integrator [Brünger *et al.* (1984)] given by the splitting: (Explicit Euler-Verlet-Implicit Euler):

$$\left\{\begin{aligned} p^{n+1/2} &= p^n - \frac{\Delta t}{2}\nabla V(q^n) - \frac{\Delta t}{2}\gamma(q^n)M^{-1}p^n + \sqrt{\frac{\Delta t}{2}}\,\sigma(q^n)\,G^n, \\ q^{n+1} &= q^n + \Delta t\; M^{-1}p^{n+1/2}, \\ p^{n+1} &= p^{n+1/2} - \frac{\Delta t}{2}\nabla V(q^{n+1}) - \frac{\Delta t}{2}\gamma(q^{n+1})M^{-1}p^{n+1} \\ &\quad + \sqrt{\frac{\Delta t}{2}}\,\sigma(q^{n+1})\,G^{n+1/2}. \end{aligned}\right. \tag{2.51}$$

In the previous algorithms $(G^0, G^{1/2}, G^1, G^{3/2}, \ldots)$ denote a sequence of i.i.d. Gaussian random vectors with zero mean and covariance Id. Notice that the last step of the numerical scheme, at first glance implicit in the variable p^{n+1}, can in fact be rewritten in an explicit manner with some straightforward algebra.

Remark 2.10 (Brownian increments). *In* (2.50) *and* (2.51), *we used independent Gaussian increments in the first and last step of the splitting procedure. It is easy to check that it is possible to use half Gaussian random variables while keeping a consistent scheme. For the BBK scheme for* *example, this variant writes:*

$$\begin{cases} p^{n+1/2} = p^n - \dfrac{\Delta t}{2}\nabla V(q^n) - \dfrac{\Delta t}{2}\gamma(q^n)M^{-1}p^n + \dfrac{1}{2}\sqrt{\Delta t}\,\sigma(q^n)\,G^n, \\ q^{n+1} = q^n + \Delta t\; M^{-1}p^{n+1/2}, \\ p^{n+1} = p^{n+1/2} - \dfrac{\Delta t}{2}\nabla V(q^{n+1}) - \dfrac{\Delta t}{2}\gamma(q^{n+1})M^{-1}p^{n+1} \\ \qquad +\dfrac{1}{2}\sqrt{\Delta t}\,\sigma(q^{n+1})\,G^{n+1}. \end{cases} \tag{2.52}$$

Such a variant is considered in [Schlick (2002)] for instance. However, the fluctuation-dissipation should be corrected in order for the kinetic temperature to be correct, see [Cancès et al. *(2007)] for more details.*

Because of the time-step error, the measures $\mu_{\Delta t}$ sampled by the numerical schemes are only approximately equal to the canonical measure μ. The advantage of considering (2.50) rather than (2.51) is that the former can be used to construct a generalized Hybrid Monte-Carlo algorithm generating a Markov chain reversible up to momentum reversal for the canonical distribution.

The Markov chains induced by the numerical schemes (2.50) and (2.51) are easily shown to be irreducible, so that they are ergodic for their actual invariant measures $\mu_{\Delta t}$ (provided the numerical schemes indeed have such an invariant measure and are not transient). The irreducibility is a consequence of the fact that two random numbers are used per time-step. For the scheme (2.50) for instance, any final configuration (q^{n+1}, p^{n+1}) can be obtained from any initial configuration (q^n, p^n) by first choosing G^n so that the final position is correct (this is done by controlling $p^{n+1/4}$ which is such that $p^{n+1/4} = M(q^{n+1} - q^n)/\Delta t + \nabla V(q^n)/2$), and then $G^{n+1/2}$ so that the final momentum is p^{n+1}.

Let us now focus on the Metropolization of the above schemes, in particular (2.50) (but similar developments can be made on (2.51)). The deterministic Verlet step alone can be corrected using a generalized Metropolis-Hastings strategy as detailed in Sections 2.1.3 and 2.1.4. The properties of the algorithm rely on the reversibility properties of the midpoint discretization of the thermostat part. Owing to [Horowitz (1991)], this splitting will be referred to as a Generalized HMC (GHMC) algorithm.

Algorithm 2.11 (Generalized Hybrid Monte-Carlo). *Consider an initial configuration* $(q^0, p^0) \in T^*\mathcal{D}$ *and iterate on* $n \geq 0$,

(1) Evolve the momenta according to midpoint Euler on a time-step of size $\Delta t/2$:

$$p^{n+1/4} = p^n - \frac{\Delta t}{4}\gamma(q^n)M^{-1}(p^n + p^{n+1/4}) + \sqrt{\frac{\Delta t}{2}}\,\sigma(q^n)G^n,$$

and compute the energy $H(q^n, p^{n+1/4})$ of the configuration $(q^n, p^{n+1/4})$.

(2) Integrate the Hamiltonian equations of motion according to the Verlet scheme:

$$\begin{cases} p^{n+1/2} = p^{n+1/4} - \dfrac{\Delta t}{2}\nabla V(q^n), \\ \tilde{q}^{n+1} = q^n + \Delta t\, M^{-1}p^{n+1/2}, \\ \tilde{p}^{n+3/4} = p^{n+1/2} - \dfrac{\Delta t}{2}\nabla V(\tilde{q}^{n+1}). \end{cases}$$

(3) Compute the energy $H(\tilde{q}^{n+1}, \tilde{p}^{n+3/4})$ of the new phase space configuration. Accept the proposal $(\tilde{q}^{n+1}, \tilde{p}^{n+3/4})$ by setting $(q^{n+1}, p^{n+3/4}) = (\tilde{q}^{n+1}, \tilde{p}^{n+3/4})$ with probability

$$\min\left\{1, \exp\left[-\beta\big(H(\tilde{q}^{n+1}, \tilde{p}^{n+3/4}) - H(q^n, p^{n+1/4})\big)\right]\right\};$$

otherwise, reject and reverse momenta by setting $(q^{n+1}, p^{n+3/4}) = (q^n, -p^{n+1/4})$;

(4) Evolve the momenta $p^{n+3/4}$ to p^{n+1} according to midpoint Euler scheme on a time-step of size $\Delta t/2$:

$$p^{n+1} = p^{n+3/4} - \frac{\Delta t}{4}\gamma(q^{n+1})M^{-1}(p^{n+1} + p^{n+3/4}) + \sqrt{\frac{\Delta t}{2}}\,\sigma(q^{n+1})\,G^{n+1/2}.$$

Since both the generalized Metropolis-Hastings step (steps (2)-(3) here, see Section 2.1.4), and the midpoint Euler discretization of the thermostat part (steps (1) and (4), see (2.49)) are reversible up to momentum flip with respect to the canonical measure, so is their composition. The aperiodic irreducibility assumption (2.6) holds for a single step of the scheme ($n_0 = 1$) since, as above, two random numbers are used per time-step so that the final position and momentum can be controlled respectively with the first and second one. As a consequence, the generalized Hybrid Monte Carlo discretization of Langevin processes arising from Algorithm 2.11 is ergodic.

Remark 2.12 (Aperiodic irreducibility). *The last step ensures the irreducibility for the momentum variables. To prove the aperiodic irreducibility, the only difficulty is to therefore check the irreducibility for the position variables. Considering step (2), this is easily obtained by choosing* *adequately $p^{n+1/2}$. If one uses more than one step of the Verlet integrator in step (2), the controllability argument on position is more involved since no simple expressions give the final position as a function of the initial one. Actually, if one uses an exact integration of the Hamiltonian equations in step (2), irreducibility does not hold anymore in general (see Section 2.1.3 and [Mackenzie (1989)]). We refer for example to [Cancès et al. (2007)] for sufficient conditions to prove irreducibility in this case.*

Remark 2.13 (Momentum reversal in the rejection step).
Using momentum reversal in the rejection step of Algorithm 2.11, namely $(q^{n+1}, p^{n+3/4}) = (q^n, -p^{n+1/4})$, destroys dynamical properties of the Langevin process on large timescales (such as time auto-correlations of observables for instance, see [Horowitz (1991); Akhmatskaya et al. (2009)]), in spite of the recent results on pathwise convergence for Metropolized dynamics [Bou-Rabee and Vanden-Eijnden (2009)]. The Metropolis correction is however necessary to sample exactly the correct measure.

2.2.4 *Overdamped limit of the Langevin dynamics*

2.2.4.1 *Limit of the stochastic processes*

Overdamped processes can be derived from Langevin processes in the so-called "overdamped regime." Let us make this precise, by considering for the ease of notation the case when the mass tensor is a scalar times identity, and the diffusion tensor (and thus the friction tensor) is also a scalar times identity which does not depend on position (so that W_t is a standard $3N$-dimensional Brownian motion):

$$M = m\,\mathrm{Id} \quad \text{and} \quad \gamma \quad \text{and} \quad \sigma \quad \text{are constant and scalar.} \tag{2.53}$$

By (2.42), we thus have $\sigma = \sqrt{2\beta^{-1}\gamma}$. A similar argument holds without these assumptions.

Let us first non-dimensionalize the Langevin equations by introducing three units:

- a unit of time t_0,
- a unit of length l_0,
- a unit of mass m_0.

Let us introduce the non-dimensional variables associated to these characteristic quantities:

$$\bar{t} = \frac{t}{t_0}, \qquad \overline{W}_{\bar{t}} = \frac{1}{\sqrt{t_0}} W_{t_0 \bar{t}},$$

$$\overline{q}_{\bar{t}} = \frac{q_t}{l_0} = \frac{q_{t_0\bar{t}}}{l_0}, \qquad \overline{p}_{\bar{t}} = \frac{p_t}{m_0 l_0 t_0^{-1}} = \frac{p_{t_0\bar{t}}}{m_0 l_0 t_0^{-1}},$$

$$\overline{V}(\overline{q}) = \beta V(q) = \beta V(l_0 \overline{q}).$$

Notice that $\overline{W}_{\bar{t}}$ is a standard $3N$-dimensional Brownian motion.

By a change of variable, the non-dimensional Langevin equation (2.41) then writes:

$$\begin{cases} d\overline{q}_{\bar{t}} = m_0 m^{-1} \overline{p}_{\bar{t}}\, d\bar{t}, \\ d\overline{p}_{\bar{t}} = -m_0^{-1} l_0^{-2} \beta^{-1} t_0^2 \nabla_{\overline{q}} \overline{V}(\overline{q}_{\bar{t}})\, d\bar{t} - \gamma m^{-1} t_0 \overline{p}_{\bar{t}}\, d\bar{t} + \sqrt{2\beta^{-1}\gamma t_0^3 m_0^{-2} l_0^{-2}}\, d\overline{W}_{\bar{t}}. \end{cases}$$

Using the following non-dimensional numbers:

$$\alpha_1 = \frac{m}{m_0}, \qquad \alpha_2 = \frac{\gamma t_0}{m_0}, \qquad \alpha_3 = \frac{\beta m_0 l_0^2}{t_0^2},$$

this equation can be rewritten as

$$\begin{cases} d\overline{q}_{\bar{t}} &= \overline{v}_{\bar{t}}\, d\bar{t}, \\ \alpha_1 d\overline{v}_{\bar{t}} &= -\dfrac{1}{\alpha_3} \nabla_{\overline{q}} \overline{V}(\overline{q}_{\bar{t}})\, d\bar{t} - \alpha_2 \overline{v}_{\bar{t}}\, d\bar{t} + \sqrt{2\dfrac{\alpha_2}{\alpha_3}}\, d\overline{W}_{\bar{t}}, \end{cases} \tag{2.54}$$

where we introduced the velocity $v_t = m^{-1} p_t$, which is non-dimensionalized (consistently with the previous non-dimensionalization) as $\overline{v}_{\bar{t}} = m_0 m^{-1} \overline{p}_{\bar{t}}$.

Consider now the following scaling for a small parameter $\varepsilon > 0$:

$$\frac{1}{\alpha_3} = \alpha_2 = \sqrt{\frac{\alpha_2}{\alpha_3}} = \frac{\alpha_1}{\varepsilon}. \tag{2.55}$$

The physical interpretation of this condition is discussed more precisely in Remark 2.16 below. Dropping the bar for the ease of notation, we get from (2.54):

$$\begin{cases} dq_t &= v_t\, dt, \\ \varepsilon dv_t &= -\nabla V(q_t)\, dt - v_t\, dt + \sqrt{2}\, dW_t. \end{cases} \tag{2.56}$$

By noticing that the second equation can be reformulated as

$$\varepsilon dv_t = -\nabla V(q_t)\, dt - dq_t + \sqrt{2}\, dW_t,$$

it is intuitively clear that in the limit $\varepsilon \to 0$, the limiting process on positions is the overdamped Langevin process presented in Section 2.2.2:

$$dq_t^0 = -\nabla V(q_t^0)\,dt + \sqrt{2}\,dW_t. \tag{2.57}$$

Let us now state a precise result concerning the limit $\varepsilon \to 0$.

Proposition 2.14. *Denote by $(q_t^\varepsilon, v_t^\varepsilon)$ the solution to (2.56), with a given initial condition $(q_0^\varepsilon, v_0^\varepsilon) = (q_{\rm init}, v_0)$, and assume that ∇V is a Lipschitz function. Then, the following pathwise convergence holds: for any time $t > 0$,*

$$\lim_{\varepsilon\to 0} \sup_{0\le s\le t} \|q_s^\varepsilon - q_s^0\| = 0 \qquad a.s.,$$

where $(q_t^0)_{t\ge 0}$ is the solution to (2.57) with the initial condition $q_0^0 = q_{\rm init}$.

Proof. It is easily seen from (2.56) that

$$\begin{aligned} v_t^\varepsilon = v_0\,\mathrm{e}^{-t/\varepsilon} - \varepsilon^{-1}\int_0^t \mathrm{e}^{-(t-s)/\varepsilon}\nabla V(q_s)\,ds \\ + \varepsilon^{-1}\sqrt{2}\int_0^t \mathrm{e}^{-(t-s)/\varepsilon}\,dW_s. \end{aligned}$$

Thus,

$$\begin{aligned} q_t^\varepsilon &= q_{\rm init} + \int_0^t v_s^\varepsilon\,ds \\ &= q_{\rm init} + \int_0^t v_0\,\mathrm{e}^{-s/\varepsilon}\,ds - \varepsilon^{-1}\int_0^t\int_0^s \mathrm{e}^{-(s-r)/\varepsilon}\nabla V(q_r^\varepsilon)\,dr\,ds \\ &\quad + \varepsilon^{-1}\sqrt{2}\int_0^t\int_0^s \mathrm{e}^{-(s-r)/\varepsilon}\,dW_r\,ds \\ &= q_{\rm init} + v_0\,\varepsilon(1-\mathrm{e}^{-t/\varepsilon}) - \varepsilon^{-1}\int_0^t\int_r^t \mathrm{e}^{-(s-r)/\varepsilon}\,ds\nabla V(q_r^\varepsilon)\,dr \\ &\quad + \varepsilon^{-1}\sqrt{2}\int_0^t\int_r^t \mathrm{e}^{-(s-r)/\varepsilon}\,ds\,dW_r \\ &= q_{\rm init} + v_0\,\varepsilon(1-\mathrm{e}^{-t/\varepsilon}) - \int_0^t \left(1-\mathrm{e}^{-(t-r)/\varepsilon}\right)\nabla V(q_r^\varepsilon)\,dr \\ &\quad + \sqrt{2}\int_0^t \left(1-\mathrm{e}^{-(t-r)/\varepsilon}\right)dW_r. \end{aligned}$$

From (2.57),

$$q_t^0 = q_{\rm init} - \int_0^t \nabla V(q_s^0)\,ds + \sqrt{2}\int_0^t dW_s$$

with the same Brownian motion, so that, finally,

$$
\begin{aligned}
q_t^\varepsilon - q_t^0 = &- \int_0^t \left(1 - \mathrm{e}^{-(t-r)/\varepsilon}\right) \left(\nabla V(q_r^\varepsilon) - \nabla V(q_r^0)\right) dr \\
&+ v_0\, \varepsilon(1 - \mathrm{e}^{-t/\varepsilon}) + \int_0^t \mathrm{e}^{-(t-r)/\varepsilon} \nabla V(q_r^0)\, dr \\
&- \sqrt{2} \int_0^t \mathrm{e}^{-(t-r)/\varepsilon}\, dW_r.
\end{aligned}
\tag{2.58}
$$

The first term is bounded by $kt \sup_{s \le t} \left| q_s^\varepsilon - q_s^0 \right|$, where k is the Lipschitz constant of ∇V. As $\varepsilon \to 0$, the second term on the right-hand side converges to zero uniformly on compact time intervals. For the third term,

$$
\left| \int_0^t \mathrm{e}^{-(t-r)/\varepsilon} \nabla V(q_r^0)\, dr \right| \le \max_{0 \le r \le t} \|\nabla V(q_r^0)\|\, \varepsilon(1 - \mathrm{e}^{-t/\varepsilon}),
$$

so that this integral also converges to zero uniformly on compact time intervals. For the last term, an integration by parts gives:

$$
\int_0^t \mathrm{e}^{-(t-r)/\varepsilon}\, dW_r = \int_0^t \frac{\mathrm{e}^{-(t-r)/\varepsilon}}{\varepsilon} (W_t - W_r)\, dr + W_t\, \mathrm{e}^{-t/\varepsilon}.
$$

By the continuity of paths of Brownian motion (and thus uniform continuity on compact intervals), the first term also goes to zero uniformly on compact interval in time, while the second one converges to zero uniformly on compact time intervals. Thus, for a fixed time t_0,

$$
\forall t \le t_0, \qquad \sup_{s \le t} \left| q_s^\varepsilon - q_s^0 \right| \le k \int_0^t \sup_{r \le s} \left| q_r^\varepsilon - q_r^0 \right| ds + r_{t_0}(\varepsilon),
$$

with $r_{t_0}(\varepsilon) \to 0$ as $\varepsilon \to 0$. An application of Gronwall's lemma yields the result. □

A variant of the previous result is the following:

Proposition 2.15. *In the limit $\varepsilon \to 0$, the probability distribution of the path $(q_t)_{t \ge 0}$ of the Langevin equation* (2.56) *converges towards the probability distribution of the path of the overdamped process $(q_t^0)_{t \ge 0}$, solution of* (2.57). *Convergence occurs in the sense of probability distributions on the Banach space of continuous trajectories endowed with the supremum norm.*

The convergence result is weaker, but we nonetheless quote it because its proof uses totally different arguments than the proof of Proposition 2.14.

Proof. We present a proof based on (partially formal) asymptotic expansions on infinitesimal generators. Let us first rewrite (2.56) as:

$$\begin{cases} dq_t^\eta = \eta^{-1} p_t^\eta \, dt, \\ dp_t^\eta = -\eta^{-1} \nabla V(q_t^\eta) \, dt - \eta^{-2} p_t^\eta \, dt + \eta^{-1} \sqrt{2} \, dW_t, \end{cases} \tag{2.59}$$

where $\eta = \sqrt{\varepsilon}$, $q_t^\eta = q_t$, and $p_t^\eta = \eta v_t$.

The interest of this rewriting appears when considering the generator associated with the Markov evolution (2.59). Defining the Hamiltonian (omitting again for simplicity the bar for non-dimensional quantities):

$$H(q,p) = \frac{|p|^2}{2} + V(q),$$

the generator of (2.59) reads

$$\begin{aligned} \mathcal{L}^\eta &= \frac{1}{\eta} \{\cdot, H\} + \frac{1}{\eta^2} \mathrm{e}^H \mathrm{div}_p (\mathrm{e}^{-H} \nabla_p \cdot) \\ &= \frac{1}{\eta} \mathcal{L}^{\mathrm{ham}} + \frac{1}{\eta^2} \mathcal{L}^{\mathrm{thm}}, \end{aligned}$$

where the operators $\mathcal{L}^{\mathrm{ham}}$ and $\mathcal{L}^{\mathrm{thm}}$ have already been encountered in Section 2.2.3.1.

Notice that the measure $\mathrm{e}^{-H(q,p)} \, dq \, dp$ is invariant for the dynamics (2.59). For a given test function φ on positions, define

$$\rho^\eta(t,q,p) = \mathbb{E}\Big[\varphi(q_t^\eta) \,\Big|\, (q_0,p_0) = (q,p) \Big].$$

By the invariance of the measure $\mathrm{e}^{-H(q,p)} \, dq \, dp$, the function ρ^η is such that, for all $\eta > 0$ and $t \geq 0$,

$$\int_{T^*\mathcal{D}} \rho^\eta(t,q,p) \, \mathrm{e}^{-H(q,p)} \, dq \, dp = \int_{T^*\mathcal{D}} \varphi(q) \, \mathrm{e}^{-H(q,p)} \, dq \, dp. \tag{2.60}$$

Besides, ρ^η satisfies the Kolmogorov equation (2.24):

$$\partial_t \rho^\eta = \mathcal{L}^\eta \rho^\eta.$$

Let us assume the following expansion on ρ^η in terms of η:

$$\rho^\eta = \sum_{n \geq 0} \eta^n \rho_n.$$

To be consistent with (2.60), a necessary condition is that, for all $t \geq 0$,

$$\int_{T^*\mathcal{D}} \rho_0(t,q,p) \, \mathrm{e}^{-H(q,p)} \, dq \, dp = \int_{T^*\mathcal{D}} \varphi(q) \, \mathrm{e}^{-H(q,p)} \, dq \, dp,$$

and, for all $n \geq 1$,

$$\int_{T^*\mathcal{D}} \rho_n(t,q,p)\,\mathrm{e}^{-H(q,p)}\,dq\,dp = 0.$$

From the Kolmogorov equation, the following hierarchy of equations is obtained by identifying terms with equal powers of η:

$$\mathcal{L}^{\mathrm{thm}}\rho_0 = 0, \tag{2.61}$$

$$\mathcal{L}^{\mathrm{thm}}\rho_1 = -\mathcal{L}^{\mathrm{ham}}\rho_0, \tag{2.62}$$

$$\partial_t\rho_0 = \mathcal{L}^{\mathrm{thm}}\rho_2 + \mathcal{L}^{\mathrm{ham}}\rho_1. \tag{2.63}$$

Equation (2.61) implies that ρ_0 is independent of the momentum variable p. Equation (2.62) then becomes

$$\mathcal{L}^{\mathrm{thm}}\rho_1 = -p^T\nabla_q\rho_0,$$

and (by the Fredholm alternative) has a unique solution such that $\int \rho_1\mathrm{e}^{-H}\,dq\,dp = 0$:

$$\rho_1(q,p) = p^T\nabla_q\rho_0(q).$$

This gives

$$\mathcal{L}^{\mathrm{ham}}\rho_1 = p^T\,\nabla_q^2\rho_0\,p - \nabla V^T\,\nabla_q\rho_0,$$

and finally, integrating equation (2.63) with respect to $\mathrm{e}^{-\frac{|p|^2}{2}}\,dp$ leads to

$$\partial_t\rho_0 = \Delta_q\rho_0 - \nabla V^T\,\nabla_q\rho_0,$$

which is the Kolmogorov equation for the overdamped process (2.57). Standard results on weak convergence on processes (see [Kushner (1984)]) allow to conclude the proof, giving a convergence result in the sense of weak convergence of processes. □

More precise results, as well as many references on the overdamped asymptotics can be found in [Freidlin (2004)].

Remark 2.16 (Physical interpretation of the scaling (2.55)**).** *Equation* (2.55) *implies that, for a fixed* ε *there is actually only one parameter to be chosen among the three* $(\alpha_1,\alpha_2,\alpha_3)$ *in terms of* ε, *the other two being then automatically fixed, since* (2.55) *is equivalent to*

$$\frac{1}{\alpha_3} = \alpha_2 = \frac{\alpha_1}{\varepsilon}.$$

(*i*) Overdamped limit*: This is the case when* α_1 *is independent of* ε*, namely*

$$\alpha_1 = 1, \quad \alpha_2 = \varepsilon^{-1} \textit{ and } \alpha_3 = \varepsilon. \tag{2.64}$$

This can be obtained in practice by setting $\gamma = \varepsilon^{-1/2}$ *and* $t_0 = \varepsilon^{-1/2}$ *(the other parameters being independent of* ε*, so that* $\overline{V}$ *does not depend on* ε*). This amounts to assuming a change of the time scaling, and a very large friction force.*

(*ii*) Zero-mass limit*: This is the case when* α_2 *(or equivalently* α_3*) is independent of* ε*:*

$$\alpha_1 = \varepsilon, \quad \alpha_2 = 1 \textit{ and } \alpha_3 = 1. \tag{2.65}$$

This can be obtained in practice by setting $m = \varepsilon$*, the other parameters being chosen independently of* ε *(again,* $\overline{V}$ *does not depend on* ε*).*

2.2.4.2 *Overdamped limit of the numerical schemes*

The Euler scheme (2.36) for overdamped processes (resp. the associated Metropolis Adjusted Langevin Algorithm 2.9) can be recovered as a special case of the BBK scheme (2.52) or the (midpoint Euler-Verlet-midpoint Euler) scheme (2.50) (resp. the Generalized Hybrid Monte-Carlo algorithm 2.11).

Let us consider again for simplicity the case of scalar and constant mass, friction and diffusion (2.53). Generalizations to the tensor case are possible.

Proposition 2.17. *Consider the (midpoint Euler-Verlet-midpoint Euler) scheme* (2.50) *with a time-step* Δt *satisfying*

$$\frac{\Delta t}{4}\gamma = m.$$

Then the scheme (2.50) *is the explicit Euler discretization of an overdamped dynamics on positions* (2.36)*:*

$$q^{n+1} = q^n - h\nabla V(q^n) + \sqrt{\frac{2h}{\beta}}\, G^n, \tag{2.66}$$

with a time-step

$$h = \frac{\Delta t^2}{2m} = \frac{2\Delta t}{\gamma}. \tag{2.67}$$

Likewise, under the same assumptions, the Generalized Hybrid Monte-Carlo Algorithm 2.11 *is the Metropolis Adjusted Langevin Algorithm* 2.9.

For the BBK scheme (2.52), *under the assumption*

$$\frac{\Delta t}{2}\gamma = m,$$

the scheme can be rewritten as (2.66) *with the time-step* h *defined in* (2.67).

Proof. Let us focus on the (midpoint Euler-Verlet-midpoint Euler) scheme (2.50), the proof being similar for the BBK scheme.

The choice $\frac{\Delta t}{4}\gamma = m$ ensures that $p^{n+1/4}$ is a Gaussian vector:

$$p^{n+1/4} = \sqrt{\frac{\Delta t}{8}}\,\sigma\, G^n.$$

Then,

$$\begin{aligned} q^{n+1} &= q^n - \frac{\Delta t^2}{2m}\nabla V(q^n) + \frac{\Delta t}{m}\sqrt{\frac{\Delta t}{8}}\,\sigma\, G^n \\ &= q^n - \frac{\Delta t^2}{2m}\nabla V(q^n) + \sqrt{\frac{\gamma \Delta t^3}{4\beta m^2}}\, G^n, \end{aligned}$$

using (2.42). Since

$$\frac{\gamma \Delta t}{4m} = 1$$

by assumption, the magnitude of the Brownian increment is finally $\Delta t(\beta m)^{-1/2} = \sqrt{2h\beta}$. This yields (2.66).

For the Generalized Hybrid Monte-Carlo algorithm, it remains to show that the acceptance probability is the same as for the MALA, which amounts to checking that

$$\begin{aligned} &\exp\left[-\beta\Big(H(\tilde{q}^{n+1},\tilde{p}^{n+3/4}) - H(q^n,p^{n+1/4})\Big)\right] \\ &= \exp\left[-\beta\big(V(\tilde{q}^{n+1}) - V(q^n)\big)\right]\frac{T(\tilde{q}^{n+1},q^n)}{T(q^n,\tilde{q}^{n+1})}, \end{aligned}$$

where T is defined by (2.37) (with Δt replaced by h). This can be checked by direct computations. □

Remark 2.18 (Hybrid Monte Carlo algorithm and MALA). *It is also possible to check that the Hybrid Monte Carlo algorithm* 2.3 *using for* Φ_τ *a single step of the Verlet algorithm with time-step* $\tau = \Delta t$ *is exactly MALA with a time-step* $h = \Delta t^2/2m$.

2.3 Convergence of sampling methods

To measure the efficiency of a Markov Chain Monte Carlo method, many criteria may be used:

- the asymptotic variance of some estimators of interest, which is related to central limit theorems for Markov chains;
- the typical time needed to leave a metastable region, which is related to large deviation estimates, see ([Ellis (1985); Dembo and Zeitouni (1998); Freidlin and Wentzell (1998)]);
- the decorrelation time over a trajectory (see [Scemama *et al.* (2006)] and references therein);
- the rate of convergence of the law of the process at time t (or of the Markov chain at the n-th step) to the canonical measure, measured in some functional norm.

In the following, we will consider the first and the third criteria in Section 2.3.1. The convergence properties in this case depend both on the method and on the average quantity to compute. We then study the fourth one in Section 2.3.2, in the particular case of the overdamped Langevin dynamics. The fourth criterium is intrinsic in the sense that it is independent of the observable to be averaged, and we will see that it is indeed relevant to characterize metastable dynamics.

2.3.1 *Sampling errors*

The aim of this section is to give some error estimates for the estimator

$$\hat{A}_N = \frac{1}{N} \sum_{n=0}^{N-1} A(q^n, p^n),$$

where (q^n, p^n) are sampled using one of the Markov chain methods presented in the previous sections (Metropolis-Hastings algorithm or discretization of a stochastic differential equation), and the initial condition (q^0, p^0) is given. It is often the case in practice that q^0 is not random (a reference geometry of the system is used as an initial condition), and in any case, the initial conditions are not distributed according to the stationary distribution of the Markov chain.

The hope is to approximate canonical averages (2.2) using the estimator $\hat{A}_N$. We have seen in the previous sections that many methods satisfy a Law of Large Numbers (LLN) under weak conditions (namely irreducibility and invariance of a probability measure for the Markov chain), so that

$$\lim_{N\to+\infty} \hat{A}_N = \int_{T^*\mathcal{D}} A(q,p)\,\widetilde{\mu}(dq\,dp) = \mathbb{E}_{\widetilde{\mu}}(A) \quad \text{a.s.}, \tag{2.68}$$

where $\widetilde{\mu}$ is the invariant probability measure of the Markov chain. Some numerical techniques may introduce time-step errors, in which case $\widetilde{\mu}$ is not the canonical measure (2.1).

It is useful to decompose the total error as the sum of some systematic error (bias), and some statistical error (related to the variance of the random variables under study). More precisely, the following equality holds:

$$\mathbb{E}\left(\left|\hat{A}_N - \mathbb{E}_\mu(A)\right|^2\right) = \left(\mathbb{E}(\hat{A}_N) - \mathbb{E}_\mu(A)\right)^2 + \mathbb{E}\left(\left|\hat{A}_N - \mathbb{E}(\hat{A}_N)\right|^2\right).$$

The first term is the square of the bias, while the second one is the square of the statistical error. Typically, the statistical error dominates the bias. We study the bias in Section 2.3.1.1, and the statistical error in Section 2.3.1.2.

2.3.1.1 *Bias*

The quantity of interest is $\mathbb{E}_\mu(A)$, where μ is the canonical measure defined in (2.1). The bias is

$$\left|\mathbb{E}(\hat{A}_N) - \mathbb{E}_\mu(A)\right| \le \left|\mathbb{E}(\hat{A}_N) - \mathbb{E}_{\widetilde{\mu}}(A)\right| + \left|\mathbb{E}_{\widetilde{\mu}}(A) - \mathbb{E}_\mu(A)\right|. \tag{2.69}$$

Perfect sampling bias. The perfect sampling bias is the second term in (2.69):

$$\left|\mathbb{E}_{\widetilde{\mu}}(A) - \mathbb{E}_\mu(A)\right|. \tag{2.70}$$

There is no bias for Metropolis-based methods since $\widetilde{\mu} = \mu$ in this case (by construction of the Metropolis-Hastings algorithm).

On the other hand, there is in general a non-zero bias for discretizations of stochastic processes without the Metropolis step. For example, for a Euler discretization (2.36) of the overdamped dynamics (2.34), the bias is typically of order Δt (under appropriate assumptions on the potential). It is possible to reduce this bias by Romberg extrapolation, see [Talay and Tubaro (1990); Bally and Talay (1995, 1996)].

Finite sampling bias. The *finite sampling bias* is the first term in (2.69). It is, for a fixed N, the difference between the average of the estimator $\mathbb{E}(\widehat{A}_N)$, and the target quantity $\mathbb{E}_\mu(A)$:

$$\left|\mathbb{E}(\widehat{A}_N) - \mathbb{E}_{\widetilde{\mu}}(A)\right|. \tag{2.71}$$

The error (2.71) is related to the fact that the initial conditions are not sampled according to the stationary measure. When $\mathbb{E}(A(q^n, p^n))$ converges exponentially fast to $\mathbb{E}_{\widetilde{\mu}}(A)$ as $n \to +\infty$, the error is of order $\mathrm{O}(N^{-1})$.

Notice that even if the bias (2.71) is zero (think of i.i.d. random variables (q^n, p^n) with law $\widetilde{\mu}$), this property is generally not conserved under a nonlinear transformation: If (2.71) is zero, then

$$\left| \mathbb{E}\left[f(\widehat{A}_N)\right] - f\left[\mathbb{E}_{\widetilde{\mu}}(A)\right] \right|$$

is different from zero in general (see also the discussion in Section 4.1.5).

2.3.1.2 *Statistical errors*

Central Limit Theorem. When the Markov chain is irreducible and admits an invariant probability measure, a Law of Large Number holds, see (2.68). Under some additional conditions on the dynamics, a Central Limit Theorem can be shown (see for instance Chapter 17 in [Meyn and Tweedie (1993)], Section 2.1.3 in [Duflo (1997)], or Chapter 4 in [Gilks *et al.* (1996)] for more details on the assumptions made on the dynamics):

$$\sqrt{N} \left| \widehat{A}_N - \mathbb{E}_{\widetilde{\mu}}(A) \right| \xrightarrow[N \to +\infty]{} \mathcal{N}(0, \sigma^2), \qquad (2.72)$$

where convergence occurs in law. The so-called asymptotic variance σ^2 is the limit as N goes to $+\infty$ of the variance of $\sqrt{N}\,\widehat{A}_N$. It may be written as the sum of the intrinsic variance (which would be obtained if the samples were i.i.d.) and an additional variance arising from the correlation between the sampled configurations:

$$\sigma^2 = \mathrm{Var}_{\widetilde{\mu}}(A) + 2 \sum_{n=1}^{+\infty} \mathbb{E}_{\widetilde{\mu}} \Big[\big(A(q^0, p^0) - \mathbb{E}_{\widetilde{\mu}}(A)\big) \big(A(q^n, p^n) - \mathbb{E}_{\widetilde{\mu}}(A)\big) \Big]. \quad (2.73)$$

This formula is detailed below. Expectations such as $\mathbb{E}_{\widetilde{\mu}}[f(q^0, p^0)\, g(q^n, p^n)]$ in the right-hand side of the above equality should be understood as an expectation over all values (q^0, p^0) distributed according to $\widetilde{\mu}$, and all possible realizations of the dynamics until the discrete time n.

It is often the case that $\sigma^2 \geq \mathrm{Var}_{\widetilde{\mu}}(A)$ (and actually, σ^2 is much larger than $\mathrm{Var}_{\widetilde{\mu}}(A)$), but there is no general rule since the correlation term (the infinite sum in (2.73)) has no sign *a priori*. The variance (2.73) can be estimated with the techniques presented in Section 2.3.1.3 below.

The conditions ensuring that the Central Limit Theorem (2.72) holds can be difficult to check in practice, and even in cases when it is possible to show that these conditions hold, it is usually not possible to compute the asymptotic variance σ^2 analytically. This variance may be very large, in particular for metastable systems when the observable depends directly on the metastable degrees of freedom. In any case, the Central Limit Theorem suggests that the error of the estimate $\widehat{A}_N$ is of order $N^{-1/2}$.

In practice, it is useful to compare the standard deviation σ to the estimated average $\widehat{A}_N$. Indeed, as a consequence of the Central Limit Theorem, confidence intervals can be obtained as

$$I(\alpha) = \left[\widehat{A}_N - \alpha\frac{\sigma}{\sqrt{N}},\ \widehat{A}_N + \alpha\frac{\sigma}{\sqrt{N}}\right],$$

the parameter α depending on the confidence level. For instance, a 95% confidence interval corresponds to $\alpha \simeq 1.96$: In the large N limit, $\mathbb{E}_{\widetilde{\mu}}(A) \in I(1.96)$ with probability 0.95.

There exist also non-asymptotic error bounds, such as the Berry-Esseen estimate or concentration inequalities, but they usually yield pessimistic error estimates.

Motivation of the expression for the variance. The formula (2.73) can be motivated as follows, see for instance [Duflo (1997)]. Upon considering the observable $A - \mathbb{E}_{\widetilde{\mu}}(A)$ instead of A, it can be assumed without loss of generality that $\mathbb{E}_{\widetilde{\mu}}(A) = 0$. For simplicity, the chain is supposed to start at equilibrium: $(q^0, p^0) \sim \widetilde{\mu}$. The computation below may however be generalized to arbitrary initial conditions.

With our assumption on the initial conditions, the chain is at equilibrium at all times: $(q^n, p^n) \sim \widetilde{\mu}$ for all $n \geq 0$, and the following stationarity property holds:

$$\mathbb{E}_{\widetilde{\mu}}\left[f(q^{n+k}, p^{n+k})\, g(q^k, p^k)\right] = \mathbb{E}_{\widetilde{\mu}}\left[f(q^n, p^n)\, g(q^0, p^0)\right]$$

for all $n, k \geq 0$. The variance of $\sqrt{N}\,\widehat{A}_N$ is

$$\begin{aligned} N\,\mathbb{E}_{\widetilde{\mu}}\left(\widehat{A}_N^2\right) &= \frac{1}{N}\sum_{k=0}^{N-1}\sum_{l=0}^{N-1}\mathbb{E}_{\widetilde{\mu}}\Big(A(q^k,p^k)A(q^l,p^l)\Big) \\ &= \frac{1}{N}\sum_{k=0}^{N-1}\mathbb{E}_{\widetilde{\mu}}\Big(A(q^k,p^k)^2\Big) + \frac{2}{N}\sum_{0\leq l<k\leq N-1}\mathbb{E}_{\widetilde{\mu}}\Big(A(q^k,p^k)\,A(q^l,p^l)\Big) \\ &= \mathbb{E}_{\widetilde{\mu}}(A^2) + 2\sum_{n=1}^{N-1}\left(1-\frac{n}{N}\right)\mathbb{E}_{\widetilde{\mu}}\Big(A(q^n,p^n)\,A(q^0,p^0)\Big), \end{aligned}$$

where we use the stationarity property in the last step. Therefore, by a dominated convergence argument, $N\,\mathbb{E}\left(\widehat{A}_N^2\right)$ converges to (2.73) as $N \to +\infty$ when the series on the right-hand side is summable.

Correlation length. To gain an understanding of the asymptotic variance in terms of correlation, it is useful to define a real number N_{corr}, called the *correlation length*, such that

$$\sigma^2 = \text{Var}_{\widetilde{\mu}}(A) N_{\text{corr}},$$

where σ^2 is the variance given by (2.73). Typically, N_{corr} is a large positive number. The variance of $\widehat{A}_N$ is of the order of

$$\frac{\sigma^2}{N} = \text{Var}_{\widetilde{\mu}}(A) \frac{N_{\text{corr}}}{N}. \tag{2.74}$$

Thus, a sample of size N may be seen as containing only N/N_{corr} independent samples, and, for practical purposes, samples obtained from the trajectory of a Markov chain subsampled at rate N_{corr} may be considered as almost independent. This motivates the estimation and the practical relevance of N_{corr}.

An estimate of N_{corr} can be deduced from estimates of $\text{Var}_{\widetilde{\mu}}(A)$ and of the variance (2.73) (see Section 2.3.1.3 for the latter quantity).

Analytical example. Consider the observable $A(q) = q$ and the one-dimensional Markov chain

$$q^{n+1} = (1 - \Delta t) q^n + \sqrt{2\beta^{-1}\Delta t}\, G^n \tag{2.75}$$

where $(G^n)_{n \geq 0}$ are independent and identically distributed according to $\mathcal{N}(0,1)$, and $\Delta t < 1$. The scheme (2.75) is a consistent discretization of the stochastic differential equation $dq_t = -q_t\, dt + \sqrt{2\beta^{-1}}\, dW_t$. For simplicity, we assume

$$q^0 \sim \mathcal{N}\left(0, \frac{1}{\beta(1 - \Delta t/2)}\right)$$

which is the invariant density of the Markov chain (2.75). The asymptotic variance of the estimator

$$\frac{1}{N} \sum_{n=1}^{N} q^n$$

of the average position for the Markov chain $(q^n)_{n \geq 0}$ given by (2.75) can be computed analytically since $\mathbb{E}(q^n) = \mathbb{E}(q^0) = 0$ for all $n \geq 0$, and $\mathbb{E}\left[q^n q^0\right] = (1 - \Delta t)^n \mathbb{E}\left[(q^0)^2\right]$. Therefore,

$$\sigma^2 = \mathbb{E}\left[(q^0)^2\right] \left(1 + 2 \sum_{n=1}^{+\infty} (1 - \Delta t)^n \right) = \frac{2}{\beta \Delta t}. \tag{2.76}$$

This variance is larger than the variance obtained for i.i.d. configurations, which is equal to $\beta^{-1}(1-\Delta t/2)^{-1}$. Besides,

$$N_{\mathrm{corr}} = \frac{2}{\Delta t} - 1,$$

which shows that the correlation length increases as the time-step Δt decreases (notice however that the correlation time $T_{\mathrm{corr}} = N_{\mathrm{corr}}\Delta t$ has a well-defined limit).

2.3.1.3 *Practical computation of error bars*

In this section, we give some computable estimates of the *statistical error* presented in Section 2.3.1.2.

Independent replica estimations. The first and most obvious error estimates may be obtained by considering *independent realizations* of the Markov chain. This is also a safe and rigorous approach, for which convergence results can be shown. Indexing by $1 \le m \le M$ the corresponding realizations (starting from independent initial conditions $(q^{m,0}, p^{m,0})$ and driven by independent noises), an estimate of the average $\mathbb{E}_{\widetilde{\mu}}(A)$ for N steps of the m-th realization is

$$\widehat{A}_N^m = \frac{1}{N}\sum_{n=1}^{N} A(q^{m,n}, p^{m,n}).$$

The asymptotic variance σ^2 of the estimator $\widehat{A}_N$ may then be estimated by the empirical variance over M realizations as

$$\sigma^2 = \lim_{N\to+\infty}\lim_{M\to+\infty} \Sigma_{N,M}^{\mathrm{real}}, \tag{2.77}$$

where

$$\Sigma_{N,M}^{\mathrm{real}} = \frac{N}{M}\sum_{m=1}^{M}\left(\widehat{A}_N^m - \frac{1}{M}\sum_{k=1}^{M}\widehat{A}_N^k\right)^2.$$

Block averaging. It is often the case in molecular dynamics that averages over a *single* long trajectory are performed. In this case, many practitioners find it convenient to resort to block averaging [Flyvbjerg and Petersen (1989)] (termed "batch means" in the statistics literature [Geyer (1992); Fishman (1996)]; see also [Scemama *et al.* (2006)]). This approach is less rigorous than estimates over independent realizations, but is useful when only a single long realization of the dynamics is considered. There are other methods in the statistics literature to estimate the variance from a given sample, like the bootstrap method (or the related jackknife method), see [Efron (1979)]. However, we are not aware of the use of such methods in molecular dynamics, and we therefore concentrate on the blocking method.

The idea of block averaging is to gather the data along one trajectory in blocks of similar and increasing sizes, and to estimate the asymptotic variance by computing averages within such blocks assuming that (i) the blocks are large enough so that a Central Limit Theorem holds within each block, and that (ii) the estimators over the different blocks are almost independent. Typically, it is observed that the so-obtained estimates of the variance increase monotonically with the block lengths.

More precisely, consider a trajectory $((q^0,p^0),(q^1,p^1),\dots)$ containing NM points, where M denotes the number of blocks and N the number of points within a block. The average within the k-th block is the random variable

$$A_N^k = \frac{1}{N} \sum_{i=(k-1)N+1}^{kN} A(q^i,p^i), \qquad k = 1,\dots,M.$$

The random variables $(A_N^1,\dots,A_N^M)$ are assumed to be approximately independent and distributed according to $\mathcal{N}(\overline{A},\sigma^2/N)$, where $\overline{A} = \mathbb{E}_{\widetilde{\mu}}(A)$ and σ^2 is the variance defined in (2.73). Therefore, an estimator of σ^2 is

$$\sigma^2 = \lim_{N\to+\infty} \lim_{M\to+\infty} \Sigma_{N,M}^{\text{block}}, \tag{2.78}$$

where

$$\Sigma_{N,M}^{\text{block}} = \frac{N}{M} \sum_{k=1}^{M} (A_N^k - A_{NM})^2,$$

$A_{NM} = A_{NM}^1$ being the average over the entire trajectory. Furthermore, the variance of the estimator $\Sigma_{N,M}^{\text{block}}$ is approximated under the assumption that $(A_N^k)_{k=1,\dots,M}$ are i.i.d. with law $\mathcal{N}(\overline{A},\sigma^2/N)$, which is the case when N is large enough. Then, $(A_N^k - A_{NM})^2 \sim \sigma^2 (G^k)^2/N$ where G^k are i.i.d. with law $\mathcal{N}(0,1)$, and $\Sigma_{N,M}^{\text{block}} \sim \sigma^2 M^{-1} \sum_{k=1}^{M} (G^k)^2$. The expectation of $\Sigma_{N,M}^{\text{block}}$ is σ^2, whereas its variance is asymptotically equal to $2\sigma^4/M$. Therefore,

$$\Sigma_{N,M}^{\text{block}} \sim \sigma^2 \left(1 + G\sqrt{\frac{2}{M}} \right), \tag{2.79}$$

where $G \sim \mathcal{N}(0,1)$. A confidence interval can therefore be obtained when the number of blocks M is large enough. The above derivation is valid only if the typical correlation length N_{corr} (see 2.74) is (much) shorter than the block length. It also suggests that there is a trade-off between the amount of correlation taken into account (which increases with the block size N) and the variability of the block estimates (which requires the number M of blocks to be large enough).

In practice, a long trajectory is first computed. Then, blocks of increasing sizes N are considered, starting from $N = 1$. When $N = 1$, all sampled configurations are considered to be independent, and the estimated variance is therefore often a lower bound of (2.73). As the block lengths are increased, more and more correlation is taken into account. The variance typically reaches a plateau at some point. For very large values of N, the number of blocks M is usually so small that the confidence interval on the estimated variance is very large, and the estimated value is not useful. See the example plotted in Figures 2.1 and 2.2 for typical behaviors of the estimate of the variance as a function of the block sizes.

Analytical example. Consider again the Markov chain (2.75). We plot in Figure 2.1 the resulting estimates of the standard deviation using multiple sequences (see (2.77)), and compare it to results obtained by block averaging (see (2.78)). Actually, in the case of independent replicas, we find it more convenient to plot the variance of averages obtained over N integration steps, namely

$$N^{-1}\Sigma^{\mathrm{real}}_{N,M} = \frac{1}{M}\sum_{m=1}^{M}\left(\widehat{A}_N^m - \frac{1}{M}\sum_{k=1}^{M}\widehat{A}_N^k\right)^2. \tag{2.80}$$

This quantity is expected to decrease as σ^2/N, the prefactor being given by (2.73). The results were obtained using $\beta = 1$ and $\Delta t = 0.05$. The reference value for the variance, computed with the analytical formula (2.76), is $2/(\beta\Delta t) = 40$.

Detecting metastability issues. Consider the double-well potential $V(q) = (q^2-1)^2$, and a trajectory of the overdamped Langevin dynamics at inverse temperature $\beta = 5$, with a time-step $\Delta t = 0.05$:

$$q^{n+1} = q^n - \Delta t\,\nabla V(q^n) + \sqrt{\frac{2\Delta t}{\beta}}\,G^n,$$

where (G^n) are i.i.d. one-dimensional Gaussian variables of mean 0 and variance 1. Two observables are considered: the potential energy, and the position. The average position should be 0 in view of the symmetry of the system, but the metastable behavior of the system, which switches

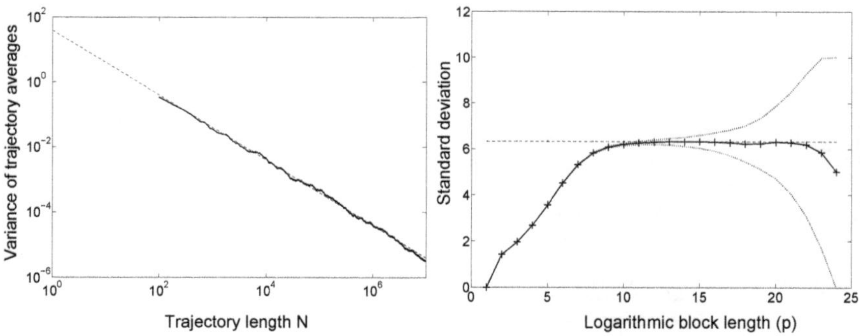

Fig. 2.1 The observable is the position for the Markov chain (2.75) with $\beta = 1$ and $\Delta t = 0.05$. Left: $N^{-1}\Sigma^{\mathrm{real}}_{N,M}$ (see (2.80)) as a function of the lengths N of the trajectories, computed using $M = 100$ independent trajectories. The linear fit (dashed line) allows to estimate the variance as $\sigma^2 \simeq 40$. Right: Estimate of the variance $\Sigma^{\mathrm{block}}_{N,M}$ given by (2.78) as a function of the block sizes $N = 2^p$ in the case $NM = 2^{24}$ (thick solid line), and 95% confidence intervals (see (2.79), dotted lines). The reference value $\sqrt{40}$ is given by the thin solid line.

from one minimum around $q = 1$ to the other minimum at $q = -1$, yields a slow convergence of the estimator $\frac{1}{N}\sum_{n=1}^{N} q^n$ of the average position. On the other hand, convergence is expected to happen much faster for an observable symmetric with respect to the change of variable $q \mapsto -q$ such as the potential energy. Figure 2.2 presents the results obtained with multiple sequences, and block averaging (in this case, only the results for the position are presented). Both techniques give very similar results. As expected, the variance on the position is much larger than the variance on the energy.

Notice however that, if no transition from one well to the other had been observed during the simulation time, the variance on the position observable would have been much smaller! The estimators of the variance presented in this section should therefore always be considered with care since they give in general only some lower bound on the actual variance.

2.3.2 *Rate of convergence for stochastic processes*

We present in this section another framework for understanding convergence properties of sampling methods, in the case of stochastic processes. We will focus on the simple gradient dynamics (2.34):

$$dq_t = -\nabla V(q_t)\,dt + \sqrt{2\beta^{-1}}\,dW_t.$$

As mentioned above, under loose assumptions on V, this dynamics is ergodic with respect to the Boltzmann-Gibbs measure ν (given by (2.3)), so

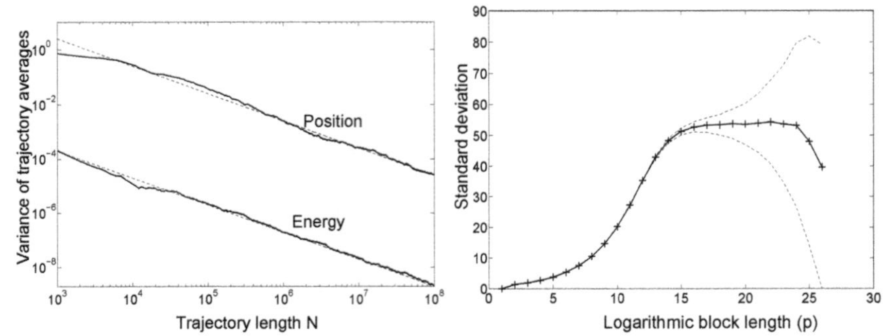

Fig. 2.2 Estimate asymptotic variances in the double-well case. Left: $N^{-1}\Sigma^{\rm real}_{N,M}$ (see (2.80)) as a function of the lengths N of the trajectories, computed using $M = 100$ independent trajectories. The linear fit (dashed line) allows to estimate the variance as $\sigma^2 \simeq 2500$ in the case when the observable is the position, and $\sigma^2 \simeq 0.20$ when the observable is the potential energy. Right: Estimate of the variance $\Sigma^{\rm block}_{N,M}$ given by (2.78) for the position observable as a function of the block sizes $N = 2^p$ in the case $NM = 2^{24}$ (thick solid line), and 95% confidence intervals (see (2.79), dotted lines). The variance estimated by block averaging is $\sigma^2 \simeq 2800$.

that it would be in principle possible to compute averages with respect to ν as ergodic averages over a trajectory $(q_t)_{t\geq 0}$. In practice, such an approach is typically not possible due to metastable states (local minima of V) in which q_t remains trapped for long times. To quantify this bad behavior, we study here the *rate of convergence of the law* of q_t to equilibrium. The random variable q_t admits a probability density function $\psi(t,\cdot)$ which satisfies the Fokker-Planck equation:

$$\partial_t \psi = \mathrm{div}(\nabla V \psi + \beta^{-1}\nabla\psi). \tag{2.81}$$

The expected longtime limit of ψ is the density of ν, namely

$$\psi_\infty = Z_\nu^{-1}\exp(-\beta V).$$

2.3.2.1 *Longtime convergence*

It is a classical result that the typical relaxation time of (2.81) is given by the spectral gap of the unbounded operator

$$\mathcal{L}^*\psi = \mathrm{div}(\nabla V \psi + \beta^{-1}\nabla\psi),$$

seen as a self-adjoint operator on the weighted $L^2(\nu)$ space. It is also well known that this relaxation time is exponentially large in the low temperature limit $\beta \to \infty$ when V has several local minima. The rate of convergence can be expressed through a variational formula given by the lowest energy barrier between any two minima of V. Classical spectral estimates, as well as their consequences on the convergence of the associated semi-group and on the exit time estimates of the associated diffusion process, can be read in [Davies (1982a, b, c); Freidlin and Wentzell (1998)].

On the other hand, a standard way to obtain a rate of convergence for (2.81) is to resort to entropy methods. The relative entropy and the Fisher information between two measures are defined as follows.

Definition 2.19 (Entropy and Fisher Information). *For two probability measures π_1 and π_2 such that π_1 is absolutely continuous with respect to π_2 (denoted $\pi_1 \ll \pi_2$ in the following), the entropy of π_1 with respect to π_2 is*

$$H(\pi_1 \,|\, \pi_2) = \int \ln\left(\frac{d\pi_1}{d\pi_2}\right) d\pi_1. \tag{2.82}$$

The Fisher information of π_1 with respect to π_2 is

$$I(\pi_1 \,|\, \pi_2) = \int \left|\nabla \ln\left(\frac{d\pi_1}{d\pi_2}\right)\right|^2 d\pi_1. \tag{2.83}$$

In the case when π_1 and π_2 are probability measures whose support is a submanifold of $\mathbb{R}^n$, the gradient operator in the definition of the Fisher information has to be understood as the gradient along the submanifold, associated to the Euclidean scalar product of the ambient space $\mathbb{R}^n$. This will be useful later on, when studying diffusions on submanifolds (see Sections 3.2.6 and 5.2).

Using the convexity of $x \mapsto x \ln x$, it is easy to check that the entropy is zero if and only if the two measures are the same. This is also true for the Fisher information.

There are many possible definitions of the entropy. Some mathematical motivations for the use of the relative entropy H defined in (2.82) can be found in [Markowich and Villani (2000)]. This particular entropy may be of interest for the following extensivity property: The relative entropy of the distribution of N independent variables (or, in less probabilistic terms, the entropy of a tensorized measure) is the sum of the relative entropies of the distributions of each random variable. This suggests that the rate of convergence to equilibrium estimated with relative entropies for weakly dependent variables may remain stable when the number of variables becomes large (see for instance the paragraph on Kac's spectral problem in Chapter 5 of [Villani (2002)]). This extensivity behavior is a consequence of the extensivity of the logarithm function involved in the definition (2.82), and has no reason to be true for other ways of measuring distances between two probability measures.

Recall that the total variation between two measures (which reduces to the L^1 norm of the difference between the two densities when the two measures are defined on $\mathbb{R}^n$ and are absolutely continuous with respect to the Lebesgue measure) can be bounded by the relative entropy. This is the Csiszár-Kullback inequality (see for example [Ané *et al.* (2000); Royer (2007)]):

$$\|\pi_1 - \pi_2\|_{TV} \leq \sqrt{2H(\pi_1 \,|\, \pi_2)}. \tag{2.84}$$

In other words, an upper bound for the entropy between π_1 and π_2 yields an upper bound for a distance between π_1 and π_2 (even though the entropy is not a distance, for example because it is not symmetric in its arguments).

We now present a way to obtain an estimate of the rate of convergence of the entropy[1] $H(\psi(t,\cdot) \,|\, \psi_\infty)$ to zero for the dynamics associated with (2.81). Notice first that (2.81) can be rewritten as

$$\partial_t \psi = \frac{1}{\beta} \text{div} \left(\psi_\infty \nabla \left(\frac{\psi}{\psi_\infty} \right) \right).$$

A straightforward computation shows that

$$\frac{d}{dt} H(\psi(t,\cdot) \,|\, \psi_\infty) = -\beta^{-1} I(\psi(t,\cdot) \,|\, \psi_\infty). \tag{2.85}$$

The exponential decay of $H(\psi(t,\cdot) \,|\, \psi_\infty)$ can then be shown provided ψ_∞ satisfies a so-called Logarithmic Sobolev Inequality (LSI).

Definition 2.20 (Logarithmic Sobolev Inequality).
A probability measure π_2 satisfies a Logarithmic Sobolev Inequality with constant $R > 0$ (in short LSI(R)) if, for all probability measures π_1 such that $\pi_1 \ll \pi_2$,

$$H(\pi_1 \,|\, \pi_2) \leq \frac{1}{2R} I(\pi_1 \,|\, \pi_2). \tag{2.86}$$

Assume that ψ_∞ satisfies[2] a LSI(R). Then the following estimate is a consequence of (2.85): $\forall t \geq 0$,

$$H(\psi(t,\cdot) \,|\, \psi_\infty) \leq H(\psi(0,\cdot) \,|\, \psi_\infty) \exp(-2\beta^{-1} R t). \tag{2.87}$$

[1] With a slight abuse of notation, for two probability density functions ψ and ψ_∞ we denote $H(\psi \,|\, \psi_\infty)$ the entropy of the probability measure $\psi(q)\,dq$ with respect to the probability measure $\psi_\infty(q)\,dq$.

[2] Again, with a slight abuse of notation, we say that a probability density function satisfies a LSI if the associated probability measure satisfies a LSI.

From the Csiszár-Kullback inequality (2.84), the exponential convergence to zero with rate $\beta^{-1}R$ of the norm $\|\psi(t,\cdot)-\psi_\infty\|_{L^1}$ is then deduced. It is easy to check that if (2.87) holds for all initial conditions $\psi(0,\cdot)$, then the LSI (2.86) is satisfied, so that the two properties (2.87) and (2.86) are actually equivalent.

For an introduction to logarithmic Sobolev inequalities, their properties and their relation to longtime behavior of solutions to PDEs, we refer to [Ané *et al.* (2000); Arnold *et al.* (2001); Villani (2003)].

2.3.2.2 *A possible quantification of metastability*

Quantifying metastability with functional inequalities. The estimate (2.87) allows us to define a metastable potential as a potential giving rise to a measure $\psi_\infty(q)\,dq$ which satisfies (2.86) for R small. Hence, the convergence of $\psi(t,\cdot)$ to ψ_∞ is slow. The interest of such a definition is that we can *quantify* the metastability, depending on how small R is.

This is a sensible measure of metastability since it is known that the constant R is small if there exist metastable states separated by large energy barriers. Links between spectral gap analysis (mentioned above) and log-Sobolev constants is a classical topic, and degeneracy of spectral gap generally implies degeneracy of log-Sobolev constant (the converse being untrue) [Davies (1983); Bakry (1997)].

In Chapter 5, we will show how the rate of convergence for the simple gradient dynamics (2.34) can be enhanced thanks to an adaptive importance sampling method (see Equation (5.68)). The main feature in this enhancement arises from convergence properties of stochastic processes with well-chosen constraints, see Section 3.2.6.

Transfer operator approach. Other definitions of metastability have been suggested. For instance, in the approach of [Schütte (1999); Huisinga (2001); Huisinga and Schmidt (2006)], a set is called metastable if the probability to leave the set for a given microscopic time span τ is small. For a given integer $m \geq 1$, the state space $\mathcal{E}$ can be decomposed as the union of sets $\mathcal{E}_1,\dots,\mathcal{E}_m$. These sets should be chosen such that the probability $p(\mathcal{E}_i)$ to leave the set $\mathcal{E}_i$ during the time span τ, for initial conditions sampled according to the canonical measure at hand, is as small as possible. Metastability is then quantified using the largest eigenvalues of the transfer operator $P^\tau = \mathrm{e}^{\tau\mathcal{L}}$, where $\mathcal{L}$ is the infinitesimal generator of the diffusion process. When the transfer operator has m eigenvalues close to 1, separated from the remained of the spectrum (see [Huisinga (2001)] for conditions ensuring this structure of the spectrum), a decomposition maximizing $p(\mathcal{E}_1) + \cdots + p(\mathcal{E}_m)$ is obtained when the first m eigenvectors of the transfer operator are as constant as possible on each set $\mathcal{E}_i$. These conditions can be used algorithmically to decompose the phase space into a union of metastable sets, see [Schütte (1999)] and the references therein.

2.3.2.3 *Obtaining logarithmic Sobolev inequalities*

Let us first mention a stability result under tensorization, due to [Gross (1975)]: When $\psi_\infty = \prod_{i=1}^M \psi_\infty^i$ and each measure $\psi_\infty^i(q)\,dq$ satisfies a LSI with constant ρ_i, then ψ_∞ satisfies a LSI with constant $\rho = \min\{\rho_1, \ldots, \rho_M\}$.

There are two classical criteria to prove that ψ_∞ satisfies a LSI, namely the Bakry-Emery criterium [Bakry and Emery (1985)], and the Holley-Stroock perturbative criterium [Holley and Stroock (1987)].

Proposition 2.21 (Bakry-Emery criterium). *Let $V : \mathbb{R}^n \to \mathbb{R}$ be an α-convex function, i.e. a function such that there exists $\alpha > 0$,*

$$\forall x \in \mathbb{R}^n, \quad \forall u \in \mathbb{R}^n, \qquad u^T \nabla^2 V(x) u \geq \alpha |u|^2,$$

where $\nabla^2 V$ is the Hessian matrix of V. Then, the probability density function $\psi_\infty \propto \exp(-V)$ satisfies a LSI(R) with $R \geq \alpha$.

Proposition 2.22 (Holley-Stroock criterium). *Let V be a function such that the probability density function $\psi_\infty \propto \exp(-V)$ satisfies a LSI(ρ) for some positive ρ. Let $\widetilde{V}$ be a bounded function, and consider the probability density function $\widetilde{\psi}_\infty \propto \exp(-V + \widetilde{V})$. Then, $\widetilde{\psi}_\infty$ satisfies a LSI($\widetilde{\rho}$) with*

$$\widetilde{\rho} \geq \rho \exp\big(-\operatorname{osc}(\widetilde{V})\big),$$

where $\operatorname{osc}(\widetilde{V}) = \sup(\widetilde{V}) - \inf(\widetilde{V})$.

Using these two criteria, it is possible to check that, for many potentials V used in practice, ψ_∞ indeed satisfies a LSI.

2.4 Methods for alchemical free energy differences

We present a first application of the sampling methods discussed in this chapter to the computation of free energy differences. The techniques presented in this section are restricted to alchemical transitions (see Section 1.3.2). Recall that alchemical transitions are described by a Hamiltonian indexed by a parameter $\lambda \in [0,1]$, and the free energy difference to be computed reads

$$\Delta F(\lambda) = F(\lambda) - F(0) = -\frac{1}{\beta}\ln\left(\frac{Z_\lambda}{Z_0}\right) = -\frac{1}{\beta}\ln\left(\frac{\displaystyle\int_{T^*\mathcal{D}} \mathrm{e}^{-\beta H_\lambda(q,p)}\,dq\,dp}{\displaystyle\int_{T^*\mathcal{D}} \mathrm{e}^{-\beta H_0(q,p)}\,dq\,dp}\right).$$

When only the potential energy depends on a parameter λ,

$$\Delta F(\lambda) = -\frac{1}{\beta}\ln\left(\frac{\displaystyle\int_{\mathcal{D}} \mathrm{e}^{-\beta V_\lambda(q)}\,dq}{\displaystyle\int_{\mathcal{D}} \mathrm{e}^{-\beta V_0(q)}\,dq}\right).$$

We will consider two methods to compute alchemical free energy differences:

(i) The first one, *free energy perturbation* (see Section 2.4.1), recasts the free energy difference as the average of some observable with respect to a canonical measure, and this average is then estimated by sampling this canonical distribution, using the methods presented in Sections 2.1 and 2.2.

(ii) The second one, *bridge estimators*, is described in Section 2.4.2. This method estimates ratios of partition functions by sampling two canonical measures.

2.4.1 *Free energy perturbation*

2.4.1.1 *General idea of the method*

Free energy perturbation (FEP) is a conceptually simple and standard technique to compute free energy differences, which can be applied to alchemical transitions (see Section 1.3.2). It was introduced in [Zwanzig (1954)]. We refer for instance to Chapter 2 in [Chipot and Pohorille (2007a)] for more physical background material. It may also be seen as a limiting case of more general nonequilibrium switchings when the switching time goes to 0, see Section 4.1 for further precision.

Free energy perturbation is based on the identity

$$\mathrm{e}^{-\beta\Delta F(\lambda)} = \mathbb{E}_{\mu_0}\left[\mathrm{e}^{-\beta(H_\lambda - H_0)}\right] = \int_{T^*\mathcal{D}} \mathrm{e}^{-\beta(H_\lambda(q,p)-H_0(q,p))}\,\mu_0(dq\,dp), \tag{2.88}$$

where the canonical measures μ_λ are defined as

$$\mu_\lambda(dq\,dp) = Z_\lambda^{-1} \mathrm{e}^{-\beta H_\lambda(q,p)}\, dq\, dp, \qquad Z_\lambda = \int_{T^*\mathcal{D}} \mathrm{e}^{-\beta H_\lambda(q,p)}\, dq\, dp. \quad (2.89)$$

An approximation of $\Delta F(1)$ (simply denoted ΔF below) is thus obtained by generating configurations (q^i, p^i) distributed according to μ_0 (using for instance the techniques presented in Sections 2.1 and 2.2) and averaging the corresponding quantities $\mathrm{e}^{-\beta(H_1 - H_0)(q^i,p^i)}$, leading to the estimator

$$\widehat{\Delta F}_M = -\frac{1}{\beta} \ln \left(\frac{1}{M} \sum_{i=1}^{M} \mathrm{e}^{-\beta(H_1(q^i,p^i) - H_0(q^i,p^i))} \right), \qquad (q^i, p^i) \sim \mu_0. \quad (2.90)$$

Alternatively, the free energy difference can be recast as an average with respect to the measure μ_λ:

$$\mathrm{e}^{\beta \Delta F(\lambda)} = \mathbb{E}_{\mu_\lambda}\left[\mathrm{e}^{\beta(H_\lambda - H_0)} \right] = \int_{T^*\mathcal{D}} \mathrm{e}^{\beta(H_\lambda(q,p) - H_0(q,p))}\, \mu_\lambda(dq\, dp). \quad (2.91)$$

In this case, $\Delta F(1)$ can still be estimated as a simple canonical average:

$$\widehat{\Delta F}_M = \frac{1}{\beta} \ln \left(\frac{1}{M} \sum_{i=1}^{M} \mathrm{e}^{\beta(H_1(q^i,p^i) - H_0(q^i,p^i))} \right), \qquad (q^i, p^i) \sim \mu_1. \quad (2.92)$$

Notice that the ratio Z_0/Z_1 is now being estimated, hence the sign change in the exponential compared to (2.88).

Motivation for the name of the method. The terminology "free energy perturbation" is reminiscent of the early days of the method, and may be motivated as follows. The above expressions (2.88) and (2.91) of $\mathrm{e}^{-\beta \Delta F(1)}$ can be recast as averages over the possible values U of the energy difference $H_1 - H_0$. Define the distributions $P_0(dU), P_1(dU)$ of energy differences as the image measures of the canonical measures μ_0 and μ_1 by $H_1 - H_0$, *i.e.*, for any test function g,

$$\int_{\mathbb{R}} g(U)\, P_0(dU) = \int_{T^*\mathcal{D}} g\Big(H_1(q,p) - H_0(q,p)\Big) \frac{\mathrm{e}^{-\beta H_0(q,p)}}{Z_0}\, dq\, dp, \quad (2.93)$$

and

$$\int_{\mathbb{R}} g(U)\, P_1(dU) = \int_{T^*\mathcal{D}} g\Big(H_1(q,p) - H_0(q,p)\Big) \frac{\mathrm{e}^{-\beta H_1(q,p)}}{Z_1}\, dq\, dp. \quad (2.94)$$

Equations (2.88) and (2.91) can then be rewritten as

$$\mathrm{e}^{-\beta \Delta F(1)} = \int_{\mathbb{R}} \mathrm{e}^{-\beta U}\, P_0(dU) = \left(\int_{\mathbb{R}} \mathrm{e}^{\beta U}\, P_1(dU) \right)^{-1}. \quad (2.95)$$

Initially, the method was a perturbative technique, in which $e^{-\beta U}$ was expanded as a power series in U. Therefore, $e^{-\beta \Delta F(1)}$ was expanded in terms of sums of moments (see (2.96) below), which were truncated to obtain estimates for the free energy difference. Nowadays, progresses in computer science and algorithms make it possible to use numerical strategies based on the exact expressions (2.88) and (2.91), but the name of the method remains. It may however still be argued that, in (2.88) for instance, the free energy of H_1 is extrapolated from the knowledge of H_0, and that, in some sense, $H_1 - H_0$ is a small perturbation of H_0.

It is already clear from (2.95), that the numerical convergence of the method may be slow since the measure proportional to $e^{-\beta U} P_0(dU)$ and the reference measure $P_0(dU)$ used to carry out the Monte-Carlo integration in general hardly overlap.

Numerical application. We consider the Widom insertion model described in Section 1.3.2.3, with an inverse temperature $\beta = 1$, for $N = 25$ particles at a density $\rho = a^{-2} = 0.8264$ (*i.e.* using a 2D square domain of side lengths $a\sqrt{N}$ with $a = 1.1$). The parameters for the smoothed Lennard-Jones potential (1.63) are $\varepsilon = 1$ and $\sigma = 1$. The distribution μ_0 is sampled with a Langevin dynamics discretized with the BBK scheme (2.51), using $\gamma = 1$ and $\Delta t = 0.005$. The position of the additional particle q^{N+1} is sampled at random in the simulation domain. The estimator (2.90) is used since it is known that Widom insertion (going from $\lambda = 0$ to $\lambda = 1$) should be preferred to Widom deletion (going from $\lambda = 1$ to $\lambda = 0$), see the discussion at the end of Section 2.4.1.4.

The results for the chemical potential (1.62) are presented in Figure 2.3. As can be seen, most of the energy difference values do not contribute much to the average, which depends heavily on the lower tail of the distribution. This explains also the sawtooth convergence of the current average. For later purposes, a very long simulation ($T = 50,000$) with a very small time-step $\Delta t = 0.0005$ allows to obtain a reference value $\mu^{\text{ex}} = 1.317 \pm 0.001$ for the chemical potential.

2.4.1.2 *Expansions using the distribution of energy differences*

In this section, we motivate the name "free energy perturbation" by presenting the fomal expansion of the free energy difference as a sum of moments of the work distribution, starting with the Gaussian case.

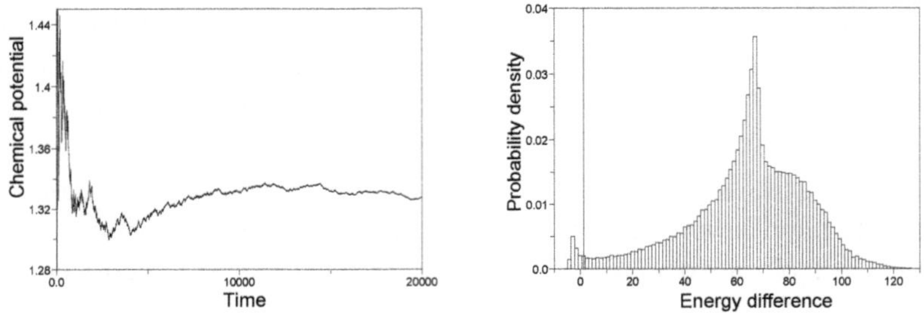

Fig. 2.3 Left: Estimate of the chemical potential as a function of time, using the estimator (2.90). Right: Distribution $P_0(dU)$ of the insertion energies. The vertical line represents the final estimate of the chemical potential.

Gaussian case. We first assume that the distribution P_0 of energy differences $U = H_1 - H_0$ (defined in (2.93)) is a Gaussian distribution, with mean $\overline{U}$ and variance σ^2:

$$P_0(dU) = \frac{1}{\sqrt{2\pi\sigma^2}} \exp\left(-\frac{(U-\overline{U})^2}{2\sigma^2}\right) dU.$$

The quantity

$$\mathbb{E}_{\mu_0}\left[\mathrm{e}^{-\beta(H_1-H_0)}\right] = \int_{\mathbb{R}} \mathrm{e}^{-\beta U} P_0(dU)$$

can then be computed explicitly since

$$\begin{aligned}
\int_{\mathbb{R}} \mathrm{e}^{-\beta U} P_0(dU) &= \mathrm{e}^{-\beta\overline{U}} \int_{\mathbb{R}} \mathrm{e}^{-\beta(U-\overline{U})} \frac{1}{\sqrt{2\pi\sigma^2}} \exp\left(-\frac{(U-\overline{U})^2}{2\sigma^2}\right) dU \\
&= \mathrm{e}^{-\beta\overline{U}} \int_{\mathbb{R}} \mathrm{e}^{-\beta x} \frac{1}{\sqrt{2\pi\sigma^2}} \exp\left(-\frac{x^2}{2\sigma^2}\right) dx \\
&= \mathrm{e}^{-\beta\overline{U}} \mathrm{e}^{\beta^2\sigma^2/2} \int_{\mathbb{R}} \frac{1}{\sqrt{2\pi\sigma^2}} \exp\left(-\frac{(x+\sigma^2\beta)^2}{2\sigma^2}\right) dx \\
&= \exp\left(-\beta\overline{U} + \frac{\beta^2}{2}\sigma^2\right).
\end{aligned}$$

Therefore,

$$\Delta F(1) = \mathbb{E}_{\mu_0}\Big(H_1 - H_0\Big) - \frac{\beta}{2}\mathrm{Var}_{\mu_0}\Big(H_1 - H_0\Big).$$

This gives an expression of the free energy difference as a sum of moments of the energy differences.

Let us discuss the efficiency of the free energy estimator (2.90) in the Gaussian case at hand. In the limit $M \to +\infty$, the asymptotic variance of $\exp\left(-\beta\widehat{\Delta F}_M\right)$ satisfies

$$\begin{aligned} M\,\mathrm{Var}\left[\exp\left(-\beta\widehat{\Delta F}_M\right)\right] &\longrightarrow \mathbb{E}_{\mu_0}\left[\mathrm{e}^{-2\beta(H_1-H_0)}\right] - \left(\mathbb{E}_{\mu_0}\left[\mathrm{e}^{-\beta(H_1-H_0)}\right]\right)^2 \\ &= \exp\left(-2\beta\overline{U} + 2\beta^2\sigma^2\right) - \exp\left(-2\beta\overline{U} + \beta^2\sigma^2\right) \\ &= \exp(-2\beta\Delta F)\Big(\exp(\beta^2\sigma^2) - 1\Big), \end{aligned}$$

which is very large when σ^2 is large. This motivates the fact that P_0 should be peaked, otherwise the quality of the estimator is likely to being poor. See also Section 4.1.4 for similar discussions in the context of work distributions for nonequilibrium switching dynamics.

General work distributions. For general (non-Gaussian) distributions $P_0(dU)$ (defined in (2.93)), the free energy difference can be recast as an infinite series involving the moments of P_0, as first derived in [Zwanzig (1954)]. To obtain this expression, the exponential $\mathrm{e}^{-\beta x}$ in formula (2.88) is formally expanded as a power expansion in x, using the so-called generating function:

$$\mathrm{e}^{-\beta\Delta F} = \mathbb{E}_{\mu_0}\left[\mathrm{e}^{-\beta(H_1-H_0)}\right] = 1 + \sum_{n=1}^{+\infty}(-\beta)^n\,\mathbb{E}_{\mu_0}\Big[(H_1-H_0)^n\Big].$$

Applying $-\beta^{-1}\ln$ (and also expanding the logarithm function in power series),

$$\Delta F(1) = \sum_{n=1}^{+\infty}\frac{(-\beta)^{n-1}}{n!}\,C_n, \tag{2.96}$$

where C_n are the so-called cumulants of the distribution P_0. The cumulants can be expressed as linear combinations of the moments of P_0. The first cumulants are

$$C_1 = \mathbb{E}_{\mu_0}\Big(H_1 - H_0\Big),$$

$$C_2 = \mathrm{Var}_{\mu_0}(H_1-H_0) = \mathbb{E}_{\mu_0}\Big((H_1-H_0)^2\Big) - \Big[\mathbb{E}_{\mu_0}\Big(H_1-H_0\Big)\Big]^2.$$

Approximations of the free energy difference $\Delta F(1)$ can now be obtained by truncating the series on the right-hand side of (2.96). In the Gaussian case, $C_n = 0$ for $n \geq 3$, so that the expansion (2.96) is exact when $n \geq 2$. However, it is difficult in general to control the accuracy of such approximations (*i.e.* the magnitude of the perfect sampling bias of the associated estimators).

2.4.1.3 *Staging*

It is often the case that the initial and the final distributions μ_0 and μ_1 hardly overlap, so that the straightforward estimators (2.90) or (2.92) are not efficient: Large sampling errors may plague the results.

In order to reduce the discrepancy between the initial and the final measures, the transformation can be performed using intermediate steps, which is called *staging* in the statistics literature. The free energy change is decomposed using $n-1$ intermediate steps λ_i $(1 \le i \le n-1)$ such that

$$0 = \lambda_0 < \lambda_1 < \cdots < \lambda_{n-1} < \lambda_n = 1.$$

The associated elementary free energy differences $\Delta F_i = F(\lambda_{i+1}) - F(\lambda_i)$ are

$$\Delta F_i = -\beta^{-1} \ln \frac{Z_{\lambda_{i+1}}}{Z_{\lambda_i}} = -\beta^{-1} \ln \int_{T^*\mathcal{D}} \mathrm{e}^{-\beta(H_{\lambda_{i+1}} - H_{\lambda_i})} d\mu_{\lambda_i}, \tag{2.97}$$

where μ_λ and Z_λ are defined in (2.89). Finally, the total free energy difference is recovered as

$$\Delta F = F(1) - F(0) = \Delta F_0 + \cdots + \Delta F_{n-1}.$$

When the number of intermediate steps n is large enough, it is expected that the overlap between μ_i and μ_{i+1} is sufficient to estimate accurately (2.97) with a straightforward estimator such as

$$\widehat{\Delta F}_{i,N_i} = -\frac{1}{\beta} \ln \left(\frac{1}{N_i} \sum_{j=1}^{N_i} \mathrm{e}^{-\beta(H_{\lambda_{i+1}}(q^j,p^j) - H_{\lambda_i}(q^j,p^j))} \right), \qquad (q^j,p^j) \sim \mu_{\lambda_i},$$

for values N_i moderately large. When many intermediate stages are considered, the method is reminiscent of thermodynamic integration (see Section 3.1), though the methods differ by the way elementary free energy differences are computed: In free energy perturbation, nonlinear averages involving logarithms and exponentials are considered (see the expression of $\widehat{\Delta F}_{i,N_i}$), and the error scales exponentially with the dimension. This can be shown rigorously in some cases, see the discussion at the end of Section 4.1.5 which is valid both for nonequilibrium methods and free energy perturbation (upon replacing work distributions by energy differences distributions). Thermodynamic integration on the other hand uses simple arithmetic averages, so that the error scales linearly with the dimension. The scaling of the statistical error with respect to the number of degrees of freedom in the system at hand is therefore in favor of thermodynamic integration.

2.4.1.4 *Umbrella sampling*

In order to improve the accuracy of free energy perturbation, some importance sampling strategy should be used in addition to (possibly, in replacement of) the staging procedure described in Section 2.4.1.3. We first review some basic facts about importance sampling, before turning to the specific case of free energy computations.

Importance sampling strategies. Before describing the Umbrella sampling method, let us first review some basic facts on importance sampling strategies. Suppose we want to estimate

$$\mathcal{A} = \int_{\mathcal{S}} A(x)\psi(x)\,dx$$

for some probability distribution $\psi(x)\,dx$. A first strategy is to sample the measure $\psi(x)\,dx$ (when this is possible), and estimate $\mathcal{A}$ as

$$\frac{1}{M}\sum_{i=1}^{M} A(x^i), \qquad x^i \sim \psi(x)\,dx. \tag{2.98}$$

We assume, for simplicity, that the configurations x^i are i.i.d. Then, the variance of the estimator (2.98) scales asymptotically (in the large M limit) as

$$\frac{1}{M}\mathrm{Var}_{\psi}(A) = \frac{1}{M}\left(\int_{\mathcal{S}} A(x)^2\psi(x)\,dx - \mathcal{A}^2\right).$$

Importance sampling is a strategy to reduce the variance of the estimator. It consists in rewriting $\mathcal{A}$ as

$$\mathcal{A} = \int_{\mathcal{S}} A(x)\frac{\psi(x)}{\phi(x)}\,\phi(x)\,dx,$$

where $\phi(x)\,dx$ is some probability measure:

$$\phi(x) \geq 0, \qquad \int_{\mathcal{S}} \phi = 1.$$

We do not worry at the moment on other conditions that ϕ should satisfy in order for the above rewriting of $\mathcal{A}$ to make sense. The function ϕ is the *importance sampling function.* The above equality suggests to resort to an estimator where configurations x^i are sampled according to the probability measure $\phi(x)\,dx$, and $\mathcal{A}$ is estimated as

$$\frac{1}{M}\sum_{i=1}^{M} A(x^i)\frac{\psi(x^i)}{\phi(x^i)}, \qquad x^i \sim \phi(x)\,dx. \tag{2.99}$$

When the configurations x^i are i.i.d. with law $\phi(x)\,dx$, the Central Limit Theorem shows that the variance of the estimator (2.99) scales asymptotically as

$$\frac{1}{M}\mathrm{Var}_\phi\left(\frac{A\psi}{\phi}\right) = \frac{1}{M}\left(\int_S \left(A(x)\frac{\psi(x)}{\phi(x)}\right)^2 \phi(x)\,dx - \mathcal{A}^2\right). \qquad (2.100)$$

As a conclusion, an importance sampling function should be such that (i) it is easy to sample from (in order to compute easily (2.99)); (ii) it enhances the accuracy of the sampling, in the sense that the asymptotic variance (2.100) of the estimator (2.99) is lower than the asymptotic variance of the simple estimator (2.98).

The question is now whether there exists an optimal importance sampling function, for which (2.100) is minimal. This amounts to solving the following optimization problem:

$$\inf\left\{ \int_S \left(\frac{A(x)\,\psi(x)}{\phi(x)}\right)^2 \phi(x)\,dx, \;\middle|\; \phi(x) \geq 0, \; \int_S \phi = 1, \; \mathrm{supp}\,\psi \subset \mathrm{supp}\,\phi \right\}. \qquad (2.101)$$

An application of the Cauchy-Schwarz inequality on $L^2(\phi)$ shows that

$$\begin{aligned}\int_S \left(\frac{A(x)\,\psi(x)}{\phi(x)}\right)^2 \phi(x)\,dx &\geq \left(\int_S \left|\frac{A(x)\,\psi(x)}{\phi(x)}\right| \phi(x)\,dx\right)^2 \\ &= \left(\int_S |A(x)\,\psi(x)|\,dx\right)^2,\end{aligned}$$

and the equality is attained for the choice

$$\phi^*(x) = \frac{|A(x)\,\psi(x)|}{\displaystyle\int_S |A\,\psi|}.$$

The latter function is therefore the optimal importance sampling function. Notice that it depends on ψ, but also on the function A whose average is to be estimated. Besides, there is no reason why it would be easy to sample from $\phi^*(x)\,dx$ in general. These considerations show that the optimal importance sampling function is in general not useful directly for practical purposes.

Remark 2.23 (Variance reduction methods).
Importance sampling is one instance of so-called variance reduction methods. Other methods which are widely used in other contexts than molecular dynamics include the antithetic variates method, the stratified sampling,

Rao-Blackwellization, and the control variates method, see [Liu (2001); Rubinstein (1981); Fishman (1996); Lapeyre et al. *(2003)]. As an example, let us briefly explain the principle of the control variates method. Suppose again we want to estimate $\mathcal{A} = \mathbb{E}(x)$, where x is a random variable. A control variate y for x is a random variable which is correlated to x, and such that its average $\mathbb{E}(y)$ is known. For simplicity, we assume it is zero: $\mathbb{E}(y) = 0$. The idea is to use as an estimate of $\mathcal{A}$ the following quantity:*

$$\frac{1}{M} \sum_{i=1}^{M} (x^i - \alpha y^i)$$

where α is a real parameter to be fixed and (x^i, y^i) are identically distributed with the same law as (x, y). Assuming that the samples are independent, this estimator will be better than the standard one (for which $\alpha = 0$) if

$$\mathrm{Var}(x - \alpha y) < \mathrm{Var}(x).$$

A simple computation gives $\mathrm{Var}(x-\alpha y) = \mathrm{Var}(x) - 2\alpha \,\mathrm{Cov}(x, y) + \alpha^2 \mathrm{Var}(y)$. The value of α which minimizes the variance is

$$\alpha^* = \frac{\mathrm{Cov}(x, y)}{\mathrm{Var}(y)},$$

and the minimum variance is then

$$\mathrm{Var}(x - \alpha^* y) = \mathrm{Var}(x) \left(1 - \frac{\mathrm{Cov}(x, y)}{\mathrm{Var}(x)\mathrm{Var}(y)}\right).$$

The value α^* is approximated in practice by

$$\frac{\displaystyle\sum_{i=1}^{M} \left(x^i - \frac{1}{M}\sum_{i=1}^{M} x^i\right) y^i}{\displaystyle\sum_{i=1}^{M} \left(y^i\right)^2}.$$

Hence, the more y and x are correlated, the greater the reduction in variance. A typical example of control variate is $x = A(z)$ and $y = A_0(z)$ (where z is the same random variable in x and y) for a function A_0 close to A.

Importance sampling for free energy perturbation. The free energy difference $\Delta F = F(1) - F(0)$ (or the elementary free energy differences considered in Section 2.4.1.3) can be computed more efficiently using some importance sampling technique, known as umbrella sampling [Torrie and Valleau (1974, 1977)] in this context. It relies on the following identity:

$$\Delta F = -\beta^{-1} \ln\left(\frac{Z_1}{Z_0}\right) = -\beta^{-1} \ln\left(\frac{\displaystyle\int_{T^*\mathcal{D}} \frac{\mathrm{e}^{-\beta H_1}}{\phi}\,\phi}{\displaystyle\int_{T^*\mathcal{D}} \frac{\mathrm{e}^{-\beta H_0}}{\phi}\,\phi}\right), \tag{2.102}$$

where $\phi(q,p)\,dq\,dp$ is an arbitrary probability distribution, which should be chosen so that: (i) it is easy to sample from $\phi(q,p)\,dq\,dp$; (ii) the variance of the estimator based on (2.102) is as small as possible. Equation (2.93) (resp. Equation (2.94)) is recovered with the choice $\phi = Z_0^{-1}\mathrm{e}^{-\beta H_0}$ (resp. $\phi = Z_1^{-1}\mathrm{e}^{-\beta H_1}$). A possible estimator of ΔF, based on (2.102), is

$$\widehat{R}_M = -\frac{1}{\beta} \ln\left(\frac{\displaystyle\sum_{i=1}^{M} \frac{\mathrm{e}^{-\beta H_1(q^i,p^i)}}{\phi(q^i,p^i)}}{\displaystyle\sum_{i=1}^{M} \frac{\mathrm{e}^{-\beta H_0(q^i,p^i)}}{\phi(q^i,p^i)}}\right), \qquad (q^i,p^i) \sim \phi(q,p)\,dq\,dp. \tag{2.103}$$

Notice that ϕ needs to be known only up to a multiplicative constant.

This approach is interesting and efficient when a good importance sampling function ϕ can be devised. The associated probability measure should heuristically be "in between" μ_0 and μ_1, *i.e.* have an appreciable overlap with both μ_0 and μ_1. In this case indeed, both the numerator and the denominator can be approximated correctly (up to an unknown but common multiplicative constant) by estimates such as $(k = 0,1)$:

$$\int_{T^*\mathcal{D}} \mathrm{e}^{-\beta H_k}\phi \simeq \frac{1}{M}\sum_{i=1}^{M} \frac{\mathrm{e}^{-\beta H_k(q^i,p^i)}}{\phi(q^i,p^i)}, \qquad (q^i,p^i) \sim \phi(q,p)\,dq\,dp.$$

The bridging property of the importance sampling function motivated the name *umbrella sampling.*[3]

Some possible choices for the umbrella function are

$$\phi(q,p) = Z_\theta^{-1}\,\mathrm{e}^{-\beta H_\theta(q,p)}, \tag{2.104}$$

where $H_\theta = (1-\theta)H_0 + \theta H_1$ with $0 < \theta < 1$, or

$$\phi(q,p) = \widetilde{Z}_\theta^{-1}\left((1-\theta)\,\mathrm{e}^{-\beta H_0(q,p)} + \theta\,\mathrm{e}^{-\beta H_1(q,p)}\right).$$

As usual in importance sampling techniques, the efficiency of the biasing potential crucially depends on the problem at hand.

[3]Quoting [Torrie and Valleau (1977)]: "The sampling distribution [...] should cover simultaneously the regions of configuration space relevant to two or more physical systems. We call this umbrella sampling."

Optimal umbrella sampling potential. It is also possible to derive the expression of an optimal umbrella function minimizing the asymptotic variance of the free energy estimator (2.103). The following computations are standard, and the reader interested by more mathematical rigor may have a look at [Chen and Shao (1997)], for instance. The basic mathematical tool underpinning the proof is the following convergence result generalizing the Central Limit Theorem for i.i.d. random variables Z_i.

Lemma 2.24. *Consider a sequence of i.i.d. random variables* $(Z_i)_{i\geq 1}$ *such that* $\mathrm{Var}(Z_1)$ *is finite, and a function* $\varphi : \mathbb{R} \to \mathbb{R}$, *differentiable at* $\mathbb{E}(Z_1)$. *Then, the following convergence in law holds:*

$$\sqrt{M}\left[\varphi\left(\frac{1}{M}\sum_{i=1}^{M} Z_i\right) - \varphi(\mathbb{E}(Z_1))\right] \longrightarrow \mathcal{N}\Big(0, \big(\varphi'[\mathbb{E}(Z_1)]\big)^2 \mathrm{Var}(Z_1)\Big). \tag{2.105}$$

Such a result is known as the delta method in the statistics literature (see for instance [van der Vaart (1998)]).

Proof. We rewrite the left-hand side of (2.105) as

$$\frac{\varphi\left(\dfrac{1}{M}\displaystyle\sum_{i=1}^{M} Z_i\right) - \varphi(\mathbb{E}(Z_1))}{\dfrac{1}{M}\displaystyle\sum_{i=1}^{M} Z_i - \mathbb{E}(Z_1)} \cdot \sqrt{M}\left[\frac{1}{M}\sum_{i=1}^{M} Z_i - \mathbb{E}(Z_1)\right].$$

By the Law of Large Numbers, the first term on the right-hand side converges almost surely to the constant value $\varphi'(\mathbb{E}(Z_1))$. The second term converges in law to $\mathcal{N}(0, \mathrm{Var}(Z_1))$ by the Central Limit Theorem. Slutsky's lemma therefore gives the expected result. □

Coming back to the estimator (2.103) of ΔF, we first rewrite it as

$$\widehat{R}_M = -\frac{1}{\beta}\ln\left(\frac{\overline{Y}_M}{\overline{X}_M}\right),$$

where

$$\overline{Y}_M = \frac{1}{M}\sum_{i=1}^{M} Y_i, \qquad \overline{X}_M = \frac{1}{M}\sum_{i=1}^{M} X_i$$

with

$$Y_i = \frac{\mathrm{e}^{-\beta H_1(q^i,p^i)}}{\phi(q^i,p^i)}, \qquad X_i = \frac{\mathrm{e}^{-\beta H_0(q^i,p^i)}}{\phi(q^i,p^i)}.$$

The configurations (q^i, p^i) are assumed to be i.i.d. with respect to the measure $\phi(q,p)\,dq\,dp$ (which is however usually not the case since Markov chain techniques are used in practice to sample $\phi(q,p)\,dq\,dp$, but it may be approximately true upon subsampling the chain, see (2.74)). In the sequel, we denote by $\mathbb{E}_\phi$ and Var_ϕ the expectation and the variance with respect to the probability measure $\phi(q,p)\,dq\,dp$. Note that

$$\mathbb{E}_\phi(X_1) = \mathbb{E}_\phi(\overline{X}_M) = Z_0, \qquad \mathbb{E}_\phi(Y_1) = \mathbb{E}_\phi(\overline{Y}_M) = Z_1.$$

Besides, rewriting

$$\overline{X}_M = Z_0\left(1 + \varepsilon_M^X\right), \qquad \varepsilon_M^X = \frac{1}{M}\sum_{i=1}^M \left(\frac{X_i}{Z_0} - 1\right),$$

$$\overline{Y}_M = Z_1\left(1 + \varepsilon_M^Y\right), \qquad \varepsilon_M^Y = \frac{1}{M}\sum_{i=1}^M \left(\frac{Y_i}{Z_1} - 1\right),$$

the Law of Large Numbers shows that ε_M^X and ε_M^Y converge almost surely to 0 in the limit $M \to +\infty$, and the asymptotic variance of these random variables is given by a Central Limit Theorem. Note however that ε_M^X and ε_M^Y are *not* independent since the same configurations (q^i, p^i) are used in X_i and Y_i. Using the expansion

$$-\ln\left(\frac{1+\varepsilon_Y}{1+\varepsilon_X}\right) \simeq \varepsilon_X - \varepsilon_Y,$$

the asymptotic variance of the estimator $\widehat{R}_M$ can be computed as

$$\begin{aligned}
\lim_{M\to+\infty} M\,\mathrm{Var}_\phi\left(\widehat{R}_M\right) &= \lim_{M\to+\infty} \frac{M}{\beta^2}\,\mathrm{Var}_\phi\left(\varepsilon_M^X - \varepsilon_M^Y\right) \\
&= \frac{1}{\beta^2}\mathrm{Var}_\phi\left(\frac{X_1}{Z_0} - \frac{Y_1}{Z_1}\right) \\
&= \frac{1}{\beta^2}\int_{T^*\mathcal{D}} \left|\frac{\mathrm{e}^{-\beta H_0}}{Z_0\,\phi} - \frac{\mathrm{e}^{-\beta H_1}}{Z_1\,\phi}\right|^2 \phi.
\end{aligned}$$

Note that, as a consequence of the correlation between X_i and Y_i, $\mathrm{Var}_\phi(\varepsilon_M^X - \varepsilon_M^Y) \neq \mathrm{Var}_\phi(\varepsilon_M^X) + \mathrm{Var}_\phi(\varepsilon_M^Y)$ (in contrast with the case of bridge sampling, see Section 2.4.2.2).

In the case $H_1 \neq H_0$, computations similar to the ones done for the minimization of (2.101) show that the optimal umbrella function is

$$\phi^*(q,p) = Z_*^{-1}\left|Z_1^{-1}\,\mathrm{e}^{-\beta H_1(q,p)} - Z_0^{-1}\,\mathrm{e}^{-\beta H_0(q,p)}\right|, \tag{2.106}$$

with

$$Z_* = \int_{T^*\mathcal{D}} \left|Z_0^{-1}\,\mathrm{e}^{-\beta H_0(q,p)} - Z_1^{-1}\,\mathrm{e}^{-\beta H_1(q,p)}\right|\,dq\,dp.$$

However, this function cannot be used as such in practice since it requires the knowledge of the free energy difference $\mathrm{e}^{-\beta\Delta F} = Z_1/Z_0$ beforehand in order to compute the relative contributions of the initial and final measures.

Numerical application. We apply the umbrella sampling technique to the Widom insertion problem. Importance sampling functions of the form (2.104) are considered, for $\theta = 0.05$ and $\theta = 0.1$. The importance sampling measure is sampled using a Langevin dynamics, and the same parameters as in Section 2.4.1.1 are used.

The results, presented in Figure 2.4, show that the choice of the umbrella sampling function is crucial: Unfortunate choices lead to a slower convergence compared to the case when no importance sampling function is used ($\theta = 0$ in this context). More precisely, it seems that the convergence to the reference value is slower when θ increases, and that the variance of the free energy difference estimate is larger (though these claims should be backed up by more systematic studies, instead of results from a single realization). It seems therefore that $\theta \simeq 0$ should be considered. This is related to the fact that Widom insertion is known to give more reliable estimates than the Widom deletion, see for instance the discussion in [Lu and Kofke (2001)] or the end of Section 2.4.2.2. Therefore, it is expected that the best estimator corresponds to values of θ close to 0 since the case $\theta = 0$ is the standard Widom insertion procedure (Widom deletion corresponding to $\theta = 1$).

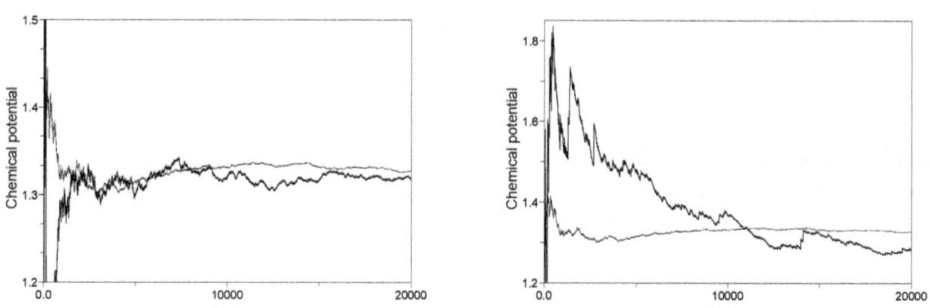

Fig. 2.4 Left: Comparison of the current estimates of the chemical potential for umbrella sampling with $\theta = 0.05$ (thick solid line), and standard FEP method (thin line). Right: Same comparison with $\theta = 0.1$.

There are however cases where umbrella sampling is more successful — see for instance the toy example in Section 2.4.2.4.

2.4.2 *Bridge sampling*

2.4.2.1 *Presentation of the method*

When it is possible to sample from both distributions μ_0 and μ_1, all sample points can be combined to obtain an estimate of the ratio of partition functions. Since this technique can also be used in other contexts than free energy perturbation methods (in particular for nonequilibrium work distributions, see Section 4.1.5), we describe it in a general framework with abstract notation.

Consider two probability measures absolutely continuous with respect to the Lebesgue measure, and with common support $\Omega \subset \mathbb{R}^n$:

$$\pi_i(x)\,dx = \frac{1}{Z_i} f_i(x)\,dx, \qquad i = 1, 2,$$

and a function α such that $\alpha f_1 f_2 \in L^1(\Omega)$ with

$$\int_\Omega \alpha f_1 f_2 \neq 0. \tag{2.107}$$

In particular, this requires that there is some overlap between π_1 and π_2, namely that $\operatorname{supp} f_1 \cap \operatorname{supp} f_2 \neq \emptyset$. Since

$$\frac{\displaystyle\int_\Omega \alpha(x) f_1(x) \pi_2(x)\,dx}{\displaystyle\int_\Omega \alpha(x) f_2(x) \pi_1(x)\,dx} = \frac{Z_1}{Z_2},$$

the ratio r of normalizing constants can be estimated as

$$r = \frac{Z_1}{Z_2} = \frac{\mathbb{E}_{\pi_2}(\alpha f_1)}{\mathbb{E}_{\pi_1}(\alpha f_2)}, \tag{2.108}$$

where the denominator is non-zero by assumption (see (2.107)). The identity (2.108) defines a family of estimators parametrized by the choice of the function α:

$$\widehat{r}_\alpha^{\,n_1,n_2} = \frac{\displaystyle\frac{1}{n_2}\sum_{j=1}^{n_2} f_1(x^{2,j})\,\alpha(x^{2,j})}{\displaystyle\frac{1}{n_1}\sum_{j=1}^{n_1} f_2(x^{1,j})\,\alpha(x^{1,j})}, \tag{2.109}$$

where $(x^{1,1}, \dots, x^{1,n_1})$ and $(x^{2,1}, \dots, x^{2,n_2})$ are assumed to be i.i.d. with respect to π_1 and π_2 respectively.

Estimates of the free energy difference can be obtained with the formula $\Delta F = -\beta^{-1} \ln r = -\beta^{-1} \ln(Z_1/Z_0)$ with the choices

$$f_1 = \mathrm{e}^{-\beta H_1}, \qquad f_2 = \mathrm{e}^{-\beta H_0}.$$

The free energy perturbation method is recovered in the case when $\alpha = 1/f_1$ or $\alpha = 1/f_2$.

Note that for a general function α, and in contrast with the umbrella sampling case (see Section 2.4.1.4), the numerator and denominator in the expression of $\widehat{r}_\alpha^{n_1,n_2}$ are independent random variables (which has an impact on the expression of the asymptotic variance of the estimator, see Section 2.4.2.2).

2.4.2.2 *Derivation of the optimal function α*

The question is whether an optimal estimator, minimizing the asymptotic variance of the free energy difference estimator, can be constructed. This question was first addressed in [Bennett (1976)] (hence the name "Bennett acceptance ratio" (BAR) in the physics and chemistry literature). Subsequent works gave more rigorous mathematical foundations to this result [Meng and Wong (1996)], or reinterpreted the original derivation, for instance by recasting the obtained estimator as maximum likelihood estimator [Shirts *et al.* (2003)].

As usual in such problems, the optimal function α depends on the quantity to compute, namely Z_1/Z_2 here. However, in contrast with the optimal umbrella sampling function (2.106), it turns out that the optimal function α^*, though not computable, is suitable for a direct numerical strategy, see (2.113) below.

More precisely, the problem can be reformulated as minimizing the asymptotic variance (where the random variable entering the variance has been centered in order for the subsequent computations to be simpler):

$$\lim_{n\to+\infty} n \operatorname{Var}\Big(\ln \widehat{r}_\alpha^{n_1,n_2} \Big) = \lim_{n\to+\infty} n \operatorname{Var}\left[\ln \left(\frac{\widehat{r}_\alpha^{n_1,n_2}}{r} \right) \right]$$

in the limit when the total number of samples $n = n_1 + n_2$ goes to infinity, with a non-vanishing fraction of samples of each kind:

$$s_1 = \frac{n_1}{n} \to \theta, \qquad s_2 = \frac{n_2}{n} \to 1 - \theta, \tag{2.110}$$

with $\theta \in (0,1)$. As in Section 2.4.1.4, we rewrite the quantity to be estimated as

$$\widehat{r}_\alpha^{n_1,n_2} = \frac{\overline{Y}_{n_2}}{\overline{X}_{n_1}},$$

with

$$\overline{X}_{n_1} = \frac{1}{n_1} \sum_{i=1}^{n_1} X_i, \qquad \overline{Y}_{n_2} = \frac{1}{n_2} \sum_{i=1}^{n_2} Y_i,$$

where $X_i = f_2(x^{1,i})\,\alpha(x^{1,i})$ and $Y_i = f_1(x^{2,i})\,\alpha(x^{2,i})$. It holds

$$\mathbb{E}\left[\overline{X}_{n_1}\right] = \mathbb{E}\left[X_1\right] = Z_2 \int_\Omega \pi_1 \alpha \pi_2, \qquad \mathbb{E}\left[\overline{Y}_{n_2}\right] = \mathbb{E}\left[Y_1\right] = Z_1 \int_\Omega \pi_1 \alpha \pi_2,$$

$$\mathrm{Var}\left[\overline{X}_{n_1}\right] = \frac{1}{n_1}\mathrm{Var}\left[X_1\right] = \frac{Z_2^2}{n_1}\left(\int_\Omega \pi_1 \alpha^2 \pi_2^2 - \left(\int_\Omega \pi_1 \alpha \pi_2\right)^2\right),$$

$$\mathrm{Var}\left[\overline{Y}_{n_2}\right] = \frac{1}{n_2}\mathrm{Var}\left[Y_1\right] = \frac{Z_1^2}{n_2}\left(\int_\Omega \pi_1^2 \alpha^2 \pi_2 - \left(\int_\Omega \pi_1 \alpha \pi_2\right)^2\right).$$

Rewrite

$$\overline{X}_{n_1} = \mathbb{E}(X_1)\,(1+\varepsilon_X^{n_1}), \qquad \overline{Y}_{n_2} = \mathbb{E}(Y_1)\,(1+\varepsilon_Y^{n_2}),$$

where the random variables $\varepsilon_X^{n_1}$ and $\varepsilon_Y^{n_2}$ are independent, and such that

$$n_1 \mathrm{Var}\left(\varepsilon_X^{n_1}\right) \longrightarrow \frac{\mathrm{Var}(X_1)}{\mathbb{E}(X_1)^2}, \qquad n_2 \mathrm{Var}\left(\varepsilon_Y^{n_2}\right) \longrightarrow \frac{\mathrm{Var}(Y_1)}{\mathbb{E}(Y_1)^2}. \tag{2.111}$$

Therefore, computations similar to the ones performed in Section 2.4.1.4 to obtain the optimal umbrella sampling function give

$$\begin{aligned}
\lim_{n\to+\infty} n\,\mathrm{Var}\left[\ln\left(\frac{\widehat{r}_\alpha^{n_1,n_2}}{r}\right)\right] &= \lim_{n\to+\infty} n\,\mathrm{Var}\left[\ln\left(\frac{1+\varepsilon_Y^{n_2}}{1+\varepsilon_X^{n_1}}\right)\right] \\
&= \lim_{n\to+\infty} n\,\mathrm{Var}\left[\varepsilon_Y^{n_2} - \varepsilon_X^{n_1}\right] \\
&= \lim_{n\to+\infty} n\left(\mathrm{Var}\left[\varepsilon_Y^{n_2}\right] + \mathrm{Var}\left[\varepsilon_X^{n_1}\right]\right),
\end{aligned}$$

since $\varepsilon_X^{n_1}$ and $\varepsilon_Y^{n_2}$ are independent. Now, (2.110) and (2.111) imply that

$$\lim_{n\to+\infty} n\left(\mathrm{Var}\left[\varepsilon_Y^{n_2}\right] + \mathrm{Var}\left[\varepsilon_X^{n_1}\right]\right) = \frac{1}{1-\theta}\frac{\mathrm{Var}(Y_1)}{\mathbb{E}(Y_1)^2} + \frac{1}{\theta}\frac{\mathrm{Var}(X_1)}{\mathbb{E}(X_1)^2} = I(\alpha),$$

with

$$I(\alpha) = \frac{1}{1-\theta}\frac{\displaystyle\int_\Omega \pi_1^2\alpha^2\pi_2}{\left(\displaystyle\int_\Omega \pi_1\alpha\pi_2\right)^2} + \frac{1}{\theta}\frac{\displaystyle\int_\Omega \pi_1\alpha^2\pi_2^2}{\left(\displaystyle\int_\Omega \pi_1\alpha\pi_2\right)^2} - \left(\frac{1}{\theta} + \frac{1}{1-\theta}\right).$$

We first keep θ fixed, and study the minimization of $I(\alpha)$ with respect to α. This problem is equivalent to the minimization of the following functional:

$$J(\alpha) = \frac{\displaystyle\int_\Omega \alpha^2 \widetilde{\pi}_1\widetilde{\pi}_2(\widetilde{\pi}_1 + \widetilde{\pi}_2)}{\left(\displaystyle\int_\Omega \alpha\widetilde{\pi}_1\widetilde{\pi}_2\right)^2},$$

with $\widetilde{\pi}_1 = \theta\pi_1$ and $\widetilde{\pi}_2 = (1-\theta)\pi_2$. A Cauchy-Schwarz inequality implies that

$$\left(\int_\Omega \alpha\widetilde{\pi}_1\widetilde{\pi}_2\right)^2 \leq \int_\Omega \alpha^2\widetilde{\pi}_1\widetilde{\pi}_2(\widetilde{\pi}_1+\widetilde{\pi}_2)\int_\Omega \frac{\widetilde{\pi}_1\widetilde{\pi}_2}{\widetilde{\pi}_1+\widetilde{\pi}_2},$$

with equality if and only if α is proportional to $(\widetilde{\pi}_1+\widetilde{\pi}_2)^{-1}$. This shows that

$$J(\alpha) \geq \left(\int_\Omega \frac{\widetilde{\pi}_1\widetilde{\pi}_2}{\widetilde{\pi}_1+\widetilde{\pi}_2}\right)^{-1},$$

the minimum being attained with the choice

$$\alpha_\theta(x) = \frac{1}{\widetilde{\pi}_1(x)+\widetilde{\pi}_2(x)} = \frac{1}{\theta\dfrac{f_1(x)}{Z_1}+(1-\theta)\dfrac{f_2(x)}{Z_2}}. \tag{2.112}$$

The function $1/\alpha_\theta$ is therefore a mixture of π_1 and π_2 with mixture proportions determined by the sample sizes of each distribution. Even though the optimal function α_θ requires *a priori* the knowledge of the ratio Z_1/Z_0, it turns out that the associated numerical procedure, presented in Section 2.4.2.3 below, does not require any prior knowledge on this ratio.

For further purposes, we denote the value of the minimum by $M(\theta)$:

$$M(\theta) = J(\alpha_\theta) = \frac{1}{\theta(1-\theta)}\left(\frac{1}{F(\theta)}-1\right),$$

with

$$F(\theta) = \int_\Omega \frac{\pi_1\,\pi_2}{\pi_2+\theta(\pi_1-\pi_2)}.$$

In practice, the choice $n_1 = n_2$, hence $\theta = 1/2$, is advocated in [Bennett (1976)] as a robust choice yielding good results even when no knowledge of the distributions is available. There exist however refined strategies allowing to optimize the parameter θ on the fly in order to obtain estimates as accurate as possible. The remarkable mathematical fact underpinning such methods is that the optimal asymptotic variance $M(\theta)$ is a convex function of the parameter θ, see [Hahn and Then (2009)]. Therefore, the optimization with respect to $\theta \in [0,1]$ leads to a unique minimum, which is in the open interval $(0,1)$ when $M'(0) < 0$ and $M'(1) > 0$, but may also be attained at the boundary values 0 or 1 for some distributions π_1 and π_2. This result is helpful in determining whether bridge sampling is more efficient than one-sided estimators obtained in the limits $\theta \to 0$ or $\theta \to 1$. For the computation of chemical potentials with the model of Section 1.3.2.3, it is found in [Hahn and Then (2009)] that values of θ close to 0 should be chosen, which corresponds to favoring insertions.

2.4.2.3 *Numerical strategy*

Equation (2.109) and the expression of the optimal function (2.112) rewritten as

$$\alpha_\theta(x) = Z_1 \frac{n_1 + n_2}{n_1 f_1(x) + n_2 r f_2(x)},$$

with $\theta = n_1/n$, suggest to construct an estimator $\widehat{r}_\alpha^{n_1,n_2}$ from the relation

$$\widehat{r}_\alpha^{n_1,n_2} = \frac{\displaystyle \frac{1}{n_2} \sum_{j=1}^{n_2} \frac{f_1(x^{2,j})}{n_1 f_1(x^{2,j}) + n_2 \widehat{r}_\alpha^{n_1,n_2} f_2(x^{2,j})}}{\displaystyle \frac{1}{n_1} \sum_{j=1}^{n_1} \frac{f_2(x^{1,j})}{n_1 f_1(x^{1,j}) + n_2 \widehat{r}_\alpha^{n_1,n_2} f_2(x^{1,j})}}. \tag{2.113}$$

The simplest strategy to find an estimate of the ratio r is to resort to an iterative approach based on a fixed-point strategy: given some initial guess r_0 (for instance, $r_0 = 1$), the estimate r_{k+1} is computed from r_k as

$$r_{k+1} = \frac{\displaystyle \frac{1}{n_2} \sum_{j=1}^{n_2} \frac{f_1(x^{2,j})}{n_1 f_1(x^{2,j}) + n_2 r_k f_2(x^{2,j})}}{\displaystyle \frac{1}{n_1} \sum_{j=1}^{n_1} \frac{f_2(x^{1,j})}{n_1 f_1(x^{1,j}) + n_2 r_k f_2(x^{1,j})}}. \tag{2.114}$$

It is then expected that $r_k \to \widehat{r}_\alpha^{n_1,n_2}$ as $k \to +\infty$.

Remark 2.25 (Comparison with adaptive umbrella sampling). *Let us emphasize that such an iterative procedure uses a given set of samples, obtained from the two distributions π_1 and π_2. In contrast, an iterative estimate of the optimal umbrella sampling function* (2.106) *would require the sampling of new configurations at each step since the measure from which configurations are sampled is updated at every step.*

Other practical aspects of the method are discussed in [Bennett (1976)]. When correlated samples from π_i are used (as is the case with usual Markov chain techniques), the quantities n_1 and n_2 should be replaced by some appropriate effective sample size taking into account the correlation among sample points, or some undersampled Markov chain trajectory should be considered.

2.4.2.4 *Numerical illustration*

We present an application for a toy model here, the case of Widom insertion being postponed to Section 4.2.3, where bridge estimation on work distributions will be performed.

Consider the problem of estimating the ratio of normalizing constants when π_1 and π_2 are Gaussian distributions possibly having a small overlap. This case is reminiscent of standard free energy perturbation computations where the distributions of backward and forward energy differences hardly overlap. Table 2.1 compares several methods in the case when $\pi_1 \sim \mathcal{N}(0,1)$ and $\pi_2 \sim \mathcal{N}(d,1)$, for increasing values of the distance d. The ratio $r = Z_1/Z_2$ is equal to 1 in all cases. The following estimator, termed "forward" in the sequel (since it requires sampling from the starting distribution π_1 only),

$$\widehat{r}^n_{\text{fwd}} = \left(\frac{1}{n}\sum_{i=1}^{n}\frac{f_2(x^i)}{f_1(x^i)}\right)^{-1}, \qquad x^i \sim \pi_1$$

is based on the importance sampling identity

$$\frac{Z_2}{Z_1} = \mathbb{E}_{\pi_1}\left(\frac{f_2}{f_1}\right).$$

Symmetrically, the estimator termed "backward"

$$\widehat{r}^n_{\text{bck}} = \frac{1}{n}\sum_{i=1}^{n}\frac{f_1(x^i)}{f_2(x^i)}, \qquad x^i \sim \pi_2,$$

relies on a similar identity. Finally, an importance sampling strategy based on the intermediate distribution $\pi_{\text{IS}} \sim \mathcal{N}(d/2,1)$ may be used (see Section 2.4.1.4), in which case the corresponding "IS" estimator is

$$\widehat{r}^n_{\text{IS}} = \frac{\displaystyle\frac{1}{n}\sum_{i=1}^{n} f_2(x^i)/f_{\text{IS}}(x^i)}{\displaystyle\frac{1}{n}\sum_{i=1}^{n} f_1(x^i)/f_{\text{IS}}(x^i)}, \qquad x^i \sim \pi_{\text{IS}},$$

where $f_{\text{IS}}(x) = \exp\left(-\frac{1}{2}(x-d/2)^2\right)$ is proportional to π_{IS}. The last estimator we consider is $\widehat{r}_\alpha^{n/2,n/2}$, the bridge estimator (2.113) (computed with the scheme (2.114)), in the case when $n_1 = n_2 = n/2$. We observed that 10 to 20 iterations were in general sufficient for the convergence of the fixed-point algorithm (2.114) with a tolerance $\varepsilon = 10^{-8}$ for the difference

$|r_{k+1} - r_k|$. In any case, for real applications, the post-processing cost associated with these iterations is usually negligible compared to the cost of generating the sample points needed for the algorithm.

The results show that the bridge estimator $\widehat{r}_\alpha^{n/2,n/2}$ always perform better than simple (unidirectional) estimators $\widehat{r}_{\rm fwd}^n$ and $\widehat{r}_{\rm bck}^n$, and is also slightly better than the importance sampling estimator $\widehat{r}_{\rm IS}^n$ for distributions with small overlaps (the variance of the estimator being smaller).

Table 2.1 Comparison of methods for estimating ratios of normalizing constants. For the bridge estimators, 1000 sample points for each distribution are used, while $n = 2000$ sample points are generated when only the forward, backward or importance sampling distributions are considered (in order to compare the methods for a fixed total number of sample points). The results are presented under the form "average (standard deviation)," and the standard deviation has been obtained by performing 10,000 independent realizations.

distance d	forward $\widehat{r}_{\rm fwd}^n$	backward $\widehat{r}_{\rm bck}^n$	IS $\widehat{r}_{\rm IS}^n$	bridge $\widehat{r}_\alpha^{n/2,n/2}$
1	1.00 (0.03)	1.00 (0.03)	1.00 (0.02)	1.00 (0.02)
2	1.02 (0.14)	1.02 (0.13)	1.00 (0.05)	1.00 (0.05)
3	1.27 (0.47)	1.26 (0.46)	1.00 (0.10)	1.00 (0.09)
4	2.80 (1.83)	2.82 (1.82)	1.02 (0.21)	1.01 (0.17)
5	14.4 (13.7)	14.5 (13.8)	1.09 (0.55)	1.05 (0.35)
6	196 (255)	191 (248)	1.30 (2.32)	1.27 (0.98)

2.5 Histogram methods

We present in this section a second application of sampling methods to the computation of free energy differences, in the reaction coordinate case this time. More precisely, we consider histogram methods, in which configurations over the whole configuration space are obtained by concatenating configurations sampled in the vicinity of given values of the reaction coordinate (upon sampling a modified potential which is the sum of the physical energy and some biasing potential centered on the target value of the reaction coordinate). An estimate of the free energy profile can then be obtained by gathering consistently these configurations.

2.5.1 *Principle of histogram methods*

For simplicity, we restrict ourselves to reaction coordinates with values in a one-dimensional space $\mathcal{M}$, but the generalization to reaction coordinates with values in higher dimensional spaces is straightforward.

2.5.1.1 *Free energy as an approximated canonical average*

Histogram techniques are based on the following mathematical remark. An approximation of the free energy at a given value z of the reaction coordinate is obtained by computing the average of the observable

$$\chi_{z,\varepsilon}(q) = \frac{1}{\varepsilon\sqrt{2\pi}} \exp\left(-\frac{|\xi(q)-z|^2}{2\varepsilon^2}\right) \tag{2.115}$$

with respect to the canonical measure (2.1). Indeed, in the limit $\varepsilon \to 0$,

$$\begin{aligned} -\frac{1}{\beta} &\ln \mathbb{E}_\mu(\chi_{z,\varepsilon}) \\ &= -\frac{1}{\beta}\ln\left(\frac{1}{Z_\mu}\int_{T^*\mathcal{D}} \frac{1}{\varepsilon\sqrt{2\pi}} \exp\left(-\frac{|\xi(q)-z|^2}{2\varepsilon^2}\right) \mathrm{e}^{-\beta H(q,p)}\, dq\, dp\right) \\ &\longrightarrow -\frac{1}{\beta}\ln\left(\frac{1}{Z_\mu}\int_{T^*\mathcal{D}} \mathrm{e}^{-\beta H(q,p)}\, \delta_{\xi(q)-z}(dq)\, dp\right) = F(z), \end{aligned}$$

where the partition function Z_μ is given in (2.1). A more detailed proof of this limit is given in Lemma 5.3. This suggests to compute averages of the functions

$$\chi_{z_i,\Delta z_i}(q) = \frac{1}{\Delta z_i\sqrt{2\pi}} \exp\left(-\frac{|\xi(q)-z_i|^2}{2(\Delta z_i)^2}\right)$$

for a discretization $z_0 < z_1 < \cdots < z_I$ of the reaction coordinate space, and obtain a free energy difference profile by interpolation. Of course, it is possible to use other functions, of the general form

$$\chi_{z,\varepsilon}(q) = \frac{1}{\varepsilon}\chi\left(\frac{\xi(q)-z}{\varepsilon}\right),$$

where χ is a nonnegative function such that $\int_{\mathcal{M}} \chi = 1$ (for instance, bin indicator functions when the reaction coordinate space is split as the union of intervals). In any case, even for perfect sampling, there is some intrinsic numerical error related to the fact that $\varepsilon > 0$ is non-zero.

A direct computation of $\mathbb{E}_\mu(\chi_{z,\varepsilon})$ is however usually not possible since metastable features of the potential prevent an efficient direct sampling of the whole configurational space by standard techniques such as the ones presented in Sections 2.1 and 2.2. In order for the free energy profile to be computed accurately, there should be many sample points in the vicinity of each value of the reaction coordinate.

2.5.1.2 *Combining partial samples*

A way to overcome the sampling difficulty mentioned above is to perform several local samplings, centered around a given value of the reaction coordinate, and to gather them consistently. Consider a sequence of sampling windows with restrained potentials[4]

$$V_i(q) = V(q) + \frac{1}{2\varepsilon_i^2}\left(\xi(q) - z_i\right)^2, \tag{2.116}$$

where $z_1, \ldots, z_I$ are the centers of the restraining potentials. The name "restraining potentials" comes from the fact that the potential increases strongly as the value of the reaction coordinate departs from z_i, so that the configurations obtained by sampling the canonical measure associated with (2.116) fall in the neighborhood of the submanifold $\{(q,p) \in T^*\mathcal{D} \mid \xi(q) = z_i\}$.

The constants $\varepsilon_1, \ldots, \varepsilon_I$ are small enough to effectively constrain the values of the reaction coordinate to remain close from the target values z_i; though not too small in order to allow sufficient overlap between neighboring distributions and to keep the efficiency of the associated sampling methods (for instance, in order to keep a time-step not too small when discretizing a Langevin dynamics driven by V_i). Sampling the canonical measure associated with the potential (2.116) raises no metastability issues when the metastability arises in the direction of the reaction coordinate (*i.e.* provided the reaction coordinate fully and accurately describes the metastable features), and the restraining potential is strong enough to localize the sampling around a fixed reaction coordinate value. This discussion shows that the choice of the constants ε_i and centers z_i may be cumbersome in practice.

It is theoretically possible to unbias the sample points obtained by sampling

$$\nu_i(dq) = Z_i^{-1}\,\mathrm{e}^{-\beta V_i}, \qquad Z_i = \int_{\mathcal{D}} \mathrm{e}^{-\beta V_i}, \tag{2.117}$$

since

$$\nu(dq) = Z_\nu^{-1}\mathrm{e}^{-\beta V(q)}\,dq = \frac{Z_i}{Z_\nu}\exp\left(\frac{\beta}{2\varepsilon_i^2}\left(\xi(q) - z_i\right)^2\right)\nu_i(dq). \tag{2.118}$$

[4] Often (and abusively) called "umbrella potentials" in the literature. As far as we understand, the initial spirit of the umbrella sampling method (see Section 2.4.1.4) is rather to find a bridging probability distribution between two probability measures, not to restrict the sampling to a given phase space region.

Averages with respect to the initial measure ν can therefore be computed from averages with respect to the biased measure ν_i, upon unbiasing according to (2.118): for any observable A,

$$\mathbb{E}_\nu(A) = \frac{\mathbb{E}_{\nu_i}\left[A \exp\left(\frac{\beta}{2\varepsilon_i^2}(\xi - z_i)^2\right)\right]}{\mathbb{E}_{\nu_i}\left[\exp\left(\frac{\beta}{2\varepsilon_i^2}(\xi - z_i)^2\right)\right]} \simeq \frac{\sum_{j=1}^{n} A(q^{i,j}) \exp\left(\frac{\beta}{2\varepsilon_i^2}\left(\xi(q^j) - z_i\right)^2\right)}{\sum_{j=1}^{n} \exp\left(\frac{\beta}{2\varepsilon_i^2}\left(\xi(q^{i,j}) - z_i\right)^2\right)}, \tag{2.119}$$

where the configurations q^j are distributed according to the probability measure $\nu_i(dq)$. Note that the (unknown) factor Z_i/Z_ν in (2.118) is unimportant as long as canonical averages are computed using only the information from one of the restrained measures ν_i, since it is sufficient to know the measure to sample from up to a multiplicative constant.

The estimation of canonical averages according to (2.119) will however work only for observables with support around z_i. In order to obtain statistics over the whole configurational space, a natural idea is to combine the estimates obtained with the restrained measures ν_i by writing ν as a mixture of the probability measures ν_i:

$$\nu(dq) = \sum_{i=1}^{I} w_i \frac{Z_i}{Z_\nu} \exp\left(\frac{\beta}{2\varepsilon_i^2}\beta(\xi(q) - z_i)^2\right) \nu_i(dq), \tag{2.120}$$

where the mixture parameters $\{w_i\}_{i=1,\dots,I}$ satisfy the constraints

$$w_i \geq 0, \qquad \sum_{i=1}^{I} w_i = 1.$$

Then, $\mathbb{E}_\nu(A)$ can be computed as

$$\mathbb{E}_\nu(A) = \frac{\sum_{i=1}^{I} w_i \frac{Z_i}{Z_1} \mathbb{E}_{\nu_i}\left[A \exp\left(\frac{\beta}{2\varepsilon_i^2}(\xi - z_i)^2\right)\right]}{\sum_{i=1}^{I} w_i \frac{Z_i}{Z_1} \mathbb{E}_{\nu_i}\left[\exp\left(\frac{\beta}{2\varepsilon_i^2}(\xi - z_i)^2\right)\right]},$$

when for instance the first partition function Z_1 is chosen as a reference (but of course any other could be chosen instead), and Z_ν disappears from the formulation. Note that this rewriting of the canonical average is useless as such since the ratios of partition functions Z_i/Z_1 are unknown. If some approximation of these ratios were known, the canonical average of A could then be estimated as

$$\mathbb{E}_\nu(A) \simeq \frac{\displaystyle\sum_{i=1}^{I} w_i \frac{Z_i}{Z_1} \frac{1}{n_i} \sum_{n=1}^{n_i} A(q^{i,n}) \exp\left(\frac{\beta}{2\varepsilon_i^2}\left(\xi(q^{i,n}) - z_i\right)^2\right)}{\displaystyle\sum_{i=1}^{I} w_i \frac{Z_i}{Z_1} \frac{1}{n_i} \sum_{n=1}^{n_i} \exp\left(\frac{\beta}{2\varepsilon_i^2}\left(\xi(q^{i,n}) - z_i\right)^2\right)}, \qquad (2.121)$$

where $(q^{i,1}, \dots, q^{i,n_i})$ are distributed according to $\nu_i(dq)$.

The extended bridge sampling estimator described in Section 2.5.2 aims precisely at estimating these ratios, allowing to resort in practice to approximations such as (2.121). The mathematical study of this algorithm will also give some useful hints to choose the weights w_i required in (2.120), which are undetermined at this stage (see (2.131) below).

2.5.2 *Extended bridge sampling*

2.5.2.1 *Presentation of the method*

We now present a practical method to estimate ratios of partition functions, as required in (2.120). In essence, the problem is to optimally use sample points from several distributions. This situation is an extension of bridge sampling to the case when more than two distributions are involved, and therefore generalizes Bennett's approach described in Section 2.4.2 — hence the name "multistate BAR" (MBAR) proposed in [Shirts and Chodera (2008)]. The method is based on several works in statistics [Geyer (1994); Meng and Wong (1996); Kong *et al.* (2003); Tan (2004)], and seems theoretically more sound to us than historic histogram methods [Ferrenberg and Swendsen (1989)] or the celebrated Weighted Histogram Analysis Method (WHAM) [Kumar *et al.* (1992)] (also presented in a more pedagogical fashion in [Frenkel and Smit (2002)]). It is also less expensive than WHAM, see the discussion in [Shirts and Chodera (2008)]. We therefore focus on the MBAR method.

Abstract setting. To make the parallel with bridge sampling as clear as possible, we return to the abstract notation of Section 2.4.2, and consider I measures $\pi_1, \dots, \pi_I$ with $\pi_i(dx) = Z_i^{-1} f_i(x)\, dx$, the partition functions $\{Z_i\}_{i=1,\dots,I}$ being unknown. The aim is to estimate ratios of partition functions Z_i/Z_j. These quantities are readily available when all ratios $y_i = Z_i/Z_1$ with respect to a reference partition function are known since

$Z_j/Z_i = y_j/y_i$. Here, the partition function Z_1 is considered as the common normalization, but any other could be considered. Bridge sampling methods could be used to estimate independently all the ratios Z_i/Z_1, but this procedure turns out to be suboptimal. This is related to the fact that the overlap of π_1 and π_i for i large enough is small.

To derive a more global estimator for the ratios of partition functions, notice first that, for any function α_{ij} such that $f_i f_j \alpha_{ij}$ is integrable,

$$Z_i \mathbb{E}_{\pi_i}\left[\alpha_{ij} f_j\right] = \int_\Omega f_i(x)\,\alpha_{ij}(x)\,f_j(x)\,dx = Z_j \mathbb{E}_{\pi_j}\left[\alpha_{ij} f_i\right].$$

Denoting $Y = (y_2, \dots, y_I)^T$, the above property implies

$$A(\alpha)Y = B(\alpha), \tag{2.122}$$

with

$$A(\alpha) = \begin{pmatrix} a_2 & -b_{23} & \dots & -b_{2I} \\ -b_{32} & a_3 & \dots & -b_{3I} \\ \vdots & & \ddots & \vdots \\ -b_{I2} & -b_{I3} & \dots & a_I \end{pmatrix}, \qquad B(\alpha) = \begin{pmatrix} b_{21} \\ b_{31} \\ \vdots \\ b_{I1} \end{pmatrix},$$

where the entries of the above matrices are

$$b_{ij} = \mathbb{E}_{\pi_j}\left[\alpha_{ij} f_i\right], \quad a_i = \sum_{j=1,\, j\neq i}^{I} \mathbb{E}_{\pi_i}\left[\alpha_{ij} f_j\right].$$

An estimator of Y is then obtained by replacing all quantities appearing in the above expressions for A and B by their empirical counterparts. This is done by sampling n_j configurations $(x^{j,1}, \dots, x^{j,n_j})$ for each measure π_j, assumed to be independent for simplicity, and approximating $b_{ij} = \mathbb{E}_{\pi_j}\left[\alpha_{ij} f_i\right]$ by

$$\widehat{b}_{ij} = \frac{1}{n_j} \sum_{n=1}^{n_j} \alpha_{ij}(x^{j,n}) f_i(x^{j,n}), \qquad x^{j,n} \sim \pi_j. \tag{2.123}$$

An estimate of a_i is obtained in a similar fashion:

$$\widehat{a}_i = \sum_{j=1,\, j\neq i}^{I} \frac{1}{n_i} \sum_{n=1}^{n_i} \alpha_{ij}(x^{i,n}) f_j(x^{i,n}). \tag{2.124}$$

This gives an equation of the form

$$\widehat{A}(\widehat{\alpha})\widehat{Y} = \widehat{B}(\widehat{\alpha}), \tag{2.125}$$

for which we assume that there exists a unique solution $\widehat{Y}$ (which is the case when $\widehat{A}(\alpha)$ is invertible). Note that all distributions need not have the same number of sample points.

Optimal choice of functions. It is desirable to have estimates $\widehat{Y}$ for which the asymptotic covariance matrix is minimal (for the order on positive definite matrices). This is the case for the choice

$$\alpha_{ij}^{n_1,\dots,n_I}(x) = \frac{n_j Z_j^{-1}}{\sum_{l=1}^I n_l Z_l^{-1} f_l(x)} \tag{2.126}$$

for $1 \le i, j \le I$ and $i \neq j$. In practice, this expression is approximated as

$$\widehat{\alpha}_{ij}^{n_1,\dots,n_I}(x) = \frac{n_j \widehat{y_j}^{-1}}{\sum_{l=1}^I n_l \widehat{y_l}^{-1} f_l(x)}.$$

In this case, the equation to be solved to obtain $\widehat{Y} = (\widehat{y}_2, \dots, \widehat{y}_I)$ is

$$\widehat{A}(\widehat{\alpha}_{ij}^{n_1,\dots,n_I})\widehat{Y} = \widehat{B}(\widehat{\alpha}_{ij}^{n_1,\dots,n_I}). \tag{2.127}$$

The proof presented in [Tan (2004)] uses maximum likelihood estimates, and error estimates (*i.e.* an expression for the asymptotic covariance matrix of $\widehat{Y}$) are also provided. Roughly speaking, the proof is an extension of the proof of optimality of the bridge sampler (see Section 2.4.2), in a matrix version.

In fact, (2.127) can be rewritten in a more explicit way as follows:

Lemma 2.26. *Consider the pooled sample of size $n = n_1 + n_2 + \cdots + n_I$ obtained by concatenation of all available sample points:*

$$(X^1, \dots, X^n) = (x^{1,1}, \dots, x^{1,n_1}, x^{2,1}, \dots, x^{2,n_2} \dots, x^{I,1}, \dots, x^{I,n_I}).$$

With the choice (2.126)*, the equation* (2.127) *is then equivalent to the following set of nonlinear equations:*

$$\forall i = 2, \dots, I, \qquad \widehat{y}_i = \sum_{m=1}^n \frac{f_i(X^m)}{\sum_{j=1}^I n_j \widehat{y_j}^{-1} f_j(X^m)}, \tag{2.128}$$

with the normalization condition $\widehat{y}_1 = 1$.

A comparison with (2.113) shows that, in the case $I = 2$, the estimator of $y_2 = Z_2/Z_1$ obtained from (2.128) is the same as the one obtained with the bridge sampling method (see Section 2.1 in [Tan (2004)]).

Proof. For $i = 2, \dots, I$, the i-th line in (2.125) can be rewritten as

$$\widehat{a}_i \widehat{y}_i = \sum_{j=1,\, j \neq i}^I \widehat{b}_{ij}\, \widehat{y}_j, \tag{2.129}$$

since $\widehat{y}_1 = 1$. The equation (2.127) has to be solved for $\widehat{\alpha}_{ij}^{n_1,\dots,n_I}$ given by (2.126):

$$\widehat{\alpha}_{ij}^{n_1,\dots,n_I}(x) = \frac{n_j \widehat{y_j}^{-1}}{A(x)}, \qquad A(x) = \sum_{l=1}^{I} n_l \widehat{y_l}^{-1} f_l(x).$$

According to (2.124), the empirical counterpart of $a_i y_i$ is

$$\begin{aligned}
\widehat{a}_i \widehat{y}_i &= \widehat{y}_i \sum_{j=1, j\neq i}^{I} \frac{1}{n_i} \sum_{k=1}^{n_i} \widehat{\alpha}_{ij}^{n_1,\dots,n_I}(x^{i,k}) f_j(x^{i,k}) \\
&= \frac{\widehat{y}_i}{n_i} \sum_{k=1}^{n_i} \frac{\sum_{j=1,\, j\neq i}^{I} n_j \widehat{y_j}^{-1} f_j(x^{i,k})}{A(x^{i,k})} \\
&= \frac{\widehat{y}_i}{n_i} \sum_{k=1}^{n_i} \left(1 - \frac{n_i \widehat{y_i}^{-1} f_i(x^{i,k})}{A(x^{i,k})} \right) \\
&= \widehat{y}_i - \sum_{k=1}^{n_i} \frac{f_i(x^{i,k})}{A(x^{i,k})}.
\end{aligned}$$

Besides,

$$\sum_{j=1,\, j\neq i}^{I} \widehat{b}_{ij}\, \widehat{y}_j = \sum_{j=1,\, j\neq i}^{I} \sum_{k=1}^{n_j} \frac{f_i(x^{j,k})}{A(x^{j,k})},$$

so that finally (2.129) can be rewritten as

$$\widehat{y}_i - \sum_{k=1}^{n_i} \frac{f_i(x^{i,k})}{A(x^{i,k})} = \sum_{j=1,\, j\neq i}^{I} \sum_{k=1}^{n_j} \frac{f_i(x^{j,k})}{A(x^{j,k})},$$

which is (2.128). □

Numerical implementation. Practical implementation strategies to solve the nonlinear equation (2.128), with a focus on free energy problems, are presented in [Meng and Wong (1996); Shirts and Chodera (2008)]. For instance, a fixed-point strategy may be used. Starting from some initial guess $(y_2^0, \dots, y_I^0)$ (for instance, $y_i^0 = 1$ for all $2 \leq i \leq I$), the fixed-point strategy is the following iterative procedure:

$$y_i^{k+1} = \sum_{m=1}^{n} \frac{f_i(X^m)}{\sum_{j=1}^{I} n_j \left(y_j^k\right)^{-1} f_j(X^m)}, \tag{2.130}$$

where $y_1^k = 1$ for all $k \geq 0$. It is expected that, for all $i = 2, \dots, I$, $y_i^k \to \widehat{y}_i$ given by (2.128).

2.5.2.2 *Recovering canonical averages*

We now return more explicitly to the framework of free energy computations. In this case, the measures ν_i are the restrained measures (2.117), while their unnormalized densities are $f_i = \mathrm{e}^{-\beta V_i}$.

In order to compute the potential of mean force, canonical averages of the form

$$-\frac{1}{\beta} \ln \mathbb{E}_\mu(\chi_{z,\varepsilon})$$

have to be computed, for functions $\chi_{z,\varepsilon}$ such as (2.115).

All the ratios of normalizing functions $y_i = Z_i/Z_1$ should first be estimated according to (2.128). It is then possible to use estimators such as (2.121), based on some mixture formula for ν in terms of the restraining measures ν_i. At this stage, it is still an open question on how to weigh the different contributions in the mixture, *i.e.* how to choose the weights w_i in (2.120)-(2.121). A simple idea is to set $w_i = 1/I$ for all $1 \le i \le I$.

Another possibility is to resort to estimates such as

$$\mathbb{E}_\nu(A) \simeq \frac{\displaystyle\sum_{m=1}^{n} A(q^m)/f_{\mathrm{mix}}(q^m)}{\displaystyle\sum_{m=1}^{n} 1/f_{\mathrm{mix}}(q^m)} \tag{2.131}$$

where $(q^m)_{m=1,\dots,n}$ is the pooled sample obtained by concatenating samples from the measures $Z_i^{-1} f_i(q)\,dq$, and the mixture weight is

$$f_{\mathrm{mix}}(q) = \sum_{i=1}^{I} \frac{n_i}{n} \widehat{y_i}^{-1} f_i(q). \tag{2.132}$$

This technique is advised in [Tan (2004); Shirts and Chodera (2008)] since it can be shown to be optimal in some sense. It also has the advantage of not introducing any new parameter in the estimate. In essence, (2.131) is some importance sampling strategy based on $f_{\mathrm{mix}}(q)\,dq$. The pooled sample $(q^m)_{m=1,\dots,n}$ can indeed be considered as drawn from the measure with density $Z_{\mathrm{mix}}^{-1} f_{\mathrm{mix}}$, since[5] the mixture formula (2.132) means that there is a probability n_j/n to draw a configuration from the j-th probability measure $Z_j^{-1} f_j(q)\,dq$.

[5]This interpretation is actually not completely correct since sampling the measure with density $Z_{\mathrm{mix}}^{-1} f_{\mathrm{mix}}$ would involve having a random number of configurations drawn from $Z_i^{-1} f_i(q)\,dq$, whereas in the pooled samples considered here, this number is deterministic and equal to n_i.

Note that for sufficiently stiff restraining potentials (ε_i small enough in (2.116)), it holds

$$f_{\text{mix}}(q) \simeq \frac{n_{i_0}}{n} \widehat{y_{i_0}}^{-1} f_{i_0}(q),$$

where q is sampled according to ν_{i_0}. The method based on (2.131)-(2.132) is therefore almost equivalent to the simple strategy which consists in setting $w_i = 1/I$ in (2.121).

2.5.2.3 *Application to the model problem*

We consider the dimer in a WCA solvent, described in Section 1.3.2.4, at an inverse temperature $\beta = 1$, with $N = 100$ particles ($N - 2$ solvent particles and the dimer). The solvent density is

$$\rho = (1 - 2/N)a^{-2} = 0.436,$$

since there are $N - 2$ solvent particles in a square box of side length $L = a\sqrt{N}$ with $a = 1.5$. The parameters describing the WCA interactions are set to $\sigma = 1$ and $\varepsilon = 1$, and the additional parameters for the dimer are $w = 2$ and $h = 2$.

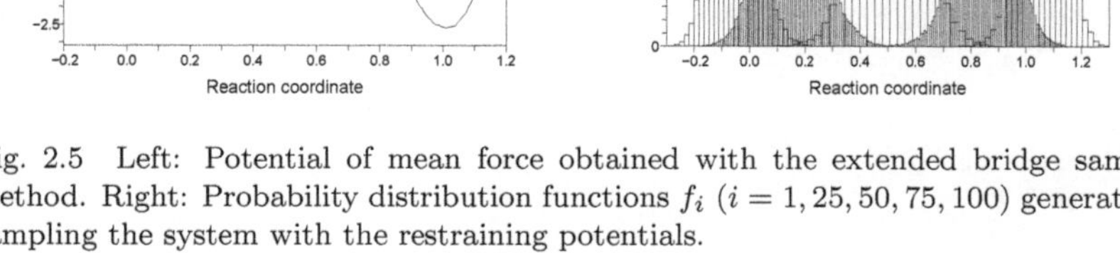

Fig. 2.5 Left: Potential of mean force obtained with the extended bridge sampling method. Right: Probability distribution functions f_i ($i = 1, 25, 50, 75, 100$) generated by sampling the system with the restraining potentials.

The restraining potentials use $\varepsilon_i = 0.1$, and $I = 100$ nodes uniformly spaced in the interval $\mathcal{M} = [-0.2, 1.2]$. The free energy is then computed after binning $\mathcal{M}$ into $N_z = 200$ bins of equal sizes. The dynamics used for the sampling is a Langevin dynamics with friction $\gamma = 1$, and integrated with a time-step $\Delta t = 0.005$. For each node z_i, 50,000 values of the reaction coordinate are stored by subsampling the trajectory every 10 steps (which represents a simulation time of 2500 per node). The final estimation of ratios of partition functions is done using the simple fixed-point

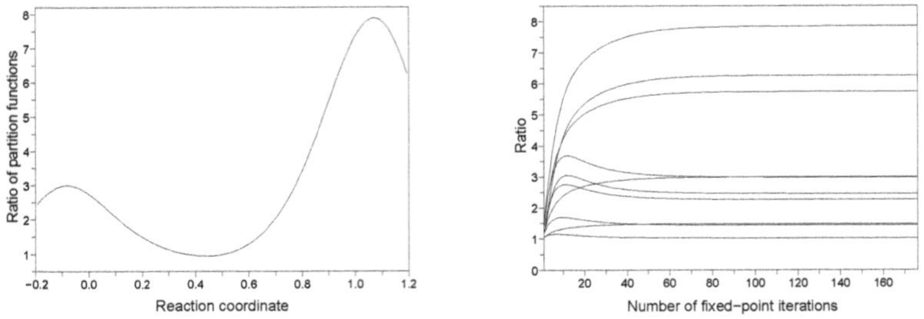

Fig. 2.6 Left: Logarithms of ratios of partition functions obtained by the fixed-point iterations. Right: Convergence of some ratios as a function of the number of fixed-point iterations.

strategy (2.130), the common denominator for the ratios being $Z_{I/2}$. The results are presented in Figures 2.5 and 2.6. Note that the logarithms of ratios of partition functions in Figure 2.6 can be seen as some smoothened free energy profile (which arises from a convolution with a Gaussian kernel, see (5.26) in the proof of Lemma 5.3).

Chapter 3

Thermodynamic integration and sampling with constraints

From a mathematical point of view, computation of free energy differences by thermodynamic integration relies on a sampling by homogeneous Markov chains.

The principle of thermodynamic integration for free energy computation is to write the free energy difference as the integral of the derivative of the free energy. Since the free energy is the logarithm of a partition function, its derivative is the average of some functional with respect to a probability measure at a fixed value of the alchemical parameter or the reaction coordinate. This technique was first introduced in [Kirkwood (1935)] in the alchemical case, and then extended to the reaction coordinate case, with the so-called Blue Moon sampling [Carter *et al.* (1989)], further refined in [den Otter and Briels (1998); Sprik and Ciccoti (1998)] with the derivation of a simpler expression for the local mean force.

The outline of this chapter is the following. In Section 3.1, the principle of the method is presented in the simple alchemical setting. Section 3.2 is then devoted to the thermodynamic integration method in the reaction coordinate case, using sampling methods in the configurational space $\mathcal{D}$. This is finally extended in Section 3.3 with sampling methods in the phase space $T^*\mathcal{D}$. The conceptual and numerical difficulty of the thermodynamic integration method in the reaction coordinate case is that it requires the sampling of conditional probability measures, namely probability measures with the support of a submanifold of the configurational or phase space.

3.1 Introduction: The alchemical setting

The alchemical setting is a pedagogical starting point. Let us recall that in this case, the *free energy* is defined by

$$F(\lambda) = -\beta^{-1} \ln \int_{T^*\mathcal{D}} \mathrm{e}^{-\beta H_\lambda(q,p)} \, dq \, dp, \tag{3.1}$$

where λ is a parameter in $\mathbb{R}$. Typically, $H_\lambda = H_0 + \lambda(H_1 - H_0)$ is a Hamiltonian built as an arbitrary interpolation between two given Hamiltonians $H_0(q,p) = \frac{1}{2}p^T M^{-1} p + V_0(q)$ and $H_1 = \frac{1}{2}p^T M^{-1} p + V_1(q)$:

$$H_\lambda(q,p) = \frac{1}{2} p^T M^{-1} p + V_\lambda(q)$$

with

$$V_\lambda = V_0 + \lambda(V_1 - V_0).$$

The quantity of interest is the free energy difference $F(1) - F(0)$.

3.1.1 *General strategy*

The derivative of F with respect to λ is the canonical average

$$F'(\lambda) = \frac{\displaystyle\int_{T^*\mathcal{D}} \frac{\partial H_\lambda}{\partial \lambda} \mathrm{e}^{-\beta H_\lambda}}{\displaystyle\int_{T^*\mathcal{D}} \mathrm{e}^{-\beta H_\lambda}} = \mathbb{E}_{\mu_\lambda}\left(\frac{\partial H_\lambda}{\partial \lambda}\right), \tag{3.2}$$

where μ_λ is the canonical measure defined by

$$\mu_\lambda(dq\,dp) = \frac{\mathrm{e}^{-\beta H_\lambda(q,p)} \, dq \, dp}{\displaystyle\int_{T^*\mathcal{D}} \mathrm{e}^{-\beta H_\lambda(q,p)} \, dq \, dp}.$$

In the case when the Hamiltonian $H_\lambda(q,p) = \frac{1}{2}p^T M^{-1} p + V_\lambda(q)$ depends on λ only through the potential part V_λ, (3.2) rewrites

$$F'(\lambda) = \frac{\displaystyle\int_{\mathcal{D}} \frac{\partial V_\lambda}{\partial \lambda} \mathrm{e}^{-\beta V_\lambda}}{\displaystyle\int_{\mathcal{D}} \mathrm{e}^{-\beta V_\lambda}} = \mathbb{E}_{\nu_\lambda}\left(\frac{\partial V_\lambda}{\partial \lambda}\right), \tag{3.3}$$

where ν_λ is the configurational canonical measure defined by

$$\nu_\lambda(dq) = \frac{\mathrm{e}^{-\beta V_\lambda(q)} \, dq}{\displaystyle\int_{\mathcal{D}} \mathrm{e}^{-\beta V_\lambda(q)} \, dq}.$$

Thermodynamic integration then consists in writing

$$F(1) - F(0) = \int_0^1 F'(\lambda)\, d\lambda,$$

and discretizing the integral on the right-hand side by a numerical integration formula. Using for example a Riemann sum, the following approximation is thus obtained:

$$F(1) - F(0) \simeq \sum_{i=1}^{n} (\lambda_i - \lambda_{i-1}) F'(\lambda_i), \tag{3.4}$$

where $(\lambda_0 = 0, \lambda_1, \dots, \lambda_{n-1}, \lambda_n = 1)$ is a given set of quadrature points. The estimation of the free energy difference is then obtained by estimating the derivatives $F'(\lambda_i)$, which can be done by sampling the Boltzmann-Gibbs measure μ_{λ_i} (resp. ν_{λ_i}), following formula (3.2) (resp. formula (3.3)). We refer to Chapter 2 for efficient sampling methods of the canonical measures μ_λ and ν_λ.

Let us briefly discuss the errors in the estimate (3.4). For simplicity, let us assume that each of the derivative values $F'(\lambda_i)$ is obtained (independently from the others) by a perfect sampling of μ_{λ_i} using an empirical mean over a fixed number of i.i.d. random variables. The errors are then of two types: (i) the deterministic error due to the numerical integration of $F'(\lambda)$ (bias) and (ii) the statistical error due to the variance of the estimators of $F'(\lambda_i)$. Concerning (ii), it has been noted (see [Schlitter (1991); Blondel (2004)]) that the quadrature points $(\lambda_0, \dots, \lambda_n)$ can be optimized in order to reduce the variance of the estimator.

Indeed, assuming a uniform repartition of the quadrature points ($\lambda_i = i\,\Delta\lambda$ and $\Delta\lambda = 1/n$), an estimate of the variance of (3.4) is (in the limit of large n, (q^i, p^i) denoting independent random variables with law μ_{λ_i}):

$$\mathrm{Var}\left(\Delta\lambda \sum_{i=1}^{n} \frac{\partial H_\lambda}{\partial \lambda}(q^i, p^i)\right) = \frac{1}{n^2} \sum_{i=1}^{n} \mathbb{E}_{\mu_\lambda}\left(\left(\frac{\partial H_\lambda}{\partial \lambda} - \mathbb{E}_{\mu_\lambda}\left(\frac{\partial H_\lambda}{\partial \lambda}\right)\right)^2\right)\Bigg|_{\lambda=\lambda_i}$$

$$\simeq \frac{1}{n} \int_0^1 \mathbb{E}_{\mu_\lambda}\left(\left(\frac{\partial H_\lambda}{\partial \lambda} - \mathbb{E}_{\mu_\lambda}\left(\frac{\partial H_\lambda}{\partial \lambda}\right)\right)^2\right) d\lambda. \tag{3.5}$$

By a change of variable, it appears that optimizing the quadrature points $(\lambda_0, \dots, \lambda_n)$ in (3.4) amounts to introducing a one-to-one increasing function $\phi : [0,1] \to [0,1]$, considering the re-parametrized Hamiltonian $H_\lambda^\phi = H_{\phi(\lambda)}$ instead of H_λ in the former formulas, and minimizing the variance (3.5) over the reparametrization functions ϕ. In other words, the problem is to minimize

$$\mathrm{Var}(\phi) = \int_0^1 \phi'(\lambda)^2 \sigma^2(\phi(\lambda))\, d\lambda$$

where

$$\sigma^2(\lambda) = \mathbb{E}_{\mu_\lambda}\left(\left(\frac{\partial H_\lambda}{\partial \lambda} - \mathbb{E}_{\mu_\lambda}\left(\frac{\partial H_\lambda}{\partial \lambda}\right)\right)^2\right).$$

The Euler-Lagrange equation associated with this minimization problem writes $(\sigma \circ \phi\, \phi')' = 0$ (together with the boundary conditions $\phi(0) = 0$ and $\phi(1) = 1$), so that the solution is

$$\phi(\lambda) = \Sigma^{-1}(\kappa\lambda),$$

where $\Sigma(s) = \int_0^s \sigma(r)\, dr$ and $\kappa = \Sigma(1)$. The variance is thus optimized by choosing $\lambda_i = \phi(i\Delta\lambda)$, which amounts to introducing more quadrature points in the regions where σ is large (see Figure 3.1). Of course, in practice, the function σ is unknown. It is however possible to obtain an estimate of the variance by empirical means, and to add adaptively new points of computations λ_i in regions where σ is large (see for example [Schlitter (1991)]). Note that in the linear case ($H_\lambda = H_0 + \lambda(H_1 - H_0)$), the variance is $\sigma^2(\lambda) = -\beta^{-1}F''(\lambda)$ and can thus be estimated by finite differences approximation of the second derivative $F''(\lambda)$.

For a general discussion of such optimizations, and generalizations to higher dimensional alchemical reaction coordinates, see [Gelman and Meng (1998)].

3.1.2 *Numerical application*

We here present some results for Widom insertion (see Section 1.3.2.3). The sampling at a fixed value of the alchemical parameter is performed with Langevin dynamics. The parameters of the system and of the dynamics are the same as in Section 2.4.1.1 (in particular, $\Delta t = 0.005$ and $\gamma = 1$ for the Langevin dynamics). A preliminary thermalization for a time $t_{\rm thm} = 0.5$ is performed every time the alchemical parameter λ is changed, before the mean force is estimated by averaging the local mean force $\Delta_N V(q^N, q)$ (see (1.61) for the definition of V_λ) over a time $T = 10\,000$. The mean force is estimated at equally spaced points $\lambda_i = i\Delta\lambda$, with spacing $\Delta\lambda = 0.01$. Figure 3.2 shows some running estimates of the mean force at some values of λ as a function of the sampling time, as well as the mean force profile. The free energy difference profile (Figure 3.3, Left) is recovered by the trapezoidal rule

$$F(\lambda_{i+1}) = F(\lambda_i) + \frac{F'(\lambda_{i+1}) + F'(\lambda_i)}{2}\,\Delta\lambda.$$

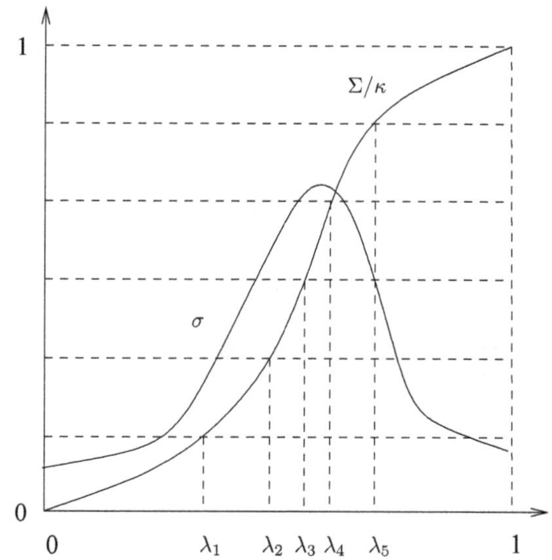

Fig. 3.1 Optimal position of the quadrature points for thermodynamic integration are obtained as the image of equally-spaced quadrature points by the inverse of the application Σ/κ.

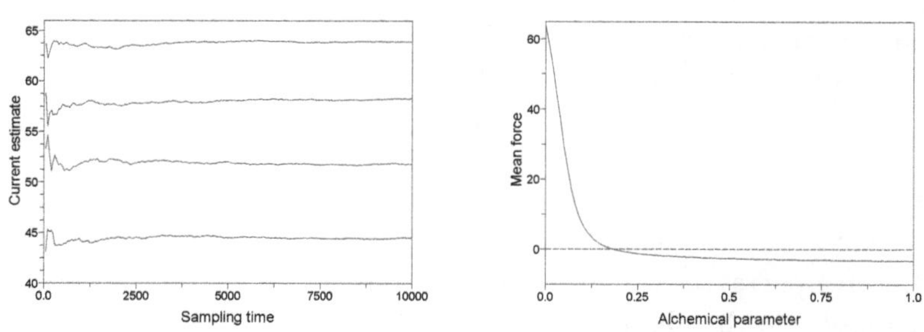

Fig. 3.2 left: Running estimate of the mean force as a function of time, for (from top to bottom) $\lambda = 0, 0.01, 0.02, 0.03$. right: Estimated mean force.

The estimated free energy difference $F(1)-F(0)$ is 1.316, in good agreement with the reference value $\mu^{\text{ex}} = 1.317 \pm 0.001$ obtained in Section 2.4.1.1. Error bars on the final value could be obtained by integrating the (independent) sampling errors for each value λ_i, which in turn are obtained as described in Section 2.3.1. Finally, a plot of $F''(\lambda)$ (estimated by finite differences) is presented in Figure 3.3 (Right). It suggests that the quadra-

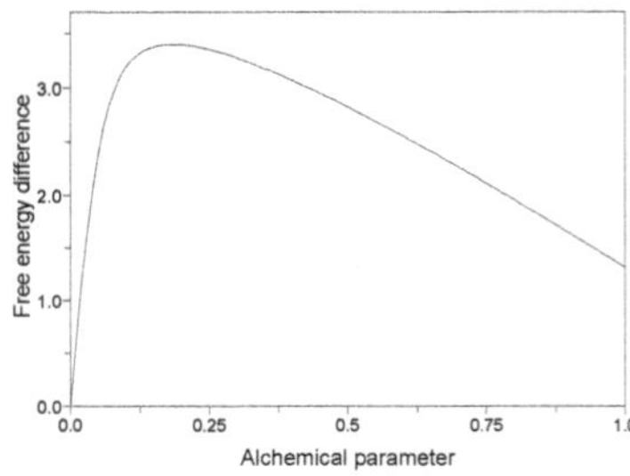

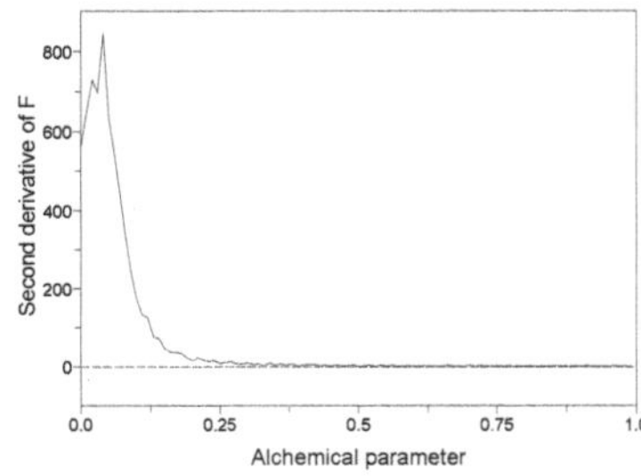

Fig. 3.3 Left: Estimated free energy difference as a function of the alchemical parameter. Right: Plot of $F''(\lambda)$, estimated by finite differences.

ture points should be more concentrated around $\lambda = 0$. This observation is consistent with the discussion at the end of Section 2.4.1.4, which shows that the critical values for the parameter λ in Widom insertion are the ones around 0.

3.2 The reaction coordinate case: configurational space sampling

Let us now discuss thermodynamic integration in the reaction coordinate case, by restricting first the discussion to the case when the Boltzmann Gibbs measure on the configurational space $\mathcal{D} = \mathbb{R}^n$ are of interest:

$$\nu(dq) = Z_\nu^{-1} \mathrm{e}^{-\beta V(q)}\, dq, \tag{3.6}$$

where $Z_\nu = \int_{\mathbb{R}^n} \mathrm{e}^{-\beta V(q)}\, dq$. This is not an important restriction in practice since the reaction coordinates are usually only functions of the position, so that the free energy can be fully defined considering only the measure ν. The extension to the phase space will be the subject of Section 3.3. Most of this section is based on the paper [Ciccotti *et al.* (2008)].

3.2.1 *Reaction coordinate and free energy*

A reaction coordinate is an application defined on the configurational space

$$\xi : \mathbb{R}^n \to \mathbb{R}^m$$

(where $n = 3N$, N being the number of particles) which indexes the transformation of interest. For example, it can be a dihedral angle in a molecule, which measures the change of conformation of the molecule. It may also be a distance between two (groups of) atoms, which measures a binding energy. In any case, it is meant to be a function with value in a small dimensional space compared to the dimension of the configurational space ($m < n$). In all the following, we assume that ξ is a smooth function such that

$$\mathrm{rank}(\nabla\xi) = m \tag{3.7}$$

where $\mathrm{rank}(\nabla\xi)$ is the rank of the $n \times m$ matrix $\left(\frac{\partial \xi_\alpha}{\partial q_i}\right)_{i,\alpha}$. The notation convention in the following is that Latin (respectively Greek) indices vary between 1 and n (respectively m).

Let us also introduce the $m \times m$ Gram matrix:

$$G = (\nabla\xi)^T \nabla\xi, \tag{3.8}$$

which writes componentwise: $G_{\alpha,\beta} = \nabla\xi_\alpha \cdot \nabla\xi_\beta$. The non-degeneracy assumption (3.7) is equivalent to the fact that $\det G > 0$. We will also need the following technical assumption: $\forall q \in \mathbb{R}^n$,

$$\sup_{1\le\alpha\le m} \left| \sum_{\zeta=1}^{m} G^{-1}_{\alpha,\zeta}(q) \nabla\xi_\zeta(q) \right| < \infty. \tag{3.9}$$

All the results we present below may be generalized to other configuration spaces $\mathcal{D}$, see Remark 1.1. Of course, the assumptions (3.7) and (3.9) need to be checked only for $q \in \mathcal{D}$.

3.2.1.1 *Marginal and conditional probability measures*

Two probability measures can be associated with the Boltzmann Gibbs measure ν on the configuration space:

- The image of the measure ν by ξ (also called the marginal of ν in ξ) which is denoted by $\nu^\xi(dz)$.
- The probability measure $\nu^\xi(dq|z)$ which is the measure ν conditioned to a fixed value z of the reaction coordinate.

These two measures are defined by the following conditioning formula:

Definition 3.1 (Marginal and conditional probability measures). *The measures ν^ξ and $\nu^\xi(\cdot|z)$ are defined by: for any bounded measurable functions $f : \mathbb{R}^m \to \mathbb{R}$ and $g : \mathbb{R}^n \to \mathbb{R}$,*

$$\int_{\mathbb{R}^n} f(\xi(q))\, g(q)\, \nu(dq) = \int_{\mathbb{R}^m} f(z) \left(\int_{\Sigma(z)} g\, d\nu^\xi(\cdot|z) \right) \nu^\xi(dz). \tag{3.10}$$

We recall that

$$\Sigma(z) = \left\{q, \xi(q) = z\right\}$$

is the submanifold of codimension m of $\mathbb{R}^n$ corresponding to positions q at a fixed value z of the reaction coordinate. The support of the measure $\nu^\xi(\cdot|z)$ is $\Sigma(z)$.

Let us now give some expressions for these two probability measures in terms of the conditional measure $\delta_{\xi(q)-z}(dq)$ (also called delta measure) which is defined as: for any test functions $f : \mathbb{R}^m \to \mathbb{R}$ and $g : \mathbb{R}^n \to \mathbb{R}$,

$$\int_{\mathbb{R}^n} f(\xi(q))\, g(q)\, dq = \int_{\mathbb{R}^m} f(z) \left(\int_{\Sigma(z)} g\, \delta_{\xi(q)-z}(dq) \right) dz,$$

or, in short,

$$dq = \delta_{\xi(q)-z}(dq)\, dz. \tag{3.11}$$

It is easy to check that

$$\nu^\xi(dz) = Z_\nu^{-1} \left(\int_{\Sigma(z)} \mathrm{e}^{-\beta V(q)} \delta_{\xi(q)-z}(dq) \right) dz$$

and

$$\nu^\xi(dq|z) = \frac{\mathrm{e}^{-\beta V(q)} \delta_{\xi(q)-z}(dq)}{\displaystyle\int_{\Sigma(z)} \mathrm{e}^{-\beta V(q)} \delta_{\xi(q)-z}(dq)}.$$

Other expressions for the two probability measures ν^ξ and $\nu^\xi(\cdot|z)$ in terms of the surface measure on $\Sigma(z)$ will be useful. To derive these expressions, a very important formula is needed, which will be used many times in the following: the co-area formula.

Lemma 3.2 (Co-area formula). *For any smooth function $f : \mathbb{R}^n \to \mathbb{R}$,*

$$\int_{\mathbb{R}^n} f(q)(\det G)^{1/2}(q)\, dq = \int_{\mathbb{R}^m} \int_{\Sigma(z)} f\, d\sigma_{\Sigma(z)}\, dz, \tag{3.12}$$

where $\sigma_{\Sigma(z)}$ denotes the surface measure on $\Sigma(z)$, namely the Lebesgue measure on $\Sigma(z)$ induced by the Lebesgue measure in the ambient Euclidean space $\mathbb{R}^n$. In the case $m = 1$, Equation (3.12) writes:

$$\int_{\mathbb{R}^n} f(q)|\nabla \xi|(q)\, dq = \int_{\mathbb{R}} \int_{\Sigma(z)} f\, d\sigma_{\Sigma(z)}\, dz. \tag{3.13}$$

In view of (3.12) and (3.11), it holds:

$$\delta_{\xi(q)-z}(dq) = (\det G)^{-1/2} d\sigma_{\Sigma(z)}. \tag{3.14}$$

A corollary of the co-area formula (3.12) is the following:

Corollary 3.3. *If a continuous random variable Q has law $\psi(q)\,dq$ in $\mathbb{R}^n$, then $\xi(Q)$ has law*

$$\left(\int_{\Sigma(z)} \psi\,(\det G)^{-1/2}\, d\sigma_{\Sigma(z)} \right) dz,$$

and the law of Q conditioned to a fixed value z of $\xi(Q)$ is

$$\frac{\psi\,(\det G)^{-1/2}\, d\sigma_{\Sigma(z)}}{\displaystyle\int_{\Sigma(z)} \psi\,(\det G)^{-1/2}\, d\sigma_{\Sigma(z)}}.$$

In particular,

$$\nu^{\xi}(dz) = \left(\int_{\Sigma(z)} Z_{\nu}^{-1} \mathrm{e}^{-\beta V}\,(\det G)^{-1/2}\, d\sigma_{\Sigma(z)} \right) dz, \tag{3.15}$$

and

$$d\nu^{\xi}(\cdot|z) = \frac{\mathrm{e}^{-\beta V}\,(\det G)^{-1/2}\, d\sigma_{\Sigma(z)}}{\displaystyle\int_{\Sigma(z)} \mathrm{e}^{-\beta V}\,(\det G)^{-1/2}\, d\sigma_{\Sigma(z)}}. \tag{3.16}$$

Proof. For any bounded functions f and g, using the co-area formula (3.12),

$$\begin{aligned}
\mathbb{E}\Big(f(\xi(Q))\, g(Q) \Big) &= \int_{\mathbb{R}^n} f(\xi(q))\, g(q)\, \psi(q)\, dq \\
&= \int_{\mathbb{R}^m} \int_{\Sigma(z)} (f \circ \xi)\, g\, \psi\, (\det G)^{-1/2} d\sigma_{\Sigma(z)}\, dz \\
&= \int_{\mathbb{R}^m} f(z) \frac{\displaystyle\int_{\Sigma(z)} g\, \psi\, (\det G)^{-1/2} d\sigma_{\Sigma(z)}}{\displaystyle\int_{\Sigma(z)} \psi\, (\det G)^{-1/2} d\sigma_{\Sigma(z)}} \left(\int_{\Sigma(z)} \psi\, (\det G)^{-1/2} d\sigma_{\Sigma(z)} \right) dz,
\end{aligned}$$

which concludes the proof in view of the definition (3.10). □

Remark 3.4 (The surface measure $\sigma_{\Sigma(z)}$). *The surface measure $\sigma_{\Sigma(z)}$ on the submanifold $\Sigma(z) \subset \mathbb{R}^n$ may be defined as follows. For any open set B of $\Sigma(z)$, consider the open set (in $\mathbb{R}^n$)*

$$B^{\varepsilon} = \{q + \lambda n(q),\, q \in B, \lambda \in (-\varepsilon, \varepsilon)\}$$

where $n(q)$ is the unit normal to $\Sigma(z)$ at point q (for the usual Euclidean scalar product). Then $\sigma_{\Sigma(z)}(B) = \lim_{\varepsilon \to 0} \frac{|B^\varepsilon|}{2\varepsilon}$, where $|B^\varepsilon|$ here denotes the Lebesgue measure (in $\mathbb{R}^n$) of B^ε.

More rigorously, if $\varphi : U \to \mathbb{R}^n$ denotes a (local) parametrization of $\Sigma(z)$ (U is an open subset of $\mathbb{R}^{n-m}$), then, for any $B \subset \varphi(U)$, $\sigma_{\Sigma(z)}(B) = \displaystyle\int_{\varphi^{-1}(B)} \left(\det(\nabla\varphi \, \nabla\varphi^T)\right)^{1/2}(x)\, dx$, where dx denotes the Lebesgue measure in $\mathbb{R}^{n-m}$ and $(\nabla\varphi)_{i,j} = \left(\frac{\partial \varphi_j}{\partial q_i}\right)$ is the Jacobian matrix (in $\mathbb{R}^{(n-m)\times n}$) of φ.

All this holds of course under regularity assumptions on $\Sigma(z)$ ($\Sigma(z)$ should be C^1). We refer for example to Section VII.5.4 in [Laudenbach (2001)].

Let us now prove Lemma 3.2.

Proof. Classical proofs of the co-area formula can be found in the reference textbooks [Ambrosio *et al.* (2000); Evans and Gariepy (1992)]. These proofs are however quite involved since they assume only Lipschitz regularity for ξ. For the sake of completeness, we give here an elementary proof in the case of a smooth ξ.

For any $q \in \mathbb{R}^n$, we introduce the matrices $\overline{A}(q)$ and $\widetilde{A}(q)$, respectively with dimensions $m \times m$ and $m \times (n-m)$ such that $(\nabla\xi)^T = \left[\overline{A}\middle|\widetilde{A}\right]$. It holds $G = (\nabla\xi)^T\nabla\xi = \overline{A}\,\overline{A}^T + \widetilde{A}\widetilde{A}^T$. Note that, using a partition of unity, it is sufficient to prove (3.12) for f with support in an open set such that $\det\overline{A} \neq 0$ (since $\mathrm{rank}(\nabla\xi) = m$). We can therefore suppose without loss of generality that the first m columns of the matrix $(\nabla\xi)^T$ are linearly independent. Let us introduce the global change of variable $\Phi(q) = (\xi(q), q_{m+1}, \ldots, q_n)$. We use the notation $z = \xi(q)$ and $y = (q_{m+1}, \ldots, q_n)$. Note that

$$(\nabla\Phi)^T = \left[\begin{array}{c|c} \overline{A} & \widetilde{A} \\ \hline 0 & \mathrm{Id}_{n-m} \end{array}\right], \text{ and } (\nabla\Phi)^{-T} = \left((\nabla\Phi)^T\right)^{-1} = \left[\begin{array}{c|c} \overline{A}^{-1} & -\overline{A}^{-1}\widetilde{A} \\ \hline 0 & \mathrm{Id}_{n-m} \end{array}\right],$$

where Id_k denotes the $k \times k$ identity matrix. This implies that

$$\left|\mathrm{Jac}(\Phi^{-1})\right| = \left|\det\overline{A}^{-1}\right| \circ \Phi^{-1}.$$

Therefore, using the change of variable $(z,y) = \Phi(q)$ and Fubini theorem, it holds:

$$\int_{\mathbb{R}^n} f(\det G)^{1/2} \tag{3.17}$$
$$= \int_{\mathbb{R}^m} \int_{\mathbb{R}^{n-m}} \left(f \left[\det \left(\overline{A}\,\overline{A}^T + \widetilde{A}\widetilde{A}^T \right) \right]^{1/2} \left| \det \overline{A}^{-1} \right| \right) \circ \Phi^{-1}(z,y)\, dy\, dz.$$

Besides, for a fixed $z \in \mathbb{R}^m$, by definition of the Lebesgue measure on $\Sigma(z)$ (see Remark 3.4) and since $y \mapsto \Phi^{-1}(z,y)$ is a (local) parametrization of $\Sigma(z)$,

$$\int f d\sigma_{\Sigma(z)} \tag{3.18}$$
$$= \int_{\mathbb{R}^{n-m}} f \circ \Phi^{-1}(z,y) \left(\det \left(\nabla \Phi^{-1}(z,.)(\nabla \Phi^{-1}(z,.))^T \right) \right)^{1/2} (y)\, dy.$$

Note that (for a fixed $z \in \mathbb{R}^m$), $\nabla \Phi^{-T}(z,.) = \left[\frac{-\overline{A}^{-1}\widetilde{A}}{\mathrm{Id}_{n-m}} \right] \circ \Phi^{-1}(z,.)$ (the gradient being only with respect to the y variable) so that

$$\nabla \Phi^{-1}(z,.)(\nabla \Phi^{-1}(z,.))^T = \left(\mathrm{Id}_{n-m} + \widetilde{A}^T \overline{A}^{-T} \overline{A}^{-1} \widetilde{A} \right) \circ \Phi^{-1}(z,.).$$

Thus, considering (3.17) and (3.18), the proof of the co-area formula is completed provided that

$$\det \left(\overline{A}\,\overline{A}^T + \widetilde{A}\widetilde{A}^T \right) \left(\det \overline{A}^{-1} \right)^2 = \det \left(\mathrm{Id}_{n-m} + \widetilde{A}^T \overline{A}^{-T} \overline{A}^{-1} \widetilde{A} \right)$$

which is equivalent to showing

$$\det \left(\mathrm{Id}_m + BB^T \right) = \det \left(\mathrm{Id}_{n-m} + B^T B \right), \tag{3.19}$$

where B denotes the $m \times (n-m)$ matrix:

$$B = \overline{A}^{-1} \widetilde{A}.$$

The identity (3.19) is a consequence of Lemma 3.5 below, which thus concludes the proof. □

Lemma 3.5. *Let $A \in \mathbb{R}^{k \times l}$ and $B \in \mathbb{R}^{l \times k}$ denote two matrices. Then*

$$\det(\mathrm{Id}_k - AB) = \det(\mathrm{Id}_l - BA).$$

Proof. The proof relies on the two identities:

$$\left[\begin{array}{c|c} \mathrm{Id}_k & A \\ \hline B & \mathrm{Id}_l \end{array} \right] \left[\begin{array}{c|c} \mathrm{Id}_k & -A \\ \hline 0 & \mathrm{Id}_l \end{array} \right] = \left[\begin{array}{c|c} \mathrm{Id}_k & 0 \\ \hline B & \mathrm{Id}_l - BA \end{array} \right] \tag{3.20}$$

and

$$\left[\begin{array}{c|c} \mathrm{Id}_k & A \\ \hline B & \mathrm{Id}_l \end{array}\right] \left[\begin{array}{c|c} \mathrm{Id}_k & 0 \\ \hline -B & \mathrm{Id}_l \end{array}\right] = \left[\begin{array}{c|c} \mathrm{Id}_k - AB & A \\ \hline 0 & \mathrm{Id}_l \end{array}\right]. \tag{3.21}$$

Now, note that the left-hand sides of (3.20) and (3.21) have the same determinant. The proof then follows on noticing that the determinant of the right-hand side of (3.20) (respectively of (3.21)) is $\det(\mathrm{Id}_l - BA)$ (respectively $\det(\mathrm{Id}_k - AB)$). □

Remark 3.6 (On the notation δ). *In the mathematical literature, the notation δ_Σ (where Σ is a smooth submanifold in $\mathbb{R}^n$) is sometimes used to denote the Dirac distribution defined by: for any smooth test function ϕ,*

$$\langle \delta_\Sigma, \phi \rangle = \int_\Sigma \phi \, d\sigma_\Sigma,$$

where σ_Σ is the surface measure and $\langle \cdot, \cdot \rangle$ here denotes the distribution bracket. The factor $(\det G)^{-1/2}$ does not appear in this formula (compare with (3.14)). The Dirac distribution δ_Σ should not be confused with the notation $\delta_{\xi(q)-z}(dq)$. The Dirac distribution is typically introduced as the derivative of a characteristic function. Namely, for any smooth domain $\Omega \in \mathbb{R}^n$ with boundary $\partial\Omega = \Sigma$ and unit outward normal n, the distribution δ_Σ satisfies: for any smooth test function ϕ,

$$\langle \delta_\Sigma, \phi \cdot n \rangle = \int_\Omega \mathrm{div}(\phi),$$

which can be rewritten (in the sense of distributions):

$$\nabla 1_\Omega = -n \, \delta_\Sigma,$$

where 1_Ω denotes the characteristic function of the domain Ω.

3.2.1.2 *The free energy*

We can now introduce the *free energy* F associated with the reaction coordinate ξ, which may be seen as an effective energy for the coarse-grained variable $\xi(q)$.

Definition 3.7 (Free energy). *The free energy F is defined by*

$$\mathrm{e}^{-\beta F(z)} \, dz = \nu^\xi(dz),$$

or equivalently

$$\begin{aligned} F(z) &= -\beta^{-1} \ln \left(\int_{\Sigma(z)} Z_\nu^{-1} \mathrm{e}^{-\beta V} \, (\det G)^{-1/2} \, d\sigma_{\Sigma(z)} \right) \\ &= -\beta^{-1} \ln \left(\int_{\Sigma(z)} Z_\nu^{-1} \mathrm{e}^{-\beta V} \, \delta_{\xi(q)-z}(dq) \right). \end{aligned} \tag{3.22}$$

The free energy is therefore the logarithm of the partition function associated with the conditional probability measure $\nu^\xi(\cdot|z)$ (up to the multiplicative factor $-\beta^{-1}$).

Let us recall that the free energy is actually defined up to an additive constant (like the potential function V, for example), so that only differences of the free energy $F(z)-F(z_0)$ make sense, for a given reference value of the reaction coordinate z_0.

Remark 3.8 (The alchemical case). *The so-called alchemical case considered in Section 3.1 can be recovered in the case when $\xi(q) = q_1$, the alchemical parameter λ being then the first component q_1 of the position vector q, see also Section 1.3.2.*

3.2.1.3 *The case of a non-standard scalar product*

In Section 3.3, a generalization of the co-area formula when the ambient space $\mathbb{R}^n$ is equipped with a non-Euclidean scalar product is needed. Let us explain how the co-area formula (3.12) writes in this context.

Assume that $\mathbb{R}^n$ has a general Euclidean structure defined by the following scalar product (instead of the standard Euclidean scalar product $q_1^T q_2$): for two vectors $q_1, q_2 \in \mathbb{R}^n$,

$$\langle q_1, q_2 \rangle_M = q_1^T M q_2, \tag{3.23}$$

where M is a given $n \times n$ positive definite symmetric matrix which is supposed to be constant for simplicity. This new structure has three consequences. First, the gradient[1] on $\mathbb{R}^n$ is $\nabla_M = M^{-1}\nabla$, so that the Gram matrix becomes:

$$G_M = \langle \nabla_M \xi, \nabla_M \xi \rangle = \nabla \xi^T M^{-1} \nabla \xi.$$

Second, the surface measure on $\Sigma(z)$ is now defined as follows (using the notation of Remark 3.4 for a parametrization of $\Sigma(z)$): for a local parametrization $\varphi : U \to \mathbb{R}^n$ ($U \subset \mathbb{R}^{n-m}$),

$$\sigma^M_{\Sigma(z)}(B) = \int_{\varphi^{-1}(B)} \left(\det(\nabla\varphi M \nabla\varphi^T)\right)^{1/2}(x)\, dx.$$

Third, the volume element on $\mathbb{R}^n$ is $(\det M)^{1/2} dq$.

[1]Let us recall that, for a given scalar product $\langle \cdot, \cdot \rangle$, the gradient $\mathrm{grad}(\phi) : \mathbb{R}^n \to \mathbb{R}^n$ of a function $\phi : \mathbb{R}^n \to \mathbb{R}$ is defined by the formula: for any $q \in \mathbb{R}^n$ and any $\delta q \in \mathbb{R}^n$, $\langle \mathrm{grad}\phi(q), \delta q \rangle = \lim_{\varepsilon \to 0} \frac{\phi(q+\varepsilon\delta q)-\phi(q)}{\varepsilon}$. The notation ∇ denotes the usual gradient associated with the Euclidean scalar product.

Thus, in this new setting, the co-area formula writes: for any smooth function $f : \mathbb{R}^n \to \mathbb{R}$,

$$\int_{\mathbb{R}^n} f\,(\det G_M)^{1/2}\,(\det M)^{1/2} dq = \int_{\mathbb{R}^m} \int_{\Sigma(z)} f d\sigma^M_{\Sigma(z)}\, dz. \tag{3.24}$$

In view of (3.24) and (3.11), we have:

$$\delta_{\xi(q)-z}(dq) = (\det M)^{-1/2} (\det G_M)^{-1/2} d\sigma^M_{\Sigma(z)}. \tag{3.25}$$

Note that the right-hand side does not depend on M, since $\delta_{\xi(q)-z}(dq)$ is defined independently of M.

Formula (3.24) can be checked by rewriting the proof of Lemma 3.2 (presented below) with the new scalar product. Alternatively, a simple way to obtain (3.24) is to start from the classical co-area formula (3.12) and use the change of variable $\tilde{q} = M^{-1/2} q$. Note first that for two vectors,

$$q_1^T q_2 = \tilde{q}_1^T M \tilde{q}_2$$

which is consistent with the scalar product (3.23).

Now, the left-hand side of the classical co-area formula:

$$\int_{\mathbb{R}^n} f (\det G)^{1/2} dq = \int_{\mathbb{R}^m} \int_{\Sigma(z)} f\, d\sigma_{\Sigma(z)}\, dz$$

can be rewritten in the new coordinates as:

$$\begin{aligned}\int_{\mathbb{R}^n} f(q)(\det G)^{1/2}(q)\, dq &= \int_{\mathbb{R}^n} f(M^{1/2}\tilde{q})(\det G)^{1/2}(M^{1/2}\tilde{q})(\det M)^{1/2}\, d\tilde{q} \\ &= \int_{\mathbb{R}^n} \tilde{f}(\tilde{q})(\det \tilde{G})^{1/2}(\tilde{q})(\det M)^{1/2}\, d\tilde{q},\end{aligned}$$

with $\tilde{f}(\tilde{q}) = f(q)$, $G = \nabla\xi^T \nabla\xi$, $\tilde{\xi}(\tilde{q}) = \xi(q)$, $\nabla_{\tilde{q}}\tilde{\xi}(\tilde{q}) = M^{1/2}\nabla\xi(q)$, $\tilde{G}(\tilde{q}) = G(q)$. Thus, since $G(q) = \nabla_q \xi^T \nabla_q \xi(q) = \nabla_{\tilde{q}}\tilde{\xi}^T M^{-1} \nabla_{\tilde{q}}\tilde{\xi}(\tilde{q})$, it follows

$$\tilde{G} = \nabla_{\tilde{q}}\tilde{\xi}^T M^{-1} \nabla_{\tilde{q}}\tilde{\xi},$$

which is consistent with the above definition of G_M. For the right-hand side, using again the notation of Remark 3.4, it holds:

$$\begin{aligned}\int_{\Sigma(z)} f\, d\sigma_{\Sigma(z)} &= \int_U f \circ \varphi(x)(\det(\nabla_x \varphi \nabla_x \varphi^T)^{1/2})(x)\, dx \\ &= \int_U f(M^{1/2}\tilde{\varphi}(x))(\det(\nabla_x \tilde{\varphi} M \nabla_x \tilde{\varphi}^T)^{1/2})(x)\, dx \\ &= \int_{\Sigma(z)} \tilde{f}\, d\tilde{\sigma}_{\Sigma(z)},\end{aligned}$$

where $\tilde{\varphi} = M^{-1/2}\varphi$ (and thus $\nabla_x \tilde{\varphi} = \nabla_x \varphi M^{-1/2}$), and $\tilde{\sigma}_{\Sigma(z)}$ has a consistent definition with $\sigma^M_{\Sigma(z)}$ above. This yields (3.24).

3.2.2 *The mean force*

As already mentioned in Section 3.1, the thermodynamic integration method is based on the formula $F(1) - F(0) = \int_0^1 F'(z)\,dz$, at least in the case of a one-dimensional reaction coordinate ($m = 1$). More generally, the free energy difference between two values z_0 and z_1 of the reaction coordinate is computed as

$$F(z_1) - F(z_0) = \int_0^1 \nabla F(\phi(\lambda)) \cdot \phi'(\lambda)\,d\lambda \tag{3.26}$$

where $\phi : [0,1] \to \mathbb{R}^m$ denotes a smooth path in the reaction coordinate space such that $\phi(0) = z_0$ and $\phi(1) = z_1$. The computation of this integral requires the calculation of the so-called *mean force* ∇F.

The following lemma gives a fundamental formula for computing $\nabla F(z)$ (see [Sprik and Ciccoti (1998); Ciccotti *et al.* (2005); den Otter and Briels (1998)]):

Lemma 3.9. *The mean force is the conditional canonical average*

$$\nabla F(z) = \int_{\Sigma(z)} f\,d\nu^{\xi}(\cdot|z), \tag{3.27}$$

where the so-called local mean force f *is a vector with components* $(f_\alpha)_{1\le\alpha\le m}$*:*

$$f_\alpha = \sum_{\gamma=1}^{m} G_{\alpha,\gamma}^{-1} \nabla\xi_\gamma \cdot \nabla V - \beta^{-1} \mathrm{div}\left(\sum_{\gamma=1}^{m} G_{\alpha,\gamma}^{-1} \nabla\xi_\gamma\right), \tag{3.28}$$

where $G_{\alpha,\gamma}^{-1}$ *denotes the* (α,γ)*-component of the inverse of the matrix* G.

The local mean force is the difference of two terms: The first one can be seen as the force exerted on the system along the reaction coordinate, and the second one is related to the curvature of the submanifolds (see Formula (3.82) below). In general, the second term is smaller than the first one, because of the β^{-1} factor, but it may also have an important contribution for some reaction coordinate (think for example of the case $\xi = V$, for which the first term is constant).

In the literature, what is called the mean force is sometimes the opposite of (3.27), in accordance with the convention that a force is the opposite of the gradient of the associated potential. In this monograph, ∇F will be called the mean force.

The proof of Lemma 3.9 is based on the following general differentiation formula:

Lemma 3.10. *Let $\psi : \mathbb{R}^n \to \mathbb{R}$ be a smooth function, and define*

$$\psi^\xi(z) = \int_{\Sigma(z)} \psi(\det G)^{-1/2} d\sigma_{\Sigma(z)} = \int \psi(q)\delta_{\xi(q)-z}(dq).$$

The derivative of ψ^ξ reads: $\forall \alpha \in \{1, \dots, m\}$,

$$\partial_{z_\alpha}\psi^\xi(z) = \int_{\Sigma(z)} \sum_{\gamma=1}^{m} \left(G^{-1}_{\alpha,\gamma}\nabla\xi_\gamma \cdot \nabla\psi + \mathrm{div}\left(G^{-1}_{\alpha,\gamma}\nabla\xi_\gamma\right)\psi\right)(\det G)^{-1/2} d\sigma_{\Sigma(z)}.$$

Proof. For a fixed $\alpha \in \{1, \dots, m\}$ and any smooth test function $g : \mathbb{R}^m \to \mathbb{R}$, the co-area formula (3.12) together with an integration by parts gives[2]:

$$\begin{aligned} \int_{\mathbb{R}^m} g\,\partial_{z_\alpha}\psi^\xi &= -\int_{\mathbb{R}^m} \psi^\xi\,\partial_{z_\alpha} g \\ &= -\int_{\mathbb{R}^n} \psi\left((\partial_{z_\alpha} g)\circ\xi\right). \end{aligned}$$

Using the fact that $(\partial_{z_\alpha} g)\circ\xi = G^{-1}_{\alpha,\gamma}\nabla\xi_\gamma \cdot \nabla(g\circ\xi)$, it holds

$$\begin{aligned} \int_{\mathbb{R}^m} g\,\partial_{z_\alpha}\psi^\xi &= -\int_{\mathbb{R}^n} \psi\, G^{-1}_{\alpha,\gamma}\nabla\xi_\gamma \cdot \nabla(g\circ\xi) \\ &= \int_{\mathbb{R}^n} \mathrm{div}\left(\psi\, G^{-1}_{\alpha,\gamma}\nabla\xi_\gamma\right) g\circ\xi \\ &= \int_{\mathbb{R}^m} g(z) \int_{\Sigma(z)} \mathrm{div}\left(\psi\, G^{-1}_{\alpha,\gamma}\nabla\xi_\gamma\right)(\det G)^{-1/2} d\sigma_{\Sigma(z)}\, dz, \end{aligned}$$

and the result follows from the identity:

$$\mathrm{div}\left(\psi\, G^{-1}_{\alpha,\gamma}\nabla\xi_\gamma\right) = G^{-1}_{\alpha,\gamma}\nabla\xi_\gamma \cdot \nabla\psi + \mathrm{div}\left(G^{-1}_{\alpha,\gamma}\nabla\xi_\gamma\right)\psi.$$

□

We are now in position to prove Lemma 3.9:

Proof. For any $\alpha \in \{1, \dots, m\}$, the definition (3.22) of the free energy gives

$$\partial_{z_\alpha} F(z) = -\beta^{-1} \frac{\partial_{z_\alpha}\left(\displaystyle\int_{\Sigma(z)} e^{-\beta V}(\det G)^{-1/2}\, d\sigma_{\Sigma(z)}\right)}{\displaystyle\int_{\Sigma(z)} e^{-\beta V}(\det G)^{-1/2}\, d\sigma_{\Sigma(z)}}.$$

[2] In all the following proofs, we use the summation convention on repeated Greek indices going from 1 to m.

Now, using Lemma 3.10 with $\psi = \mathrm{e}^{-\beta V}$, the numerator reads

$$\begin{aligned}
&\partial_{z_\alpha}\left(\int_{\Sigma(z)} \mathrm{e}^{-\beta V} (\det G)^{-1/2}\, d\sigma_{\Sigma(z)}\right) \\
&= \int_{\Sigma(z)} \sum_{\gamma=1}^{m} \left(G^{-1}_{\alpha,\gamma}\nabla\xi_\gamma \cdot \nabla \mathrm{e}^{-\beta V} + \mathrm{div}\left(G^{-1}_{\alpha,\gamma}\nabla\xi_\gamma\right)\mathrm{e}^{-\beta V}\right)(\det G)^{-1/2} d\sigma_{\Sigma(z)} \\
&= -\beta \int_{\Sigma(z)} f_\alpha\, \mathrm{e}^{-\beta V}(\det G)^{-1/2} d\sigma_{\Sigma(z)},
\end{aligned}$$

which yields the result, in view of the expression (3.16) of the measure $d\nu^\xi(\cdot|z)$. □

Remark 3.11 (More general formulas for the mean force).
As noted in Section 4.4 *in [Chipot and Pohorille (2007b)] (see also Equation* (17) *in [den Otter (2000)] and Equation* (4) *in [Ciccotti* et al. *(2005)]), there exist generalizations of formula* (3.27) *for the mean force. Indeed, for any smooth function $W : \mathbb{R}^n \to \mathbb{R}^{n\times m}$ such that the $m \times m$ matrix G^W defined by*

$$(G^W)_{\alpha,\beta} = W_\alpha \cdot \nabla\xi_\beta$$

is invertible ($W_\alpha \in \mathbb{R}^n$ being the α-th column of W), another expression for the derivative of ψ^ξ in Lemma 3.10 *is:*

$$\partial_{z_\alpha}\psi^\xi(z) = \int_{\Sigma(z)} \sum_{\gamma=1}^{m} \left((G^W)^{-1}_{\alpha,\gamma} W_\gamma \cdot \nabla\psi + \mathrm{div}\left((G^W)^{-1}_{\alpha,\gamma} W_\gamma\right)\psi\right)(\det G)^{-1/2} d\sigma_{\Sigma(z)},$$

where $(G^W)^{-1}_{\alpha,\gamma}$ denotes the (α,γ)-component of the inverse of the matrix G^W. Using this formula, the mean force writes:

$$\nabla F(z) = \int_{\Sigma(z)} f^W d\nu^\xi(\cdot|z), \tag{3.29}$$

where f^W is defined by: $\forall \alpha \in \{1,\ldots,m\}$,

$$f^W_\alpha = \sum_{\gamma=1}^{m} (G^W)^{-1}_{\alpha,\gamma} W_\gamma \cdot \nabla V - \beta^{-1}\mathrm{div}\left(\sum_{\gamma=1}^{m}(G^W)^{-1}_{\alpha,\gamma} W_\gamma\right). \tag{3.30}$$

Formula (3.27) *is obtained for the specific choice $W = \nabla\xi$. The interest of* (3.29)–(3.30) *is that the function W may be seen as a parameter to be tuned to enhance the numerical methods. For example, it is used in some specific situations to simplify the computation of the local mean force (in particular when the reaction coordinate involves some degrees of freedom on which molecular constraints are applied).*

In view of (3.26) and (3.27), the algorithm to get an estimate of the free energy difference by the thermodynamic integration method is clear: Discretize the integral in (3.26) and estimate ∇F using (3.27) by sampling the conditional measures $\nu^\xi(\cdot|z)$. What makes formula (3.27) useful for computational purposes is that ∇F is expressed as an average of f with respect to the measure $\nu^\xi(\cdot|z)$. The only remaining question is then how to sample the measure $\nu^\xi(\cdot|z)$.

Note that the conditional measure $\nu^\xi(\cdot|z)$ can be written as

$$d\nu^\xi(\cdot|z) = \frac{\mathrm{e}^{-\beta V^\xi}\, d\sigma_{\Sigma(z)}}{\int_{\Sigma(z)} \mathrm{e}^{-\beta V^\xi}\, d\sigma_{\Sigma(z)}}, \tag{3.31}$$

where the modified potential V^ξ is defined by

$$V^\xi = V + \frac{1}{2\beta} \ln(\det G). \tag{3.32}$$

The question is then: How to sample a measure with density $\mathrm{e}^{-\beta V^\xi}$ with respect to the Lebesgue measure on $\Sigma(z)$? The aim of the next two sections, 3.2.3 and 3.2.4, is to give an answer to this question.

We close this section with three remarks.

Remark 3.12 (One-dimensional reaction coordinate). *In the codimension* 1 *situation* ($m = 1$), *formulas are simpler. The local mean force is*

$$f = \frac{\nabla\xi \cdot \nabla V}{|\nabla\xi|^2} - \beta^{-1}\mathrm{div}\left(\frac{\nabla\xi}{|\nabla\xi|^2}\right).$$

And the modified potential V^ξ is:

$$V^\xi = V + \beta^{-1} \ln|\nabla\xi|.$$

Remark 3.13 (Computing canonical averages by conditioning). *Let us anticipate on the following sections and assume that we have a method to sample the probability measures $\nu^\xi(\cdot|z)$. Then, a method to compute the mean of any functional ϕ with respect to the Boltzmann-Gibbs measure ν is to use the conditioning formula* (3.10)*:*

$$\int_{\mathcal{D}} \phi\, d\nu = \frac{\displaystyle\int_{\mathbb{R}^m}\int_{\Sigma(z)} \phi\, d\nu^\xi(\cdot|z) \exp(-\beta F(z))\, dz}{\displaystyle\int_{\mathbb{R}^m} \exp(-\beta F(z))\, dz}. \tag{3.33}$$

The free energy F is obtained from ∇F (see Remark 3.14 *below) up to an additive constant which does not intervene in* (3.33)*, and, as mentioned* *above, the computation of* ∇F *can be done by sampling the measures* $\nu^\xi(\cdot|z)$. *Likewise, the computation of the integrals* $\int_{\Sigma(z)} \phi \, d\nu^\xi(\cdot|z)$ *again requires the sampling of the measures* $\nu^\xi(\cdot|z)$. *Thus, under the assumption that* m *is sufficiently small so that direct numerical integrations with respect to the reaction coordinate value* z *is possible (say,* $1 \leq m \leq 4$*), we have a method to compute the average of any functional* ϕ *with respect to* ν. *The efficiency of such a method, compared to a direct sampling of the canonical measure* ν *by methods such as those presented in Chapter* 2, *is of course related to the efficiency of the sampling procedure for the measures* $\nu^\xi(\cdot|z)$, *compared to the sampling procedure for the measure* ν. *It is clear that this efficiency crucially depends on the choice of the reaction coordinate* ξ *(see Section* 3.2.6 *for more details).*

Remark 3.14 (Reconstructing free energies from mean forces). *Let us assume that* $m \geq 2$. *In practice, based on the formula* (3.27), *approximations of* ∇F *are obtained typically at some points of a grid of the reaction coordinate space. A natural question is then how to reconstruct the whole free energy profile:* $z \mapsto F(z)$ *(up to an additive constant). Let us assume that an approximation* U *of* ∇F *is known on some bounded open set* $\mathcal{M}$ *of* $\mathbb{R}^m$. *Note that in practice,* U *is not even a gradient in general (due to sampling and interpolation errors).*

A natural algorithm to reconstruct F *from* U *consists in solving the Poisson problem*

$$-\Delta F = -\mathrm{div}(U) \text{ on } \mathcal{M}, \tag{3.34}$$

with appropriate boundary conditions, typically Neumann boundary conditions

$$\frac{\partial F}{\partial n} = U \cdot n \text{ on } \partial\mathcal{M},$$

where n *denotes the unit outward normal to* $\mathcal{M}$. *Periodic boundary conditions could also be used if some components of* ξ *have values in a periodic domain (angles for example). To solve this Poisson problem, standard methods such as finite element methods, finite difference methods or fast Fourier transform for example, may be used. The solution* F *is defined up to an additive constant, which may be imposed in practice by assuming that* $\int_{\mathcal{M}} F = 0$, *for example.*

Note that Equation (3.34) *is the Euler Lagrange equation associated with the minimization problem:*

$$F = \operatorname*{argmin}_{G \in H^1(\mathcal{M})} \int_{\mathcal{M}} |\nabla G - U|^2$$

so that F can be interpreted as the function such that its gradient is the closest to U.

In dimension $m = 2$ or $m = 3$, solving (3.34) amounts to computing the so-called Helmholtz (or Hodge) decomposition of the vector field U as:

$$U = \nabla F + \operatorname{curl} W,$$

where, if $m = 3$, W is a vector field and $\operatorname{curl} W = \nabla \times W$, *and if $m = 2$, W has values in $\mathbb{R}$ and* $\operatorname{curl} W = (\partial_y W, -\partial_x W)$.

3.2.3 *Sampling measures on submanifolds of* $\mathbb{R}^n$

The aim of this section is to explain how to sample a measure with support on a submanifold of $\mathbb{R}^n$, with application to the sampling of the conditional measure $\nu^\xi(\cdot|z)$ defined by (3.16). In view of (3.31)–(3.32), this amounts to answering the following question: How to sample a measure

$$d\tilde{\nu}_\Sigma = \frac{e^{-\beta\tilde{V}} d\sigma_\Sigma}{\displaystyle\int_\Sigma e^{-\beta\tilde{V}} d\sigma_\Sigma} \tag{3.35}$$

with support a smooth submanifold $\Sigma = \Sigma(0) = \{q \in \mathbb{R}^n, \, \xi(q) = 0\}$ of $\mathbb{R}^n$ which is defined as the zero level set of a smooth function $\xi : \mathbb{R}^n \to \mathbb{R}^m$. We denote $\Sigma = \Sigma(0)$ in this section for the ease of notation.

Note that we may assume without loss of generality that $z = 0$ in (3.31)–(3.32) (all the formulas given in this section (3.2.3) and in the next section (3.2.4) remain the same whatever z), and that V^ξ is actually any potential $\tilde{V} : \mathbb{R}^n \to \mathbb{R}$ (such that $\int_\Sigma e^{-\beta\tilde{V}} d\sigma_\Sigma < \infty$).

In this section, a (continuous in time) dynamics which is ergodic with respect to $\tilde{\nu}_\Sigma$ is derived. The numerical discretization of this dynamics is the subject of the next Section 3.2.4.

One idea to sample the measure $\tilde{\nu}_\Sigma$ is to use a projection on Σ of the simple gradient (or overdamped Langevin) dynamics:

$$dq_t = -\nabla\tilde{V}(q_t)\, dt + \sqrt{2\beta^{-1}}\, dW_t, \tag{3.36}$$

which is known to sample the Boltzmann-Gibbs measure $\dfrac{e^{-\beta\tilde{V}(q)}\, dq}{\int_{\mathbb{R}^n} e^{-\beta\tilde{V}}}$. Of course, there are many ways to project such a dynamics on a submanifold, and we choose a projected dynamics which indeed samples $\tilde{\nu}_\Sigma$ and admits a very natural discretization (see Section 3.2.4.2).

3.2.3.1 *Geometrical notation*

It is convenient to introduce the orthogonal projection operator[3] at a point $q \in \Sigma$ onto the tangent space $T_q\Sigma$ of Σ:

$$P(q) = \mathrm{Id} - \sum_{\alpha,\zeta=1}^{m} G^{-1}_{\alpha,\zeta}(q)\nabla\xi_\alpha(q) \otimes \nabla\xi_\zeta(q). \tag{3.37}$$

To check that $P(q)$ is indeed the orthogonal projector onto $T_q\Sigma$, note that for any $1 \leq \gamma \leq m$,

$$\begin{aligned} P\nabla\xi_\gamma &= \nabla\xi_\gamma - \sum_{\alpha,\zeta=1}^{m} G^{-1}_{\alpha,\zeta}\nabla\xi_\alpha\nabla\xi_\zeta \cdot \nabla\xi_\gamma \\ &= \nabla\xi_\gamma - \sum_{\alpha,\zeta=1}^{m} G^{-1}_{\alpha,\zeta}G_{\zeta,\gamma}\nabla\xi_\alpha = 0, \end{aligned}$$

while, for any vector u such that $\forall 1 \leq \zeta \leq m$, $u \cdot \nabla\xi_\zeta = 0$, we have

$$Pu = u - \sum_{\alpha,\zeta=1}^{m} G^{-1}_{\alpha,\zeta}\nabla\xi_\alpha\nabla\xi_\zeta \cdot u = u.$$

Since $P(q)$ is an orthogonal projector, it is an idempotent symmetric matrix:

$$P(x)P(x) = P(x) \text{ and } P(x)^T = P(x).$$

Note also that in the special case of orthogonal constraints, namely if $\nabla\xi_\alpha \cdot \nabla\xi_\zeta = \delta_{\alpha,\zeta}|\nabla\xi_\alpha|^2$, (3.37) simplifies into:

$$P(q) = \mathrm{Id} - \sum_{\alpha=1}^{m} n_\alpha(q) \otimes n_\alpha(q), \tag{3.38}$$

where the normal n_α is defined by

$$n_\alpha(q) = \frac{\nabla\xi_\alpha(q)}{|\nabla\xi_\alpha(q)|}, \qquad 1 \leq \alpha \leq m. \tag{3.39}$$

To define the projected dynamics, we also need to introduce the mean curvature vector of the submanifold Σ (see also Remark 3.17 below).

Lemma 3.15. *The mean curvature vector $\mathcal{H}(q) \in \mathbb{R}^n$ at point q of the submanifold Σ is:*

$$\mathcal{H} = -\sum_{\alpha=1}^{m} \kappa_\alpha n_\alpha, \tag{3.40}$$

[3]Note that for any $z \in \mathbb{R}^m$ (and not only $z = 0$) and for any point $q \in \Sigma(z)$, $P(q)$ is actually the orthogonal projection operator onto the tangent space $T_q\Sigma(z)$ to $\Sigma(z)$ at a point q.

where the curvature κ_α along the normal n_α (given by (3.39)) is defined by

$$\kappa_\alpha = |\nabla\xi_\alpha| \sum_{\gamma=1}^{m} G^{-1}_{\gamma,\alpha} \left(\Delta\xi_\gamma - \nabla^2\xi_\gamma : \left(\sum_{\zeta,\delta=1}^{m} G^{-1}_{\delta,\zeta} \nabla\xi_\delta \otimes \nabla\xi_\zeta \right) \right). \quad (3.41)$$

Other expressions for $\mathcal{H}$ in terms of the projection operator P are:

$$\mathcal{H} = \left(\sum_{j,k=1}^{n} P_{j,k} \nabla_j P_{i,k} \right)_{1 \le i \le n} \quad (3.42)$$

$$= \left((\det G)^{-1/2} \sum_{j=1}^{n} \nabla_j \left[(\det G)^{1/2} P_{i,j} \right] \right)_{1 \le i \le n}. \quad (3.43)$$

Finally, the curvature κ_α also writes:

$$\kappa_\alpha = |\nabla\xi_\alpha| (\det G)^{-1/2} \mathrm{div} \left((\det G)^{1/2} \sum_{\gamma=1}^{m} G^{-1}_{\alpha,\gamma} \nabla\xi_\gamma \right). \quad (3.44)$$

Note that since the vector $\mathcal{H}$ is orthogonal to the submanifold Σ, it holds

$$(\mathrm{Id} - P(q))\mathcal{H}(q) = \mathcal{H}(q). \quad (3.45)$$

Another formula which will be useful in the following is:

$$-\nabla\xi_\gamma \cdot \mathcal{H} = P : \nabla^2\xi_\gamma. \quad (3.46)$$

This is a direct consequence of (3.40) and (3.41):

$$\begin{aligned}
-\nabla\xi_\gamma \cdot \mathcal{H} &= \kappa_\alpha \nabla\xi_\gamma \cdot n_\alpha \\
&= G_{\gamma,\alpha} |\nabla\xi_\alpha|^{-1} \kappa_\alpha \\
&= \Delta\xi_\gamma - \nabla^2\xi_\gamma : \left(\sum_{\zeta,\delta=1}^{m} G^{-1}_{\delta,\zeta} \nabla\xi_\delta \otimes \nabla\xi_\zeta \right) \\
&= P : \nabla^2\xi_\gamma.
\end{aligned}$$

Remark 3.16 (One-dimensional reaction coordinate). *In the codimension 1 situation ($m = 1$), formulas are simpler. The projection operator P simply writes*

$$P = \mathrm{Id} - n \otimes n$$

where $n = \dfrac{\nabla\xi}{|\nabla\xi|}$ and the mean curvature vector is $\mathcal{H} = -\kappa n$ with $\kappa =$ $\mathrm{div}(n)$.

Let us now prove Lemma 3.15.

Proof. Equations (3.40) and (3.41) define the mean curvature vector, and we need to prove that equations (3.42), (3.43) and (3.44) are indeed equivalent definitions. We will discuss in Remark 3.17 the link between these definitions and another one which can be found in other textbooks.

Let us start with (3.42). We have (using again the summation convention on repeated indices: for Greek letters between 1 and m, and for Latin letters between 1 and $n = 3N$):

$$\begin{aligned}
P_{j,k}\nabla_j P_{i,k} &= \nabla_j(P_{j,k}P_{i,k}) - P_{i,k}\nabla_j(P_{j,k}) \\
&= \nabla_j(P_{i,j}) - P_{i,k}\nabla_j(P_{j,k}) \\
&= (\delta_{i,k} - P_{i,k})\nabla_j(P_{j,k}) \\
&= -G^{-1}_{\alpha,\zeta}\nabla_i\xi_\alpha\nabla_k\xi_\zeta\nabla_j(G^{-1}_{\gamma,\delta}\nabla_j\xi_\gamma\nabla_k\xi_\delta) \\
&= -G^{-1}_{\alpha,\zeta}\nabla_i\xi_\alpha\nabla_k\xi_\zeta\Big(\nabla_j G^{-1}_{\gamma,\delta}\nabla_j\xi_\gamma\nabla_k\xi_\delta + G^{-1}_{\gamma,\delta}\Delta\xi_\gamma\nabla_k\xi_\delta \\
&\quad + G^{-1}_{\gamma,\delta}\nabla_j\xi_\gamma\nabla_j\nabla_k\xi_\delta\Big) \\
&= -\nabla_i\xi_\delta\nabla_j G^{-1}_{\gamma,\delta}\nabla_j\xi_\gamma - \nabla_i\xi_\delta G^{-1}_{\gamma,\delta}\Delta\xi_\gamma \\
&\quad - G^{-1}_{\alpha,\zeta}\nabla_i\xi_\alpha\nabla_k\xi_\zeta G^{-1}_{\gamma,\delta}\nabla_j\xi_\gamma\nabla_j\nabla_k\xi_\delta.
\end{aligned} \tag{3.47}$$

Let us now compute $\nabla_j(G^{-1}_{\gamma,\delta})$. We have

$$0 = \nabla_j(G_{\alpha,\zeta}G^{-1}_{\zeta,\delta}) = \nabla_j(G_{\alpha,\zeta})G^{-1}_{\zeta,\delta} + G_{\alpha,\zeta}\nabla_j(G^{-1}_{\zeta,\delta}),$$

so that

$$\begin{aligned}
\nabla_j(G^{-1}_{\gamma,\delta}) &= -G^{-1}_{\gamma,\alpha}\nabla_j(G_{\alpha,\zeta})G^{-1}_{\zeta,\delta} \\
&= -G^{-1}_{\gamma,\alpha}G^{-1}_{\zeta,\delta}\left(\nabla_j\nabla_k\xi_\alpha\nabla_k\xi_\zeta + \nabla_j\nabla_k\xi_\zeta\nabla_k\xi_\alpha\right).
\end{aligned} \tag{3.48}$$

Therefore, the first term in (3.47) is

$$\begin{aligned}
&-\nabla_i\xi_\delta\nabla_j G^{-1}_{\gamma,\delta}\nabla_j\xi_\gamma \\
&= \nabla_i\xi_\delta\nabla_j\xi_\gamma G^{-1}_{\gamma,\alpha}G^{-1}_{\zeta,\delta}\left(\nabla_j\nabla_k\xi_\alpha\nabla_k\xi_\zeta + \nabla_j\nabla_k\xi_\zeta\nabla_k\xi_\alpha\right) \\
&= \nabla_i\xi_\delta\nabla_j\xi_\gamma G^{-1}_{\gamma,\alpha}G^{-1}_{\zeta,\delta}\nabla_j\nabla_k\xi_\alpha\nabla_k\xi_\zeta + \nabla_i\xi_\delta\nabla_j\xi_\gamma G^{-1}_{\gamma,\alpha}G^{-1}_{\zeta,\delta}\nabla_j\nabla_k\xi_\zeta\nabla_k\xi_\alpha \\
&= \nabla_i\xi_\alpha\nabla_j\xi_\gamma G^{-1}_{\gamma,\delta}G^{-1}_{\zeta,\alpha}\nabla_j\nabla_k\xi_\delta\nabla_k\xi_\zeta + \nabla_i\xi_\delta\nabla_j\xi_\zeta G^{-1}_{\zeta,\alpha}G^{-1}_{\gamma,\delta}\nabla_j\nabla_k\xi_\gamma\nabla_k\xi_\alpha,
\end{aligned}$$

where, in the last line, we have swapped α and δ in the first term and we have swapped ζ and γ in the second term. Note now that the first term in the last line and the last term in (3.47) cancel, so that:

$$\begin{aligned}
P_{j,k}\nabla_j P_{i,k} &= \nabla_i\xi_\delta\, G^{-1}_{\gamma,\delta}\left(-\Delta\xi_\gamma + G^{-1}_{\zeta,\alpha}\nabla_j\xi_\zeta\nabla_j\nabla_k\xi_\gamma\nabla_k\xi_\alpha\right), \\
&= \nabla_i\xi_\delta\, G^{-1}_{\gamma,\delta}\left(-\Delta\xi_\gamma + \nabla^2\xi_\gamma : (G^{-1}_{\alpha,\zeta}\nabla\xi_\alpha \otimes \nabla\xi_\zeta)\right), \\
&= -\kappa_\delta|\nabla\xi_\delta|^{-1}\nabla_i\xi_\delta,
\end{aligned}$$

which proves (3.42).

Let us now consider (3.43). We will need the well-known Jacobi's formula: for a given tensor M,

$$\nabla \ln(\det M) = \sum_{\alpha,\zeta} M^{-1}_{\alpha,\zeta} \nabla M_{\zeta,\alpha}. \tag{3.49}$$

Using (3.49) and (3.48), we have:

$$\begin{aligned}
&(\det G)^{-1/2}\nabla_j((\det G)^{1/2}P_{i,j})\\
&= (\det G)^{-1/2}\nabla_j((\det G)^{1/2})P_{i,j} + \nabla_j P_{i,j}\\
&= \frac{1}{2}\nabla_j \ln(\det G)P_{i,j} - \nabla_j(G^{-1}_{\gamma,\delta}\nabla_i\xi_\gamma\nabla_j\xi_\delta),\\
&= \frac{1}{2}G^{-1}_{\alpha,\zeta}\nabla_j G_{\alpha,\zeta}P_{i,j} - \nabla_j G^{-1}_{\gamma,\delta}\nabla_i\xi_\gamma\nabla_j\xi_\delta\\
&\quad - G^{-1}_{\gamma,\delta}\nabla_j\nabla_i\xi_\gamma\nabla_j\xi_\delta - G^{-1}_{\gamma,\delta}\nabla_i\xi_\gamma\Delta\xi_\delta\\
&= \frac{1}{2}G^{-1}_{\alpha,\zeta}\nabla_i G_{\alpha,\zeta} - \frac{1}{2}G^{-1}_{\alpha,\zeta}\nabla_j G_{\alpha,\zeta}G^{-1}_{\gamma,\delta}\nabla_i\xi_\gamma\nabla_j\xi_\delta\\
&\quad + G^{-1}_{\gamma,\alpha}G^{-1}_{\zeta,\delta}\left(\nabla_j\nabla_k\xi_\alpha\nabla_k\xi_\zeta + \nabla_j\nabla_k\xi_\zeta\nabla_k\xi_\alpha\right)\nabla_i\xi_\gamma\nabla_j\xi_\delta\\
&\quad - G^{-1}_{\gamma,\delta}\nabla_j\nabla_i\xi_\gamma\nabla_j\xi_\delta - G^{-1}_{\gamma,\delta}\nabla_i\xi_\gamma\Delta\xi_\delta\\
&= G^{-1}_{\alpha,\zeta}\nabla_i\nabla_j\xi_\alpha\nabla_j\xi_\zeta - G^{-1}_{\alpha,\zeta}\nabla_j\nabla_k\xi_\alpha\nabla_k\xi_\zeta G^{-1}_{\gamma,\delta}\nabla_i\xi_\gamma\nabla_j\xi_\delta\\
&\quad + G^{-1}_{\gamma,\alpha}G^{-1}_{\zeta,\delta}\nabla_j\nabla_k\xi_\alpha\nabla_k\xi_\zeta\nabla_i\xi_\gamma\nabla_j\xi_\delta + G^{-1}_{\gamma,\alpha}G^{-1}_{\zeta,\delta}\nabla_j\nabla_k\xi_\zeta\nabla_k\xi_\alpha\nabla_i\xi_\gamma\nabla_j\xi_\delta\\
&\quad - G^{-1}_{\gamma,\delta}\nabla_j\nabla_i\xi_\gamma\nabla_j\xi_\delta - G^{-1}_{\gamma,\delta}\nabla_i\xi_\gamma\Delta\xi_\delta.
\end{aligned}$$

The first and the fifth terms cancel, so that

$$\begin{aligned}
&(\det G)^{-1/2}\nabla_j((\det G)^{1/2}P_{i,j})\\
&= -G^{-1}_{\zeta,\alpha}\nabla_j\nabla_k\xi_\zeta\nabla_k\xi_\alpha G^{-1}_{\gamma,\delta}\nabla_i\xi_\gamma\nabla_j\xi_\delta\\
&\quad + G^{-1}_{\gamma,\delta}G^{-1}_{\zeta,\alpha}\nabla_j\nabla_k\xi_\delta\nabla_k\xi_\zeta\nabla_i\xi_\gamma\nabla_j\xi_\alpha + G^{-1}_{\gamma,\delta}G^{-1}_{\zeta,\alpha}\nabla_j\nabla_k\xi_\zeta\nabla_k\xi_\delta\nabla_i\xi_\gamma\nabla_j\xi_\alpha\\
&\quad - G^{-1}_{\gamma,\delta}\nabla_i\xi_\gamma\Delta\xi_\delta,
\end{aligned}$$

where, in the last expression we have swapped α and ζ in the first term and α and δ in the third term. Now note that the third term cancels with the first term (swapping j and k) so that we obtain:

$$\begin{aligned}
(\det G)^{-1/2}\nabla_j((\det G)^{1/2}P_{i,j}) &= \nabla_i\xi_\gamma G^{-1}_{\gamma,\delta}\left(G^{-1}_{\zeta,\alpha}\nabla_j\nabla_k\xi_\delta\nabla_k\xi_\zeta\nabla_j\xi_\alpha - \Delta\xi_\delta\right)\\
&= -\kappa_\gamma\nabla_i\xi_\gamma|\nabla\xi_\gamma|^{-1},
\end{aligned}$$

which completes the proof of (3.43).

Let us finally prove (3.44). For a fixed $\alpha \in \{1, \ldots, m\}$, we have

$$\begin{aligned}(\det G)^{-1/2}\mathrm{div}\left((\det G)^{1/2}G^{-1}_{\alpha,\gamma}\nabla\xi_\gamma\right)\\ = G^{-1}_{\alpha,\gamma}\nabla\xi_\gamma \cdot \nabla \ln\left((\det G)^{1/2}\right) + \nabla G^{-1}_{\alpha,\gamma}\cdot\nabla\xi_\gamma + G^{-1}_{\alpha,\gamma}\Delta\xi_\gamma.\end{aligned} \tag{3.50}$$

Using (3.49), the first term in (3.50) is:

$$\begin{aligned}G^{-1}_{\alpha,\gamma}\nabla\xi_\gamma\cdot\nabla\ln(\det G)^{1/2} &= \frac{1}{2}G^{-1}_{\alpha,\gamma}\nabla\xi_\gamma\cdot G^{-1}_{\zeta,\delta}\nabla G_{\delta,\zeta}\\ &= \frac{1}{2}G^{-1}_{\alpha,\gamma}\nabla\xi_\gamma\cdot G^{-1}_{\zeta,\delta}\left(\nabla^2\xi_\delta\nabla\xi_\zeta + \nabla^2\xi_\zeta\nabla\xi_\delta\right)\\ &= G^{-1}_{\alpha,\gamma}G^{-1}_{\zeta,\delta}\nabla^2\xi_\delta : \nabla\xi_\zeta\otimes\nabla\xi_\gamma.\end{aligned}$$

Using (3.48) the second term in (3.50) is

$$\begin{aligned}\nabla G^{-1}_{\alpha,\gamma}\cdot\nabla\xi_\gamma &= -G^{-1}_{\gamma,\delta}G^{-1}_{\zeta,\alpha}\left(\nabla^2\xi_\delta\nabla\xi_\zeta + \nabla^2\xi_\zeta\nabla\xi_\delta\right)\cdot\nabla\xi_\gamma\\ &= -G^{-1}_{\gamma,\delta}G^{-1}_{\zeta,\alpha}\left(\nabla^2\xi_\delta : \nabla\xi_\zeta\otimes\nabla\xi_\gamma + \nabla^2\xi_\zeta : \nabla\xi_\delta\otimes\nabla\xi_\gamma\right).\end{aligned}$$

Therefore, we have:

$$\begin{aligned}(\det G)^{-1/2}\mathrm{div}\left((\det G)^{1/2}G^{-1}_{\alpha,\gamma}\nabla\xi_\gamma\right)\\ = -G^{-1}_{\zeta,\alpha}\nabla^2\xi_\zeta : \left(G^{-1}_{\gamma,\delta}\nabla\xi_\delta\otimes\nabla\xi_\gamma\right) + G^{-1}_{\alpha,\gamma}\Delta\xi_\gamma,\end{aligned}$$

which yields (3.44), using the definition (3.41) of κ_α. □

We end this paragraph on geometrical notation by a general remark on the mean curvature vector.

Remark 3.17 (The mean curvature vector). *The aim of this remark is to make a link between the definition we used for the mean curvature vector, and the one found in some textbooks. We prove that the vector* $-\sum_{\alpha=1}^{m}\kappa_\alpha n_\alpha$ *(see (3.40)) is indeed the so-called mean curvature vector* $\mathcal{H}$ *defined as [Ambrosio and Soner (1996a)]*[4]

$$\mathcal{H} = -\sum_{\alpha=1}^{m}\mathrm{div}_\Sigma(\nu_\alpha)\nu_\alpha, \tag{3.51}$$

where div_Σ *is the tangential divergence and* $(\nu_1, \ldots, \nu_m)(q)$ *denotes a smooth orthonormal vector field generating the space normal to* Σ *at point* q*. The geometric interpretation of* $\mathcal{H}$ *is that it points in the direction where the area of* Σ *decreases most, when* Σ *is moved along that direction. This*

[4]Depending on the textbook, the mean curvature vector is defined as $\pm\sum_{\alpha=1}^{m}\kappa_\alpha n_\alpha$, or as $m^{-1}\sum_{\alpha=1}^{m}\kappa_\alpha n_\alpha$. The vector $\mathcal{H}$ defined by (3.51) is also sometimes called the additive curvature vector.

vector intervenes in mean curvature flows or in the divergence theorem on manifolds (3.55) *(see [Ambrosio and Soner (1996a, b)]). This vector only depends on the geometry of the surface* Σ *as a submanifold of* $\mathbb{R}^n$. *In other words, the dynamics* (3.52) *is intrinsic, like the measure* $\tilde{\nu}_\Sigma$ *it samples.*

To derive the expression $\mathcal{H} = -\sum_{\alpha=1}^m \kappa_\alpha n_\alpha$ *from* (3.51), *note first that this definition does not depend on the choice of the vector field* $(\nu_1, ..., \nu_m)$. *Thus, the mean curvature vector is characterized by the fact that, for any vector* ν *in the normal space to* Σ *at point* q, $\mathcal{H} \cdot \nu = -\mathrm{div}_\Sigma(\nu)$. *On the one hand, for* $1 \le \alpha_0 \le m$,

$$\begin{aligned}
&-\kappa_\alpha n_\alpha \cdot n_{\alpha_0} \\
&= -|\nabla \xi_\alpha| G^{-1}_{\gamma,\alpha} \left(\Delta \xi_\gamma - \nabla^2 \xi_\gamma : (G^{-1}_{\delta,\zeta} \nabla \xi_\delta \otimes \nabla \xi_\zeta)\right) \frac{\nabla \xi_\alpha}{|\nabla \xi_\alpha|} \cdot \frac{\nabla \xi_{\alpha_0}}{|\nabla \xi_{\alpha_0}|} \\
&= -G^{-1}_{\gamma,\alpha} \left(\Delta \xi_\gamma - \nabla^2 \xi_\gamma : (G^{-1}_{\delta,\zeta} \nabla \xi_\delta \otimes \nabla \xi_\zeta)\right) G_{\alpha,\alpha_0} |\nabla \xi_{\alpha_0}|^{-1} \\
&= -\left(\Delta \xi_{\alpha_0} - \nabla^2 \xi_{\alpha_0} : (G^{-1}_{\delta,\zeta} \nabla \xi_\delta \otimes \nabla \xi_\zeta)\right) |\nabla \xi_{\alpha_0}|^{-1}.
\end{aligned}$$

On the other hand, for $1 \le \delta \le m$,

$$\begin{aligned}
\mathrm{div}_\Sigma(n_\delta) &= \mathrm{Id} : (P \nabla (\nabla \xi_\delta / |\nabla \xi_\delta|)) \\
&= \delta_{i,j} \left(\delta_{i,k} - G^{-1}_{\alpha,\zeta} \nabla_i \xi_\alpha \nabla_k \xi_\zeta\right) \nabla_k \left(\nabla_j \xi_\delta |\nabla \xi_\delta|^{-1})\right) \\
&= \left(\delta_{i,k} - G^{-1}_{\alpha,\zeta} \nabla_i \xi_\alpha \nabla_k \xi_\zeta\right) \left(\nabla_k \nabla_i \xi_\delta |\nabla \xi_\delta|^{-1} - \nabla_i \xi_\delta \nabla_k \xi_\delta |\nabla \xi_\delta|^{-2}\right) \\
&= \Delta \xi_\delta |\nabla \xi_\delta|^{-1} - G^{-1}_{\alpha,\zeta} \nabla_i \xi_\alpha \nabla_k \xi_\zeta \nabla_k \nabla_i \xi_\delta |\nabla \xi_\delta|^{-1} - 1 \\
&\quad + G^{-1}_{\alpha,\zeta} \nabla_i \xi_\alpha \nabla_k \xi_\zeta \nabla_i \xi_\delta \nabla_k \xi_\delta |\nabla \xi_\delta|^{-2} \\
&= \Delta \xi_\delta |\nabla \xi_\delta|^{-1} - \nabla^2 \xi_\delta : (G^{-1}_{\alpha,\zeta} \nabla \xi_\alpha \otimes \nabla \xi_\zeta) |\nabla \xi_\delta|^{-1}.
\end{aligned}$$

Therefore, for any $1 \le \alpha_0 \le m$,

$$\sum_{\alpha=1}^m -\kappa_\alpha n_\alpha \cdot n_{\alpha_0} = -\mathrm{div}_\Sigma(n_{\alpha_0}).$$

Since $(n_1, ..., n_m)(q)$ *generates the space normal to* Σ *at point* q, *this proves that* $\mathcal{H} = -\sum_{\alpha=1}^m \kappa_\alpha n_\alpha$.

3.2.3.2 *Projected dynamics*

Using the operator P and the mean curvature vector $\mathcal{H}$, we can now define the projected dynamics (written here in Itô form):

$$dq_t = P(q_t)\left(-\nabla \tilde{V}(q_t)\,dt + \sqrt{2\beta^{-1}}\,dW_t\right) + \beta^{-1} \mathcal{H}(q_t)\,dt. \tag{3.52}$$

In the following, this dynamics is called the projected (or constrained) overdamped Langevin (or gradient) dynamics.

The solution to the projected dynamics (3.52) leaves on the submanifold $\Sigma(\xi(q_0))$. In particular, if $\xi(q_0) = 0$, then $q_t \in \Sigma$.

Lemma 3.18. *Let q_t be a solution to* (3.52). *Then $\forall t \geq 0$, almost surely,*

$$\xi(q_t) = \xi(q_0).$$

Proof. This is the consequence of an Itô calculus: for $\gamma \in \{1, \dots, m\}$,

$$d\xi_\gamma(q_t) = \nabla \xi_\gamma(q_t) \cdot dq_t + \beta^{-1} PP^T : \nabla^2 \xi_\gamma(q_t)\, dt.$$

We recall that the notation $PP^T : \nabla^2 \xi_\gamma$ means $P_{i,k} P_{j,k} \frac{\partial^2 \xi_\gamma}{\partial x_i \partial x_j}$. Using the fact that $P = P^T$ and $PP = P$, it holds:

$$\begin{aligned} d\xi_\gamma(q_t) &= \nabla \xi_\gamma(q_t) \cdot P(q_t) \left(-\nabla \tilde{V}(q_t)\, dt + \sqrt{2\beta^{-1}}\, dW_t \right) \\ &\quad + \beta^{-1} \nabla \xi_\gamma(q_t) \cdot \mathcal{H}(q_t)\, dt + \beta^{-1} P : \nabla^2 \xi_\gamma(q_t)\, dt. \end{aligned}$$

Now, note that

$$\begin{aligned} &\nabla \xi_\gamma(q_t) \cdot P(q_t) \left(-\nabla \tilde{V}(q_t)\, dt + \sqrt{2\beta^{-1}}\, dW_t \right) \\ &= P(q_t) \nabla \xi_\gamma(q_t) \cdot \left(-\nabla \tilde{V}(q_t)\, dt + \sqrt{2\beta^{-1}}\, dW_t \right) \\ &= 0. \end{aligned}$$

The fact that $d\xi_\gamma(q_t) = 0$ is then a direct consequence of (3.46). □

The dynamics (3.52) may also be written using a Stratonovitch product:

Lemma 3.19. *The dynamics* (3.52) *writes equivalently:*

$$dq_t = -P(q_t) \nabla \tilde{V}(q_t)\, dt + \sqrt{2\beta^{-1}} P(q_t) \circ dW_t, \tag{3.53}$$

where $\circ$ denotes the Stratonovitch product.

Proof. The Stratonovitch product rewrites in terms of Itô product as (see (2.19)): for a fixed $i \in \{1, \dots, n\}$,

$$\begin{aligned} P_{i,k}(q_t) \circ dW_{k,t} &= \frac{1}{2} P_{j,k} \nabla_j P_{i,k}(q_t)\, dt + P_{i,k}(q_t) dW_{k,t} \\ &= \frac{1}{2} \mathcal{H}_i(q_t)\, dt + P_{i,k}(q_t) dW_{k,t}, \end{aligned}$$

where we denote by $(W_{k,t})_{1 \leq k \leq n}$ the components of W_t, and we used (3.42). □

3.2.3.3 *Ergodicity of the projected dynamics*

The main result of this section 3.2.3 is that the projected dynamics (3.52) indeed samples $\tilde{\nu}_\Sigma$ defined by (3.35).

Proposition 3.20. *Assume that* $\xi(q_0) = 0$, *and consider* (q_t) *solution to* (3.52). *The distribution* $d\tilde{\nu}_\Sigma = \dfrac{e^{-\beta \tilde{V}}\, d\sigma_\Sigma}{\displaystyle\int_\Sigma e^{-\beta \tilde{V}}\, d\sigma_\Sigma}$ *is the unique equilibrium distribution of the diffusion* q_t.

The uniqueness of the equilibrium distribution implies that (q_t) is ergodic with respect to $\tilde{\nu}_\Sigma$. Thus, by the Birkhoff ergodic theorem, Proposition 3.20 implies that if ϕ is in $L^p(\tilde{\nu}_\Sigma)$ (with $p > 1$),

$$\lim_{T\to\infty} \frac{1}{T} \int_0^T \phi(q_t)\, dt = \frac{\displaystyle\int_\Sigma \phi\, e^{-\beta \tilde{V}}\, d\sigma_\Sigma}{\displaystyle\int_\Sigma e^{-\beta \tilde{V}}\, d\sigma_\Sigma} = \int_\Sigma \phi\, d\tilde{\nu}_\Sigma, \tag{3.54}$$

where (q_t) is a solution of (3.52), and the convergence is almost sure and in L^p (with respect to the initial condition $q_0 \in \Sigma$).

The proof of Proposition 3.20 relies on the divergence theorem on manifolds:

Lemma 3.21. *Let* Σ *be a smooth submanifold of* $\mathbb{R}^n$. *For all functions* $\phi \in C^1_c(\mathbb{R}^n, \mathbb{R}^n)$,

$$\int_\Sigma \mathrm{div}_\Sigma(\phi)\, d\sigma_\Sigma = -\int_\Sigma \mathcal{H} \cdot \phi\, d\sigma_\Sigma, \tag{3.55}$$

where div_Σ *denotes the surface divergence:*

$$\mathrm{div}_\Sigma(\phi) = \mathrm{tr}(P\nabla\phi) = \sum_{i,j} P_{i,j} \nabla_j \phi_i, \tag{3.56}$$

where P *is the orthogonal projection operator defined by* (3.38).

Note that even if ϕ may be defined in a neighborhood of Σ, $\mathrm{div}_\Sigma \phi$ only depends on the values of ϕ on Σ.

In case $\Sigma = \mathbb{R}^n$, we recall that the usual divergence theorem writes $\int_{\mathbb{R}^n} \mathrm{div}(\phi) = 0$. The right-hand side in (3.55) is due to the "non-flatness" of Σ. This term is zero in two cases: (i) Σ is a vector space (flat submanifold) so that $\mathcal{H} = 0$, or (ii) ϕ is with value in the tangent space to Σ so that $\mathcal{H} \cdot \phi = 0$.

Using (3.42) and the fact that, from (3.43),

$$\mathrm{div} P + P\nabla \ln\left((\det G)^{1/2}\right) = \mathcal{H},$$

it is easy to check that this lemma is equivalent to the following, which is closest in its formulation to the usual divergence theorem:

Lemma 3.22. *Let Σ be a smooth submanifold of $\mathbb{R}^n$. For all functions $\phi \in \mathcal{C}^1_c(\mathbb{R}^n, \mathbb{R}^n)$,*

$$\int_\Sigma \mathrm{div}_\Sigma(P\phi)\, d\sigma_\Sigma = 0, \tag{3.57}$$

where div_Σ *denotes the surface divergence* (3.56). *Equivalently, for all functions $\phi \in \mathcal{C}^1_c(\mathbb{R}^n, \mathbb{R}^n)$,*

$$\int_\Sigma \mathrm{div}(P\phi)\, (\det G)^{-1/2}\, d\sigma_\Sigma = 0, \tag{3.58}$$

which rewrites using the δ notation: $\int_\Sigma \mathrm{div}(P\phi)\, \delta_{\xi(q)}(dq) = 0.$

Let us prove Lemma 3.21.

Proof. This is a well-known result (see for example [Ambrosio and Soner (1996a, b)]), but we give here a short proof based on the co-area formula (3.12) for the sake of completeness.

We may assume without loss of generality that the submanifold Σ is actually the zero level set $\Sigma(0)$ of some function $\xi : \mathbb{R}^n \to \mathbb{R}^m$. Consider $\phi \in \mathrm{C}^1_c(\mathbb{R}^n, \mathbb{R}^n)$ and $f \in \mathrm{C}^\infty_c(\mathbb{R}^m, \mathbb{R})$. We have, using the co-area formula (3.12) and an integration by parts (the notation $\circ$ here denotes the composition operator):

$$\begin{aligned}
\int_{\mathbb{R}^m} f(z) \int_{\Sigma(z)} \mathrm{div}_{\Sigma(z)}(\phi)\, d\sigma_{\Sigma(z)}\, dz &= \int_{\mathbb{R}^n} (f\circ\xi)\, \mathrm{div}_{\Sigma(z)}(\phi)(\det G)^{1/2} \\
&= \int_{\mathbb{R}^n} (f\circ\xi)\, P_{i,j}\nabla_j(\phi_i)(\det G)^{1/2} \\
&= -\int_{\mathbb{R}^n} \phi_i \nabla_j \left((f\circ\xi)\, P_{i,j}(\det G)^{1/2}\right) \\
&= -\int_{\mathbb{R}^n} \phi_i \nabla_j \left(P_{i,j}(\det G)^{1/2}\right)\ (f\circ\xi),
\end{aligned}$$

where we used the fact that $P\nabla(f\circ\xi) = P\nabla\xi_\alpha(\nabla_\alpha f)\circ\xi = 0$. Using again the co-area formula and then formula (3.43) for the mean curvature vector, it holds:

$$\begin{aligned}
&\int_{\mathbb{R}^m} f(z) \int_{\Sigma(z)} \mathrm{div}_{\Sigma(z)}(\phi)\, d\sigma_{\Sigma(z)}\, dz \\
&\qquad = -\int_{\mathbb{R}^m} f(z) \int_{\Sigma(z)} \phi_i \nabla_j \left(P_{i,j}(\det G)^{1/2}\right) (\det G)^{-1/2}\, d\sigma_{\Sigma(z)}\, dz \\
&\qquad = -\int_{\mathbb{R}^m} f(z) \int_{\Sigma(z)} \phi \cdot \mathcal{H}\, d\sigma_{\Sigma(z)}\, dz.
\end{aligned}$$

The fact that this equality holds for all functions f concludes the proof. □

We are now in position to prove Proposition 3.20.

Proof. The transition probability function for the Markov process (q_t) from Σ to Σ is strictly positive so that any invariant measure is equivalent to the Lebesgue measure σ_Σ, which implies the uniqueness of the invariant measure (see Proposition 6.1.9 in [Duflo (1997)]).

It therefore suffices to prove that $\tilde{\nu}_\Sigma$ is an invariant measure for (3.52). To this end, define $u(t,q) = \mathbb{E}_q(f(q_t))$, where q_t satisfies (3.52) and $\mathbb{E}_q$ denotes the expectation over this process conditional on $q_0 = q$. Then $\tilde{\nu}_\Sigma$ is an invariant measure for (3.52) if

$$\int_\Sigma u(t,q)\, \tilde{\nu}_\Sigma(dq) = \int_\Sigma u(0,q)\, \tilde{\nu}_\Sigma(dq). \tag{3.59}$$

To check (3.59), note that u satisfies the backward Kolmogorov equation

$$\frac{\partial u}{\partial t} = -P\nabla\tilde{V} \cdot \nabla u + \beta^{-1}\mathcal{H} \cdot \nabla u + \beta^{-1} P : \nabla^2 u,$$

where we recall the notation $P : \nabla^2 u = \sum_{i,j=1}^n P_{i,j} \frac{\partial^2 u}{\partial x_i \partial x_j}$. It follows that

$$\begin{aligned}
&\frac{d}{dt}\int_\Sigma u\, d\tilde{\nu}_\Sigma \\
&\quad = Z_\Sigma^{-1} \int_\Sigma \Big(- P\nabla\tilde{V} \cdot \nabla u + \beta^{-1}\mathcal{H} \cdot \nabla u + \beta^{-1} P : \nabla^2 u \Big) \exp(-\beta\tilde{V})\, d\sigma_\Sigma \\
&\quad = \beta^{-1} Z_\Sigma^{-1} \int_\Sigma \Big(\mathrm{div}_\Sigma(\nabla u \exp(-\beta\tilde{V})) + \mathcal{H} \cdot \nabla u \exp(-\beta\tilde{V}) \Big)\, d\sigma_\Sigma \\
&\quad = 0,
\end{aligned}$$

where $Z_\Sigma = \int_\Sigma \exp(-\beta\tilde{V})\, d\sigma_\Sigma$, div_Σ denotes the surface divergence (see (3.56)), and we used the divergence theorem on manifolds (see Lemma 3.21). This shows that (3.59) holds, which concludes the proof. □

Remark 3.23 (Reversibility with respect to $\tilde{\nu}_\Sigma$). *The diffusion* (q_t) *solution to* (3.52) *is actually reversible with respect to* $\tilde{\nu}_\Sigma$. *Indeed, using* (3.42), *it is easy to check that the infinitesimal generator* $\mathcal{L}$ *of the diffusion writes: for any smooth test function* u,

$$\begin{aligned}\mathcal{L}u &= -P\nabla\tilde{V}\cdot\nabla u + \beta^{-1}\mathcal{H}\cdot\nabla u + \beta^{-1}P : \nabla^2 u \\ &= \beta^{-1}\exp(\beta\tilde{V})\,\mathrm{div}_\Sigma\left(\exp(-\beta\tilde{V})\nabla_\Sigma u\right).\end{aligned}$$

Therefore, using the divergence theorem on manifolds (3.57) *and the notation*

$$\nabla_\Sigma = P\nabla \tag{3.60}$$

to denote the surface gradient on Σ, *it holds: for any smooth test functions* u *and* v,

$$\int_\Sigma (\mathcal{L}u)v\,d\tilde{\nu}_\Sigma = -\beta^{-1}\int_\Sigma \nabla_\Sigma u\cdot\nabla_\Sigma v\,d\tilde{\nu}_\Sigma = \int_\Sigma (\mathcal{L}v)u\,d\tilde{\nu}_\Sigma,$$

which is exactly the reversibility property (2.30). *This implies the stationarity of* $\tilde{\nu}_\Sigma$.

Note that even if u *may be defined in a neighborhood of* Σ, $\nabla_\Sigma u$ *depends only on the values of* u *on* Σ.

3.2.3.4 *Softly and rigidly constrained dynamics*

We used one particular method to project the gradient dynamics (3.36) onto Σ. Another natural method would be to consider the penalized dynamics

$$dq_t^\eta = -\nabla\left(\tilde{V} + \frac{1}{2\eta}\sum_{\alpha=1}^m \xi_\alpha^2\right)(q_t^\eta)\,dt + \sqrt{2\beta^{-1}}\,dW_t, \tag{3.61}$$

where $\eta > 0$ is a (small) parameter. The additional term involving $\nabla(\xi_\alpha^2)$ in (3.61) is a penalty term which constraints q_t in the vicinity of $\Sigma = \{x, \xi(x) = 0\}$. Letting $\eta \to 0$ amounts to imposing the constraint that $q_t \in \Sigma$ almost surely. In fact, it can be shown (see for example Appendix C in [Ciccotti *et al.* (2008)] for the case $m = 1$) that the limit process (q_t) of (q_t^η) when $\eta \to 0$ is solution of the stochastic differential equation:

$$\begin{aligned}dq_t = P(q_t)\left(-\nabla\left(\tilde{V} + \frac{1}{2\beta}\ln(\det G)\right)(q_t)\,dt + \sqrt{2\beta^{-1}}\,dW_t\right) \\ + \beta^{-1}\mathcal{H}(q_t)\,dt.\end{aligned} \tag{3.62}$$

This equation is different from (3.52). In particular, (3.62) samples $\nu^\xi(\cdot|z)$ with $\tilde{V} = V$ (and not $\tilde{V} = V^\xi$, as for (3.52)).

Note that, for a given potential $\tilde{V}$, the measure $\tilde{\nu}_\Sigma$ depends on ξ only through its zero set (which defines Σ). The values of ξ around Σ are irrelevant. In this sense, $\tilde{\nu}_\Sigma$ is an intrinsic quantity. Accordingly the stochastic differential equation (3.52) (and in particular the mean curvature vector $\mathcal{H}$, see Remark 3.17) can be defined knowing only Σ. In contrast, the measure $\nu^\xi(\cdot|0)$ also depends on the values of ξ around Σ (since its expression involves $\nabla\xi$). In this sense, it is a non-intrinsic quantity. Because the constraints are softly imposed in (3.61) (and not rigidly as in (3.52)), in the limit as $\eta \to 0$ the limiting process (q_t) still "sees" the variation of ξ around Σ, through the term $\frac{1}{2\beta}\ln(\det G)$ in (3.62). This term is called the Fixman correction.

The fact that the equilibrium measure associated with the rigidly constrained dynamics (3.52), namely

$$\tilde{\nu}_\Sigma = \frac{e^{-\beta\tilde{V}}\,d\sigma_\Sigma}{\int_\Sigma e^{-\beta\tilde{V}}\,d\sigma_\Sigma}$$

and the equilibrium measure associated with the softly constrained dynamics (3.62), namely

$$\frac{e^{-\beta\tilde{V}+\frac{1}{2}\ln(\det G)}d\sigma_\Sigma}{\int_\Sigma e^{-\beta\tilde{V}+\frac{1}{2}\ln(\det G)}d\sigma_\Sigma}$$

are not identical is related to an apparent paradox which has often been discussed in the literature, about the different statistics at equilibrium of rigid and stiff bonds in bead spring models: see p. 228 in [Öttinger (1995)], Section 4.6 in [Peters (2000)], [Hinch (1994)], [Fixman (1974)], [van Kampen (1981)], or paragraph 3 in [Morse (2004)]. We here exhibit the difference between statistical properties of rigidly and softly constrained dynamics in the framework of overdamped Langevin dynamics, but this question has also been discussed either in the framework of Hamiltonian systems at equilibrium (in the canonical ensemble) or in the non-zero-mass Langevin dynamics framework (see e.g. [Sprik and Ciccoti (1998); Darve *et al.* (2002); Hartmann and Schütte (2005a, b)]).

3.2.4 *Sampling measures on submanifolds of $\mathbb{R}^n$: discretization*

The aim of this section is to derive a discretization scheme for the projected dynamics (3.52), based on a rewriting of this dynamics using Lagrange multipliers. We recall that we may assume without loss of generality a projection on the submanifold $\Sigma = \Sigma(0)$. All the results given in this section remain the same for a projection on the submanifold $\Sigma(z)$ whatever z.

The bottom line of the proposed discretization methods is that, since we intend to use the stochastic processes to perform trajectorial averages in order to compute averages with respect to $\tilde{\nu}_\Sigma$ (defined by (3.35)), the presented numerical discretizations are such that the constraint $\xi(q_t) = 0$ is exactly satisfied (for the discretized process). This restriction rules out standard dicretization schemes of (3.52), like Euler schemes for which the constraint $\xi(q_t) = 0$ would not be satisfied over long times.

3.2.4.1 *Rewriting of the projected dynamics* (3.52) *using Lagrange multipliers*

In this section, a projection of the dynamics (3.36) using Lagrange multipliers associated with the constraint $\xi(q_t) = 0$ is introduced, and shown to be formally equivalent to (3.52). Consider the following projection of the dynamics (3.36):

$$\begin{cases} dq_t = -\nabla \tilde{V}(q_t)\, dt + \sqrt{2\beta^{-1}} dW_t + \displaystyle\sum_{\alpha=1}^{m} \nabla \xi_\alpha(q_t)\, d\lambda_{\alpha,t}, \\ \text{with } \lambda_t \in \mathbb{R}^m \text{ an adapted process such that } \xi(q_t) = 0. \end{cases} \tag{3.63}$$

The vector $\lambda_t \in \mathbb{R}^m$ has components $(\lambda_{\alpha,t})_{1 \leq \alpha \leq m}$. Since we suppose that $\xi(q_0) = 0$, we set $\lambda_0 = 0$. In the following, we will denote by

$$dY_t = \sum_{\alpha=1}^{m} \nabla \xi_\alpha(q_t)\, d\lambda_{\alpha,t} \tag{3.64}$$

the $\mathbb{R}^n$-valued process associated with the constraints $\xi(q_t) = 0$. It can be checked that (q_t) solution to (3.63) is actually solution to (3.52). Equivalently, Y_t satisfies $Y_0 = 0$ and

$$dY_t = (P(q_t) - \mathrm{Id})\Big(- \nabla \tilde{V}(q_t)\, dt + \sqrt{2\beta^{-1}} dW_t\Big) + \beta^{-1} \mathcal{H}(q_t)\, dt. \tag{3.65}$$

The interest of this reformulation is twofold: First, it suggests natural consistent discretization schemes and this is the subject of this Section 3.2.4; second, in the case $\tilde{V} = V^\xi$, it can be checked that the mean force can actually be obtained by averaging the Lagrange multiplier Y_t, and this is the subject of next Section 3.2.5.

Before giving a numerical scheme for discretizing (3.52) based on the equivalent reformulation (3.63), let us explain why these two dynamics are indeed the same, assuming $m = 1$ for simplicity. The fact that $(\lambda_t)_{t\geq 0}$ is adapted with respect to the filtration of the Brownian motion $(W_t)_{t\geq 0}$ means that at a given time t, the random variable λ_t only depends on the past, *i.e.* is measurable with respect to $(W_s)_{0\leq s\leq t}$. This implies that $(\lambda_t)_{t\geq 0}$ may be decomposed into two parts:

$$\lambda_t = \lambda_t^f + \lambda_t^m$$

where $(\lambda_t^f)_{t\geq 0}$ is a stochastic process with finite variation, and λ_t^m is a martingale which may be written as:

$$\lambda_t^m = \int_0^t S_s \cdot dW_s,$$

with (S_t) a stochastic process with values in $\mathbb{R}^n$. Computing $d\xi(q_t)$, it holds:

$$\begin{aligned} d\xi(q_t) &= \nabla\xi(q_t) \cdot \left(-\nabla\tilde{V}(q_t)\,dt + \nabla\xi(q_t)d\lambda_t^f\right) \\ &+ \frac{1}{2}\nabla^2\xi(q_t) : \left[\left(\sqrt{2\beta^{-1}}\mathrm{Id} + \nabla\xi(q_t)\otimes S_t\right)\left(\sqrt{2\beta^{-1}}\mathrm{Id} + \nabla\xi(q_t)\otimes S_t\right)^T\right] dt \\ &+ \left(\sqrt{2\beta^{-1}}\nabla\xi(q_t) + |\nabla\xi|^2(q_t)S_t\right) \cdot dW_t. \end{aligned}$$

Therefore, for $d(\xi(q_t))$ to be zero, it is necessary that the martingale part is zero, which gives

$$S_t = -\sqrt{2\beta^{-1}}\,\frac{\nabla\xi}{|\nabla\xi|^2}(q_t)$$

and thus, using the fact that the bounded variation part should also be zero,

$$\begin{aligned} d\lambda_t^f &= \frac{\nabla\xi\cdot\nabla\tilde{V}}{|\nabla\xi|^2}(q_t)\,dt - \beta^{-1}\frac{1}{|\nabla\xi|^2}\nabla^2\xi : P(q_t)\,dt \\ &= \frac{\nabla\xi\cdot\nabla\tilde{V}}{|\nabla\xi|^2}(q_t)\,dt + \beta^{-1}\frac{\nabla\xi\cdot\mathcal{H}}{|\nabla\xi|^2}(q_t)\,dt, \end{aligned}$$

where we used (3.46). Using (3.45), we thus recover (3.65), and as a result, (3.52).

3.2.4.2 *Discretization of the projected dynamics* (3.52)

As shown in the previous section, the dynamics (3.52) may be rewritten in terms of Lagrange multipliers associated with the constraint $\xi(q_t) = 0$, see (3.63). This suggests natural (predictor-corrector like) discretization schemes of (3.52), namely:

$$\begin{cases} q^{n+1} = q^n - \nabla\tilde{V}(q^n)\,\Delta t + \sqrt{2\beta^{-1}}\Delta W^n + \sum_{\alpha=1}^{m} \lambda_\alpha^n \nabla\xi_\alpha(q^{n+1}), \\ \text{with } (\lambda_\alpha^n)_{1\le\alpha\le m} \text{ such that } \xi(q^{n+1}) = 0, \end{cases} \tag{3.66}$$

and

$$\begin{cases} q^{n+1} = q^n - \nabla\tilde{V}(q^n)\,\Delta t + \sqrt{2\beta^{-1}}\Delta W^n + \sum_{\alpha=1}^{m} \lambda_\alpha^n \nabla\xi_\alpha(q^{n}), \\ \text{with } (\lambda_\alpha^n)_{1\le\alpha\le m} \text{ such that } \xi(q^{n+1}) = 0, \end{cases} \tag{3.67}$$

where $\Delta W^n = W_{t_{n+1}} - W_{t_n}$ denotes the Brownian increment. We assume that the time-step size Δt is fixed, so that $t_n = n\Delta t$ and q^n is meant to be an approximation of q_{t_n} (where q_t satisfies (3.52)).

The implicit scheme in (3.66) can in fact be rewritten as a variational problem as follows:

$$\begin{cases} q^{n+1,\star} = q^n - \nabla\tilde{V}(q^n)\,\Delta t + \sqrt{2\beta^{-1}}\Delta W^n, \\ q^{n+1} = \arg\min_{y\in\mathbb{R}^n} \left\{ |q^{n+1,\star} - y|^2, \xi(y) = 0 \right\}. \end{cases} \tag{3.68}$$

In this case, the numbers $(\lambda_\alpha^n)_{1\le\alpha\le m}$ can be interpreted as scalar Lagrange multipliers associated with the constraint $\xi(q^{n+1}) = 0$. It may be expected that the scheme (3.66) exhibits better stability properties than the scheme (3.67) since it admits the variational interpretation (3.68). In other words, the existence of solution to (3.66) is ensured under milder assumptions than for (3.67).

In practice, in order for (3.66), (3.67) or (3.68) to be well-posed, Δt has to be sufficiently small so that $q^{n+1,\star}$ is not "too far" from the submanifold Σ. To solve (3.66), classical methods for optimization problems with constraints can be used, see [Glowinski and Tallec (1989)] for a presentation of the classical Uzawa algorithm, and to [Bonnans *et al.* (2002)] for more advanced methods. To solve (3.67), classical methods for nonlinear problems (like Newton method, see [Bonnans *et al.* (2002)]) can be used. We also refer to Chapter 7 of [Leimkuhler and Reich (2005)] where similar problems are discussed.

In the following, we will check that the schemes (3.66), (3.67) and (3.68) are consistent with (3.52), and we will admit that they are well posed and indeed convergent (in the mean square sense for example), namely that the trajectory $(q_0, \dots, q_M)$ where $M = T/\Delta t$ converges when $\Delta t \to 0$ (for a fixed T) to $(q_s)_{0\leq s\leq T}$ which satisfies (3.52). Since $(q_t)_{t\geq 0}$ is ergodic with respect to $\tilde{\nu}_\Sigma$ (see Proposition 3.20), the schemes (3.66)

and (3.67) can be used to sample $\tilde{\nu}_\Sigma$ and to compute quantities such as (3.54). We refer to Section 3.2.4.4 where the ergodicity of the discretized scheme and the error due to the time discretization are discussed.

We will come back below to the more specific discussion of the computation of free energy differences and mean forces (see Section 3.2.5).

3.2.4.3 *Consistency of the predictor-corrector schemes.*

The main result of this section is the following:

Proposition 3.24. *The two schemes* (3.66) *and* (3.67) *are consistent discretization, in the limit* $\Delta t \to 0$, *of* (3.52).

The proof of Proposition 3.24 relies on the following lemma which gives expansions in Δt of the λ_α^n appearing in (3.66) and (3.67).

Lemma 3.25. *Let q^n be the solution of* (3.66) *or* (3.67). *Then λ_α^n is such that:*

$$\lambda_\alpha^n = \lambda_\alpha^{0,n}\sqrt{\Delta t} + \lambda_\alpha^{1,n}\Delta t + \mathrm{o}(\Delta t), \tag{3.69}$$

with

$$\lambda_\alpha^{0,n} = -\sqrt{2\beta^{-1}}\sum_{\zeta=1}^{m} G_{\alpha,\zeta}^{-1}\nabla\xi_\zeta(q^n)\cdot w^n, \tag{3.70}$$

where $w^n = \Delta W^n/\sqrt{\Delta t}$ are i.i.d. Gaussian variables in $\mathbb{R}^n$ with zero mean and variance Id, *and*

$$\begin{aligned}\lambda_\alpha^{1,n} = &\sum_{\zeta=1}^{m} G_{\alpha,\zeta}^{-1}\nabla\xi_\zeta\cdot\nabla\tilde{V}(q^n)\\ &+\beta^{-1}\sum_{\zeta,\delta=1}^{m} G_{\alpha,\zeta}^{-1}G_{\gamma,\delta}^{-1}\nabla^2\xi_\zeta : \nabla\xi_\gamma\otimes\nabla\xi_\delta(q^n)\\ &-\beta^{-1}\sum_{\zeta=1}^{m} G_{\alpha,\zeta}^{-1}\Delta\xi_\zeta(q^n).\end{aligned} \tag{3.71}$$

Proof. For the sake of brevity we only present the proof of Lemma 3.25 for the scheme (3.66). The proof for the scheme (3.67) is similar.

The Lagrange multipliers λ_α^n are obtained by requiring that $\xi(q^{n+1}) = 0$, with $\xi(q^n) = 0$. Using (3.66) as well as the *a priori* expansion (3.69) of λ_α^n, this is equivalent to requiring that: for any $1 \le \zeta \le m$,

$$
\begin{aligned}
0 &= \xi_\zeta(q^{n+1}) \\
&= \nabla\xi_\zeta(q^n) \cdot \Big(-\nabla\tilde{V}(q^n)\,\Delta t + \sqrt{2\beta^{-1}}\Delta W^n \\
&\qquad\qquad + (\lambda_\alpha^{0,n}\sqrt{\Delta t} + \lambda_\alpha^{1,n}\Delta t)\nabla\xi_\alpha(q^{n+1}) \Big) \\
&\quad + \frac{1}{2}(K^n)^T \nabla^2\xi_\zeta(q^n) K^n + \mathrm{o}(\Delta t),
\end{aligned}
\tag{3.72}
$$

where $K^n = \sqrt{2\beta^{-1}}\Delta W^n + \lambda_\alpha^{0,n}\sqrt{\Delta t}\nabla\xi_\alpha(q^{n+1})$. Since

$$
\nabla\xi_\alpha(q^{n+1}) = \nabla\xi_\alpha(q^n) + \nabla^2\xi_\alpha(q^n)\overline{K}^n + \mathrm{o}(\sqrt{\Delta t}),
$$

where $\overline{K}^n = \sqrt{2\beta^{-1}}\Delta W^n + \lambda_\alpha^{0,n}\sqrt{\Delta t}\nabla\xi_\alpha(q^n)$, equating terms of equal orders in Δt in (3.72) gives

$$
\begin{cases}
0 = \sqrt{2\beta^{-1}}\nabla\xi_\zeta(q^n)\cdot\Delta W^n + \lambda_\alpha^{0,n}\sqrt{\Delta t}\,G_{\alpha,\zeta}(q^n), \\
0 = -\nabla\xi_\zeta(q^n)\cdot\nabla\tilde{V}(q^n)\Delta t + \lambda_\alpha^{1,n} G_{\alpha,\zeta}(q^n)\Delta t \\
\qquad + \sqrt{\Delta t}\lambda_\alpha^{0,n}(\nabla\xi_\zeta)^T\nabla^2\xi_\alpha(q^n)\overline{K}^n + \dfrac{1}{2}(\overline{K}^n)^T\nabla^2\xi_\zeta(q^n)\overline{K}^n.
\end{cases}
$$

From this, we obtain the formula (3.70) for $\lambda_n^{0,\alpha}$ and the following expression for $\lambda_n^{1,\alpha}$:

$$
\begin{aligned}
\lambda_\alpha^{1,n} &= G_{\alpha,\zeta}^{-1}\nabla\xi_\zeta\cdot\nabla\tilde{V}(q^n) \\
&\quad - G_{\alpha,\zeta}^{-1}\lambda_\gamma^{0,n}\lambda_\delta^{0,n}\nabla^2\xi_\gamma : \nabla\xi_\zeta\otimes\nabla\xi_\delta(q^n) \\
&\quad - \tfrac{1}{2}G_{\alpha,\zeta}^{-1}\lambda_\gamma^{0,n}\lambda_\delta^{0,n}\nabla^2\xi_\zeta : \nabla\xi_\gamma\otimes\nabla\xi_\delta(q^n) \\
&\quad - \sqrt{2\beta^{-1}}\,G_{\alpha,\zeta}^{-1}\,\lambda_\gamma^{0,n}\,\nabla^2\xi_\gamma(q^n) : \nabla\xi_\zeta(q^n)\otimes w^n \\
&\quad - \sqrt{2\beta^{-1}}\,G_{\alpha,\zeta}^{-1}\,\lambda_\gamma^{0,n}\,\nabla^2\xi_\zeta(q^n) : \nabla\xi_\gamma(q^n)\otimes w^n \\
&\quad - \beta^{-1}\,G_{\alpha,\zeta}^{-1}\nabla^2\xi_\zeta(q^n) : w^n\otimes w^n.
\end{aligned}
$$

We now use (3.70) in this expression together with the fact that in the limit as $\Delta t \to 0$, $w^n \otimes w^n = \mathrm{Id}$ since w^n is always multiplied by $\mathcal{F}_{n\Delta t}$ measurable functions, namely functions which only depends on the filtration generated by the Brownian motion up to time $n\Delta t$.[5] For example, in the limit $\Delta t \to 0$,

$$
\begin{aligned}
\lambda_\gamma^{0,n}\lambda_\delta^{0,n} &= 2\beta^{-1}\Delta t^{-1}G_{\gamma,\gamma'}^{-1}\nabla\xi_{\gamma'}(q^n)\cdot\Delta W^n G_{\delta,\delta'}^{-1}\nabla\xi_{\delta'}(q^n)\cdot\Delta W^n \\
&= 2\beta^{-1}\,G_{\gamma,\gamma'}^{-1}\,G_{\delta,\delta'}^{-1}\,G_{\gamma',\delta'}(q^n) + \mathrm{o}(1) = 2\beta^{-1}\,G_{\gamma,\delta}^{-1}(q^n) + \mathrm{o}(1).
\end{aligned}
$$

[5] This is the cornerstone of Itô's formula, see for example the proof of Theorem 3.3 in [Karatzas and Shreve (1988)].

We thus obtain the following expression for $\lambda_\alpha^{1,n}$:

$$\begin{aligned}\lambda_\alpha^{1,n} &= G^{-1}_{\alpha,\zeta}\nabla\xi_\zeta\cdot\nabla\tilde{V}(q^n)\\ &\quad -2\beta^{-1}G^{-1}_{\alpha,\zeta}G^{-1}_{\gamma,\delta}\nabla^2\xi_\gamma:\nabla\xi_\zeta\otimes\nabla\xi_\delta(q^n)\\ &\quad -\beta^{-1}G^{-1}_{\alpha,\zeta}G^{-1}_{\gamma,\delta}\nabla^2\xi_\zeta:\nabla\xi_\gamma\otimes\nabla\xi_\delta(q^n)\\ &\quad +2\beta^{-1}G^{-1}_{\alpha,\zeta}G^{-1}_{\gamma,\delta}\nabla^2\xi_\gamma:\nabla\xi_\delta\otimes\nabla\xi_\zeta(q^n)\\ &\quad +2\beta^{-1}G^{-1}_{\alpha,\zeta}G^{-1}_{\gamma,\delta}\nabla^2\xi_\zeta:\nabla\xi_\delta\otimes\nabla\xi_\gamma(q^n)\\ &\quad -\beta^{-1}G^{-1}_{\alpha,\zeta}\Delta\xi_\zeta(q^n)+\mathrm{o}(1),\end{aligned}$$

from which we deduce (3.71). □

We are now in position to prove Proposition 3.24.

Proof. Let us first consider the scheme (3.66) (or, equivalently, (3.68)). To check its consistency with (3.52), we compute $\lambda_\alpha^n\nabla\xi_\alpha(q^{n+1})\Delta t$ using the expressions for $\lambda_\alpha^{0,n}$ and $\lambda_\alpha^{1,n}$ given in Lemma 3.25, and the property that $\Delta W^n\otimes\Delta W^n = \mathrm{Id}\,\Delta t$ in the limit as $\Delta t\to 0$, since ΔW^n is always multiplied by $\mathcal{F}_{n\Delta t}$ measurable functions. This gives (using the expression (3.41) for κ_α)

$$\begin{aligned}&\lambda_\alpha^n\nabla\xi_\alpha(q^{n+1})\\ &= \left(\lambda_\alpha^{0,n}\sqrt{\Delta t}+\lambda_\alpha^{1,n}\Delta t\right)\nabla\xi_\alpha(q^n)\\ &\quad +\lambda_\alpha^{0,n}\sqrt{\Delta t}\nabla^2\xi_\alpha(q^n)\left(\sqrt{2\beta^{-1}}\Delta W^n+\lambda_\gamma^{0,n}\sqrt{\Delta t}\nabla\xi_\gamma(q^n)\right)+\mathrm{o}(\Delta t)\\ &= -\sqrt{2\beta^{-1}}G^{-1}_{\alpha,\zeta}\nabla\xi_\alpha\nabla\xi_\zeta(q^n)\cdot\Delta W^n + G^{-1}_{\alpha,\zeta}\nabla\xi_\alpha\nabla\xi_\zeta\cdot\nabla\tilde{V}(q^n)\Delta t\\ &\quad +\beta^{-1}G^{-1}_{\alpha,\zeta}\nabla\xi_\alpha G^{-1}_{\gamma,\delta}\nabla^2\xi_\zeta:\nabla\xi_\gamma\otimes\nabla\xi_\delta(q^n)\Delta t\\ &\quad -\beta^{-1}G^{-1}_{\alpha,\zeta}\nabla\xi_\alpha\Delta\xi_\zeta(q^n)\Delta t-2\beta^{-1}G^{-1}_{\alpha,\zeta}\nabla^2\xi_\alpha\nabla\xi_\zeta(q^n)\Delta t\\ &\quad +2\beta^{-1}G^{-1}_{\gamma,\alpha}\nabla^2\xi_\alpha\nabla\xi_\gamma(q^n)\Delta t+\mathrm{o}(\Delta t),\\ &= (P(q^n)-\mathrm{Id})\left(-\nabla\tilde{V}(q^n)\Delta t+\sqrt{2\beta^{-1}}\Delta W^n\right)\\ &\quad -\beta^{-1}\nabla\xi_\alpha G^{-1}_{\alpha,\zeta}\left(-G^{-1}_{\gamma,\delta}\nabla^2\xi_\zeta:\nabla\xi_\gamma\otimes\nabla\xi_\delta(q^n)+\Delta\xi_\zeta(q^n)\right)\Delta t+\mathrm{o}(\Delta t)\\ &= (P(q^n)-\mathrm{Id})\left(-\nabla\tilde{V}(q^n)\Delta t+\sqrt{2\beta^{-1}}\Delta W^n\right)\\ &\quad -\beta^{-1}\kappa_\alpha n_\alpha(q^n)\Delta t+\mathrm{o}(\Delta t).\end{aligned}$$

This shows that (3.66) can be rewritten as

$$\begin{aligned}q^{n+1} = q^n + P(q^n)\left(-\nabla\tilde{V}(q^n)\Delta t+\sqrt{2\beta^{-1}}\Delta W^n\right)\\ -\beta^{-1}\kappa_\alpha n_\alpha(q^n)\Delta t+\mathrm{o}(\Delta t),\end{aligned}\tag{3.73}$$

which is a consistent discretization of (3.52).

We now consider the scheme (3.67). In this case, using the expressions for $\lambda_\alpha^{0,n}$ and $\lambda_\alpha^{1,n}$ given in Lemma 3.25, we obtain

$$\begin{aligned}
&\lambda_\alpha^n \nabla\xi_\alpha(q^n) \\
&= \left(\lambda_\alpha^{0,n}\sqrt{\Delta t} + \lambda_\alpha^{1,n}\Delta t\right) \nabla\xi_\alpha(q^n) + \mathrm{o}(\Delta t) \\
&= -\sqrt{2\beta^{-1}} G^{-1}_{\alpha,\zeta} \nabla\xi_\alpha \nabla\xi_\zeta(q^n) \cdot \Delta W^n + G^{-1}_{\alpha,\zeta} \nabla\xi_\alpha \nabla\xi_\zeta \cdot \nabla\tilde{V}(q^n)\Delta t \\
&\quad + \beta^{-1} G^{-1}_{\alpha,\zeta} \nabla\xi_\alpha G^{-1}_{\gamma,\delta} \nabla^2\xi_\zeta : \nabla\xi_\gamma \otimes \nabla\xi_\delta(q^n)\Delta t \\
&\quad - \beta^{-1} G^{-1}_{\alpha,\zeta} \nabla\xi_\alpha \Delta\xi_\zeta(q^n)\Delta t + \mathrm{o}(\Delta t) \\
&= (P(q^n) - \mathrm{Id})\left(-\nabla\tilde{V}(q^n)\Delta t + \sqrt{2\beta^{-1}}\Delta W^n\right) \\
&\quad - \beta^{-1}\kappa_\alpha n_\alpha(q^n)\Delta t + \mathrm{o}(\Delta t),
\end{aligned}$$

which shows that (3.67) is also of the form (3.73) and proves that this scheme is consistent with (3.52). □

3.2.4.4 *Ergodicity of the numerical schemes and time discretization error*

We end this section by mentioning a result from [Faou and Lelièvre (2009)] concerning the ergodicity of the numerical schemes (3.66) and (3.67). Under suitable assumptions, it can be shown that the dynamics (3.66) and (3.67) are indeed ergodic with respect to a measure $\tilde{\nu}_\Sigma^{\Delta t}$ with the support of the submanifold Σ (compare with (3.54)):

$$\lim_{M\to\infty} \frac{1}{M} \sum_{n=1}^{M} \phi(q^n) = \int_\Sigma \phi \, d\tilde{\nu}_\Sigma^{\Delta t}. \tag{3.74}$$

A natural question is then: How different from $\tilde{\nu}_\Sigma$ is $\tilde{\nu}_\Sigma^{\Delta t}$? This question has been addressed in [Talay and Tubaro (1990); Talay (1991)] in the case of ergodic dynamics in $\mathbb{R}^n$. In [Faou and Lelièvre (2009)], these results are extended to the case of diffusions with constraint (for more general stochastic differential equations and numerical scheme than those considered here). More precisely, the following result is proven: There exists a positive constant Δt_0 such that for all positive $\Delta t < \Delta t_0$ and for all smooth function $\phi : \Sigma \to \mathbb{R}$, there exists $C > 0$ such that:

$$\left| \int_\Sigma \phi \, d\tilde{\nu}_\Sigma^{\Delta t} - \int_\Sigma \phi \, d\tilde{\nu}_\Sigma \right| \le C\Delta t.$$

We refer to [Faou and Lelièvre (2009)] for a precise statement of the general result, and some numerical experiments.

Remark 3.26 (Metropolis-Hastings algorithm in phase space). *A natural question at this point is whether it is possible to use the Metropolis-Hastings algorithm to eliminate the error in the sampled measure due to the time discretization, as for MALA (see Algorithm* 2.9*). The problem is that the transition kernel corresponding to a single step of the numerical schemes* (3.66) *and* (3.67) *has no simple analytical expression due to the nonlinear projection step. It is however possible to circumvent this difficulty, using algorithms in the phase space (see Section* 3.3.5.5*).*

3.2.5 *Computing the mean force*

In this section, various methods to compute the mean force $\nabla F(z)$ at some fixed value z of the reaction coordinate are discussed. Free energy differences are then obtained by numerical integration, as explained above (see (3.26) and (3.27)).

3.2.5.1 *Methods based on the sampling of the conditional measures $\nu^\xi(\cdot|z)$*

In view of (3.27)–(3.28), a natural algorithm to estimate $\nabla F(z)$ consists in computing the average of the local mean force f with respect to the conditional probability measures $\nu^\xi(\cdot|z)$. The last two sections 3.2.3 and 3.2.4 have been devoted to the sampling of such measures. More precisely, it has been shown that the dynamics (for q_0 such that $\xi(q_0) = z$)

$$dq_t = P(q_t)\left(-\nabla V^\xi(q_t)\,dt + \sqrt{2\beta^{-1}}\,dW_t\right) + \beta^{-1}\mathcal{H}(q_t)\,dt, \tag{3.75}$$

where $V^\xi = V + \beta^{-1}\ln|\det G|^{1/2}$ (see (3.32)) is ergodic with respect to $\nu^\xi(\cdot|z)$. Thus, for every smooth test function $\phi : \mathbb{R}^n \to \mathbb{R}$, for $\nu^\xi(\cdot|z)$-almost every $q_0 \in \Sigma(z)$,

$$\lim_{T\to\infty} \frac{1}{T}\int_0^T \phi(q_t)\,dt = \int_{\Sigma(z)} \phi\,d\nu^\xi(\cdot|z), \tag{3.76}$$

where q_t satisfies (3.75). Note that, for q_t to sample $\nu^\xi(\cdot|z)$, the potential used in the projected dynamics is V^ξ and not V.

Let us now explain two methods to compute $\nabla F(z)$. For simplicity, we here concentrate on continuous in time dynamics. Discretization aspects are discussed in the next section 3.2.5.2.

The first method simply consists in using the formula

$$\lim_{T\to\infty} \frac{1}{T}\int_0^T f(q_t)dt = \nabla F(z), \tag{3.77}$$

where q_t satisfies (3.75) and f is the local mean force. Let us recall its expression for convenience (see (3.28)):

$$f_\alpha = \sum_{\gamma=1}^{m} G^{-1}_{\alpha,\gamma} \nabla \xi_\gamma \cdot \nabla V - \beta^{-1} \mathrm{div} \left(\sum_{\gamma=1}^{m} G^{-1}_{\alpha,\gamma} \nabla \xi_\gamma \right).$$

This expression involves second derivatives of ξ which may be difficult to compute. The next method is a way to circumvent this difficulty.

The second method is based on the fact that $\nabla F(z)$ can be obtained by averaging the Lagrange multiplier Y_t associated with the constraint $\xi(q_t) = z$ (see (3.81) below). Let us explain this in detail. As mentioned in Section 3.2.4, the projected dynamics (3.52) can be rewritten using Lagrange multipliers, see (3.63). In the remainder of this section, we consider the dynamics (3.63) projected on $\Sigma(z)$ with $\tilde{V} = V^\xi$, which we rewrite for convenience:

$$\begin{cases} dq_t = -\nabla V^\xi(q_t)\, dt + \sqrt{2\beta^{-1}} dW_t + \displaystyle\sum_{\alpha=1}^{m} \nabla \xi_\alpha(q_t)\, d\lambda_{\alpha,t}, \\ \text{with } \lambda_t \in \mathbb{R}^m \text{ an adapted process such that } \xi(q_t) = z, \end{cases} \quad (3.78)$$

and

$$dY_t = \sum_{\alpha=1}^{m} \nabla \xi_\alpha(q_t)\, d\lambda_{\alpha,t}. \quad (3.79)$$

We have shown that the two dynamics (3.75) and (3.78) are equivalent, and are ergodic with respect to $\nu^\xi(\cdot|z)$. Let us recall that the Lagrange multiplier Y_t also writes:

$$dY_t = (P(q_t) - \mathrm{Id}) \left(-\nabla V^\xi(q_t)\, dt + \sqrt{2\beta^{-1}} dW_t \right) + \beta^{-1} \mathcal{H}(q_t)\, dt. \quad (3.80)$$

The remainder of this section is devoted to the proof of the proposition:

Proposition 3.27. *Consider the processes* (q_t) *solution to* (3.75) *with* $\xi(q_0) = z$ *and the associated Lagrange multiplier* Y_t *defined by* (3.80). *Then for* $\alpha \in \{1, \ldots, m\}$,

$$\lim_{T\to\infty} \frac{1}{T} \int_0^T \sum_{\zeta=1}^{m} G^{-1}_{\alpha,\zeta} \nabla \xi_\zeta(q_t) \cdot dY_t = \nabla_\alpha F(z), \quad (3.81)$$

almost surely and in L^p, $p \geq 1$.

To prove this result, another expression for the mean force is needed.

Lemma 3.28. *The local mean force f defined by (3.28) can be rewritten as:* $\forall \alpha \in \{1, \dots, m\}$,

$$f_\alpha = \sum_{\gamma=1}^{m} G^{-1}_{\alpha,\gamma} \nabla \xi_\gamma \cdot \left(\nabla V^\xi + \beta^{-1} \mathcal{H} \right), \tag{3.82}$$

where the modified potential V^ξ is defined by (3.32) and $\mathcal{H}$ is the mean curvature vector. Thus, the mean force can be written as:

$$\nabla_\alpha F(z) = \frac{\displaystyle\int_{\Sigma(z)} \sum_{\gamma=1}^{m} G^{-1}_{\alpha,\gamma} \nabla \xi_\gamma \cdot \left(\nabla V^\xi + \beta^{-1} \mathcal{H} \right) \exp(-\beta V^\xi) \, d\sigma_{\Sigma(z)}}{\displaystyle\int_{\Sigma(z)} \exp(-\beta V^\xi) \, d\sigma_{\Sigma(z)}}. \tag{3.83}$$

Proof. These results are obtained by straightforward computations, using the fact that (by (3.40)), for $\alpha \in \{1, \dots, m\}$,

$$\sum_{\gamma=1}^{m} G^{-1}_{\alpha,\gamma} \nabla \xi_\gamma \cdot \mathcal{H} = |\nabla \xi_\alpha|^{-1} \kappa_\alpha,$$

that (by (3.44)),

$$|\nabla \xi_\alpha|^{-1} \kappa_\alpha = \operatorname{div} \left(\sum_{\gamma=1}^{m} G^{-1}_{\alpha,\gamma} \nabla \xi_\gamma \right) + \frac{1}{2} \sum_{\gamma=1}^{m} G^{-1}_{\alpha,\gamma} \nabla \xi_\gamma \cdot \nabla \ln(\det G),$$

and that (from the definition (3.32) of V^ξ)

$$\exp(-\beta V^\xi) = \exp(-\beta V) |\det G|^{-1/2}.$$

□

We are now in a position to prove Proposition 3.27.

Proof. The fact that the potential in the dynamics (3.75) is $\tilde{V}$ has two consequences. First, by Proposition 3.20, the measure sampled by (q_t) solution to (3.75) (and thus to (3.78)) is indeed $\nu^\xi(\cdot|z)$.

Second, the process Y_t can be expressed as (3.80). Hence, since $(P(q) - \mathrm{Id}) \, G^{-1}_{\alpha,\gamma} \nabla \xi_\gamma(q) = -G^{-1}_{\alpha,\gamma} \nabla \xi_\gamma(q)$, it holds

$$\begin{aligned} G^{-1}_{\alpha,\gamma} \nabla \xi_\gamma(q_t) \cdot dY_t = {} & G^{-1}_{\alpha,\gamma} \nabla \xi_\gamma \cdot \left(\nabla V^\xi + \beta^{-1} \mathcal{H} \right)(q_t) \, dt \\ & - \sqrt{2\beta^{-1}} G^{-1}_{\alpha,\gamma} \nabla \xi_\gamma(q_t) \cdot dW_t. \end{aligned}$$

In the bounded variation part, we recover the expression (3.82) of the mean force. Now, (3.81) follows from the ergodicity of q_t (see (3.76)) and the fact that

$$\lim_{T \to \infty} \frac{1}{T} \int_0^T G^{-1}_{\alpha,\gamma} \nabla \xi_\gamma(q_t) \cdot dW_t = 0,$$

by (3.9) (see for example Theorem 1.3.15 in [Duflo (1997)]). □

Remark 3.29 (Cost interpretation of the Lagrange multiplier). *The fact that averaging the Lagrange multiplier indeed yields the mean force may seem miraculous. In this remark, we would like to mention another context where the Lagrange multipliers are related to the variation of some quantity with respect to a constraint, in the framework of optimization.*

For simplicity, we assume $m = 1$. Let us consider the following problem:

$$A(z) = \inf_{x \in \mathbb{R}^n} \left\{ \tilde{V}(x),\, \xi(x) = z \right\}.$$

Under suitable assumptions (see for example [Bonnans et al. *(2002)]), it is known that, for a given z, the solution $x(z)$ to this constrained optimization problem satisfies the Euler-Lagrange equations: there exists $\lambda(z) \in \mathbb{R}$ such that*

$$\begin{cases} \nabla \tilde{V}\big(x(z)\big) = \lambda(z) \nabla \xi\big(x(z)\big), \\ \xi\big(x(z)\big) = z, \end{cases} \tag{3.84}$$

and $A(z) = \tilde{V}\big(x(z)\big)$. It can then be checked that

$$A'(z) = \lambda(z).$$

Indeed, using the Euler-Lagrange equations (3.84)*:*

$$\begin{aligned} A'(z) &= \nabla \tilde{V}\big(x(z)\big) \cdot x'(z) \\ &= \lambda(z) \nabla \xi\big(x(z)\big) \cdot x'(z) \\ &= \lambda(z) \frac{d}{dz}\Big(\xi\big(x(z)\big)\Big) \\ &= \lambda(z). \end{aligned}$$

This basic fact is somewhat related to what we showed in this section, since formally, in the limit of zero temperature, the dynamics (3.78) *indeed samples the minimizers of $\tilde{V} = V^\xi$ under the constraint $\xi(x) = 0$.*

3.2.5.2 *Methods based on the sampling of the conditional measures $\nu^\xi(\cdot|z)$: discretization*

We would like to discuss in this section practical aspects of the computation of the mean force using one of the two methods introduced in the last section: A trajectory average of the local mean force (3.77), or a trajectory average of the Lagrange multiplier (3.81). The discretization methods are based on the numerical schemes (3.66) and (3.67).

First, for both methods and as mentioned in the last section, in order to sample the right distribution (namely $\nu^\xi(\cdot|z)$), the potential in schemes (3.66) and (3.67) should be $\tilde{V} = V^\xi = V + \beta^{-1} \ln|\det G|^{1/2}$ (see (3.32)). A remark is in order about the computation of ∇V^ξ.

Remark 3.30 (Computation of ∇V^ξ by finite differences). *In the numerical schemes, the computation of ∇V^ξ requires the computation of:*

$$\beta^{-1}\nabla \ln\left(|\det G|^{-1/2}\right)(q^n)\Delta t = -\frac{1}{2}\beta^{-1}\sum_{\alpha,\zeta}\left(G^{-1}_{\alpha,\zeta}\nabla G_{\alpha,\zeta}\right)(q^n)\Delta t$$
$$= -\beta^{-1}\sum_{\alpha,\zeta}\left(G^{-1}_{\alpha,\zeta}\nabla^2\xi_\alpha\nabla\xi_\zeta\right)(q^n)\Delta t, \tag{3.85}$$

where we used Jacobi's formula (3.49). *In order to avoid the computation of $\nabla^2\xi_\alpha(q^n)$, which may be cumbersome,* (3.85) *can be approximated by:*

$$-\beta^{-1}\sum_{\alpha,\zeta}\left(G^{-1}_{\alpha,\zeta}\nabla^2\xi_\alpha\nabla\xi_\zeta\right)(q^n)\Delta t$$
$$= -\beta^{-1}\sum_{\alpha,\zeta=1}^{m} G^{-1}_{\alpha,\zeta}(q^n)\Big(\nabla\xi_\alpha\big(q^n+\Delta t\nabla\xi_\zeta(q^n)\big)-\nabla\xi_\alpha(q^n)\Big)+\mathrm{o}(\Delta t). \tag{3.86}$$

We also refer to Section 3.2.5.3 *for alternative approaches avoiding the use of the modified potential V^ξ (and thus the computation of second derivatives of ξ).*

Let us now discuss the discretization of the two methods presented in Section 3.2.5.1 to compute the mean force.

Concerning the method based on a trajectory average of the local mean force (3.77), the idea is simply to use as an estimate of the mean force the trajectory average:

$$\frac{1}{M}\sum_{n=1}^{M} f(q^n)$$

where $M = T/\Delta t$ and q^n is the solution to (3.66) or to (3.67), with $\tilde{V}$ replaced by V^ξ, and the constraint $\xi(q^n) = z$ (rather than $\xi(q^n) = 0$). A problem with this method is that the expression for f involve second derivatives of ξ:

$$f_\alpha = \sum_{\gamma=1}^{m} G^{-1}_{\alpha,\gamma}\nabla\xi_\gamma\cdot\nabla V - \beta^{-1}\mathrm{div}\left(\sum_{\gamma=1}^{m} G^{-1}_{\alpha,\gamma}\nabla\xi_\gamma\right).$$

A method based on finite differences to circumvent this difficulty is the following. The first term in $\nabla F(z)$ can be obtained from

$$\begin{aligned} &\lim_{T\to\infty}\lim_{\Delta t\to 0}\frac{1}{M}\sum_{n=1}^{M}\sum_{\gamma=1}^{m}\left(\nabla V\cdot G^{-1}_{\alpha,\gamma}\nabla\xi_\gamma\right)(q^n) \\ &\quad=\sum_{\gamma=1}^{m}\int_{\Sigma(0)}\nabla V\cdot G^{-1}_{\alpha,\gamma}\nabla\xi_\gamma\,d\nu^\xi(\cdot|z). \end{aligned} \tag{3.87}$$

For the second term,

$$\begin{aligned} -\beta^{-1}\lim_{T\to\infty}\lim_{\Delta t\to 0}\frac{1}{M\Delta t}\sum_{n=1}^{M}\sum_{\gamma=1}^{m}&\Big((G^{-1}_{\alpha,\gamma}\nabla\xi_\gamma)(q^n+\Delta W^n) \\ &-(G^{-1}_{\alpha,\gamma}\nabla\xi_\gamma)(q^n)\Big)\cdot\Delta W^n \end{aligned} \tag{3.88}$$

$$=-\beta^{-1}\sum_{\gamma=1}^{m}\int_\Sigma\nabla\cdot(G^{-1}_{\alpha,\gamma}\nabla\xi_\gamma)\,d\nu^\xi(\cdot|z),$$

where we used the fact that $\Delta W^n\otimes\Delta W^n=\mathrm{Id}\,\Delta t$ in the limit as $\Delta t\to 0$. Equations (3.87) and (3.88) (together with the approximation (3.86)) allow us to estimate $\nabla_\alpha F(z)$ without having to compute $\nabla^2\xi_\alpha$. Of course, this simplification comes at the price of a larger variance, see Remark 3.34.

The second method consists in averaging the Lagrange multiplier associated with the constraints, see Proposition 3.27. More precisely, as in the continuous in time case (see (3.81)), the mean force $\nabla_\alpha F(z)$ may be computed by averaging the Lagrange multipliers λ^n_α entering the algorithms (3.66) or (3.67).

Proposition 3.31. *Let (q^n) be the solution to (3.66) or to (3.67), with $\tilde V$ replaced by $V^\xi=V+\beta^{-1}\ln|\det G|^{1/2}$, and the constraint $\xi(q^n)=z$ (rather than $\xi(q^n)=0$). Then,*

$$\lim_{T\to\infty}\lim_{\Delta t\to 0}\frac{1}{M\Delta t}\sum_{n=1}^{M}\lambda^n_\alpha=\nabla_\alpha F(z), \tag{3.89}$$

where $M=T/\Delta t$.

This method (together with the approximation (3.86)) thus allows us to estimate $\nabla_\alpha F(z)$ without having to compute $\nabla^2\xi_\alpha$. Again, this simplification comes at the price of a larger variance, see Remark 3.34. Note that the Lagrange multipliers involved in (3.89) are actually needed to build the dynamics (q^n), so that, in contrast with the previous method, this method yields the mean force without any additional cost.

Proposition 3.31 is a direct consequence of Proposition 3.27 and the following lemma.

Lemma 3.32. *Let (q^n) be the solution to (3.66) or to (3.67), with $\tilde{V}$ replaced by $V^\xi = V + \beta^{-1} \ln|\det G|^{1/2}$. Assume moreover that (3.9) holds. Then, for $1 \le \alpha \le m$, λ_α^n is such that*

$$\lim_{\Delta t \to 0} \frac{1}{M\Delta t} \sum_{n=1}^{M} \lambda_\alpha^n = \frac{1}{T} \int_0^T \sum_{\zeta=1}^{m} G_{\alpha,\zeta}^{-1} \nabla \xi_\zeta(q_t) \cdot dY_t \tag{3.90}$$

with Y_t defined by (3.80) and where $T = M\Delta t$ is fixed (so that $M \to \infty$).

Proof. Using Lemma 3.25, we have

$$\lim_{\Delta t \to 0} \frac{1}{M\Delta t} \sum_{n=1}^{M} \lambda_\alpha^n$$

$$= \frac{1}{T} \int_0^T G_{\alpha,\zeta}^{-1} \left(\nabla \xi_\zeta \cdot \nabla V^\xi + \beta^{-1} G_{\gamma,\delta}^{-1} \nabla^2 \xi_\zeta : \nabla \xi_\gamma \otimes \nabla \xi_\delta - \beta^{-1} \Delta \xi_\zeta \right)(q_t)\, dt$$

$$- \sqrt{2\beta^{-1}} \frac{1}{T} \int_0^T G_{\alpha,\zeta}^{-1} \nabla \xi_\zeta(q_t) \cdot dW_t.$$

Then, (3.90) is obtained using the fact that, for a fixed $\alpha \in \{1, \dots, m\}$, $(P(q) - \mathrm{Id})\, G_{\alpha,\zeta}^{-1} \nabla \xi_\zeta(q) = -G_{\alpha,\zeta}^{-1} \nabla \xi_\zeta(q)$ and, by (3.40)–(3.41), again for a fixed $\alpha \in \{1, \dots, m\}$,

$$-\kappa_\alpha |\nabla \xi_\alpha|^{-1} = G_{\alpha,\gamma}^{-1} \nabla \xi_\gamma \cdot \mathcal{H} = G_{\alpha,\zeta}^{-1} G_{\gamma,\delta}^{-1} \nabla^2 \xi_\zeta : \nabla \xi_\gamma \otimes \nabla \xi_\delta - G_{\alpha,\zeta}^{-1} \Delta \xi_\zeta. \qquad \square$$

We end this paragraph concerning the mean force computation by two remarks concerning the variance of the methods.

Remark 3.33 (A variance reduction method). *In the numerical scheme we have described to compute the mean force by averaging the Lagrange multipliers (see formula (3.89)), there are three sources of error: The time discretization error ($\Delta t \to 0$), the longtime limit error ($T \to \infty$) and the statistical error due to the fact that we use a stochastic process. In this remark, we focus on the statistical error. It is linked to the variance of the result. Let us consider the case $m = 1$ for simplicity. It holds:*

$$\frac{1}{M\Delta t} \sum_{n=1}^{M} \lambda^n = -\sqrt{2\beta^{-1}} \frac{1}{T} \sum_{n=1}^{M} \frac{\nabla \xi}{|\nabla \xi|^2}(q^n) \cdot \Delta W^n + \frac{1}{T} \sum_{n=1}^{M} \lambda^{1,n} \Delta t + \mathrm{o}(1).$$

In the limit $\Delta t \to 0$, the first term in the right-hand side converges to the martingale part

$$-\sqrt{2\beta^{-1}}\frac{1}{T}\int_0^T \frac{\nabla\xi}{|\nabla\xi|^2}(q_s)\cdot dW_s$$

of the constraining force Y_t, while the second part converges to the bounded variation part of Y_t (see Equation (3.80)). It is the limit, when T goes to infinity, of the second term which yields the mean force $F'(0)$. The first term goes to 0 in the limit $T \to \infty$ but this term is responsible for a large variance of the result.

Therefore, a natural idea to reduce the variance is to eliminate the first term. It is possible to directly compute this term (which is the projection of the Brownian increment) and to subtract it from the Lagrange multiplier. Alternatively, the following scheme, which is very easy to implement, may be used. We consider that $t = t_n$ and we denote by $\lambda(\Delta W^n)$ the Lagrange multiplier obtained from (3.66) or (3.67) with a Brownian increment ΔW^n. The next position q^{n+1} is defined by (3.66) or (3.67), but the Lagrange multiplier used in formula (3.89) is now defined as:

$$\lambda^n = \frac{1}{2}\left(\lambda(\Delta W^n) + \lambda(-\Delta W^n)\right). \tag{3.91}$$

It can be checked that this does not change the value of the bounded variation part $\lambda^{1,n}$ of λ^n, but "eliminates" the martingale part $\lambda^{0,n}$. This method reminds us of the antithetic variables variance reduction method classically used in Monte Carlo methods (see for example [Lapeyre et al. *(2003)]). In practice, this method seems to be very efficient (see [Lelièvre* et al. *(2007a)]). This idea can be straightforwardly generalized to the case $m > 1$.*

Remark 3.34 (Approximating the local mean force). *The methods presented above to circumvent the computation of second derivatives of ξ in the local mean force f (see Equations (3.88) and (3.89)) amount to approximating the local mean force by some finite differences formula involving the Brownian increment. For a fixed $q^n = q$, the bottom line of the argument is indeed to note that*

$$\lim_{M\to\infty}\lim_{\Delta t\to 0}\frac{1}{M\Delta t}\sum_{m=1}^{M}\left[\xi\left(q+\sqrt{\Delta t}\,G^m\right) - 2\xi(q) + \xi\left(q-\sqrt{\Delta t}\,G^m\right)\right] = \Delta\xi(q),$$

where G^m are independent n-dimensional Gaussian vectors with covariance Id. *When passing to the limit $\Delta t \to 0$, it holds:*

$$\lim_{M\to\infty}\frac{1}{M}\sum_{m=1}^{M} G^m\cdot\nabla^2\xi(q)\,G^m = \Delta\xi(q).$$

The estimator $\frac{1}{M}\sum_{m=1}^{M} G^m \cdot \nabla^2\xi(q)\, G^m$ *is unbiased (its expectation is* $\Delta\xi(q)$*) but has a finite non-zero variance* $\frac{2}{M}\nabla^2\xi(q) : \nabla^2\xi(q)$*. Thus, the stochastic approximation of the local mean force by finite differences (which avoids the sometimes cumbersome computation of second derivatives of* ξ*) has a price in terms of variance. Using the analytic expression for the local mean force (like* (3.28)*) is better in terms of variance.*

3.2.5.3 *Methods based on the sampling of the measures* $\nu_{\Sigma(z)}$

We have already discussed the importance of using the modified potential V^ξ instead of V in the constrained diffusion in order to sample the measure $\nu^\xi(\cdot|z)$ and to compute the correct mean force. We also mentioned that this induces computational difficulties related to the calculation of ∇V^ξ which contains second order derivative of the reaction coordinate ξ (see Remark 3.30). In this section, we would like to discuss a method which circumvent this difficulty by (i) computing first another "free energy" called the rigid free energy, and then (ii) recovering the correct free energy by an additional computation.

(i) Computing "generalized free energies", and in particular the rigid free energy. Let us first discuss the computation of "generalized free energies". It is interesting to note that all the former computations may be generalized to the following "generalized free energy":

$$F_g(z) = -\beta^{-1} \ln \int_{\Sigma(z)} \exp(-\beta V)\, g\, d\sigma_{\Sigma(z)}, \tag{3.92}$$

where g is a given positive function[6] such that $\int_{\Sigma(z)} \exp(-\beta V)\, g\, d\sigma_{\Sigma(z)} < \infty$. Indeed, in this case, the expression of the gradient of F_g is given by the following formula (which is a generalization of (3.83))

$$\nabla_\alpha F_g(z) = \frac{\displaystyle\int_{\Sigma(z)} \sum_{\gamma=1}^{m} G^{-1}_{\alpha,\gamma} \nabla\xi_\gamma \cdot (\nabla V_g + \beta^{-1}\mathcal{H}) \exp(-\beta V_g)\, d\sigma_{\Sigma(z)}}{\displaystyle\int_{\Sigma(z)} \exp(-\beta V_g)\, d\sigma_{\Sigma(z)}}, \tag{3.93}$$

where the modified potential V_g is defined by:

$$V_g = V - \beta^{-1} \ln g. \tag{3.94}$$

Suppose now that in the numerical schemes (3.66) or (3.67), we take $\tilde{V} = V_g$. Then, following the arguments of the former sections, it is easy

[6] With this notation, the free energy defined by (3.22) is $F_{(\det G)^{-1/2}}$, up to an additive constant. Note that this constant does not intervene in the mean force.

to check that (in the limit $\Delta t \to 0$), (q^n) is ergodic with respect to the probability measure

$$\frac{\exp(-\beta V_g)\, d\sigma_{\Sigma(z)}}{\displaystyle\int_{\Sigma(z)} \exp(-\beta V_g)\, d\sigma_{\Sigma(z)}}.$$

Moreover, as explained above, an approximation of $\nabla F_g(z)$ can then be obtained in two ways: (i) by averaging the "generalized local mean force" $\sum_{\gamma=1}^m G^{-1}_{\alpha,\gamma} \nabla\xi_\gamma \cdot (\nabla V_g + \beta^{-1}\mathcal{H})$ over a trajectory (q_n), or (ii) by averaging the Lagrange multipliers associated with the constraint $\xi(q^n) = z$: for $1 \leq \alpha \leq m$,

$$\lim_{T\to\infty} \lim_{\Delta t \to 0} \frac{1}{M\Delta t} \sum_{n=1}^{M} \lambda_\alpha^n = \nabla_\alpha F_g(0), \tag{3.95}$$

where $M = T/\Delta t$.

An important particular case is when $g = 1$. This amounts to considering the former algorithms, with $\tilde{V} = V$. In this case, the process (q^n) samples (in the limit $\Delta t \to 0$)

$$d\nu_{\Sigma(z)} = \frac{\exp(-\beta V)\, d\sigma_{\Sigma(z)}}{\displaystyle\int_{\Sigma(z)} \exp(-\beta V)\, d\sigma_{\Sigma(z)}}.$$

By averaging over a trajectory (q^n) the so-called *rigid local mean force*:

$$f_{\rm rgd} = \sum_{\gamma=1}^{m} G^{-1}_{\alpha,\gamma} \nabla\xi_\gamma \cdot \left(\nabla V + \beta^{-1}\mathcal{H}\right),$$

or the Lagrange multiplier associated with the constraint $\xi(q^n) = z$, the so-called *rigid mean force* is obtained:

$$\nabla F_{\rm rgd}(z) = \frac{\displaystyle\int_{\Sigma(z)} f_{\rm rgd} \exp(-\beta V)\, d\sigma_{\Sigma(z)}}{\displaystyle\int_{\Sigma(z)} \exp(-\beta V)\, d\sigma_{\Sigma(z)}}.$$

Using Lemma 3.10 and (3.44), it is easy to check that the rigid mean force is the gradient of the *rigid free energy*:

$$F_{\rm rgd}(z) = -\beta^{-1} \ln\left(\int_{\Sigma(z)} \exp(-\beta V)\, d\sigma_{\Sigma(z)}\right).$$

We use the name "rigid free energy" because it is associated with the probability measure $\nu_{\Sigma(z)}$ which is naturally sampled by the rigidly constrained stochastic process (3.52) with $\tilde{V} = V$, see Section 3.2.3.4. In the literature, F_{rgd} is sometimes called the "gauge invariant free energy" (see [E and Vanden-Eijnden (2004)]), since $F_{\text{rgd}}(z)$ is a quantity which depends on the reaction coordinate ξ only through its level set $\Sigma(z)$, contrary to the standard free energy F which involves some gradient of ξ on $\Sigma(z)$. In other words, the free energy difference $F_{\text{rgd}}(1) - F_{\text{rgd}}(0)$ only depends on the submanifolds $\Sigma(0)$ and $\Sigma(1)$, and not on the parametrization of the foliation in-between $z = 0$ and $z = 1$ (see also Remark 1.3).

(ii) Recovering the standard free energy from the rigid free energy. As mentioned above, the interest of considering the rigid free energy F_{rgd} rather than the standard free energy F is that the associated mean force ∇F_{rgd} can be computed by sampling the probability measure $\nu_{\Sigma(z)}$, namely using the numerical schemes (3.66) or (3.67) with $\tilde{V} = V$. For the standard mean force ∇F, one has to use $\tilde{V} = V^{\xi}$ which may be cumbersome since this requires the computation of second derivatives of ξ (see however Remark 3.30).

A natural question is then: Is it possible to recover the standard free energy from the rigid free energy? This is indeed the case by noticing that:

$$\begin{aligned}
F(z) &= F_{\text{rgd}}(z) + F(z) - F_{\text{rgd}}(z) \\
&= F_{\text{rgd}}(z) - \beta^{-1} \ln \left(\frac{\displaystyle\int_{\Sigma(z)} \exp(-\beta V)(\det G)^{-1/2} d\sigma_{\Sigma(z)}}{\displaystyle\int_{\Sigma(z)} \exp(-\beta V) d\sigma_{\Sigma(z)}} \right) \\
&= F_{\text{rgd}}(z) - \beta^{-1} \ln \left(\int_{\Sigma(z)} (\det G)^{-1/2} d\nu_{\Sigma(z)} \right).
\end{aligned}$$

Thus, to compute F, it is possible to use the rigidly constrained dynamics (3.52) with $\tilde{V} = V$ (which samples $\nu_{\Sigma(z)}$) in order to (i) compute F_{rgd}, for example by averaging the Lagrange multipliers to get ∇F_{rgd} as explained above, and (ii) compute the difference between F and F_{rgd} by averaging $(\det G)^{-1/2}$ with respect to the probability measure $\nu_{\Sigma(z)}$. This yields a method to compute the free energy without computing second order derivatives of ξ. We refer for example to Equation (4.32) in [Chipot and Pohorille (2007b)], and references therein.

This remark of course applies to any energy F_g by changing $(\det G)^{-1/2}$ to g in the previous computations.

3.2.5.4 *A numerical illustration*

We consider the model system (dimer in a solvent) described in Section 1.3.2.4, with the same parameters as in Section 2.5.2.3. The mean force is estimated at the values $z_i = z_{\min} + i\Delta z$, with $z_{\min} = 0.1$ and $\Delta z = 0.012$, by ergodic averages obtained with the projected dynamics (3.66) integrated up to a time $T = 125$ with a step size $\Delta t = 2.5 \times 10^{-4}$. The resulting mean force profile, obtained by ergodic averages of the local mean force (which can be analytically computed in this case), is presented in Figure 3.4, together with the associated free energy profile. Note that the latter is very similar to the profile obtained in Section 2.5.2.3.

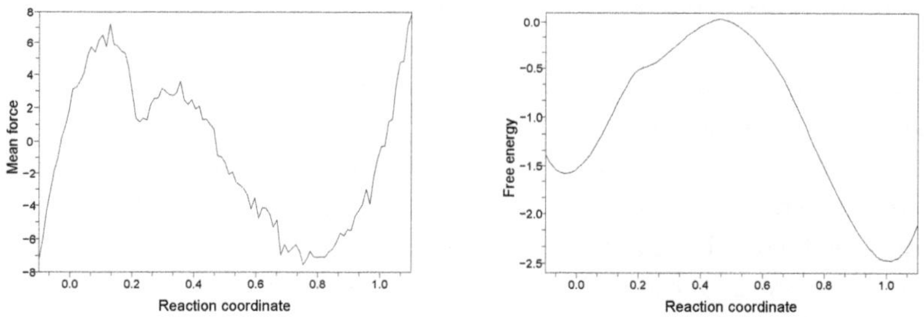

Fig. 3.4 Left: Estimated mean force. Right: Corresponding free energy.

It is interesting to study more carefully the variance of the results depending on the estimator used to compute the mean force: ergodic average of an analytical formula of the local mean force, ergodic average of the Lagrange multiplier or ergodic average of the Lagrange multiplier with variance reduction (see Remark 3.33). To this end, computations at a fixed value $z = -0.1$ of the reaction coordinate are performed. On Figure 3.5, the analytical local mean force, the full Lagrange multipliers (3.69), and the Lagrange multipliers (3.71) where the martingale part has been subtracted are plotted as a function of time in the case $\Delta t = 2.5 \times 10^{-5}$. We did not plot the results obtained with the variance reduction method (3.91), since they are almost indistinguishable from those obtained with (3.71).

The results show that the variance reduction has a dramatic impact since the curves obtained with the Lagrange multipliers for which the martingale part have been subtracted are very close to the ones obtained with the analytical local mean force. Note however that there are regions where the values obtained with (3.71) are noticeably different from the local mean

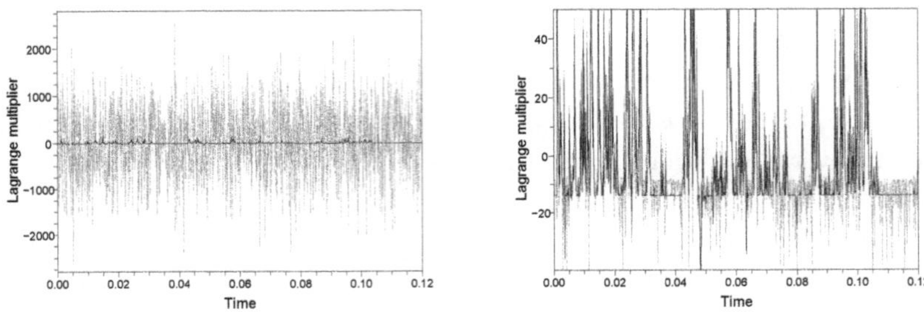

Fig. 3.5 Left: Comparison of the full Lagrange multipliers (3.69) (grey line) and the local mean force (black line). Right: Comparison of the Lagrange multipliers with the martingale part subtracted (3.71) (grey line), and the local mean force (black line). Note that the scales of the vertical axis are very different on the two pictures.

force. For example, on Figure 3.5, Right, on the interval $[0.10, 0.12]$, the local mean force is almost constant, while its approximation by (3.71) varies. These variations are intrinsic, and do not decrease in magnitude as the time-step is decreased. This is a numerical illustration of the fact that results obtained with the analytical formula of the local mean force have lower variance than those obtained using Lagrange multipliers, see Remark 3.34.

3.2.6 *On the efficiency of constrained sampling*

In this section, we would like to discuss the efficiency of constraining methods, by looking at the rate of convergence of the law of q_t solution to the rigidly constrained dynamics (3.52), which we recall for convenience:

$$dq_t = P(q_t)\left(-\nabla\tilde{V}(q_t)\,dt + \sqrt{2\beta^{-1}}\,dW_t\right) + \beta^{-1}\mathcal{H}(q_t)\,dt.$$

The equilibrium measure $\tilde{\nu}_\Sigma$ reads:

$$d\tilde{\nu}_\Sigma = Z_\Sigma^{-1}\exp(-\beta\tilde{V})\,d\sigma_\Sigma, \qquad Z_\Sigma = \int_\Sigma \exp(-\beta\tilde{V})\,d\sigma_\Sigma.$$

The question we ask is the following: Why do these methods perform better (to compute the free energy) than a naive algorithm which would consist in sampling the canonical measure ν (typically by the gradient dynamics (2.34)), and then computing the free energy by looking at the image of the measure ν by ξ (see Section 1.3.4)?

We use the entropy techniques presented in Section 2.3.2 to prove the following result:

Proposition 3.35. *Assume that the probability measure $\tilde{\nu}_\Sigma$ satisfies a logarithmic Sobolev inequality with constant ρ: for any function ϕ such that $\phi\, d\sigma_\Sigma$ is a probability measure,*

$$H\big(\phi\, d\sigma_\Sigma \,|\, \tilde{\nu}_\Sigma\big) \le \frac{1}{2\rho} I\big(\phi\, d\sigma_\Sigma \,|\, \tilde{\nu}_\Sigma\big)$$

where the entropy is:

$$H\big(\phi\, d\sigma_\Sigma \,|\, \tilde{\nu}_\Sigma\big) = \int_\Sigma \ln\left(\frac{\phi}{Z_\Sigma^{-1}\,\mathrm{e}^{-\beta\tilde{V}}}\right)\phi\, d\sigma_\Sigma$$

and the Fisher information writes:

$$I\big(\phi\, d\sigma_\Sigma \,|\, \tilde{\nu}_\Sigma\big) = \int_\Sigma \left|\nabla_\Sigma \ln\left(\frac{\phi}{Z_\Sigma^{-1}\,\mathrm{e}^{-\beta\tilde{V}}}\right)\right|^2 \phi\, d\sigma_\Sigma,$$

where $\nabla_\Sigma = P\nabla$ is defined by (3.60)*. Then the law of q_t solution to* (3.52) *converges to its equilibrium value exponentially fast with rate $\beta^{-1}\rho$. More precisely, if $\psi(t,\cdot)\, d\sigma_\Sigma$ denotes the law of q_t, then: $\forall t \ge 0$*

$$H\big(\psi(t,\cdot) d\sigma_\Sigma \,|\, \tilde{\nu}_\Sigma\big) \le H\big(\psi(0,\cdot) d\sigma_\Sigma \,|\, \tilde{\nu}_\Sigma\big)\, \exp(-2\beta^{-1}\rho t),$$

and thus

$$\big\|\psi(t,\cdot) d\sigma_\Sigma - \tilde{\nu}_\Sigma\big\|_{\mathrm{TV}} \le \sqrt{H\big(\psi(0,\cdot) d\sigma_\Sigma \,|\, \tilde{\nu}_\Sigma\big)}\, \exp(-\beta^{-1}\rho t).$$

Proof. Let us first write the Fokker-Planck equation associated with the dynamics (3.52). Since ψ is the density of q_t with respect to the measure σ_Σ, it holds for any smooth test function u,

$$\mathbb{E}(u(q_t)) = \int_\Sigma u(q)\psi(t,q)\, \sigma_\Sigma(dq).$$

The infinitesimal generator $\mathcal{L}$ of the dynamics (3.52) writes (see Remark 3.23):

$$\begin{aligned}\mathcal{L}u &= -P\nabla\tilde{V}\cdot\nabla u + \beta^{-1}\mathcal{H}\cdot\nabla u + \beta^{-1}P : \nabla^2 u \\ &= \beta^{-1}\mathrm{e}^{\beta\tilde{V}}\,\mathrm{div}_\Sigma\left(\mathrm{e}^{-\beta\tilde{V}}\nabla_\Sigma u\right),\end{aligned}$$

where the surface divergence div_Σ is defined by (3.56). Besides, by Itô's calculus:

$$\frac{d}{dt}\mathbb{E}(u(q_t)) = \mathbb{E}(\mathcal{L}u(q_t)).$$

In terms of ψ, this rewrites:

$$\begin{aligned}\frac{d}{dt}\int_\Sigma u\,\psi(t,\cdot)d\sigma_\Sigma &= \int_\Sigma (\mathcal{L}u)\,\psi(t,\cdot)d\sigma_\Sigma \\ &= \beta^{-1}\int_\Sigma \mathrm{e}^{\beta\tilde{V}}\mathrm{div}_\Sigma\left(\mathrm{e}^{-\beta\tilde{V}}\nabla_\Sigma u\right)\psi(t,\cdot)d\sigma_\Sigma \\ &= -\beta^{-1}\int_\Sigma \nabla_\Sigma u\cdot\nabla_\Sigma\left(\psi(t,\cdot)\mathrm{e}^{\beta\tilde{V}}\right)\mathrm{e}^{-\beta\tilde{V}}\,d\sigma_\Sigma,\end{aligned}$$

where we used the divergence theorem (3.57) to perform an integration by parts. We thus obtain the following variational formulation of the Fokker-Planck equation: for any smooth test function u,

$$\int_\Sigma u\,\partial_t\psi(t,\cdot)d\sigma_\Sigma = -\beta^{-1}\int_\Sigma \nabla_\Sigma u\cdot\nabla_\Sigma\left(\psi(t,\cdot)\mathrm{e}^{\beta\tilde{V}}\right)\mathrm{e}^{-\beta\tilde{V}}\,d\sigma_\Sigma.$$

Thus, the derivative of the relative entropy of ψ with respect to its equilibrium value $\psi_\infty = Z_\Sigma^{-1}\exp(-\beta\tilde{V})$ is (using the fact that $\frac{d}{dt}\int_\Sigma \psi(t,\cdot)d\sigma_\Sigma = 0$):

$$\begin{aligned}\frac{d}{dt}H\big(\psi(t,\cdot)d\sigma_\Sigma\,|\,\tilde{\nu}_\Sigma\big) &= \frac{d}{dt}\int_\Sigma \ln\left(\psi(t,\cdot)\mathrm{e}^{\beta\tilde{V}}\right)\psi(t,\cdot)d\sigma_\Sigma \\ &= \int_\Sigma \ln\left(\psi(t,\cdot)\mathrm{e}^{\beta\tilde{V}}\right)\partial_t\psi(t,\cdot)d\sigma_\Sigma \\ &= -\beta^{-1}\int_\Sigma \nabla_\Sigma\left(\ln(\psi(t,\cdot))\mathrm{e}^{\beta\tilde{V}}\right)\cdot\nabla_\Sigma\left(\psi(t,\cdot)\mathrm{e}^{\beta\tilde{V}}\right)\mathrm{e}^{-\beta\tilde{V}}\,d\sigma_\Sigma \\ &= -\beta^{-1}\int_\Sigma \left|\nabla_\Sigma\left(\ln(\psi(t,\cdot))\mathrm{e}^{\beta\tilde{V}}\right)\right|^2\psi(t,\cdot)d\sigma_\Sigma \\ &= -\beta^{-1}I\big(\psi(t,\cdot)d\sigma_\Sigma\,|\,\tilde{\nu}_\Sigma\big) \\ &\leq -2\beta^{-1}\rho\,H\big(\psi(t,\cdot)d\sigma_\Sigma\,|\,\tilde{\nu}_\Sigma\big),\end{aligned}$$

which yields the result. □

Let us go back to the counterparts of these results for free energy computation. This convergence result should be compared with (2.87), which tells that the simple gradient dynamics

$$dq_t = -\nabla V(q_t)\,dt + \sqrt{2\beta^{-1}}dW_t$$

converges exponentially fast to equilibrium with rate $\beta^{-1}R$, where R is the logarithmic Sobolev constant associated with the canonical probability measure ν. On the other hand, the constrained dynamics (3.52) (with $\tilde{V} = V^\xi$ and $\xi(q_0) = z$) converges exponentially fast to equilibrium with rate $\beta^{-1}\rho(z)$, where $\rho(z)$ is the logarithmic Sobolev constant associated with the conditional probability measures $\nu^\xi(\cdot|z)$. Thus, the efficiency of thermodynamic integration is related to a good choice of ξ, such that, for all z, $\rho(z)$ is significantly larger than R. The reader is invited to consider the simple examples mentioned in the introduction (see Sections 1.3.3.1 and 1.3.2.4) to convince himself that in those cases, the constrained dynamics (which essentially consists in the sampling of a convex potential) indeed goes faster to equilibrium than the simple overdamped dynamics (which contains a double-well potential along the reaction coordinate).

3.3 The reaction coordinate case: Phase space sampling

Constrained systems can be used as a sampling tool for free energy computations, by constraining the value of a given reaction coordinate to compute the mean force. This is the so-called thermodynamic integration method, as explained in Section 3.2 and the present section. In many situations of interest, molecular dynamics simulations are also performed for mechanical systems described in the phase space, with some rigidly constrained degrees of freedom, such as covalent bonds, or bond angles. Considering such constrained systems in phase space, rather than in configuration space only as in Section 3.2, is important for at least two reasons. First, the most natural physical description of molecular systems includes masses and momenta through Newton's equation of motion. Second, discretizing a phase space description may lead to some numerical advantages; for instance, it will be shown below how to construct Metropolis-Hasting schemes sampling *exactly* probability distributions supported by a submanifold which is for example the level set of some reaction coordinate.

We start with an introduction to mechanically constrained systems in Section 3.3.1, where many useful concepts are motivated. Phase space measures, and in particular canonical measures for constrained systems, are made precise in Section 3.3.2. Poisson brackets and the generator of the dynamics are defined in Section 3.3.3. Section 3.3.4 presents the case of mechanically constrained Langevin processes; and Section 3.3.5 their numerical discretization. Finally, thermodynamic integration with constrained Langevin processes is detailed in Section 3.3.6.

3.3.1 *Constrained mechanical systems*

The theory of constrained dynamical systems is a classical topic of Hamiltonian theory of motion (see [Arnol'd (1989); Abraham and Marsden (1978); Cohen *et al.* (2006); Hairer *et al.* (2006)]). A rigorous sense can be given to the loose idea that constrained mechanical systems (solution to (3.97) or (3.106) below), have a "Hamiltonian structure" associated with the Hamiltonian H and a phase space $T^*\Sigma(z) \subset T^*\mathcal{D}$. Roughly speaking, this amounts to finding a set of internal coordinates such that the usual Hamiltonian formalism is conserved. Some recent mathematical work on constrained mechanical systems and free energy computations in molecular dynamics simulation can be found in [Hartmann and Schütte (2005b); Hartmann (2008)].

3.3.1.1 *Definition of the dynamics*

Consider a Hamiltonian (mechanical) system $(q,p) \in T^*\mathcal{D} = \mathcal{D} \times \mathbb{R}^{3N}$ subjected to an m-dimensional constraint

$$\xi(q) = \Big(\xi_1(q), \ldots, \xi_m(q)\Big)^T = z,$$

where $\xi : \mathcal{D} \to \mathbb{R}^m$, and denote by

$$\Sigma(z) = \Big\{ q \in \mathcal{D} \;\Big|\; \xi(q) = z \Big\} \subset \mathbb{R}^{3N}$$

the submanifold of co-dimension m defined by the constraints on the positions of the system. The gradient of the constraint is by convention

$$\nabla\xi(q) = \Big(\nabla\xi_1(q), \ldots, \nabla\xi_m(q)\Big) \in \mathbb{R}^{3N\times m}.$$

By the implicit function theorem, the submanifold $\Sigma(z)$ is non-degenerate when the mass-matrix dependent Gram matrix

$$\begin{aligned} G_M(q) &= \nabla\xi(q)^T M^{-1}\,\nabla\xi(q) \\ &= \Big(\partial_{q_i}\xi_\alpha(q)\,M^{-1}_{i,j}\,\partial_{q_j}\xi_\beta(q)\Big)_{\alpha,\beta=1,\ldots,m} \in \mathbb{R}^{m\times m}, \end{aligned} \tag{3.96}$$

where ∂_{q_i} denotes the partial derivative with respect to the i-th degree of freedom ($1 \le i \le 3N$), is invertible in the neighborhood of any $q \in \Sigma(z)$. Note that the latter condition is equivalent to the invertibility of the Gram matrix $G(q) = \nabla\xi(q)^T\,\nabla\xi(q)$. A constrained mechanical system is the solution of the system:

$$\begin{cases} \dfrac{dq(t)}{dt} = M^{-1}p(t), \\ \dfrac{dp(t)}{dt} = -\nabla V(q(t)) + \nabla\xi(q(t))\,\dfrac{d\lambda(t)}{dt}, \\ \xi(q(t)) = z, \end{cases} \tag{3.97}$$

where

$$\nabla \xi(q(t)) \, \frac{d\lambda(t)}{dt} = \sum_{\alpha=1}^{m} \nabla \xi_\alpha(q_t) \frac{d\lambda_\alpha(t)}{dt} \in \mathbb{R}^{3N}$$

is the mechanical force constraining the evolution of the system to remain on the submanifold $\Sigma(z)$. The coordinates of $\lambda(t) \in \mathbb{R}^m$ are the Lagrange multipliers, which are additional degrees of freedom associated with the mechanical constraints. It will be shown below (see Lemma 3.44) that the energy of the system is preserved, so that the constraining force does not exchange work with the system, in accordance with D'Alembert principle.

The constraint $\xi(q) = z$ on the positions implicitly poses additional constraints on the momenta. Indeed, a time-derivation of the position constraint leads to

$$p(t)^T M^{-1} \nabla \xi(q(t)) = 0$$

along a trajectory. This constraint on the momenta is often termed a "hidden velocity constraint". Note that it involves the mass-matrix M. A constrained system therefore evolves on a so-called cotangent bundle, which is the submanifold of phase space $T^*\Sigma(z) \subset T^*\mathcal{D}$ defined, using the tangent bundle $T\Sigma(z) \subset T\mathcal{D} = \mathcal{D} \times \mathbb{R}^{3N}$

$$T\Sigma(z) = \left\{ (q,v) \in T\mathcal{D} \;\middle|\; \xi(q) = z, \; \nabla \xi^T(q)\, v = 0 \right\},$$

as

$$T^*\Sigma(z) = \left\{ (q,p) \in T^*\mathcal{D} \;\middle|\; (q, M^{-1}p) \in T\Sigma(z) \right\}. \tag{3.98}$$

For a given $q \in \Sigma(z)$, the set of tangent velocities and cotangent momenta are also denoted respectively by

$$T_q\Sigma(z) = \left\{ v \in \mathbb{R}^{3N} \;\middle|\; \nabla \xi(q)^T\, v = 0 \right\} \tag{3.99}$$

and

$$T_q^*\Sigma(z) = \left\{ p \in \mathbb{R}^{3N} \;\middle|\; M^{-1}p \in T_q\Sigma(z) \right\}. \tag{3.100}$$

Remark 3.36 (Highly oscillatory systems). *Constrained dynamics are not, except in the case of affine constraints, limits of highly oscillatory Hamiltonian systems. For instance, the infinite stiffness limit ($\varepsilon \to 0$) of Hamiltonian systems with potential energies including some "soft" constraints of the form*

$$V_\varepsilon(q) = V(q) + \frac{1}{\varepsilon^2} \left| \xi(q) - z \right|^2,$$

has been studied in several works [Rubin and Ungar (1957); Takens (1980); van Kampen (1985); Bornemann and Schütte (1992); Reich (1995, 2000)]. The limiting dynamics can be fully characterized when the highly oscillatory potential has a non-resonant harmonic behavior (at least almost everywhere on the trajectory, see [Takens (1980)]). The latter dynamics is then described by the so-called adiabatic regime where the ratio of the couple energy/frequency of each highly oscillatory eigenmode is conserved over time. This description involves an effective potential depending on the initial energies and frequencies of the highly oscillatory degrees of freedom. See also [Cohen et al. *(2006); Le Bris and Legoll (2007)] for recent works on related numerical issues.*

In the case of stochastically perturbed highly oscillatory motions, the so-called Fixman entropic potential accounts for the correction between the case of rigid and soft constraints, see Section 3.2.3.4 *and Remark* 3.51.

3.3.1.2 *Explicit expression of the Lagrange multipliers*

The explicit expression of the Lagrange multipliers in (3.97) can be found by differentiating the momentum constraint

$$p(t)^T M^{-1}\nabla\xi(q(t)) = \Big[p(t)^T M^{-1}\nabla\xi_1(q(t)), \dots, p(t)^T M^{-1}\nabla\xi_m(q(t))\Big] = 0$$

with respect to time. Indeed,

$$\begin{aligned}\frac{d}{dt}\left[p(t)^T M^{-1}\nabla\xi(q(t))\right] &= \left(-\nabla V(q(t)) + \nabla\xi(q(t))\,\frac{d\lambda(t)}{dt}\right)^T M^{-1}\nabla\xi(q(t)) \\ &\quad + \Big[p(t)^T M^{-1}\nabla^2\xi_1(q(t))M^{-1}p(t), \dots, p(t)^T M^{-1}\nabla^2\xi_m(q(t))M^{-1}p(t)\Big],\end{aligned}$$

so that

$$\begin{aligned}\frac{d\lambda(t)}{dt} &= G_M^{-1}(q(t))\Big[-\mathrm{Hess}_{q(t)}(\xi)\big(M^{-1}p(t), M^{-1}p(t)\big) \\ &\qquad + \nabla\xi(q(t))^T M^{-1}\nabla V(q(t))\Big], \\ &= f_{\mathrm{rgd}}^M(q(t), p(t))\end{aligned} \tag{3.101}$$

with the notation

$$\mathrm{Hess}_q(\xi)\big(v_1, v_2\big) = \Big[v_1^T\nabla^2\xi_1(q)v_2, \dots, v_1^T\nabla^2\xi_m(q)v_2\Big]. \tag{3.102}$$

In (3.101), we have introduced the "constraining force field", a vector in $\mathbb{R}^m$ defined by:

$$f_{\mathrm{rgd}}^M(q,p) = G_M^{-1}(q)\nabla\xi(q)^T M^{-1}\nabla V(q) - G_M^{-1}(q)\mathrm{Hess}_q(\xi)(M^{-1}p, M^{-1}p). \tag{3.103}$$

The expression of the Lagrange multiplier can be used to rewrite the constrained dynamics in a more explicit way. It is convenient to this end to introduce the orthogonal projection of the momenta on the tangent momenta space $T_q^*\Sigma(z)$ given by (3.100). The projection is orthogonal with respect to the scalar product induced by M^{-1} (*i.e.* $\langle p, \tilde{p}\rangle_{M^{-1}} = p^T M^{-1}\tilde{p}$), so that

$$P_M(q) = \mathrm{Id} - \nabla\xi(q)\, G_M^{-1}(q)\nabla\xi(q)^T M^{-1}. \tag{3.104}$$

The matrix elements of this operator are

$$(P_M)_{i,k}(q) = \delta_{i,k} - \partial_{q_i}\xi_\alpha(q)\left[G_M^{-1}(q)\right]_{\alpha,\beta}\partial_{q_j}\xi_\beta(q) M_{j,k}^{-1}.$$

It is easily checked that P_M is indeed a projector property since $P_M(q)^2 = P_M(q)$. Besides,

$$M^{-1}P_M(q) = P_M(q)^T M^{-1}, \tag{3.105}$$

which shows the orthogonality of the projector with respect to the scalar product induced by M^{-1}. Thus, $P_M(q)p = p$ when $p \in T_q^*\Sigma(z)$ and, if $p \in \mathbb{R}^{3N}$ is such that $\langle p, \tilde{p}\rangle_{M^{-1}} = 0$ for all $\tilde{p} \in T_q^*\Sigma(z)$, then $P_M(q)p = 0$.

A more explicit reformulation of the constrained dynamics (3.97) is finally

$$\begin{cases} \dfrac{dq(t)}{dt} = M^{-1}p(t), \\ \dfrac{dp(t)}{dt} = -P_M(q(t))\nabla V(q(t)) \\ \qquad\qquad -\nabla\xi(q(t))\, G_M^{-1}(q(t))\, \mathrm{Hess}_{q(t)}(\xi)\big(M^{-1}p(t), M^{-1}p(t)\big), \end{cases} \tag{3.106}$$

or equivalently for the equation on momenta,

$$\frac{dp(t)}{dt} = -\nabla V(q(t)) + \nabla\xi(q(t)) f_{\mathrm{rgd}}^M(q(t), p(t)).$$

Note that the physical force contraining the system is thus $\nabla\xi(q(t)) f_{\mathrm{rgd}}^M(q(t), p(t))$.

3.3.1.3 *Generalization of the constraints*

An important concept to study constrained Hamiltonian systems (in particular, to use the co-area formula in phase space, as well as the Poisson bracket formulation of the Liouville equation) is the effective velocity v_ξ and the effective momentum p_ξ associated with the constrained degrees of freedom ξ:

$$v_\xi(q, p) = \nabla\xi(q)^T M^{-1}p \in \mathbb{R}^m, \tag{3.107}$$

and

$$p_\xi(q,p) = G_M^{-1}(q)\, v_\xi(q,p) = G_M^{-1}(q)\nabla\xi(q)^T M^{-1} p \in \mathbb{R}^m. \qquad (3.108)$$

The expression of the effective velocity is obtained by deriving the constraint ξ along an unconstrained trajectory, as

$$\frac{d\xi(q(t))}{dt} = v_\xi(q(t),p(t)).$$

The hidden velocity constraint corresponds to $v_\xi(q,p) = 0$, or equivalently $p_\xi(q,p) = 0$ (since $G_M(q)$ is assumed to be invertible). The term $G_M^{-1}(q)$ in the expression (3.108) of the effective momentum may be interpreted as the effective mass of the constrained degrees of freedom. This can be motivated by a decomposition of the kinetic energy of the system into a tangential and orthogonal part, using the projector (3.104) for a given position $q \in \mathcal{D}$:

$$\begin{aligned} E_{\rm kin}(p) =& \frac{1}{2} p^T M^{-1} p \\ =& \frac{1}{2} p^T P_M(q)^T M^{-1} P_M(q) p \\ &+ \frac{1}{2} p^T \left(\mathrm{Id} - P_M(q)\right)^T M^{-1} \left(\mathrm{Id} - P_M(q)\right) p. \end{aligned}$$

For an unconstrained trajectory $t \mapsto q(t)$, the orthogonal part can be rewritten as

$$\begin{aligned} E_{\rm kin}^{\perp}(q(t),p(t)) &= \frac{1}{2} p(t)^T \Big(\mathrm{Id} - P_M(q(t))\Big)^T M^{-1} \Big(\mathrm{Id} - P_M(q(t))\Big) p(t) \\ &= \frac{1}{2} v_\xi(q(t),p(t))^T G_M^{-1}(q(t))\, v_\xi(q(t),p(t)). \end{aligned}$$

The constraints on a mechanical system can now be rewritten under the general form

$$\Xi(p,q) = \zeta \in \mathbb{R}^{2m} \qquad (3.109)$$

where either the effective momentum is constrained, in which case

$$\Xi(q,p) = \begin{pmatrix} \xi(q) \\ p_\xi(q,p) \end{pmatrix}, \qquad \zeta = \begin{pmatrix} z \\ p_z \end{pmatrix}, \qquad (3.110)$$

or the effective velocity is constrained:

$$\Xi(q,p) = \begin{pmatrix} \xi(q) \\ v_\xi(q,p) \end{pmatrix}, \qquad \zeta = \begin{pmatrix} z \\ v_z \end{pmatrix}. \qquad (3.111)$$

Stationary mechanical constraints are *defined* by the specific choice $p_z = 0$ and $v_z = 0$ in (3.110) and (3.111) respectively. The phase space associated with such constraints is denoted, for velocity constraints (3.107)

$$\Sigma_{\xi,v_\xi}(z,v_z) = \Big\{ (q,p) \in T^*\mathcal{D} \ \Big|\ \xi(q) = z,\ v_\xi(q,p) = v_z \Big\}, \qquad (3.112)$$

and for momenta constraints (3.108)

$$\Sigma_{\xi,p_\xi}(z,p_z) = \left\{(q,p) \in T^*\mathcal{D} \;\middle|\; \xi(q) = z,\ p_\xi(q,p) = p_z\right\}. \tag{3.113}$$

When the type of generalized constraints need not be precised,

$$\Sigma_\Xi(\zeta) = \left\{(q,p) \in T^*\mathcal{D} \;\middle|\; \Xi(q,p) = \zeta\right\}. \tag{3.114}$$

In the same way, $q \in \Sigma(z)$ being given, the affine space of constrained velocities is

$$\Sigma_{v_\xi(q,\cdot)}(v_z) = \left\{p \in \mathbb{R}^{3N} \;\middle|\; v_\xi(q,p) = v_z\right\},$$

while the affine space of constrained momenta is

$$\Sigma_{p_\xi(q,\cdot)}(p_z) = \left\{p \in \mathbb{R}^{3N} \;\middle|\; p_\xi(q,p) = p_z\right\}.$$

The phase space of mechanical constraints, defined by (3.98), is then simply

$$T^*\Sigma(z) = \Sigma_{\xi,v_\xi}(z,0) = \Sigma_{\xi,p_\xi}(z,0),$$

with co-tangent space:

$$T_q^*\Sigma(z) = \Sigma_{v_\xi(q,\cdot)}(0) = \Sigma_{p_\xi(q,\cdot)}(0).$$

As will be detailed in Section 3.3.2, a practical motivation for introducing such generalized constraints (3.109) is to use the co-area formula on the one hand, and divergence theorems for transport operators of constraints systems on the other hand. This will be of paramount importance for non-equilibrium methods where the constraints evolve in time according to a predefined schedule, see Section 4.3.2.

3.3.2 *Phase space measures for constrained systems*

The goal of this section is to rigorously define the canonical distribution of constrained mechanical systems, namely

$$\begin{cases} \mu_{T^*\Sigma(z)}(dq\,dp) = Z_{z,0}^{-1}\,\mathrm{e}^{-\beta H(q,p)}\,\sigma_{T^*\Sigma(z)}(dq\,dp), \\ Z_{z,0} = \displaystyle\int_{T^*\Sigma(z)} \mathrm{e}^{-\beta H(q,p)}\,\sigma_{T^*\Sigma(z)}(dq\,dp), \end{cases} \tag{3.115}$$

where $\sigma_{T^*\Sigma(z)}(dq\,dp)$ is the phase space Liouville measure (see Section 3.3.2.1). The marginal distribution with respect to position variables of the canonical distribution $\mu_{T^*\Sigma(z)}(dq\,dp)$ is

$$\begin{cases} \nu_{\Sigma(z)}^M(dq) = \dfrac{1}{Z_z^M}\,\mathrm{e}^{-\beta V(q)}\,\sigma_{\Sigma(z)}^M(dq), \\ Z_z^M = \displaystyle\int_{\Sigma(z)} \mathrm{e}^{-\beta V(q)}\,\sigma_{\Sigma(z)}^M(dq), \end{cases} \tag{3.116}$$

where $\sigma_{\Sigma(z)}^M(dq)$ is the surface measure induced by the scalar product associated with M, see Sections 3.2.1.3 and 3.3.2.1.

Remark 3.37 (Mass matrices and scalar products).
Throughout this book (see also Remark 3.4 and Section 3.2.1.3), the following convention is used for the mass matrix M:

- *When phase space dynamics are considered (Langevin or Hamiltonian dynamics), the position space $\mathbb{R}^{3N}$ is endowed with a scalar product $\langle q_1, q_2 \rangle_M = q_1^T M q_2$ given by the mass matrix M and $\sigma^M_{\Sigma(z)}(dq)$ refers to surface measures induced by the latter scalar product. In the same way, the momentum space is endowed with a scalar product $\langle p_1, p_2 \rangle_{M^{-1}} = p_1^T M^{-1} p_2$, and $\sigma^{M^{-1}}_{\Sigma_{p_\xi(q,\cdot)}(p_z)}(dp)$ is the surface measure on the momentum space $\Sigma_{p_\xi(q,\cdot)}(p_z)$ induced by the latter scalar product.*
- *For dynamics in position space $\mathbb{R}^{3N}$ only (such as the overdamped Langevin process), the system is endowed with the usual Euclidean scalar product $q_1^T q_2$ induced by $M = \mathrm{Id}$, and $\sigma_{\Sigma(z)}(dq)$ refers to surface measures induced by the latter.*

Note however that the mass matrix can be easily eliminated in all formulas by working in mass-weighted coordinates. This is done with the following change of coordinates:

$$\begin{cases} \tilde{q} = M^{1/2}\, q, \\ \tilde{p} = M^{-1/2}\, p. \end{cases} \tag{3.117}$$

The equation of motion $\dfrac{d^2 q(t)}{dt^2} = -M^{-1} \nabla V(q(t))$ *can then be rewritten as*

$$\frac{d^2 \tilde{q}(t)}{dt^2} = -M^{-1/2}\, \nabla V(q(t)) = -M^{-1/2}\, \nabla V\left(M^{-1/2}\tilde{q}(t)\right) = -\nabla \widetilde{V}(\tilde{q}(t)),$$

with $\widetilde{V}(\tilde{q}) = V(M^{-1/2}\tilde{q})$, *and the system is described by the Hamiltonian*

$$\widetilde{H}(\tilde{q},\tilde{p}) = \frac{1}{2}\, |\tilde{p}|^2 + \widetilde{V}(\tilde{q}).$$

Masses can be reintroduced on demand by the inverse of the change of variables (3.117). *This yields a systematic and easy way to deal with mass matrices.*

3.3.2.1 *Surface measures*

The surface measure of a submanifold in Euclidean spaces is introduced in Section 3.2 (see Remark 3.4 and Section 3.2.1.3). The phase space measure (also termed Liouville measure) on the phase space $T^*\Sigma(z)$ of constrained mechanical systems is not induced by a scalar product (defined by a positive definite symmetric 2-form) but by the *symplectic*, or skew-symmetric 2-form on $T^*\mathcal{D}$ defined by the canonical skew-symmetric matrix (1.10) in $\mathbb{R}^{6N}$, namely

$$J := J_{6N\times 6N} := \begin{pmatrix} 0 & \mathrm{Id}_{3N} \\ -\mathrm{Id}_{3N} & 0 \end{pmatrix}. \tag{3.118}$$

Consider a state $(q,p) \in \mathbb{R}^{6N}$, and a set of $2n \leq 6N$ linearly independent vectors $u(q,p) = (u_1(q,p),\ldots,u_{2n}(q,p)) \in \mathbb{R}^{6N\times 2n}$ representing a tangent space of a sub-manifold of dimension $2n$. The $2n$-dimensional phase space volume of the parallelepiped

$$\begin{aligned}&\mathrm{Span}_{[0,1]}(u_1(q,p),\ldots,u_{2n}(q,p)) := \\ &\qquad \left\{\alpha_1 u_1(q,p) + \ldots + \alpha_{2n} u_{2n}(q,p) \,\middle|\, (\alpha_1,\ldots,\alpha_{2n}) \in [0,1]^{2n}\right\}\end{aligned}$$

spanned by the latter vectors is given by the determinant

$$\mathrm{Vol}_{2n}\Big(\mathrm{Span}_{[0,1]}(u_1(q,p),\ldots,u_{2n}(q,p))\Big) = \left|\det \mathcal{G}(u(q,p))\right|^{1/2}, \tag{3.119}$$

where the entries of the skew-symmetric Gram matrix $\mathcal{G} \in \mathbb{R}^{2n\times 2n}$ are

$$\mathcal{G}_{a,b}(u) = (u_a)^T J_{6N\times 6N}\, u_b, \qquad a,b = 1,\ldots,2n.$$

Since $\mathcal{G}$ is skew-symmetric, the above notion of volume requires an even number $2n$ of tangent vectors in order for $\mathcal{G}(u)$ to be invertible. Sub-manifolds of dimension $2n$ such that any basis of tangential vectors $(u_1(q,p),\ldots,u_{2n}(q,p))$ satisfy $\det \mathcal{G}(u(q,p)) \neq 0$ (called "symplectic sub-manifolds") are thus endowed with a non-degenerate volume form.

In the case of mechanical constraints with phase space $T^*\Sigma(z)$, the dimension of the submanifold is

$$2n = 6N - 2m,$$

and the basis of vectors $(u_1(q,p),\ldots,u_{6N-2m}(q,p))$ at a given point $(q,p) \in T^*\Sigma(z)$ is chosen to span the tangential space

$$\mathrm{Span}\Big\{u_1(q,p),\ldots,u_{6N-2m}(q,p)\Big\} = T_{(q,p)}\left(T^*\Sigma(z)\right).$$

The measure induced by (3.119) is, by definition, the phase space measure (or Liouville measure), denoted by $\sigma_{T^*\Sigma(z)}(dq\,dp)$ in the sequel. This construction can be generalized to the submanifolds $\Sigma_{\xi,p_\xi}(z,p_z)$ and $\Sigma_{\xi,v_\xi}(z,v_z)$ for non-zero (v_z,p_z), and the associated phase space measures will be denoted by $\sigma_{\Sigma_{\xi,p_\xi}(z,p_z)}(dq\,dp)$ and $\sigma_{\Sigma_{\xi,v_\xi}(z,v_z)}(dq\,dp)$ respectively. When we do not wish to make precise the measure and the phase space at hand, we will simply denote one of the two previous measures by $\sigma_{\Sigma_{\Xi}(\zeta)}(dq\,dp)$.

Note that if no constraints are considered ($m = 0$), the above definition yields the usual Liouville measure $dq\,dp$ in $T^*\mathcal{D}$. Indeed in this case, $u(q,p) \in \mathbb{R}^{6N\times 6N}$ and $\left|\det \mathcal{G}\big(u(q,p)\big)\right|^{1/2} = \left|\det u(q,p)\right|$ which is the usual Lebesgue volume formula for a parallelepiped.

Remark 3.38 (Internal coordinates). *Darboux's completion theorem (see for instance [Arnol'd (1989)]) states the existence of so-called local internal symplectic coordinates*

$$(q,p) \mapsto \Big(Q_1(q,p), \dots, Q_{3N-m}(q,p), P_1(q,p), \dots, P_{3N-m}(q,p)\Big),$$

which define a local system of coordinates of the manifolds $\Sigma_{\xi,p_\xi}(z,p_z)$ *for any given* (z,p_z) *(or* $\Sigma_{\xi,v_\xi}(z,v_z)$ *for any given* (z,v_z)*), and furthermore preserve the symplectic form, in the sense that*

$$\nabla(Q,P)^T J \nabla(Q,P) = J_{(6N-2m)\times(6N-2m)}. \tag{3.120}$$

In an internal system of symplectic coordinates, the measure on the phase space $\Sigma_{\xi,p_\xi}(z,p_z)$ *simply reads*

$$\sigma_{\Sigma_{\xi,p_\xi}(z,p_z)}(dq\,dp) = dQ\,dP.$$

A similar equality also holds true for $\Sigma_{\xi,v_\xi}(z,v_z)$. *It can be shown that the constrained motions* (3.97) *or* (3.106) *written in internal coordinates then have the usual Hamiltonian formulation, however with a generalized position dependent mass matrix. Cartesian coordinates are of primary interest in the present book (and in numerical simulations), and the use of internal coordinates will therefore be avoided in proofs.*

Let us however explain how to construct internal coordinates. For constraints on the positions of the mechanical system, a local internal coordinates can be obtained from the so-called local generalized coordinates

$$\Big(x(q), p_x(q,p)\Big) = \Big(Q(q), \xi(q), P(q,p), \tilde{p}_x(q,p)\Big),$$

where $Q : \mathcal{D} \subset \mathbb{R}^{3N} \to \mathbb{R}^{3N-m}$ *is a local completion of* ξ, *obtained with the implicit function theorem, so that (i)* $q \mapsto x(q) = (Q(q), \xi(q))$ *is a local diffeomorphism of* $\mathbb{R}^{3N}$*; (ii) the momentum* p_x *is*

$$p_x(q,p) = M_x(q)\, v_x(q,p) = (\nabla x)^{-1}(q)\, p$$

where the associated mass and velocity in generalized coordinates are respectively

$$M_x(q) = \left(\nabla x\right)^{-1}(q) M \left(\nabla x^T\right)^{-1}(q), \qquad v_x(q,p) = \nabla x^T(q) M^{-1} p.$$

The generalized coordinates are thus a local diffeomorphism of $\mathbb{R}^{6N}$. *Let us show that the latter preserves the symplectic structure, in the sense that:*

$$\nabla(x,p_x)^T J\,\nabla(x,p_x) = \begin{pmatrix} \begin{pmatrix}\nabla x\\0\end{pmatrix}^T J \begin{pmatrix}\nabla x\\0\end{pmatrix} & \begin{pmatrix}\nabla x\\0\end{pmatrix}^T J\nabla p_x \\ \nabla p_x^T J \begin{pmatrix}\nabla x\\0\end{pmatrix} & \nabla p_x^T J \nabla p_x \end{pmatrix} = J. \tag{3.121}$$

Indeed, it is easily shown that $\begin{pmatrix}\nabla x\\0\end{pmatrix}^T J \begin{pmatrix}\nabla x\\0\end{pmatrix} = 0$ *and* $\begin{pmatrix}\nabla x\\0\end{pmatrix}^T J\nabla p_x =$ Id. *Besides, for any coordinates* $i,j = 1,\dots,3N$, *the differentiation rule* (3.147) *below implies*

$$\partial_{q_i} p_{x,j} = -(\nabla x)^{-1}_{j,k}\left[\partial_{q_i}\partial_{q_k} x_l\right](\nabla x)^{-1}_{l,m}\, p_m,$$

so that

$$\left(\nabla_p\, p_{x,i}\right)^T \nabla_q\, p_{x,j} = -\mathrm{Hess}(x_l)\Big((\nabla x)^{-1}_{i,\cdot}, (\nabla x)^{-1}_{j,\cdot}\Big)(\nabla x)^{-1}_{l,m} p_m.$$

Thus, by symmetry, $(\nabla p_x)^T J \nabla p_x = 0$. *Note that* (3.120) *follows directly from* (3.121).

Let us show now that the internal coordinates (Q,P) *form a set of coordinates of the submanifolds* $\Sigma_{\xi,p_\xi}(z,p_z)$ *(the case of* $\Sigma_{\xi,v_\xi}(z,v_z)$ *could be treated in the same way). This is equivalent to showing that the modified set of coordinates:*

$$(q,p) \mapsto (Q(q), \xi(q), P(q,p), p_\xi(q,p))$$

is a local diffeomorphism. First $q \mapsto (Q(q),\xi(q))$ *is clearly a local diffeomorphism. Moreover, by definition of* p_x, $p = \nabla x(q) p_x(q,p)$ *so that*

$$p = \nabla Q(q) P(q,p) + \nabla\xi(q)\tilde{p}_x(q,p),$$

and the orthogonal projection of p *on* $T^*_q\Sigma(z)$ *is* $P_M(q)p = P_M(q)\nabla Q(q) P(q,p)$. *On the other hand, using the definition of* p_ξ, *it can be checked that* $P_M(q)p = p - \nabla\xi(q)p_\xi(q,p)$. *Thus, the following decomposition holds:*

$$p = P_M(q)\nabla Q(q) P(q,p) + \nabla\xi(q) p_\xi(q,p).$$

We have therefore, for a fixed q, *an analytical expression of the inverse of the linear transformation* $p \mapsto (P(q,p), p_\xi(q,p))$, *which is thus invertible. This shows that* (Q,P) *is indeed a set of coordinates of* $\Sigma_{\xi,p_\xi}(z,p_z)$.

3.3.2.2 *The co-area formula*

The co-area formula in phase space (see also Lemma 3.2 for the position space case) relates the phase space measures, $\sigma_{\Sigma_{\xi,p_\xi}(z,p_z)}$ or $\sigma_{\Sigma_{\xi,v_\xi}(z,v_z)}$ defined in the last section, and the conditional measure $\delta_{(\xi(q)-z,p_\xi(q,p)-p_z)}(dq\,dp)$ defined by

$$dq\,dp = \delta_{(\xi(q)-z,p_\xi(q,p)-p_z)}(dq\,dp)\,dz\,dp_z, \tag{3.122}$$

i.e. for any compactly supported smooth test function φ,

$$\int_{T^*\mathcal{D}} \varphi(q,p)\,dq\,dp = \int_{\mathbb{R}^{2m}} \int_{\Sigma_{\xi,p_\xi}(z,p_z)} \varphi(q,p)\,\delta_{(\xi(q)-z,p_\xi(q,p)-p_z)}(dq\,dp)\,dz\,dp_z.$$

The measure $\delta_{(\xi(q)-z,v_\xi(q,p)-v_z)}(dq\,dp)$ is defined in a similar way.

Proposition 3.39 (Co-area). *Let $\Sigma_\Xi(\zeta)$ be the phase space defined by generalized constraints $\Xi(p,q)=\zeta\in\mathbb{R}^{2m}$ of the form* (3.110) *or* (3.111). *Define the skew-symmetric Gram tensor of dimension $2m\times 2m$ associated with the constraints as*

$$\Gamma(q,p) = \nabla\Xi^T(q,p)\,J\,\nabla\Xi(q,p)\in\mathbb{R}^{2m\times 2m}. \tag{3.123}$$

Assume that Γ is locally non-degenerate on a neighborhood of $\Sigma_\Xi(\zeta)$. Then,

$$dq\,dp = \left|\det\Gamma(q,p)\right|^{-1/2}\sigma_{\Sigma_\Xi(\zeta)}(dq\,dp)\,d\zeta. \tag{3.124}$$

Proof. The proof could be carried out by following the steps of the proof of Lemma 3.2, using J instead of the usual Euclidean scalar product induced by the identity matrix. Alternatively, a direct proof can be made by combining the decompositions (3.129) and (3.130) given below, the explicit computation of $\det\Gamma$ (see below), and the co-area formula for general scalar product proven in Section 3.2.1.3. □

The Gram matrix Γ (also called symplectic Gram matrix) associated with the generalized constraints (3.110) or (3.111) can be explicitly computed by block. Indeed, for $\Xi=(\xi,p_\xi)$ or $\Xi=(\xi,v_\xi)$,

$$\nabla\Xi(q,p) = \begin{pmatrix}\nabla_q\Xi\\ \nabla_p\Xi\end{pmatrix} = \begin{pmatrix}\nabla\xi & \nabla_q p_\xi\\ 0 & \nabla_p p_\xi\end{pmatrix}.$$

Then, the upper left block of Γ reads:

$$\begin{pmatrix}\nabla\xi\\ 0\end{pmatrix}^T J \begin{pmatrix}\nabla\xi\\ 0\end{pmatrix} = 0\in\mathbb{R}^{m\times m}.$$

Besides,

$$\begin{pmatrix} \nabla\xi \\ 0 \end{pmatrix}^T J\,\nabla p_\xi = \nabla\xi^T\,\nabla_p\,p_\xi = \mathrm{Id} \in \mathbb{R}^{m\times m},$$

so that

$$\Gamma = \begin{pmatrix} 0 & \mathrm{Id} \\ -\mathrm{Id} & \nabla p_\xi^T J\,\nabla p_\xi \end{pmatrix} \tag{3.125}$$

$\det(\Gamma) = 1$ in the case (3.111). In the case (3.111), the upper right and lower left blocks are

$$\begin{pmatrix} \nabla\xi \\ 0 \end{pmatrix}^T J\,\nabla v_\xi = \nabla\xi^T\,\nabla_p\,v_\xi = G_M \in \mathbb{R}^{m\times m},$$

so that the Gram matrix reads

$$\Gamma(q,p) = \begin{pmatrix} 0 & G_M \\ -G_M & \nabla v_\xi^T\,J\,\nabla v_\xi \end{pmatrix}. \tag{3.126}$$

Then, $\det(\Gamma) = \det(G_M)^2$. Note that in both cases $\det(\Gamma) > 0$.

Let us conclude this section with different definitions of the phase space measure for the generalized constraints on effective momentum (3.110) or on effective velocity (3.111), equivalent to the definition presented in Section 3.3.2.1.

Proposition 3.40. *The following definitions of the phase space measures* $\sigma_{\Sigma_{\xi,p_\xi}(z,p_z)}(dq\,dp)$ *and* $\sigma_{\Sigma_{\xi,v_\xi}(z,v_z)}(dq\,dp)$ *are equivalent:*

(1) The phase space measures are given, through the volume form (3.119)*, by the Liouville measure induced on* $\Sigma_{\xi,p_\xi}(z,p_z)$ *or* $\Sigma_{\xi,v_\xi}(z,v_z)$ *by the phase space structure of* $\mathbb{R}^{6N}$ *associated with* J.

(2) The phase space measure on $\Sigma_{\xi,p_\xi}(z,p_z)$ *can be identified with the conditional measure defined in* (3.122)*:*

$$\sigma_{\Sigma_{\xi,p_\xi}(z,p_z)}(dq\,dp) = \delta_{(\xi(q)-z,\,p_\xi(q,p)-p_z)}(dq\,dp), \tag{3.127}$$

while the phase space measure on $\Sigma_{\xi,v_\xi}(z,v_z)$ *is related to the corresponding conditional measure as*

$$\sigma_{\Sigma_{\xi,v_\xi}(z,v_z)}(dq\,dp) = \det(G_M)\,\delta_{(\xi(q)-z,\,v_\xi(q,p)-v_z)}(dq\,dp). \tag{3.128}$$

(3) The phase space measures are given by the product of surface measures:

$$\sigma_{\Sigma_{\xi,p_\xi}(z,p_z)}(dq\,dp) = \sigma^{M^{-1}}_{\Sigma_{p_\xi(q,\cdot)}(p_z)}(dp)\,\sigma^{M}_{\Sigma(z)}(dq), \tag{3.129}$$

and

$$\sigma_{\Sigma_{\xi,v_\xi}(z,v_z)}(dq\,dp) = \sigma^{M^{-1}}_{\Sigma_{v_\xi(q,\cdot)}(v_z)}(dp)\,\sigma^{M}_{\Sigma(z)}(dq). \tag{3.130}$$

In these expressions, the surface measure $\sigma^M_{\Sigma(z)}(dq)$ *on* $\Sigma(z)$ *is induced by the scalar product* $\langle q, \tilde{q}\rangle_M = q^T M \tilde{q}$ *on* $\mathbb{R}^{3N}$*, while the surface measures on the affine spaces* $\Sigma_{p_\xi(q,\cdot)}(p_z)$ *or* $\Sigma_{v_\xi(q,\cdot)}(v_z)$ *(for a given* $q \in \Sigma(z)$*) are induced by the scalar product* $\langle p, \tilde{p}\rangle_{M^{-1}} = p^T M^{-1}\tilde{p}$ *on* $\mathbb{R}^{3N}$*.*

Proof. The second point is equivalent to the first one by the co-area formula (3.124), using the adequate determinant of the Gram matrices (3.125) or (3.126).

The third point is equivalent to the first one by two applications of the co-area formula in $\mathbb{R}^{3N}$ (see Lemma 3.2 and Section 3.2.1.3), for q and p variables successively. Indeed,

$$\delta_{(\xi(q)-z,\,p_\xi(q,p)-p_z)}(dq\,dp) = \delta_{p_\xi(q,p)-p_z}(dp)\,\delta_{\xi(q)-z}(dq),$$

and a similar relation holds for constraints stated in terms of effective velocity. The second conditional measure is

$$\delta_{\xi(q)-z}(dq) = \det(M)^{-1/2}\det\big(G_M(q)\big)^{-1/2}\,\sigma^M_{\Sigma(z)}(dq).$$

The conditional measure on the momenta or velocities is, q being given,

$$\delta_{p_\xi(q,p)-p_z}(dp) = \det(M)^{1/2}\det(G_M(q))^{1/2}\,\sigma^{M^{-1}}_{\Sigma_{p_\xi(q,\cdot)}(p_z)}(dp)$$

for constrained effective momentum, while

$$\delta_{v_\xi(q,p)-v_z}(dp) = \det(M)^{1/2}\det(G_M(q))^{-1/2}\,\sigma^{M^{-1}}_{\Sigma_{v_\xi(q,\cdot)}(v_z)}(dp)$$

for constrained effective velocity. □

The proof shows in particular that the measure $\sigma_{\Sigma_{\xi,v_\xi}}$ depends on the mass tensor M, which is not the case for the measure $\sigma_{\Sigma_{\xi,p_\xi}}$.

Remark 3.41 (Phase space measure in internal coordinates). *As already noted in Remark* 3.38*, the phase space measures in any internal system of symplectic (sometimes called "canonical") coordinates are given by:*

$$\sigma_{\Sigma_{\xi,p_\xi}(z,p_z)}(dq\,dp) = dQ\,dP.$$

The equivalence with the first point of Proposition 3.40 *is shown by computing the phase space volume of parallelepipeds in internal coordinates. Indeed, by construction, any tangent vector* u *to* $\Sigma_{\xi,p_\xi}(z, p_z)$ *in generalized coordinates* $(x, p_x)(q,p) = (Q(q), \xi(q), P(q,p), \tilde{p}_x(q,p))$ *(see Remark* 3.38*) is of the general form*

$$u = \begin{pmatrix} u^x \\ u^{p_x} \end{pmatrix}, \quad u^x = \begin{pmatrix} u^Q \\ 0 \end{pmatrix}, \quad u^{p_x} = \begin{pmatrix} u^P \\ u^{\tilde{p}_x} \end{pmatrix}.$$

Thus,

$$u_a^T J u_b = (u_a^x)^T u_b^{p_x} - (u_b^x)^T u_a^{p_x} = \begin{pmatrix} u_a^Q \\ u_a^P \end{pmatrix}^T J_{(6N-2m)\times(6N-2m)} \begin{pmatrix} u_b^Q \\ u_b^P \end{pmatrix}$$

where u_a, u_b *are any two tangent vectors of* $\Sigma_{\xi,p_\xi}(z,p_z)$. *Using the definition* (3.119), *the volume of a parallelepiped spanned by a basis of tangential vectors* $(u_1(q,p),\ldots,u_{6N-2m}(q,p))$ *at some point* $(q,p) \in \Sigma_{\xi,p_\xi}(z,p_z)$ *is then given by:*

$$\left|\det\left(B^T J_{(6N-2m)\times(6N-2m)} B\right)\right|^{1/2} = |\det B|,$$

where

$$B = \begin{pmatrix} u_1^Q, \ldots, u_{6N-2m}^Q \\ u_1^P, \ldots, u_{6N-2m}^P \end{pmatrix} \in \mathbb{R}^{(6N-2m)\times(6N-2m)}.$$

This yields the usual Euclidean formula for volumes of parallelepipeds with respect to the Lebesgue measure $dQ\,dP$, *in internal coordinates.*

3.3.2.3 *Canonical distributions*

Now that the phase space measure has been characterized through Proposition 3.40 of the previous section, the canonical distribution with constraints can be defined. This gives for effective momenta constraints (3.110):

$$\mu_{\Sigma_{\xi,p_\xi}(z,p_z)}(dq\,dp) := Z_{z,p_z}^{-1}\,\mathrm{e}^{-\beta H(q,p)}\,\sigma_{\Sigma_{\xi,p_\xi}(z,p_z)}(dq\,dp),$$

and for effective momenta constraints (3.111):

$$\mu_{\Sigma_{\xi,v_\xi}(z,v_z)}(dq\,dp) := Z_{z,v_z}^{-1}\,\mathrm{e}^{-\beta H(q,p)}\,\sigma_{\Sigma_{\xi,v_\xi}(z,v_z)}(dq\,dp).$$

The canonical distribution $\mu_{\Sigma_{\xi,p_\xi}(z,p_z)}$ identifies with the following conditional probability distribution:

$$\begin{aligned} \mu^{\xi,p_\xi}(dq\,dp\,|\,z,p_z) &:= Z_{z,p_z}^{-1}\,\mathrm{e}^{-\beta H(q,p)}\,\delta_{\xi(q)-z,\,p_\xi(q,p)-p_z}(dq\,dp), \\ &= \mu_{\Sigma_{\xi,p_\xi}(z,p_z)}(dq\,dp) \end{aligned} \tag{3.131}$$

where

$$Z_{z,p_z} = \int_{\Sigma_{\xi,p_\xi}(z,p_z)} \mathrm{e}^{-\beta H(q,p)}\,\sigma_{\Sigma_{\xi,p_\xi}(z,p_z)}(dq\,dp).$$

In (3.131), $\mu_{\Sigma_{\xi,p_\xi}(z,p_z)}(dq\,dp)$ is the canonical Boltzmann distribution on the phase space $\Sigma_{\xi,p_\xi}(z,p_z)$, which identifies with $\mu^{\xi,p_\xi}(dq\,dp\,|\,z,p_z)$, the original probability distribution $\mu(dq\,dp)$ conditioned by the values (z,p_z) of the functions (ξ,p_ξ) (see also (3.10) for similar notations in the position state space).

In the case when $p_z = 0$ (mechanical constraints on the positions, implying implicitly a momentum constraint), we use a simplified notation for the canonical distribution:

$$\begin{aligned}\mu_{T^*\Sigma(z)}(dq\,dp) &:= Z_{z,0}^{-1}\,\mathrm{e}^{-\beta H(q,p)}\,\sigma_{T^*\Sigma(z)}(dq\,dp),\\ &= Z_{z,0}^{-1}\,\mathrm{e}^{-\beta H(q,p)}\,\delta_{\xi(q)-z,p_\xi(q,p)}(dq\,dp),\\ &= \mu^{\xi,p_\xi}(dq\,dp\,|\,z,0).\end{aligned}$$

In the physics literature (see [Darve (2007)]), averages with respect to $\mu_{T^*\Sigma(z)}(dq\,dp)$ are often denoted: for some smooth function A,

$$\langle A\rangle_{\xi,\dot{\xi}} = \int_{T^*\Sigma(z)} A(q,p)\,\mu_{T^*\Sigma(z)}(dq\,dp),$$

where the notation $\dot{\xi}$ underlines that the momenta are constrained to zero. The latter measure may then be decomposed as:

$$\mu_{T^*\Sigma(z)}(dq\,dp) = \kappa^{M^{-1}}_{T^*_q\Sigma(z)}(dp)\nu^{M}_{\Sigma(z)}(dq), \tag{3.132}$$

where the kinetic probability distribution (which is the momentum distribution conditionally to a given position q) reads:

$$\kappa^{M^{-1}}_{T^*_q\Sigma(z)}(dp) := \left(\frac{\beta}{2\pi}\right)^{(3N-m)/2} \exp\left(-\beta\frac{p^T M^{-1} p}{2}\right)\,\sigma^{M^{-1}}_{T^*_q\Sigma(z)}(dp), \tag{3.133}$$

and the q-marginal distribution of $\mu_{T^*\Sigma(z)}$ is therefore

$$\nu^{M}_{\Sigma(z)}(dq) := \frac{1}{Z^M_z}\,\mathrm{e}^{-\beta V(q)}\sigma^{M}_{\Sigma(z)}(dq).$$

The canonical distribution with position constraints and associated hidden momentum constraints $\mu_{T^*\Sigma(z)}$ is different from the probability distribution μ conditionally to a fixed value z of ξ,

$$\mu^{\xi}(dq\,dp\,|\,z) := Z_z^{-1}\left(\frac{\beta}{2\pi}\right)^{3N/2}\mathrm{e}^{-\beta H(q,p)}\,\delta_{\xi(q)-z}(dq)\,dp$$

used to define the free energy F, since momenta are not constrained (see (1.57) and (3.22)). Recall indeed that

$$\mathrm{e}^{-\beta F(z)} = \int_{\Sigma(z)\times\mathbb{R}^{3N}} \mu^{\xi}(dq\,dp\,|\,z).$$

In the physics literature, averages with respect to this measure are usually referred to as (see for instance [den Otter and Briels (1998); Darve (2007)]): for some smooth function A,

$$\langle A \rangle_\xi = \int_{\Sigma(z)\times\mathbb{R}^{3N}} A(q,p)\, \mu^\xi(dq\, dp \,|\, z).$$

The q-marginal of $\mu^\xi(dq\, dp|z)$ is then the conditional probability distribution:

$$\nu^\xi(dq \,|\, z) = Z_z^{-1} \mathrm{e}^{-\beta V(q)}\, \delta_{\xi(q)-z}(dq),$$

and the free energy also satisfies

$$\mathrm{e}^{-\beta F(z)} = \int_{\Sigma(z)} \nu^\xi(dq \,|\, z).$$

In general $\nu^M_{\Sigma(z)}(dq) \neq \nu^\xi(dq|z)$ (even for an identity mass-matrix) since, by the co-area formula (3.24),

$$\delta_{\xi(q)-z}(dq) = \det(M)^{-1/2} \det(G_M(q))^{-1/2} \sigma^M_{\Sigma(z)}(dq) \neq \sigma^M_{\Sigma(z)}(dq).$$

Therefore, averages with or without momentum constraints are different. This has an important consequence: When free energy differences are computed with phase space dynamics where momenta are constrained, a correction term has to be taken into account, see Section 3.3.6.

3.3.3 *Hamilton and Poisson formalisms with constraints*

The canonical Poisson bracket (1.12) on $T^*\mathcal{D}$ (see Section 1.2.2.2), associated with the skew-symmetric symplectic matrix (1.10), reads

$$\{\varphi_1, \varphi_2\} = (\nabla \varphi_1)^T J \,\nabla \varphi_2 = (\nabla_q \varphi_1)^T \,\nabla_p \varphi_2 - (\nabla_p \varphi_1)^T \,\nabla_q \varphi_2,$$

for any smooth test functions $\varphi_1, \varphi_2 : T^*\mathcal{D} \to \mathbb{R}$. If $\varphi_1 : T^*\mathcal{D} \to \mathbb{R}^{n_1}$ and $\varphi_2 : T^*\mathcal{D} \to\in \mathbb{R}^{n_2}$ are two vector fields,

$$\{\varphi_1, \varphi_2\} = (\nabla \varphi_1)^T J \,\nabla \varphi_2 = \Big(\{\varphi_{1,\alpha}, \varphi_{2,\beta}\} \Big)_{\alpha=1,\dots,n_1, \beta=1,\dots,n_2} \in \mathbb{R}^{n_1 \times n_2}$$

denotes the $n_1 \times n_2$ matrix of the Poisson brackets of the components. The unconstrained Hamiltonian equations of motion can be written in the Liouvillian transport form (1.13):

$$\frac{d\varphi(q_t, p_t)}{dt} = \{\varphi, H\}\,(q_t, p_t),$$

for any test function φ.

In this section, the definition of the Poisson bracket is generalized to Hamiltonian motions constrained on submanifolds of the whole phase space. More precisely, we show that the equations of motion with constraints (3.97) or (3.106) can also be written as a Liouvillian transport equation

$$\frac{d\varphi(q_t,p_t)}{dt} = \{\varphi, H\}_\Xi (q_t,p_t) = \nabla\varphi^T J_\Xi \nabla H(q_t,p_t) \tag{3.134}$$

for a generalized Poisson bracket $\{\cdot,\cdot\}_\Xi$ associated with the full constraints (3.111) or (3.110), see (3.143) below. We will prove below by direct computations that the Liouville phase space measure is preserved by the evolution, see Corollary 3.49. Moreover, it can be shown that the notion of symplecticity associated with the constrained symplectic matrix J_Ξ still holds for Hamiltonian flows solution to (3.134), and can be proved from the Poisson structure (3.142) (see Chapter 8 in [Arnol'd (1989)], and Chapter VII.1 in [Hairer *et al.* (2006)]). In the same way, the reversibility (see the definition in Section 1.2.2.3) of the equations of motion (3.97), (3.106) or (3.135) follows from the symmetry of the Hamiltonian H under momenta flip.

3.3.3.1 *Definition of the constrained dynamics*

Using the general notation (3.109) for the constrained degrees of freedom, and denoting by $\Sigma_\Xi(\zeta)$ the associated phase space (see (3.114)) and Γ the associated Gram matrix (3.123), constrained equations of motion can be written in the following general form

$$\begin{cases} \dfrac{d}{dt}\begin{pmatrix} q(t) \\ p(t) \end{pmatrix} = J\,\nabla H(q(t),p(t)) - J\,\nabla\Xi(q(t),p(t))\,\dfrac{d\Lambda(t)}{dt}, \\ \Xi(q(t),p(t)) = \zeta, \end{cases} \tag{3.135}$$

where the Lagrange multipliers $\Lambda(t) \in \mathbb{R}^{2m}$. The system (3.135) is well posed, in the sense that the Lagrange multipliers are uniquely defined. Indeed, a time derivation of the constraint leads to

$$0 = \frac{d\,\Xi(q(t),p(t))}{dt} = \nabla\Xi^T J\nabla H(q(t),p(t)) - \nabla\Xi^T J\nabla\Xi(q(t),p(t))\frac{d\Lambda(t)}{dt},$$

which yields, in view of the invertibility of Γ,

$$\frac{d\Lambda(t)}{dt} = \Gamma^{-1}(q(t),p(t))\ \{\Xi, H\}\,(q(t),p(t)). \tag{3.136}$$

Remark 3.42 (Application to mechanical constraints).
The equations (3.97) *or* (3.106) *are recovered in the case of constrained effective velocity* $\Xi = (\xi, v_\xi)^T$ *when* $v_z = 0$ *(a similar proof holds for the*

case of constrained momenta when $p_z = 0$*). Indeed, decomposing the Lagrange multiplier as* $\Lambda(t) = (\lambda_1(t), \lambda_2(t))^T$ *with* $\lambda_i(t) \in \mathbb{R}^m$, (3.135) *yields:*

$$\begin{cases} \dfrac{dq(t)}{dt} = M^{-1}\, p(t) - \nabla_p v_\xi(q(t), p(t)) \dfrac{d\lambda_2(t)}{dt}, \\ \dfrac{dp(t)}{dt} = -\nabla V(q(t)) + \nabla \xi(q(t))\, \dfrac{d\lambda_1(t)}{dt} + \nabla_q v_\xi(q(t), p(t)) \dfrac{d\lambda_2(t)}{dt}, \\ \xi(q(t)) = z, \\ v_\xi(q(t), p(t)) = 0. \end{cases}$$

A time derivation of the position constraint leads to

$$0 = \frac{d\xi(q(t))}{dt} = v_\xi(q(t), p(t)) - G_M(q(t))\, \frac{d\lambda_2(t)}{dt}.$$

Since $v_\xi(q(t), p(t)) = 0$*, the invertibility of* G_M *implies* $\dfrac{d\lambda_2(t)}{dt} = 0$*. The classical formulation of constrained systems* (3.97) *is therefore recovered.*

An alternative way to recover the case of mechanical constraints is to start from (3.136)*. Consider the case when* $\Xi = (\xi, v_\xi)^T$*. By the same computations as the ones performed to obtain* (3.101)*,*

$$\{\Xi, H\}(q,p) = \begin{pmatrix} v_\xi(q,p) \\ \mathrm{Hess}_q(\xi)(M^{-1}p, M^{-1}p) - \nabla\xi(q)^T M^{-1} \nabla V(q) \end{pmatrix}, \quad (3.137)$$

where the Hessian operator Hess *is defined in* (3.102)*. Using*

$$\Gamma^{-1} = \begin{pmatrix} G_M^{-1}\, \nabla v_\xi^T J \,\nabla v_\xi\, G_M^{-1} & -G_M^{-1} \\ G_M^{-1} & 0 \end{pmatrix},$$

and the fact that $v_\xi(q,p) = 0$ *when mechanical constraints are considered, it follows using the notation* f_{rgd}^M *introduced in* (3.103)*,*

$$\Gamma^{-1}\{\Xi, H\} = \begin{pmatrix} f_{\mathrm{rgd}}^M \\ 0 \end{pmatrix}. \quad (3.138)$$

Thus, considering a test function $\varphi : \mathbb{R}^{6N} \to \mathbb{R}$*, and remarking that*

$$\{\varphi, \Xi\} \begin{pmatrix} a \\ 0 \end{pmatrix} = -a^T \nabla\xi^T \nabla_p \varphi$$

for any $a \in \mathbb{R}^m$*, we obtain:*

$$\{\varphi, \Xi\}\, \Gamma^{-1}\{\Xi, H\} = -\left(f_{\mathrm{rgd}}^M\right)^T \nabla\xi^T \nabla_p \varphi. \quad (3.139)$$

The equations (3.135) *with the Lagrange multiplier* (3.136) *are then the equations* (3.106)*. A similar computation holds for the constraints* $\xi(q) = z$ *and* $p_\xi(q,p) = 0$ *by remarking that*

$$\nabla_q p_\xi(q,p) = G_M^{-1}(q)\, \nabla_q v_\xi(q,p) + \nabla\big(G_M^{-1}\big)(q)\, v_\xi(q,p) = G_M^{-1}(q)\, \nabla_q v_\xi(q,p)$$

since $v_\xi(q,p) = 0$*.*

Defining the constrained symplectic matrix

$$J_\Xi(q,p) = J - J\,\nabla\Xi(q,p)\,\Gamma^{-1}(q,p)\,\nabla\Xi^T(q,p)\,J, \tag{3.140}$$

the constrained evolution (3.135) can be rewritten more compactly as

$$\frac{d}{dt}\begin{pmatrix} q(t) \\ p(t) \end{pmatrix} = J_\Xi(q(t),p(t))\,\nabla H(q(t),p(t)). \tag{3.141}$$

Note that J_Ξ is a skew-symmetric matrix since Γ^{-1} is skew-symmetric. The Poisson bracket associated with constraints can now be defined as:

Definition 3.43 (Poisson bracket for constrained systems). *For any smooth observables $\varphi_1, \varphi_2 : T^*\mathcal{D} \to \mathbb{R}$ with compact support in an open set where $\Gamma(q,p)$ is invertible, the Poisson bracket associated with generalized constraints* (3.109) *is defined by:*

$$\begin{aligned}\{\varphi_1,\varphi_2\}_\Xi &= \{\varphi_1,\varphi_2\} - \{\varphi_1,\Xi\}\,\Gamma^{-1}\,\{\Xi,\varphi_2\} \\ &= \nabla\varphi_1^T J_\Xi \nabla\varphi_2.\end{aligned} \tag{3.142}$$

Note that the definition (3.142) does not involve the particular values ζ taken by the constraints Ξ. It can be checked that $\{\cdot,\cdot\}_\Xi$ verifies the characteristic properties of Poisson brackets (see Section 1.2.2.2), namely the skew-symmetry, Jacobi's identity, and Leibniz' rule. A divergence formula with associated integration by parts will be shown below in Proposition 3.46.

3.3.3.2 *Properties of the constrained dynamics*

With the definition (3.142) at hand, it is possible to rewrite constrained Hamiltonian dynamics (3.135) or (3.141) as Liouvillian transport equation, for any smooth test function φ:

$$\frac{d\varphi(q(t),p(t))}{dt} = \{\varphi,H\}_\Xi\,(q(t),p(t)). \tag{3.143}$$

Lemma 3.44 (Preservation of energy and constraints). *The energy and the constraints are invariants of the motion.*

Proof. Since $\Gamma = \{\Xi,\Xi\}$, it follows that

$$\{H,\Xi\}_\Xi = \{H,\Xi\} - \{H,\Xi\}\,\Gamma^{-1}\,\{\Xi,\Xi\} = 0.$$

Besides,

$$\{H,H\}_\Xi = \{H,\Xi\}\,\Gamma^{-1}\,\{\Xi,H\} = 0$$

by skew-symmetry of Γ^{-1}. Therefore, the solution of (3.143) (equivalently, the solution of (3.135) or (3.141)) is such that

$$\frac{d}{dt}\Big[H(q(t),p(t))\Big]=0, \qquad \frac{d}{dt}\Big[\Xi(q(t),p(t))\Big]=0.$$

This shows that the constraints Ξ and the energy H are indeed invariants of the dynamics. □

We can now state the reversibility up to momentum reversal of the constrained dynamics in the case when $(q,p)\in T^*\Sigma(z)$:

Lemma 3.45. *The constrained evolution equations* (3.97) *or* (3.106) *(equivalent to* (3.135) *in the case when $v_z=0$ or $p_z=0$) in the state space $T^*\Sigma(z)$ are symmetric and reversible: if $S:(q,p)\mapsto(q,-p)$, then for any test function φ and for all $(q,p)\in T^*\Sigma(z)$,*

$$\{\varphi\circ S,H\}_\Xi\,(q,p)=-\,\{\varphi,H\}_\Xi\,(S(q,p)).$$

Equivalently, if ϕ_t denotes the flow associated to the dynamics (3.97),

$$S\circ\phi_t\circ S=\phi_{-t}=(\phi_t)^{-1}.$$

Proof. Reversibility up to momenta reversal is a consequence of the following invariance property: if $t\mapsto(q(t),p(t))$ is solution of (3.97), then so is $t\mapsto(q(-t),-p(-t))$. This can be double checked by remarking that for any test function $\varphi\in C^\infty(\mathbb{R}^{6N})$:

$$\{\varphi\circ S,H\}=-\,\{\varphi,H\}\circ S,$$

on the one hand; while on the other hand (3.139) yields for $(q,p)\in T^*\Sigma(z)$:

$$\{\varphi,\Xi\}\,\Gamma^{-1}\,\{\Xi,H\}=-(\nabla_p\varphi)^T\nabla\xi\, f^M_{\rm rgd}.$$

Now since $f^M_{\rm rgd}\circ S(q,p)=f^M_{\rm rgd}(q,p)$,

$$\{\varphi\circ S,\Xi\}\,\Gamma^{-1}\,\{\Xi,H\}=-\left(\{\varphi,\Xi\}\,\Gamma^{-1}\,\{\Xi,H\}\right)\circ S.$$

Combining the two computations yields $\{\varphi\circ S,H\}_\Xi=-\,\{\varphi,H\}_\Xi\circ S$. □

The key property relating the phase space measure $\sigma_{\Sigma_\Xi(\zeta)}(dq\,dp)$ defined in Section 3.3.2 with the Poisson bracket (3.142) is the following divergence theorem.

Proposition 3.46 (Divergence theorem on submanifolds). *Consider the Poisson bracket $\{\cdot,\cdot\}_\Xi$ defined by* (3.142) *in an open neighborhood $\mathcal{O}$ of the submanifold $\Sigma_\Xi(\zeta)\subset\mathbb{R}^{6N}$ where Γ is invertible. Then, for any smooth test functions $\varphi_1,\varphi_2: T^*\mathcal{D}\to\mathbb{R}$ with compact support in $\mathcal{O}$,*

$$\int_{\Sigma_\Xi(\zeta)}\{\varphi_1,\varphi_2\}_\Xi\; d\sigma_{\Sigma_\Xi(\zeta)}=0. \tag{3.144}$$

Proof. It is possible to obtain this result as a corollary of Darboux's theorem (see [Abraham and Marsden (1978)] or Chapter 8 in [Arnol'd (1989)], and Remark 3.38), which states the existence of local coordinates (Q, P) for which the phase space measure and the Poisson bracket have the canonical forms $\sigma_{\Sigma_\Xi(\zeta)}(dq\,dp) = dP\,dQ$, and $\{\phi_1, \phi_2\}_\Xi = \nabla_Q \phi_1 \cdot \nabla_P \phi_2 - \nabla_P \phi_1 \cdot \nabla_Q \phi_2$.

Nevertheless, a proof in Cartesian coordinates is instructive in the present setting. For this purpose, consider smooth test functions φ_1, φ_2 with compact support in $\mathcal{O}$. By the co-area formula (3.124),

$$\big|\det \Gamma(q,p)\big|^{-1/2} \sigma_{\Sigma_\Xi(\zeta)}(dq\,dp)\,d\zeta = dq\,dp,$$

so that proving the proposition amounts to showing that $I = 0$, with

$$\begin{aligned} I &:= \int_{\mathbb{R}^{2m}} \phi(\zeta) \int_{\Sigma_\Xi(\zeta)} \{\varphi_1, \varphi_2\}_\Xi \, d\sigma_{\Sigma_\Xi(\zeta)}\, d\zeta \\ &= \int_{\mathbb{R}^{6N}} \Big(\phi \circ \Xi\Big) \{\varphi_1, \varphi_2\}_\Xi \, |\det \Gamma|^{1/2} \, dq\,dp \end{aligned}$$

for a given test function ϕ with compact support in $\mathbb{R}^{2m}$. The invariance of the constraints (see Lemma 3.44) implies

$$\{\phi \circ \Xi, \varphi_2\}_\Xi = \sum_{a=1}^{2m} \Big(\partial_a \phi \circ \Xi\Big) \{\Xi_a, \varphi_2\}_\Xi = 0,$$

so that

$$\Big(\phi \circ \Xi\Big) \{\varphi_1, \varphi_2\}_\Xi = \{\widetilde{\varphi}_1, \varphi_2\}_\Xi$$

with $\widetilde{\varphi}_1 = \varphi_1\,(\phi \circ \Xi)$. Then, developing the Poisson bracket as (3.142),

$$\begin{aligned} I &= \int_{\mathbb{R}^{6N}} \{\tilde{\varphi}_1, \varphi_2\} \, |\det \Gamma|^{1/2} \, dq\,dp - \int_{\mathbb{R}^{6N}} \{\tilde{\varphi}_1, \Xi\}\, \Gamma^{-1} \{\Xi, \varphi_2\} \, |\det \Gamma|^{1/2} \, dq\,dp \\ &= I_1 + I_2. \end{aligned}$$

We now use the relation (3.145) proved below in Lemma 3.48, together with an integration by parts in $\mathbb{R}^{6N}$ (we use below the summation convention on repeated indices for latin letters $a, b, \ldots$ from 1 to $2m$):

$$\begin{aligned} I_2 &= -\int_{\mathbb{R}^{6N}} \{\widetilde{\varphi}_1, \Xi_a\} \, (\Gamma^{-1})_{a,b} \, \{\Xi_b, \varphi_2\} \, |\det \Gamma|^{1/2} \, dq\,dp \\ &= \int_{\mathbb{R}^{6N}} \Big\{ \, |\det \Gamma|^{1/2} \, (\Gamma^{-1})_{a,b} \, \{\Xi_b, \varphi_2\}\,,\, \Xi_a \Big\} \, \widetilde{\varphi}_1 \, dq\,dp \\ &= \int_{\mathbb{R}^{6N}} \{\{\Xi_b, \varphi_2\}\,, \Xi_a\} \, (\Gamma^{-1})_{a,b} \, |\det \Gamma|^{1/2} \, \widetilde{\varphi}_1 \, dq\,dp. \end{aligned}$$

Exchanging the indices a, b in the above sum, and using the skew-symmetry property $(\Gamma^{-1})_{a,b} = -(\Gamma^{-1})_{b,a}$,

$$I_2 = \frac{1}{2} \int_{\mathbb{R}^{6N}} \Big(\{\{\Xi_b, \varphi_2\}, \Xi_a\} - \{\{\Xi_a, \varphi_2\}, \Xi_b\} \Big) (\Gamma^{-1})_{a,b} \, |\det \Gamma|^{1/2} \, \widetilde{\varphi}_1 \, dq \, dp$$
$$= \frac{1}{2} \int_{\mathbb{R}^{6N}} \Big(\{\{\Xi_b, \varphi_2\}, \Xi_a\} + \{\{\varphi_2, \Xi_a\}, \Xi_b\} \Big) (\Gamma^{-1})_{a,b} \, |\det \Gamma|^{1/2} \, \widetilde{\varphi}_1 \, dq \, dp$$
$$= -\frac{1}{2} \int_{\mathbb{R}^{6N}} \{\Gamma_{a,b}, \varphi_2\} \, (\Gamma^{-1})_{a,b} \, |\det \Gamma|^{1/2} \, \widetilde{\varphi}_1 \, dq \, dp$$

by Jacobi's identity (1.14). Finally, the determinant variation formula (3.146) below implies

$$|\det \Gamma|^{-1/2} \left\{ |\det \Gamma|^{1/2}, \cdot \right\} = \frac{1}{2} \{\ln |\det \Gamma|, \cdot\} = \frac{1}{2} \{\Gamma_{a,b}, \cdot\} (\Gamma^{-1})_{b,a}.$$

This enables one to conclude the proof by a final integration by parts in $\mathbb{R}^{6N}$:

$$I_2 = \int_{\mathbb{R}^{6N}} \left\{ |\det \Gamma|^{1/2}, \varphi_2 \right\} \widetilde{\varphi}_1 \, dq \, dp$$
$$= -\int_{\mathbb{R}^{6N}} \{\widetilde{\varphi}_1, \varphi_2\} \, |\det \Gamma|^{1/2} \, dq \, dp = -I_1.$$

Finally $I = 0$ and the result follows. □

Remark 3.47 (Relationship with the divergence formula (3.57)). *In Lemma 3.22, we have shown the following divergence formula: for any function $\Phi : \mathbb{R}^n \to \mathbb{R}^n$,*

$$\int_{\Sigma(z)} \mathrm{div}_{\Sigma(z)}(P\Phi) \, d\sigma_{\Sigma(z)} = 0,$$

where $\mathrm{div}_{\Sigma(z)}$ denotes the surface divergence (3.56). This result is a consequence of (3.144) upon choosing $M = \mathrm{Id}$, $\Xi = (\xi, v_\xi)$, $\zeta = (z, 0)$ (so that $d\sigma_{\Sigma_\Xi(\zeta)} = d\sigma_{T^*\Sigma(z)}$), and

$$\varphi_1(q,p) = p^T P(q) \Phi(q), \qquad \varphi_2(q,p) = \exp\left(-\frac{|p|^2}{2}\right).$$

Indeed,

$$\{\varphi_1, \varphi_2\}_\Xi (q,p) = \left\{ p^T P(q) \Phi(q), \exp\left(-\frac{|p|^2}{2}\right) \right\}_\Xi$$
$$= \left\{ p^T P(q) \Phi(q), \exp\left(-\frac{|p|^2}{2}\right) \right\} - \{p^T P(q) \Phi(q), \Xi\} \, \Gamma^{-1} \left\{ \Xi, \exp\left(-\frac{|p|^2}{2}\right) \right\}$$
$$= -p^T \nabla_q \Big(P(q) \Phi(q) \Big) p \, \exp\left(-\frac{|p|^2}{2}\right),$$

where we have used (3.139) *with* $V = 0$. *Therefore, using* (3.177),

$$
\begin{aligned}
0 &= \int_{\Sigma_\Xi(\zeta)} \{\varphi_1, \varphi_2\}_\Xi \, d\sigma_{\Sigma_\Xi(\zeta)} \\
&= -\int_{T^*\Sigma(z)} p^T \nabla_q \Big(P(q)\Phi(q) \Big) p \exp\left(-\frac{|p|^2}{2} \right) \sigma_{T^*\Sigma(z)}(dq\, dp) \\
&= -\int_{\Sigma(z)} P(q) : \nabla_q \Big(P(q)\Phi(q) \Big) \, \sigma_{\Sigma(z)}(dq) \\
&= -\int_{\Sigma(z)} \mathrm{div}_{\Sigma(z)}(P\Phi) \, d\sigma_{\Sigma(z)}.
\end{aligned}
$$

An important property used in the previous proof is

Lemma 3.48. *For any* $a \in \{1, \ldots, 2m\}$ *:*

$$\sum_{b=1}^{2m} \left\{ |\det\Gamma|^{1/2} \, (\Gamma^{-1})_{a,b}, \Xi_b \right\} = 0. \tag{3.145}$$

Proof. The proof relies on the following computation rules for any family of invertible square matrices $\theta \mapsto A_\theta$:

$$\frac{d}{d\theta} \ln |\det A_\theta| = \mathrm{tr}\left(A_\theta^{-1} \frac{d}{d\theta} A_\theta \right), \tag{3.146}$$

and

$$A_\theta \frac{d}{d\theta}(A_\theta^{-1}) = -\Big(\frac{d}{d\theta} A_\theta \Big) A_\theta^{-1}. \tag{3.147}$$

First, using (3.147) with A_θ replaced by Γ and $\frac{d}{d\theta}$ replaced by $\{\cdot, \Xi_c\}$,

$$\Gamma_{a,b} \left\{ (\Gamma^{-1})_{b,c}, \Xi_c \right\} = - \left\{ \Gamma_{a,b}, \Xi_c \right\} (\Gamma^{-1})_{b,c}.$$

By the skew-symmetry of Γ^{-1},

$$
\begin{aligned}
\Gamma_{a,b} \left\{ (\Gamma^{-1})_{b,c}, \Xi_c \right\} &= - \left\{ \left\{ \Xi_a, \Xi_b \right\}, \Xi_c \right\} (\Gamma^{-1})_{b,c} \\
&= -\frac{1}{2} \Big(\left\{ \left\{ \Xi_a, \Xi_b \right\}, \Xi_c \right\} - \left\{ \left\{ \Xi_a, \Xi_c \right\}, \Xi_b \right\} \Big) (\Gamma^{-1})_{b,c} \\
&= -\frac{1}{2} \Big(\left\{ \left\{ \Xi_a, \Xi_b \right\}, \Xi_c \right\} + \left\{ \left\{ \Xi_c, \Xi_a \right\}, \Xi_b \right\} \Big) (\Gamma^{-1})_{b,c}.
\end{aligned}
$$

Jacobi's identity (1.14) and (3.146) then yield

$$
\begin{aligned}
\Gamma_{a,b} \left\{ (\Gamma^{-1})_{b,c}, \Xi_c \right\} &= \frac{1}{2} \left\{ \left\{ \Xi_b, \Xi_c \right\}, \Xi_a \right\} (\Gamma^{-1})_{b,c} = -\frac{1}{2} \left\{ \Gamma_{c,b}, \Xi_a \right\} (\Gamma^{-1})_{b,c} \\
&= -\frac{1}{2} \left\{ \ln |\det \Gamma|, \Xi_a \right\} = - |\det \Gamma|^{-1/2} \left\{ |\det \Gamma|^{1/2}, \Xi_a \right\} \\
&= - |\det \Gamma|^{-1/2} \, \Gamma_{a,b} (\Gamma^{-1})_{b,c} \left\{ |\det \Gamma|^{1/2}, \Xi_c \right\}
\end{aligned}
$$

since $\Gamma_{a,b}(\Gamma^{-1})_{b,c} = \delta_{a,c}$ where δ is the Kronecker symbol. Finally, the equality

$$\Gamma_{a,b}\left\{(\Gamma^{-1})_{b,c}, \Xi_c\right\} + |\det \Gamma|^{-1/2}\,\Gamma_{a,b}(\Gamma^{-1})_{b,c}\left\{|\det \Gamma|^{1/2}, \Xi_c\right\} = 0$$

can be rearranged with Leibniz' rule (1.15) to conclude

$$|\det \Gamma|^{-1/2}\,\Gamma_{a,b}\left\{|\det \Gamma|^{1/2}\,(\Gamma^{-1})_{b,c}, \Xi_c\right\} = 0,$$

which yields (3.145). □

We conclude this section by an important corollary to Proposition 3.46.

Corollary 3.49. *The Liouville/phase space measure $\sigma_{\Sigma_\Xi(\zeta)}(dq\,dp)$ is conserved by the Hamiltonian flow associated to constrained dynamics solution to* (3.135).

Proof. Denote by Ψ_t the flow of (3.135), and consider a smooth test function φ with compact support. Differentiating $\Psi_t = \Psi_{t-h} \circ \Psi_h$ with respect to h at $h = 0$ yields

$$0 = \frac{d}{dh}\left(\Psi_{t-h} \circ \Psi_h\right)\Big|_{h=0} = -\frac{d}{dt}\Psi_t + \{\Psi_t, H\}_\Xi\,,$$

so that

$$\frac{d}{dt}\Big(\varphi \circ \Psi_t\Big) = \{\varphi \circ \Psi_t, H\}_\Xi\,.$$

Thus, by the divergence formula (3.144),

$$\frac{d}{dt}\left(\int_{\Sigma_\Xi(\zeta)} \varphi \circ \Psi_t \, d\sigma_{\Sigma_\Xi(\zeta)}\right) = 0.$$

Since this equality is valid for any compactly supported smooth function, the preservation of $\sigma_{\Sigma_\Xi(\zeta)}$ by the flow $t \mapsto \Psi_t$ follows. □

3.3.4 *Constrained Langevin processes*

In this section, a generalization of Langevin processes (2.39) with mechanical constraints $\xi(q) = z$ is considered, by adding random terms to the Hamiltonian evolution (3.97).

3.3.4.1 *Definition of the dynamics*

A constrained Langevin process is the solution of the following stochastic differential equation:

$$\begin{cases} dq_t = M^{-1}p_t\,dt, \\ dp_t = -\nabla V(q_t)\,dt - \gamma(q_t)M^{-1}p_t\,dt + \sigma(q_t)\,dW_t + \nabla\xi(q_t)\,d\lambda_t, \\ \xi(q_t) = z, \end{cases} \tag{3.148}$$

where the adapted process[7] $t \mapsto \lambda_t$ is the Lagrange multiplier associated with the (vectorial) constraint $\xi(q) = z \in \mathbb{R}^m$. The standard fluctuation/dissipation identity (2.42) for unconstrained processes

$$\forall q \in \mathcal{D}, \qquad \sigma(q)\,\sigma^T(q) = \frac{2}{\beta}\gamma(q), \tag{3.149}$$

should still be imposed in order for the canonical measure to be invariant, see Proposition 3.53 below.

Remark 3.50 (Itô vs. Stratonovitch integration). *Since the process $t \mapsto q_t$ is of finite variation, the choice of stochastic integration is indifferent in* (3.148)*: Itô integration $\nabla\xi(q_t)\,d\lambda_t$ and $\sigma(q_t)\,dW_t$, or Stratonovitch integration $\nabla\xi(q_t)\circ d\lambda_t$ and $\sigma(q_t)\circ dW_t$ are identical, see Remark* 2.6.

As in the deterministic case (see Section 3.3.1), the Lagrange multiplier can be computed explicitly:

$$\begin{aligned} d\lambda_t = -G_M^{-1}(q_t)\Big[&\mathrm{Hess}_{q_t}(\xi)\big(M^{-1}p_t, M^{-1}p_t\big)\,dt \\ &+ \nabla\xi(q_t)^T M^{-1}\Big(- \nabla V(q_t)\,dt - \gamma(q_t)M^{-1}p_t\,dt + \sigma(q_t)\,dW_t\Big)\Big], \end{aligned}$$

where Hess is defined in (3.102). Therefore, (3.148) can be recast in a more explicit form as

$$\begin{cases} dq_t = M^{-1}p_t\,dt, \\ dp_t = P_M(q_t)\Big[- \nabla V(q_t)\,dt - \gamma(q_t)M^{-1}p_t\,dt + \sigma(q_t)\,dW_t\Big] \\ \qquad\quad -\nabla\xi(q_t)G_M^{-1}(q_t)\mathrm{Hess}_{q_t}(\xi)(M^{-1}p_t, M^{-1}p_t)\,dt, \end{cases} \tag{3.150}$$

where P_M is the projector defined in (3.104) by

$$P_M(q) = \mathrm{Id} - \nabla\xi(q)G_M^{-1}(q)\nabla\xi^T(q)M^{-1}.$$

[7] *I.e.* a random variable depending only on the past values of the Brownian motion.

Defining the projected fluctuation-dissipation matrices

$$\sigma_P(q) = P_M(q)\,\sigma, \qquad \gamma_P(q) := P_M(q)\,\gamma\,P_M(q)^T, \tag{3.151}$$

which still satisfy the fluctuation-dissipation identity

$$\sigma_P\,\sigma_P^T = \frac{2}{\beta}\gamma_P, \tag{3.152}$$

the momenta dynamics in (3.150) can be rewritten with the constraining force f_{rgd}^M (3.103) as follows:

$$dp_t = -\nabla V(q_t)\,dt + \nabla\xi(q_t)\,f_{\mathrm{rgd}}^M(q(t),p(t))\,dt - \gamma_P(q_t)M^{-1}p_t\,dt + \sigma_P(q_t)\,dW_t.$$

Remark 3.51 (Highly oscillatory systems). *As in the deterministic case (see Remark* 3.36*), constrained dynamics are not, except in the very special case of affine constraints, the limit of a highly oscillatory system with slow manifold $T^*\Sigma(z)$, for instance with potential energies of the form*

$$V_\varepsilon(q) = V(q) + \frac{1}{\varepsilon^2}\,|\xi(q) - z|^2.$$

The infinite stiffness limit ($\varepsilon \to 0$) of highly oscillatory dynamics with random perturbations has been studied in [Reich (2000)], where it is shown that adiabatic effective potentials (derived from the conservation of the ratio energy/frequency of fast modes) are still required to describe the limiting dynamics. However, a formal argument based on "overdamping" the fast modes leads to some Markovian effective dynamics, which has already been introduced in the overdamped case in Section 3.2.3.4*. This effective dynamics, called "softly constrained", is obtained by changing the potential V of the "rigidly constrained" Langevin dynamics* (3.150) *to an effective potential $V + V_{\mathrm{fix}}$. The additional term*

$$V_{\mathrm{fix}}(q) = \frac{1}{2\beta}\ln\Big(\det G_M(q)\Big), \tag{3.153}$$

sometimes called Fixman corrector, is due to [Fixman (1978)]. The difference between soft and rigid constraints therefore arises from the geometry of the constraints ($G_M \neq \mathrm{Id}$), and is the dynamical counterpart of the difference between the canonical distribution μ^ξ with constraints on position only, whose marginal on position is proportional to $\mathrm{e}^{-\beta V(q)}\delta_{\xi(q)-z}(dq) = \mathrm{e}^{-\beta(V(q)+V_{\mathrm{fix}}(q))}\,\sigma_{\Sigma(z)}^M(dq)$, and the canonical distribution $\mu_{T^*\Sigma(z)}$ with constraints on both positions and momenta, whose marginal on positions is proportional to $\mathrm{e}^{-\beta V(q)}\sigma_{\Sigma(z)}^M(dq)$ (see Section 3.3.2.3 *for more details, and Remark* 3.60 *below).*

The Markov generator of the above Langevin process is given by the following proposition.

Proposition 3.52. *For general constraints* Ξ *given by* (3.109), *the Markov generator of the constrained process* (3.148) *or* (3.150), *defined for smooth test functions with compact support on the open set where* Γ *is invertible, is*

$$\mathcal{L}_\Xi = \mathcal{L}_\Xi^{\rm ham} + \mathcal{L}_\Xi^{\rm thm} \tag{3.154}$$

where the Hamiltonian transport operator reads

$$\mathcal{L}_\Xi^{\rm ham} = \{\cdot, H\}_\Xi \,,$$

while the fluctuation-dissipation part is

$$\mathcal{L}_\Xi^{\rm thm} = \frac{1}{2}{\rm div}_p\Big(P_M\,\sigma\sigma^T P_M^T \nabla_p\,\cdot\Big) - p^T M^{-1} P_M \gamma P_M^T \nabla_p. \tag{3.155}$$

Using the fluctuation-dissipation relation (3.149), the generator $\mathcal{L}_\Xi^{\rm thm}$ can be rewritten more compactly as

$$\mathcal{L}_\Xi^{\rm thm} = \frac{1}{\beta}\,{\rm e}^{\beta H}{\rm div}_p\Big({\rm e}^{-\beta H}\,\gamma_P\,\nabla_p\,\cdot\Big), \tag{3.156}$$

where γ_P is defined by (3.151).

Proof. The expression of the generator is obtained by Itô calculus using the relation: for any smooth test function φ,

$$\frac{d}{dt}\Big(\varphi(q_t,p_t)\Big) = \Big[\mathcal{L}_\Xi(\varphi)\Big](q_t,p_t),$$

where (q_t, p_t) satisfies (3.150). The dynamics (3.150) can be seen as the deterministic constrained Hamiltonian evolution (3.106) perturbed by an additional term $P_M(q_t)(-\gamma(q_t)M^{-1}p_t\,dt + \sigma(q_t)\,dW_t)$ in the evolution of the momenta. The generator $\mathcal{L}_\Xi^{\rm ham}$ of the constrained Langevin dynamics (3.150) in the case when $\sigma, \gamma = 0$ is given by (3.143).

The additional term characterizes the fluctuation-dissipation operator. The diffusive part arises from the Brownian term, and its expression

$$\frac{1}{2}{\rm div}_p\Big(P_M\,\sigma\sigma^T P_M^T \nabla_p\,\cdot\Big)$$

is obtained directly from the standard Itô calculus. The dissipation operator

$$-\big(P_M\,\gamma\,M^{-1}p\big)^T\nabla_p = -p^T M^{-1}\gamma\,P_M^T\nabla_p$$

can be rewritten in a more symmetric way by remarking that (3.105) implies $P_M(q)^T M^{-1} p = M^{-1}p$ when $p \in T_q^*\Sigma(z)$, so that the dissipation operator finally reads

$$-p^T M^{-1} P_M \gamma P_M^T \nabla_p.$$

The addition of these two contributions gives the expression of $\mathcal{L}_\Xi^{\rm thm}$. □

With the expression of the generator at hand, it is easily checked that the canonical measure is invariant, and the ergodicity of the dynamics can be shown with some hypo-ellipticity result.

Proposition 3.53. *When the fluctuation/dissipation relation* (3.149) *holds, the constrained Langevin operator* (3.154) *associated with the dynamics* (3.148) *satisfies on* $T^*\Sigma(z)$ *the detailed balance condition up to momenta reversal* (2.32), *with stationary Boltzmann-Gibbs distribution*

$$\mu_{T^*\Sigma(z)}(dq\,dp) = Z_{z,0}^{-1}\,\mathrm{e}^{-\beta H(q,p)}\,\sigma_{T^*\Sigma(z)}(dq\,dp).$$

If $P_M(q)\gamma P_M(q)^T$ *is everywhere strictly positive in the sense of symmetric matrices, then the process* (3.148) *is ergodic: for any smooth test function* φ,

$$\lim_{T\to+\infty}\frac{1}{T}\int_0^T \varphi(q_t,p_t)\,dt = \int_{T^*\Sigma(z)} \varphi\,d\mu_{T^*\Sigma(z)} \qquad \text{a.s.}$$

Proof. The detailed balance condition up to momentum reversal to be satisfied in this context is

$$\int_{T^*\Sigma(z)} \varphi_1\,\mathcal{L}_\Xi(\varphi_2)\,d\mu_{T^*\Sigma(z)} = \int_{T^*\Sigma(z)} (\varphi_2\circ S)\,\mathcal{L}_\Xi(\varphi_1\circ S)\,d\mu_{T^*\Sigma(z)},$$

for any test functions φ_1, φ_2, and where $S:(q,p)\mapsto(q,-p)$ is the momentum flip. Note that H satisfies $H\circ S = H$.

The Hamiltonian part $\mathcal{L}_\Xi^{\mathrm{ham}}$ is the transport operator associated with the constrained evolution (3.106), which is symmetric and time reversible (see Lemma 3.45):

$$\{\varphi\circ S, H\}_\Xi = -\{\varphi, H\}_\Xi\circ S.$$

On the other hand,

$$\mathrm{e}^{-\beta H}\{\cdot, H\}_\Xi = -\frac{1}{\beta}\left\{\cdot, \mathrm{e}^{-\beta H}\right\}_\Xi,$$

so that:

$$\begin{aligned}\mathrm{e}^{-\beta H}\varphi_2\circ S\,\{\varphi_1\circ S, H\}_\Xi &= -\left(\mathrm{e}^{-\beta H}\varphi_2\,\{\varphi_1, H\}_\Xi\right)\circ S\\ &= \left(\mathrm{e}^{-\beta H}\varphi_1\,\{\varphi_2, H\}_\Xi + \left\{\varphi_2\varphi_1, \frac{\mathrm{e}^{-\beta H}}{\beta}\right\}_\Xi\right)\circ S,\end{aligned}$$

and the divergence formula (3.144) yields the balance condition for the Hamiltonian part, using the fact that the distribution $\mu_{T^*\Sigma(z)}$ is invariant under the momenta flip S.

For the fluctuation/dissipation part, it is easily checked, that

$$\mathcal{L}_{\Xi}^{\mathrm{thm}}(\varphi \circ S) = \mathcal{L}_{\Xi}^{\mathrm{thm}}(\varphi) \circ S$$

for any smooth test function φ, so that the detailed balance condition up to momenta reversal will follow from the usual detailed balance condition:

$$\int_{T^*\Sigma(z)} \varphi_1 \, \mathcal{L}_{\Xi}^{\mathrm{thm}}(\varphi_2) \, d\mu_{T^*\Sigma(z)} = \int_{T^*\Sigma(z)} \varphi_2 \, \mathcal{L}_{\Xi}^{\mathrm{thm}}(\varphi_1) \, d\mu_{T^*\Sigma(z)}.$$

To prove the latter relation, consider the decomposition of the phase space measure (3.129) or (3.130) in Proposition 3.40 (see also (3.132)) and use the divergence theorem (3.58) in the linear space $T_q^*\Sigma(z)$ for the variable p, q being fixed, namely:

$$\int_{T_q^*\Sigma(z)} \mathrm{div}_p \left(P_M(q)\phi \right) \, \sigma_{T_q^*\Sigma(z)}^{M^{-1}}(dp) = 0, \tag{3.157}$$

with

$$\phi = \gamma P_M^T \nabla_p(\varphi_2) \, \mathrm{e}^{-\beta H} \, \varphi_1.$$

Equation (3.157) is obtained from the divergence theorem (3.58) by remarking that the matrix G in (3.58) is constant for linear constraints. This yields a symmetric expression in (φ_1, φ_2), using the formula (3.156) for $\mathcal{L}_{\Xi}^{\mathrm{thm}}$:

$$\int_{T^*\Sigma(z)} \varphi_1 \, \mathcal{L}_{\Xi}^{\mathrm{thm}}(\varphi_2) \, d\mu_{T^*\Sigma(z)} = -\int_{T^*\Sigma(z)} \nabla_p^T \varphi_1 P_M \gamma P_M^T \nabla_p \varphi_2 \, d\mu_{T^*\Sigma(z)},$$

and the detailed balance condition follows.

Ergodicity comes from the hypo-ellipticity of the operator $\mathcal{L}_{\Xi}$ on $T^*\Sigma(z)$, which is itself a consequence of the fact that $P_M \gamma P_M^T$ is strictly positive on $\Sigma(z)$. The proof can be done using local coordinates and Theorem 2.7. □

3.3.5 *Numerical implementation*

In this section, we are interested in numerical methods to discretize constrained Hamilton and Langevin dynamics. The reader is referred to Section 2.2.3 for similar considerations in the unconstrained case. As for unconstrained Langevin dynamics, it is natural to build numerical schemes based on a splitting of the dynamics (Section 3.3.5.3) into a Hamiltonian transport part (Section 3.3.5.1), and a fluctuation/dissipation part acting only on the momenta (Section 3.3.5.2). These schemes can be corrected by a Metropolis rule, in order to eliminate the time-step error (Section 3.3.5.4). Finally, numerical schemes for the overdamped Langevin dynamics can be obtained in some limiting regime, which offers a way to correct the time-step error in the usual Euler schemes (Section 3.3.5.5). Application to free energy computation will be the topic of the next section (Section 3.3.6). For simplicity, we restrict ourselves to constant matrices γ and σ. Generalizations to position dependent matrices are straightforward.

3.3.5.1 *Numerical schemes for the Hamiltonian part*

The deterministic equations of motion (3.97) with position constraints $\xi(q) = z$, can be integrated by a velocity-Verlet like algorithm. This version with constraints is often called "RATTLE" in the literature and is a velocity version of the classical "SHAKE" algorithm (see Chapter VII.1 in [Hairer *et al.* (2006)], or Chapter 7 in [Leimkuhler and Reich (2005)] for more details and historical references). The constrained velocity-Verlet scheme is an explicit integrator (which requires however the solution of a nonlinear problem to enforce the constraints on position):

$$\begin{cases} p^{n+1/2} = p^n - \dfrac{\Delta t}{2}\nabla V(q^n) + \nabla\xi(q^n)\,\lambda^{n+1/2}, \\ q^{n+1} = q^n + \Delta t\, M^{-1}\, p^{n+1/2}, \\ \xi(q^{n+1}) = z, & (C_q) \\ p^{n+1} = p^{n+1/2} - \dfrac{\Delta t}{2}\nabla V(q^{n+1}) + \nabla\xi(q^{n+1})\,\lambda^{n+1}, \\ \nabla\xi^T(q^{n+1})M^{-1}p^{n+1} = 0, & (C_p) \end{cases} \tag{3.158}$$

where Δt is the time-step, $\lambda^{n+1/2} \in \mathbb{R}^m$ are the Lagrange multipliers associated with the position constraints (C_q), and $\lambda^{n+1} \in \mathbb{R}^m$ are the Lagrange multipliers associated with the velocity constraints (C_p). The nonlinear constraints (C_q) are typically enforced using a Newton algorithm (see the references and the discussion in Section 3.2.4.2). In (3.158), the (linear) momentum projection is always well defined, whereas the nonlinear projection used to enforce the position constraints (C_q) requires the following definition:

Definition 3.54 (Domain $D_{\Delta t}$). *The domain $D_{\Delta t} \subset T^*\Sigma(z)$ is defined as the set of $(q^n, p^n) \in T^*\Sigma(z)$ such that there is a unique solution (q^{n+1}, p^{n+1}) verifying* (3.158).

Solving the position constraints $\xi(q^{n+1}) = z$ consists in projecting on $\Sigma(z)$ a point in a Δt-neighborhood of q^n, thus by the implicit function theorem, the domain $D_{\Delta t}$ verifies:

$$\lim_{\Delta t \to 0} D_{\Delta t} = T^*\Sigma(z).$$

It may happen that there is no solution if the time-step is too large, and, even for small time-steps, that several projections exist, see for instance Example 2 in Chapter 7 in [Leimkuhler and Reich (2005)] where a motion constrained to a circle is considered; when the system is inside the circle, it can be projected on the circle in two different ways, but one projection is closer from the unprojected point than the other. In general, the correct solution is obtained as a continuation limit using the implicit function theorem. When the numerical trajectory remains in a compact set of phase space, and when the potential V is $C^1(\mathcal{D})$, the time-step can be chosen small enough so that the numerical scheme is well posed for all $n \geq 0$. See [Hairer *et al.* (2006); Leimkuhler and Reich (2005)] for further precision on those numerical issues. In practice, $D_{\Delta t}$ can be chosen to be the set of (q^n, p^n) such that the Newton algorithm enforcing constraints converges within a given precision threshold, in a limited number of iterations.

As for the Verlet scheme in the unconstrained case, the associated numerical flow shares three important qualitative properties with the exact flow of (3.97): it is symmetric, time reversible and symplectic (see [Leimkuhler and Skeel (1994)]).

3.3.5.2 *Fluctuation-dissipation part*

The fluctuation-dissipation part in the constrained Langevin equation (3.150) corresponds to an Ornstein-Uhlenbeck process $t \mapsto p_t$ for a fixed position q given by:

$$\begin{aligned} dp_t =& P_M(q)\left(-\gamma\, M^{-1} p_t\, dt + \sigma\, dW_t\right) \\ =& -\gamma_P(q) M^{-1} p_t\, dt + \sigma_P(q)\, dW_t, \end{aligned}$$

where the second equality comes from the fact that $P_M^T(q)M^{-1}p = M^{-1}p$ when $p \in T_q^*\Sigma(z)$. The matrices (γ_P, σ_P) are defined in (3.151) and satisfies the fluctuation-dissipation relation (3.152). This stochastic process can be integrated explicitly when the exponential of the matrix $\gamma_P(q) = P_M(q)\gamma P_M(q)^T$ can be computed. When this computation is cumbersome, a simple and efficient alternative is to resort to a generalization of the midpoint Euler scheme (2.47). For a given $q \in \Sigma_z$ and $p^n \in T_q^*\Sigma(z)$, the corresponding one-step integrator over a time interval Δt reads

$$\begin{cases} p^{n+1} = p^n - \dfrac{\Delta t}{2}\gamma\, M^{-1}(p^n + p^{n+1}) + \sqrt{\Delta t}\,\sigma\, G^n + \nabla\xi(q)\,\lambda^n, \\ \nabla\xi(q)^T M^{-1} p^{n+1} = 0, \end{cases} \tag{3.159}$$

where $(G^n)_{n\geq 0}$ are i.i.d. standard random Gaussian vectors, and $\lambda^n \in \mathbb{R}^m$ are the Lagrange multipliers associated with the momenta constraints. This yields

$$p^{n+1} = \left(\mathrm{Id} + \frac{\Delta t}{2}\gamma M^{-1}\right)^{-1}\left[\left(\mathrm{Id} - \frac{\Delta t}{2}\gamma M^{-1}\right)p^n + \sqrt{\Delta t}\,\sigma\, G^n + \nabla\xi(q)\,\lambda^n\right],$$

together with the constraints $\nabla\xi(q)^T M^{-1} p^{n+1} = 0$. The momentum p^{n+1} may be obtained by first integrating the unconstrained dynamics as

$$\widetilde{p}^{n+1} = \left(\mathrm{Id} + \frac{\Delta t}{2}\gamma M^{-1}\right)^{-1}\left[\left(\mathrm{Id} - \frac{\Delta t}{2}\gamma M^{-1}\right)p^n + \sqrt{\Delta t}\,\sigma\, G^n\right],$$

and then computing the Lagrange multiplicator λ^n by solving the linear system:

$$\nabla\xi(q)^T M^{-1}\widetilde{p}^{n+1} + \nabla\xi(q)^T M^{-1}\left(\mathrm{Id} + \frac{\Delta t}{2}\gamma M^{-1}\right)^{-1}\nabla\xi(q)\,\lambda^n = 0.$$

The invertibility of $\nabla\xi(q)^T M^{-1}\left(\mathrm{Id} + \frac{\Delta t}{2}\gamma M^{-1}\right)^{-1}\nabla\xi(q)$ is a consequence of the invertibility of $\nabla\xi(q)^T M^{-1}\nabla\xi(q)$ since M, γ are symmetric non-negative matrices. Note that the latter computation becomes particularly simple by choosing a dissipation matrix of the form $\gamma = \frac{\gamma_0}{m_0}M$, where $(m_0, \gamma_0) \in \mathbb{R}^2$ are tunable mass and friction. Moreover, a sufficient criteria for stability is, as in the unconstrained case (2.48), for order on symmetric matrices

$$\frac{\Delta t}{2}\gamma \leq M.$$

The Markov chain defined by the scheme (3.159) is reversible (both in the plain sense and up to momentum reversal) with respect to the probability distribution $\kappa^{M^{-1}}_{T^*_q\Sigma(z)}$, the momenta marginal of the constrained canonical distribution defined in (3.133). If $\gamma_P(q) := P_M(q)\,\gamma\, P_M(q)^T$ is strictly positive, it is ergodic with respect to this measure (see Proposition 2.1). Indeed, projecting both sides of (3.159) on $T^*_q\Sigma(z)$ with the operator $P_M(q)$, using $P_M(q)\nabla\xi(q) = 0$ and for $p^n, p^{n+1} \in T^*_q\Sigma(z)$

$$\begin{aligned} M^{-1}(p^n + p^{n+1}) &= M^{-1}P_M(q)(p^n + p^{n+1}) \\ &= P_M(q)^T M^{-1} P_M(q)(p^n + p^{n+1}), \end{aligned}$$

leads to a relation very similar to (2.47): p^n and p^{n+1} are in $T_q^*\Sigma(z)$ and satisfy

$$p^{n+1} = p^n - \frac{\Delta t}{2}\gamma_P(q)\, M^{-1}(p^n + p^{n+1}) + \sqrt{\Delta t}\,\sigma_P(q)\, G^n.$$

A computation similar to the one performed in the unconstrained case (see (2.49)) then allows us to conclude to the reversibility properties.

3.3.5.3 *Numerical schemes obtained by a splitting strategy*

A (midpoint Euler-Verlet-midpoint Euler) splitting strategy based on the schemes (3.159) and (3.158) leads to the following scheme:

$$\begin{cases} p^{n+1/4} = p^n - \dfrac{\Delta t}{4}\gamma\, M^{-1}(p^n + p^{n+1/4}) + \sqrt{\dfrac{\Delta t}{2}}\,\sigma\, G^n + \nabla\xi(q^n)\,\lambda^{n+1/4}, \\ \nabla\xi(q^n)^T M^{-1} p^{n+1/4} = 0, \end{cases}$$

$$\begin{cases} p^{n+1/2} = p^{n+1/4} - \dfrac{\Delta t}{2}\nabla V(q^n) + \nabla\xi(q^n)\,\lambda^{n+1/2}, \\ q^{n+1} = q^n + \Delta t\, M^{-1}\, p^{n+1/2}, \\ \xi(q^{n+1}) = z, \\ p^{n+3/4} = p^{n+1/2} - \dfrac{\Delta t}{2}\nabla V(q^{n+1}) + \nabla\xi(q^{n+1})\,\lambda^{n+3/4}, \\ \nabla\xi(q^{n+1})^T M^{-1} p^{n+3/4} = 0, \end{cases} \tag{3.160}$$

$$\begin{cases} p^{n+1} = p^{n+3/4} - \dfrac{\Delta t}{4}\gamma\, M^{-1}(p^{n+3/4} + p^{n+1}) + \sqrt{\dfrac{\Delta t}{2}}\,\sigma\, G^{n+1/2} \\ \qquad\quad + \nabla\xi(q^{n+1})\,\lambda^{n+1}, \\ \nabla\xi(q^{n+1})^T M^{-1} p^{n+1} = 0. \end{cases}$$

Alternatively, as in the unconstrained case (see (2.51) and (2.52)), the splitting based on half a step of the explicit Euler scheme, then a Verlet step, finally half a step of the implicit Euler scheme, will be referred to as the BBK algorithm with constraints. It reads, when the same random number is used for the explicit and the previous implicit Euler parts (see (2.52) for the unconstrained case):

$$
\left\{
\begin{aligned}
& p^{n+1/4} = p^n - \frac{\Delta t}{2}\gamma\, M^{-1} p^n + \frac{\sqrt{\Delta t}}{2}\,\sigma\, G^n + \nabla\xi(q^n)\,\lambda^{n+1/4}, \\
& \nabla\xi(q^n)^T M^{-1} p^{n+1/4} = 0,
\end{aligned}
\right.
$$

$$
\left\{
\begin{aligned}
& p^{n+1/2} = p^{n+1/4} - \frac{\Delta t}{2}\nabla V(q^n) + \nabla\xi(q^n)\,\lambda^{n+1/2}, \\
& q^{n+1} = q^n + \Delta t\, M^{-1}\, p^{n+1/2}, \\
& \xi(q^{n+1}) = z, \\
& p^{n+3/4} = p^{n+1/2} - \frac{\Delta t}{2}\nabla V(q^{n+1}) + \nabla\xi(q^{n+1})\,\lambda^{n+3/4}, \\
& \nabla\xi(q^{n+1})^T M^{-1} p^{n+3/4} = 0,
\end{aligned}
\right.
\tag{3.161}
$$

$$
\left\{
\begin{aligned}
& p^{n+1} = p^{n+3/4} - \frac{\Delta t}{2}\gamma\, M^{-1} p^{n+1} + \frac{\sqrt{\Delta t}}{2}\,\sigma\, G^{n+1} \\
& \qquad + \nabla\xi(q^{n+1})\,\lambda^{n+1}, \\
& \nabla\xi(q^{n+1})^T M^{-1} p^{n+1} = 0.
\end{aligned}
\right.
$$

As already explained above, in (3.160)-(3.161), the (linear) momentum projection amounts to inverting a matrix of the form $\nabla\xi(q)^T S \nabla\xi(q)$, where $S = M^{-1}\left(\mathrm{Id} + \theta\Delta t\gamma M^{-1}\right)^{-1}$ with $\theta > 0$. The invertibility of the latter matrices are equivalent to the invertibility of $G_M(q)$ which is invertible by assumption. On the contrary, the nonlinear projection used to enforce the position constraints $\xi(q^{n+1}) = z$ is well defined only when $(q^n, p^{n+1/4})$ in (3.160) or in (3.161) belongs to the domain $D_{\Delta t} \subset \mathbb{R}^{6N}$ defined in Section 3.3.5.1.

Remark 3.55 (Consistency and convergence). *We do not discuss convergence results for numerical discretizations of stochastic differential equations in this book, but standard results for unconstrained dynamics could be extended with some care to the constrained case, therefore showing the convergence in law of the path. Let us give some hints on such a proof for the splitting scheme* (3.160). *First note that the midpoint Euler scheme* (3.159) *is by construction a consistent discretization of the fluctuation/dissipation part, while (using the implicit function theorem) the scheme* (3.158) *is a consistent discretization of the Hamiltonian part. Convergence in distribution of the paths means that for a given initial condition* (q^0, p^0), *the probability distribution of the path* $t \mapsto (q^{\lfloor t/\Delta t \rfloor}, p^{\lfloor t/\Delta t \rfloor})$ *(where* $\lfloor\cdot\rfloor$ *denotes the integer part function) of the discretized process on* $[0, T]$, *converges to the distribution of the path $t \mapsto (q_t, p_t)$, the distance being measured in the supremum norm. To prove the latter assertion, consider the exit time $\tau_r^{\Delta t}$ corresponding to the first time when the discretized process reaches a (large) ball $B_r \subset D_{\Delta t} \subset \mathbb{R}^{6N}$ of radius r which contains (q^0, p^0). The implicit function theorem ensures that the projection steps associated with the nonlinear constraints on positions have a unique solution, and that the numerical flow $\Psi_{\Delta t}(q, p)$ is a smooth map defining a consistent discretization of the Hamiltonian dynamics. Now standard results of convergence of discretizations of stochastic differential equations can be applied to the process stopped at $\tau_r^{\Delta t}$ (see for instance [Ethier and Kurtz (1986)]), and weak convergence of probability distribution of paths holds for the convergence of the interpolated discretized process stopped at $\tau_r^{\Delta t}$, towards the Langevin dynamics stopped at some random time τ_r^0. Then, since τ_r^0 is bounded from below by the exit time of the Langevin process from $B_{r/2}$, and taking $r \to +\infty$ with $\Delta t \to 0$, the convergence of unstopped processes on finite time intervals follows. Such a method does not give any information on the speed of convergence.*

3.3.5.4 *Metropolization*

Usually, the invariant probability distribution of the Markov chain sampled by the numerical schemes differs from the invariant probability distribution of the associated continuous in time stochastic processes. Relying on the time-reversibility (up to momentum reversal) and the preservation of the phase space measure $\sigma_{T^*\Sigma(z)}(dq\,dp)$ by the flow of the RATTLE scheme (3.158), it is possible to eliminate the time-step error in the splitting scheme (3.160) by resorting to a Generalized Hybrid Monte-Carlo (GHMC) algorithm, similarly to what is done in the unconstrained case (see Algorithm 2.11).

Algorithm 3.56 (GHMC with constraints). *Consider an initial configuration $(q^0, p^0) \in T^*\Sigma(z)$, and a sequence $(G^n, G^{n+1/2})_{n \geq 0}$ of i.i.d. standard Gaussian vectors. Iterate on $n \geq 0$,*

(1) Evolve the momenta according to the midpoint Euler scheme

$$\begin{cases} p^{n+1/4} = p^n - \dfrac{\Delta t}{4} \gamma M^{-1}(p^n + p^{n+1/4}) + \sqrt{\dfrac{\Delta t}{2}} \sigma\, G^n + \nabla \xi(q^n)\, \lambda^{n+1/4}, \\ \nabla \xi(q^n)^T M^{-1} p^{n+1/4} = 0, \end{cases}$$

and compute the energy $E^n = H(q^n, p^{n+1/4})$ of the new configuration;

(2) Integrate the Hamiltonian part according to the RATTLE scheme:

$$\begin{cases} p^{n+1/2} = p^{n+1/4} - \dfrac{\Delta t}{2} \nabla V(q^n) + \nabla \xi(q^n)\, \lambda^{n+1/2}, \\ \widetilde{q}^{n+1} = q^n + \Delta t\, M^{-1} p^{n+1/2}, \\ \xi(\widetilde{q}^{n+1}) = z, \\ \widetilde{p}^{n+3/4} = p^{n+1/2} - \dfrac{\Delta t}{2} \nabla V(\widetilde{q}^{n+1}) + \nabla \xi(\widetilde{q}^{n+1})\, \lambda^{n+3/4}, \\ \nabla \xi(\widetilde{q}^{n+1})^T M^{-1} \widetilde{p}^{n+3/4} = 0, \end{cases}$$

and set $E^{n+1} = H(\widetilde{q}^{n+1}, \widetilde{p}^{n+3/4})$*;*

(3) Accept the proposal $(q^{n+1}, p^{n+3/4}) := (\widetilde{q}^{n+1}, \widetilde{p}^{n+3/4})$ *with probability*

$$\min\Big(\exp(-\beta(E^{n+1} - E^n)), 1 \Big).$$

Otherwise, reject and flip momentum (see Remark 2.5*):* $(q^{n+1}, p^{n+3/4}) = (q^n, -p^{n+1/4})$*;*

(4) Evolve the momenta according to the midpoint Euler scheme

$$\begin{cases} p^{n+1} = p^{n+3/4} - \dfrac{\Delta t}{4} \gamma M^{-1} (p^{n+3/4} + p^{n+1}) + \sqrt{\dfrac{\Delta t}{2}}\, \sigma G^{n+1/2} + \nabla \xi(q^{n+1}) \lambda^{n+1}, \\ \nabla \xi(q^{n+1})^T M^{-1} p^{n+1} = 0. \end{cases}$$

We implicitly assume that the projection step in (2) is well defined. In such a case, the GHMC algorithm with constraints leaves invariant the equilibrium distribution $\mu_{T^*\Sigma(z)}(dq\, dp)$. In practice however, the nonlinear projection step enforcing the position constraints $\xi(q^{n+1}) = z$ is well defined only for $(q^n, p^{n+1/4}) \in D_{\Delta t} \subset \mathbb{R}^{6N}$, where $D_{\Delta t}$ is defined in Section 3.3.5.1. It is possible to modify Algorithm 3.56 by introducing a rejection in steps (1), (2) and (4) which rejects the states which left the set $D_{\Delta t} \subset \mathbb{R}^{6N}$. By doing so, the global algorithm has an invariant equilibrium distribution given by $\mu_{T^*\Sigma(z)}(dq\, dp)$ conditioned on the set $D_{\Delta t} \subset \mathbb{R}^{6N}$. This invariant distribution can be written explicitly as follows:

$$\frac{1}{Z_{z,0,\Delta t}} \mathrm{e}^{-\beta H(q,p)} \mathbf{1}_{(q,p) \in D_{\Delta t}} \sigma_{T^*\Sigma(z)}(dq\, dp).$$

3.3.5.5 *Exact sampling of constrained overdamped processes*

We did introduce in Section 3.2 the constrained overdamped Langevin process, which satisfies the stochastic differential equation:

$$dq_t = -\nabla V(q_t)\, dt + \sqrt{\frac{2}{\beta}}\, dW_t + \nabla \xi(q_t)\, d\lambda_t, \qquad (3.162)$$

where λ_t is a stochastic process such that $\xi(q_t) = z$. Equivalently, (3.162) can be rewritten in the Stratonovitch form as

$$dq_t = -P(q_t)\nabla V(q_t)\,dt + \sqrt{\frac{2}{\beta}} P(q_t) \circ dW_t,$$

where $\circ$ denotes the Stratonovitch integration, and P is the projector defined by (3.37) (or (3.104) with the choice $M = \mathrm{Id}$, see Remark 3.37).

As in the unconstrained case (see Propositions 2.14 and 2.15, and Remark 2.16), the constrained overdamped Langevin process (3.162) may be obtained from a scaling limit of the constrained Langevin (3.148), using a formal argument as in the proof of Proposition 2.15. Moreover, as in the unconstrained case in Section 2.2.4.2, an Euler-Maruyama discretization of the overdamped process (3.162) can also be obtained as some limit of numerical schemes for the Langevin equation, such as (3.160) or (3.161). This is detailed in the following proposition.

Proposition 3.57. *Suppose that the following relation is satisfied:*

$$\frac{\Delta t}{4}\gamma = M = \mathrm{Id}. \tag{3.163}$$

Then the splitting scheme (3.160) *yields the Euler scheme* (3.67) *for constrained dynamics, with the following parameters:*

$$\begin{cases} q^{n+1} = q^n - \dfrac{\Delta t^2}{2}\nabla V(q^n) + \dfrac{\Delta t}{\sqrt{\beta}}\, G^n + \nabla\xi(q^n)\,\lambda^{n+1}, \\ \xi(q^{n+1}) = z, \end{cases} \tag{3.164}$$

where $(G^n)_{n\geq 0}$ *are i.i.d. centered and normalized Gaussian variables, and* $(\lambda^n)_{n\geq 1}$ *are the Lagrange multipliers associated with the constraint* $\xi(q^{n+1}) = z$.

The same holds true for the BBK scheme (3.161) *with the relation*

$$\frac{\Delta t}{2}\gamma = M = \mathrm{Id}.$$

Proof. The choice (3.163) in the scheme (3.160) leads to

$$p^{n+1/4} = \sqrt{\frac{\Delta t}{8}}\,\sigma\, G^n + \frac{1}{2}\nabla\xi(q^n)\,\lambda^{n+1/4},$$

where $\lambda^{n+1/4}$ is associated with the constraints $\nabla\xi(q^n)^T M^{-1} p^{n+1/4} = 0$. Then,

$$p^{n+1/2} = -\frac{\Delta t}{2}\nabla V(q^n) + \sqrt{\frac{\Delta t}{8}}\,\sigma\, G^n + \nabla\xi(q^n)\left(\frac{1}{2}\,\lambda^{n+1/4} + \lambda^{n+1/2}\right),$$

and (recall $M = \mathrm{Id}$)

$$p^{n+1} = q^n + \Delta t p^{n+1/2},$$

where $\lambda^{n+1/2}$ is associated with the position constraints $\xi(q^{n+1}) = z$. The fluctuation/dissipation relation (3.149) can be reformulated in this context as

$$\sigma\sigma^T = \frac{2}{\beta}\gamma = \frac{8}{\beta\Delta t}\,\mathrm{Id},$$

and the scheme (3.164) is recovered by taking the associated Lagrange multiplier equal to $\lambda^{n+1} = \Delta t\left(\frac{1}{2}\,\lambda^{n+1/4} + \lambda^{n+1/2}\right)$. The proof for (3.161) is similar. □

This point of view allows to construct an exact Metropolis correction to the Euler scheme (3.164), using the Generalized Hybrid Monte-Carlo scheme (see Algorithm 3.56) with the special relation (3.163). The algorithm requires a cut-off parameter $R_{\Delta t} > 0$ on the momenta, chosen so that the integrator is everywhere well defined. We thus define $R_{\Delta t} > 0$ such that, for any $q \in \mathcal{D}$ and $|p|^2 \leq R_{\Delta t}$, the state (q, p) belongs to $D_{\Delta t}$ (defined in Section 3.3.5.1).

Algorithm 3.58 (Metropolis adjusted Euler with constraints). *Consider an initial configuration $q^0 \in \Sigma(z)$. Iterate on $n \geq 0$,*

(1) Sample a random vector in the tangent space $T_{q^n}\Sigma(z)$:

$$p^n = \beta^{-1/2}\, G^n + \nabla\xi(q^n)\widetilde{\lambda}^n,$$

where the Lagrange multiplier $\widetilde{\lambda}^n$ is such that

$$\nabla\xi(q^n)^T\, p^n = 0,$$

and where $(G^n)_{n\geq 0}$ are i.i.d. standard random Gaussian variables (note that $p^n = \beta^{-1/2}\, P(q^n)G^n$), and compute the energy $E^n = \dfrac{1}{2}\,|p^n|^2 + V(q^n)$ of the configuration (q^n, p^n);

(2) If $|p^n|^2 > R_{\Delta t}$, set $E^{n+1} = +\infty$ and go to (3); otherwise perform one integration step of the RATTLE scheme with identity mass-matrix, and time-step Δt:

$$\begin{cases} p^{n+1/2} = p^n - \dfrac{\Delta t}{2}\,\nabla V(q^n) + \nabla\xi(q^n)\,\lambda^{n+1/2}, \\ \widetilde{q}^{n+1} = q^n + \Delta t\, p^{n+1/2}, \\ \xi(\widetilde{q}^{n+1}) = z, \\ \widetilde{p}^{n+1} = p^{n+1/2} - \dfrac{\Delta t}{2}\,\nabla V(\widetilde{q}^{n+1}) + \nabla\xi(\widetilde{q}^{n+1})\,\lambda^{n+1}, \\ \nabla\xi(\widetilde{q}^{n+1})^T\,\widetilde{p}^{n+1} = 0; \end{cases}$$

(3) If $\left|\widetilde{p}^{n+1}\right|^2 > R_{\Delta t}$*, set* $E^{n+1} = +\infty$*; otherwise compute the energy* $E^{n+1} = \frac{1}{2}\left|\widetilde{p}^{n+1}\right|^2 + V(\widetilde{q}^{n+1})$ *of the new phase space configuration. Accept the proposal and set* $q^{n+1} = \widetilde{q}^{n+1}$ *with probability*

$$\min\Big(\exp(-\beta(E^{n+1} - E^n)), 1\Big);$$

otherwise, reject and set $q^{n+1} = q^n$.

Again, by construction,

Proposition 3.59. *The canonical distribution* $Z_z^{-1}\,\mathrm{e}^{-\beta V(q)}\sigma_{\Sigma(z)}(dq)$ *is an invariant probability measure of the Markov chain* $(q^n)_{n\geq 0}$ *generated by Algorithm* 3.58.

Proof. Algorithm 3.58 can be seen as a variant of Algorithm 3.56 with the relation (3.163). The latter as an invariant distribution of the form:

$$Z^{-1}\,\mathrm{e}^{-\beta H_{\Delta t}(q,p)}\sigma_{T_q^*\Sigma(z)}(dp)\,\sigma_{\Sigma(z)}(dq),$$

with

$$H_{\Delta t}(q,p) = \begin{cases} +\infty & \text{if } |p|^2 > R_{\Delta t}, \\ \frac{1}{2}|p|^2 + V(q) & \text{if } |p|^2 > R_{\Delta t}. \end{cases}$$

This yields the claimed q-marginal distribution. □

3.3.6 *Thermodynamic integration with constrained Langevin processes*

3.3.6.1 *Free energy*

As throughout this book, the free energy F associated with a reaction coordinate ξ is defined, up to an additive constant, as the Gibbs energy associated with the marginal probability ν^ξ of the degrees of freedom ξ under the canonical distribution (see (3.22)):

$$\mathrm{e}^{-\beta F(z)}dz = \nu^\xi(dz).$$

In the context of full phase space, it can be rewritten up to an additive constant denoted by C in the sequel, and which may vary from line to line, as

$$F(z) = -\frac{1}{\beta}\ln\int_{\Sigma(z)\times\mathbb{R}^{3N}} \mathrm{e}^{-\beta H(q,p)}\delta_{\xi(q)-z}(dq)\,dp \tag{3.165}$$

$$= -\frac{1}{\beta}\ln\int_{\Sigma(z)} \mathrm{e}^{-\beta V(q)}(\det G_M(q))^{-1/2}\sigma^M_{\Sigma(z)}(dq) + \mathrm{C}, \tag{3.166}$$

where G_M is the Gram matrix defined in (3.96) and $\sigma_{\Sigma(z)}^M$ is the surface measure induced by the mass matrix M, as defined in Section 3.3.2, see also Section 3.2.1.3. By construction, this definition of free energy is independent of the mass matrix considered, thanks to the geometric correction introduced by $(\det G_M(q))^{-1/2}$. Now, when using constrained simulation in phase space, the momenta variables of the dynamical system are also constrained, and a modified rigid free energy without geometric correction is naturally computed (see also Remark 3.51). The latter depends on the considered mass matrix, and is defined through:

$$\begin{aligned} F_{\rm rgd}^M(z) &= -\frac{1}{\beta} \ln \int_{T^*\Sigma(z)} {\rm e}^{-\beta H(q,p)} \sigma_{T^*\Sigma(z)}(dq\, dp) \qquad (3.167) \\ &= -\frac{1}{\beta} \ln \int_{\Sigma(z)} {\rm e}^{-\beta V(q)} \sigma_{\Sigma(z)}^M(dq) + {\rm C}, \end{aligned}$$

where the effective momentum p_ξ is defined in (3.108). These two definitions of free energies are related through the identity:

$$\begin{aligned} F(z) - F_{\rm rgd}^M(z) &= -\frac{1}{\beta} \ln \int_{T^*\Sigma(z)} (\det G_M)^{-1/2} d\mu_{T^*\Sigma(z)} + {\rm C} \qquad (3.168) \\ &= -\frac{1}{\beta} \ln \int_{\Sigma(z)} (\det G_M)^{-1/2} d\nu_{\Sigma(z)}^M + {\rm C} \end{aligned}$$

where $\mu_{T^*\Sigma(z)}$ is the equilibrium distribution with constraints (3.115), and $\nu_{\Sigma(z)}^M$ the associated marginal (3.116). For any value of the reaction coordinates, the difference $F(z) - F_{\rm rgd}^M(z)$ can thus be computed by sampling the probability distribution $\mu_{T^*\Sigma(z)}$. For instance, the trajectorial average of $(\det G_M)^{-1/2}$ along the process (3.148) can be used to evaluate the right-hand side of (3.168). Note that this computation does not involve the computation of any second order derivatives of ξ. The actual profile $z \mapsto F(z)$ can then be easily reconstructed from the modified free energy profile $z \mapsto F_{\rm rgd}^M(z)$ by these means. Thus, in this section, we will focus on the computation of $F_{\rm rgd}^M$, knowing that the profile $z \mapsto F(z)$ can be deduced from $z \mapsto F_{\rm rgd}^M(z)$ using (3.168) and a sampling method.

Remark 3.60 (The method of potential modification). *The physical free energy $z \mapsto F(z)$ can also be computed directly by thermodynamic integration directly using the formulation:*

$$F(z) = -\frac{1}{\beta} \ln \int_{T^*\Sigma(z)} {\rm e}^{-\beta H^\xi(q,p)} \sigma_{T^*\Sigma(z)}(dq\, dp),$$

with the modified Hamiltonian:

$$H^\xi(q,p) = \frac{1}{2}p^T M^{-1} p + V(q) + \frac{1}{2\beta}\ln\Big(\det G_M(q)\Big).$$

This formulation however requires the computation of the gradient of $\det G_M(q)$ *in the evaluation of the forces, which may be cumbersome. This is actually the point of view adopted in Section* 3.2.1 *above for the overdamped Langevin dynamics.*

A generalized version involving the effective momentum (3.110) or the effective velocity (3.111) of the rigid free energy $F_{\rm rgd}^M$ can be defined using the generalized notation (3.109) through:

$$F_{\rm rgd}^{\Xi}(\zeta) = -\frac{1}{\beta}\ln\int_{\Sigma_\Xi(\zeta)} \mathrm{e}^{-\beta H(q,p)}\sigma_{\Sigma_\Xi(\zeta)}(dq\,dp). \tag{3.169}$$

In particular, the original rigid free energy (3.167) can be recovered through the identity:

$$F_{\rm rgd}^M(z) = F_{\rm rgd}^{\Xi}(z,0),$$

where Ξ involves either the effective momentum (3.108) or the effective velocity (3.107).

Remark 3.61 (About the generalized free energy F^{ξ,p_ξ}). *When* Ξ *involves the effective momentum, the above generalized free energy also reads:*

$$F_{\rm rgd}^{\xi,p_\xi}(z,p_z) = -\frac{1}{\beta}\ln\int_{\Sigma_{\xi,p_\xi}(z,p_z)} \mathrm{e}^{-\beta H(q,p)}\delta_{\xi(q)-z,\,p_\xi(q,p)-p_z}(dq\,dp).$$

It is thus associated to the marginal probability $\mu^{\xi,p_\xi}(dz\,dp_z)$ *of the equilibrium distribution* μ *with respect to the variables* (ξ,p_ξ)*:*

$$\frac{1}{Z_\mu}\mathrm{e}^{-\beta F_{\rm rgd}^{\xi,p_\xi}(z,p_z)}dz\,dp_z = \mu^{\xi,p_\xi}(dz\,dp_z).$$

Therefore, it would also be natural to denote the latter without the subscript "rgd":

$$F^{\xi,p_\xi} = F_{\rm rgd}^{\xi,p_\xi}.$$

This is of particular interest in order to compare $F(z)$ *with* $F_{\rm rgd}^M(z)$*, which both arise from marginal probability distributions of the canonical distribution. While the former can be identified through* (3.165) *with the marginal probability distribution in* ξ*, the latter can be identified with the marginal probability distribution in* (ξ,p_ξ)*, conditioned to zero effective momentum* $p_\xi(q,p) = 0$.

3.3.6.2 *The case of molecular constraints*

We here discuss how to generalize all the computations to systems with molecular constraints. This section can be considered as independent and may be omitted in a first reading.

In practice, many systems are subject to molecular constraints, typically fixed length for covalent bonds, or fixed angle between covalent bonds (see Section 1.2.1.2). The reader is referred to [Rapaport (1995)] for practical details for simulating molecular constraints. In the context of free energy computations, two types of constraints are thus considered: first, the molecular constraints,

$$\xi_{\rm mc}(q) = (\xi_{{\rm mc},1}(q), \ldots, \xi_{{\rm mc},\overline{m}}(q)) = 0,$$

and second the reaction coordinates denoted in the present section by $\xi_{\rm rc} \in \mathbb{R}^m$, with $\overline{m} + m < 3N$. The submanifold of molecular constraints is

$$\Sigma_{\rm mc} = \left\{ q \in \mathcal{D} \,\middle|\, \xi_{\rm mc}(q) = 0 \right\},$$

and the submanifold of reaction coordinates is

$$\Sigma_{\rm rc}(z_{\rm rc}) = \left\{ q \in \mathcal{D} \,\middle|\, \xi_{\rm rc}(q) = z_{\rm rc} \right\}.$$

It is assumed that the full Gram matrix

$$G_M^{\rm mc,rc} := \nabla \begin{pmatrix} \xi_{\rm mc} \\ \xi_{\rm rc} \end{pmatrix}^T M^{-1} \nabla \begin{pmatrix} \xi_{\rm mc} \\ \xi_{\rm rc} \end{pmatrix} \in \mathbb{R}^{(\overline{m}+m)\times(\overline{m}+m)}$$

is everywhere invertible on $\Sigma_{\rm mc} \cap \Sigma_{\rm rc}(z_{\rm rc})$. Likewise, we denote

$$G_M^{\rm rc} := \nabla \xi_{\rm rc}^T M^{-1} \nabla \xi_{\rm rc} \in \mathbb{R}^{m \times m},$$

and

$$G_M^{\rm mc} := \nabla \xi_{\rm mc}^T M^{-1} \nabla \xi_{\rm mc} \in \mathbb{R}^{\overline{m} \times \overline{m}}.$$

Assuming rigid mechanical constraints on the molecular constraints $\xi_{\rm mc}$, we are led to consider the canonical distribution

$$\begin{aligned} \mu_{T^*\Sigma_{\rm mc}}(dp dq) &= \frac{1}{Z_{\rm mc}} \mathrm{e}^{-\beta H(q,p)} \sigma_{T^*\Sigma_{\rm mc}}(dq\,dp) \\ &= \frac{1}{Z_{\rm mc}} \mathrm{e}^{-\beta H(q,p)} \delta_{\xi_{\rm mc}(q), p_{\xi_{\rm mc}}(p,q)}(dq\,dp), \end{aligned} \qquad (3.170)$$

as a model to define the canonical ensemble of systems with molecular constraints. According to Section 3.3.2, $\sigma_{T^*\Sigma_{\rm mc}}$ denotes the phase space measure, and $\delta_{\xi_{\rm mc}(q), p_{\xi_{\rm mc}}(dp\,dq)}$ denotes the conditional measure associated with the constraints $(\xi_{\rm mc}(q) = 0, p_{\xi_{\rm mc}}(q,p) = 0)$, where $p_{\xi_{\rm mc}}$ is the effective momentum (3.108) associated with $\xi_{\rm mc}$ (see Section 3.3.1). These two measures are equal, see (3.127).

Remark 3.62 (On the choice of the distribution (3.170)). *The distribution $\mu_{T^*\Sigma_{\rm mc}}$ in (3.170) is obtained by constraining rigidly $\xi_{\rm mc}(q) = 0$ and not "softly" (in which case $\delta_{\xi_{\rm mc}(q),p_{\xi_{\rm mc}}(p,q)}(dq\,dp)$ would be replaced by $\delta_{\xi_{\rm mc}(q)}(dq)\,dp$, see Remarks 3.36 and 3.51). As explained in Section 3.3.4, the distribution (3.170) is the equilibrium distribution of a Langevin process (thermostated Hamiltonian dynamics) with rigid position constraints $\xi_{\rm mc}(q) = 0$. Two remarks are in order. First, it is possible to consider the free energy associated with the soft molecular constraints rather than the rigid molecular constraints, which reads (the integration domain being $\Sigma_{\rm mc} \cap \Sigma_{\rm rc}(z_{\rm rc}) \times \mathbb{R}^{3N}$):*

$$\tilde{F}^{\rm mc}(z_{\rm rc}) = -\frac{1}{\beta} \ln \int {\rm e}^{-\beta H(q,p)} \delta_{\xi_{\rm mc}(q),\xi_{\rm rc}(q)-z_{\rm rc}}(dq)\,dp,$$

and to compute the latter up to an appropriate modification of Eq. (3.172) below. Second, we prefer to stick to the rigidly constrained potential since, for reasons coming from quantum mechanics, the system with molecular constraints typically "does not oscillate" in the vicinity of the submanifolds defined by $\xi_{\rm mc}(q) = 0$. Indeed, the characteristic energy associated with a quantum oscillator is $\hbar\omega$ where ω is the frequency of the oscillator. For molecular bonds, $\hbar\omega \gg k_B T$ at usual temperature; for instance $\hbar\omega/k_B =$ 7600 K for H-C bonds, and $\hbar\omega/k_B = 2300$ K for C-C bonds (see [Schlick (2002)], Table 8.1). As a consequence, the thermal energy is not sufficient to change the quantum mechanical energy of the oscillator, which thus remains in its fundamental state. Since the classical limit of the fundamental state is loosely speaking the minimal energy position, it is reasonable to assume that such constraints strictly remain at their fixed, equilibrium value.

Moreover, by associativity of conditional measures, which amounts to: $\forall \phi : \mathbb{R}^m \to \mathbb{R}$, $\forall \varphi : \mathbb{R}^{6N} \to \mathbb{R}$,

$$\int_{\mathbb{R}^m} \phi(z_{\rm rc}) \int_{(T^*\Sigma_{\rm mc})\cap\Sigma_{\rm rc}(z_{\rm rc})} \varphi(q,p)\, \delta_{\xi_{\rm mc}(q),p_{\xi_{\rm mc}}(q,p),\xi_{\rm rc}(q)-z_{\rm rc}}(dq\,dp)\, dz_{\rm rc}$$
$$= \int_{T^*\Sigma_{\rm mc}} \phi\circ\xi_{\rm rc}(q)\, \varphi(q,p)\, \delta_{\xi_{\rm mc}(q),p_{\xi_{\rm mc}}(q,p)}(dq\,dp),$$

the probability distribution $\mu_{T^*\Sigma_{\rm mc}}$ conditioned by a value of the reaction coordinates $\xi_{\rm rc}(q) = z_{\rm rc}$ writes, up to a normalizing factor:

$${\rm e}^{-\beta H(q,p)} \delta_{\xi_{\rm mc}(q),p_{\xi_{\rm mc}}(q,p),\xi_{\rm rc}(q)-z_{\rm rc}}(dq\,dp).$$

Therefore, the definition of the free energy associated with the reaction coordinates reads as follows (the integration domain being

$T^*\Sigma_{\rm mc} \cap (\Sigma_{\rm rc}(z_{\rm rc}) \times \mathbb{R}^{3N})$):

$$F^{\rm mc}(z_{\rm rc}) = -\frac{1}{\beta} \ln \int {\rm e}^{-\beta H(q,p)} \delta_{\xi_{\rm mc}(q), p_{\xi_{\rm mc}}(q,p), \xi_{\rm rc}(q) - z_{\rm rc}}(dq\, dp).$$

The conditional distribution can be decomposed using the co-area formula (3.24) and the definition of effective momentum (3.108):

$$\begin{aligned}\delta_{\xi_{\rm mc}(q), p_{\xi_{\rm mc}}(q,p), \xi_{\rm rc}(q) - z_{\rm rc}}(dq\, dp) &= \delta_{p_{\xi_{\rm mc}}(q,p)}(dp)\delta_{\xi_{\rm mc}(q), \xi_{\rm rc}(q) - z_{\rm rc}}(dq) \\ &= (\det G_M^{\rm mc}(q))^{1/2} \sigma^{M^{-1}}_{T_q^*\Sigma_{\rm mc}}(dp)\, (\det G_M^{\rm mc, rc}(q))^{-1/2} \sigma^M_{\Sigma_{\rm rc}(z_{\rm rc}) \cap \Sigma_{\rm mc}}(dq).\end{aligned}$$

Integrating out the momenta in the linear space $T_q^*\Sigma_{\rm mc}$ with scalar product $\langle p_1, p_2 \rangle_{M^{-1}} = p_1^T M^{-1} p_2$, the free energy can then be rewritten as (the integration domain being $\Sigma_{\rm rc}(z_{\rm rc}) \cap \Sigma_{\rm mc}$):

$$F^{\rm mc}(z_{\rm rc}) = -\frac{1}{\beta} \ln \int {\rm e}^{-\beta V(q)} \frac{(\det G_M^{\rm mc}(q))^{1/2}}{(\det G_M^{\rm mc, rc}(q))^{1/2}} \sigma^M_{\Sigma_{\rm rc}(z_{\rm rc}) \cap \Sigma_{\rm mc}}(dq) + {\rm C}.$$

As a consequence, the free energy $F^{\rm mc}$ can be computed from the following "rigid" free energy similar to (3.167) (the integration domain being $T^*(\Sigma_{\rm rc}(z_{\rm rc}) \cap \Sigma_{\rm mc})$):

$$F^{{\rm mc},M}_{\rm rgd}(z_{\rm rc}, 0) = -\frac{1}{\beta} \ln \int {\rm e}^{-\beta H(q,p)} \sigma_{T^*(\Sigma_{\rm rc}(z_{\rm rc}) \cap \Sigma_{\rm mc})}(dp\, dq), \tag{3.171}$$

using the following formula similar to (3.168) (the integration domain being $T^*(\Sigma_{\rm rc}(z_{\rm rc}) \cap \Sigma_{\rm mc})$):

$$F^{\rm mc}(z_{\rm rc}) - F^{{\rm mc},M}_{\rm rgd}(z_{\rm rc}, 0) = -\frac{1}{\beta} \ln \int \frac{(\det G_M^{\rm mc})^{1/2}}{(\det G_M^{{\rm mc},rc})^{1/2}} d\mu_{T^*(\Sigma_{\rm rc}(z_{\rm rc}) \cap \Sigma_{\rm mc})} + {\rm C}. \tag{3.172}$$

Thus the case of molecular constraints can be treated within the general framework, where computing "rigid" free energies of the type (3.171) is considered.

3.3.6.3 *The mean force*

In this section, the constraining force associated with a deterministic mechanical system (see Section 3.3.1) with rigid constraints is related to derivatives of the "rigid" free energy. This result is the parallel in full phase space to (3.28) in the overdamped case. We first need the following result, similar to Lemma 3.10 in the overdamped case.

Lemma 3.63. *For any compactly supported smooth test function φ on $\mathbb{R}^{6N}$:*

$$\nabla_\zeta \left(\int_{\Sigma_\Xi(\zeta)} \varphi \, d\sigma_{\Sigma_\Xi(\zeta)} \right) = \int_{\Sigma_\Xi(\zeta)} \Gamma^{-1} \{\Xi, \varphi\} \, d\sigma_{\Sigma_\Xi(\zeta)}.$$

Proof. The proof requires the full constraints formalism detailed in Section 3.3.3, in order to use the co-area formula in phase space. Consider a test function $\phi : \mathbb{R}^{2m} \to \mathbb{R}$, and use an integration by parts, as well as the co-area formula of Proposition 3.39 to obtain:

$$\begin{aligned} I &:= \int_{\mathbb{R}^{2m}} \phi(\zeta) \left(\nabla_\zeta \int_{\Sigma_\Xi(\zeta)} \varphi(q,p)\, \sigma_{\Sigma_\Xi(\zeta)}(dq\,dp) \right) d\zeta \\ &= - \int_{\mathbb{R}^{2m}} \nabla_\zeta \phi(\zeta) \left(\int_{\Sigma_\Xi(\zeta)} \varphi(q,p)\, \sigma_{\Sigma_\Xi(\zeta)}(dq\,dp) \right) d\zeta \\ &= - \int_{\mathbb{R}^{6N}} \Gamma^{-1} \{\Xi, \phi \circ \Xi\}\, \varphi \det(\Gamma)^{1/2}\, dq\,dp, \end{aligned}$$

where in the last line the following chain rule has been used:

$$\{\Xi, \phi \circ \Xi\}(q,p) = \{\Xi, \Xi\}(q,p) \nabla_\zeta \phi(\Xi(q,p)) = \Gamma(q,p) \nabla_\zeta \phi(\Xi(q,p)).$$

Now an integration by parts with respect to $dq\,dp$, together with the formula

$$\sum_{b=1}^{2m} \left\{ \det(\Gamma)^{1/2} \Gamma^{-1}_{\cdot,b}, \Xi_b \right\} = 0$$

proved in Lemma 3.48 yield:

$$\begin{aligned} I &= \int_{\mathbb{R}^{6N}} (\phi \circ \Xi) \left\{ \Xi_b, \det(\Gamma)^{1/2} \Gamma^{-1}_{\cdot,b} \varphi \right\} dq\,dp \\ &= \int_{\mathbb{R}^{6N}} (\phi \circ \Xi)\, \Gamma^{-1} \{\Xi, \varphi\} \det(\Gamma)^{1/2}\, dq\,dp \\ &= \int_{\mathbb{R}^{2m}} \phi(\zeta) \left(\int_{\Sigma_\Xi(\zeta)} \Gamma^{-1} \{\Xi, \varphi\}\, d\sigma_{\Sigma_\Xi(\zeta)} \right) d\zeta, \end{aligned}$$

which gives the result. □

Proposition 3.64. *The constraining force* $f^M_{\mathrm{rgd}} : T^*\Sigma(z) \to \mathbb{R}^m$, *defined in* (3.103) *by:*

$$f^M_{\mathrm{rgd}}(q,p) = G_M^{-1}(q) \nabla\xi(q)^T M^{-1} \nabla V(q) - G_M^{-1}(q) \mathrm{Hess}_q(\xi)(M^{-1}p, M^{-1}p),$$

yields on average the rigid mean force:

$$\nabla_z F^M_{\mathrm{rgd}}(z) = \int_{T^*\Sigma(z)} f^M_{\mathrm{rgd}}(q,p)\, \mu_{T^*\Sigma(z)}(dq\,dp). \tag{3.173}$$

Moreover, for non-tangential velocities ($p_\xi(q,p) = p_z \neq 0$ *or* $v_\xi(q,p) = v_z \neq 0$), *the formula can be generalized as follows:*

$$\begin{pmatrix} f^\Xi_{\mathrm{rgd}} \\ g^\Xi_{\mathrm{rgd}} \end{pmatrix} = \Gamma^{-1} \{\Xi, H\}, \tag{3.174}$$

where Γ *is defined in* (3.123), *and the rigid mean force is given by*

$$\nabla_\zeta F^{\Xi}_{\text{rgd}}(\zeta) = \frac{1}{Z_\zeta}\int_{\Sigma_\Xi(\zeta)} \begin{pmatrix} f^{\Xi}_{\text{rgd}} \\ g^{\Xi}_{\text{rgd}} \end{pmatrix} \mathrm{e}^{-\beta H}\sigma_{\Sigma_\Xi(\zeta)}(dq\,dp) \tag{3.175}$$

where $Z_\zeta = \int_{\Sigma_\Xi(\zeta)} \mathrm{e}^{-\beta H}\sigma_{\Sigma_\Xi(\zeta)}(dq\,dp)$. *When* (q,p) *verifies* $p_\xi(q,p) = v_\xi(q,p) = 0$, *then* $g^{\Xi}_{\text{rgd}}(q,p) = 0$ *and* $f^{\Xi}_{\text{rgd}}(q,p) = f^{M}_{\text{rgd}}(q,p)$.

Proof. Formulas (3.174) and (3.175) are obtained directly by replacing φ by $\mathrm{e}^{-\beta H}$ in Lemma 3.63. The fact that $(g^{\Xi}_{\text{rgd}}(q,p), f^{\Xi}_{\text{rgd}}(q,p)) = (0, f^{M}_{\text{rgd}}(q,p))$ in the tangential case is obtained with (3.138). □

The following lemma highlights the link between the constraining force f^M_{rgd} and the local mean force f introduced in the overdamped case (see (3.82)).

Lemma 3.65. *The rigid mean force* (3.173) *can be rewritten as:*

$$\nabla_z F^M_{\text{rgd}}(z) = \int_{T^*\Sigma(z)} \overline{f}^M_{\text{rgd}}(q)\,\mu_{T^*\Sigma(z)}(dq\,dp) = \int_{\Sigma(z)} \overline{f}^M_{\text{rgd}}(q)\,\nu^M_{\Sigma(z)}(dq),$$

where

$$\overline{f}^M_{\text{rgd}}(q) = G_M^{-1}(q)\nabla\xi(q)^T M^{-1}\nabla V(q) - \beta^{-1}G_M^{-1}(q)\mathrm{Hess}_q(\xi) : \left(M^{-1}P_M(q)\right). \tag{3.176}$$

It should be noted that $\overline{f}^M_{\text{rgd}} = f$ defined in (3.82) (using also (3.46)) if mass weighted coordinates ($M = \mathrm{Id}$) and a Fixman corrected potential ($V = V^\xi = V + V_{\text{fix}}$, see (3.153)) are used in (3.176).

Proof. Consider the momenta Gaussian distribution $\kappa^{M^{-1}}_{T^*_q\Sigma(z)}(dp)$ defined in (3.133), which is the marginal distribution in momenta variable of $\mu_{T^*\Sigma(z)}(dq\,dp)$ conditioned by a given $q \in \Sigma(z)$. Proving Lemma 3.65 amounts to showing that the average of the constraining force f^M_{rgd} with respect to $\kappa^{M^{-1}}_{T^*_q\Sigma(z)}(dp)$ yields:

$$\overline{f}^M_{\text{rgd}}(q) = \int_{T^*_q\Sigma(z)} f^M_{\text{rgd}}(q,p)\,\kappa^{M^{-1}}_{T^*_q\Sigma(z)}(dp).$$

First, we compute the covariance matrix

$$\Sigma := \mathrm{cov}\left(\kappa^{M^{-1}}_{T^*_q\Sigma(z)}\right)$$

of the Gaussian distribution $\kappa^{M^{-1}}_{T^*_q\Sigma(z)}(dp)$. Since $\kappa^{M^{-1}}_{T^*_q\Sigma(z)}(dp)$ is a centered Gaussian distribution, Σ satisfies, for all $p_1, p_2 \in \mathbb{R}^{3N}$,

$$\begin{aligned} &p_1^T M^{-1}\Sigma M^{-1} p_2 \\ &:= \int_{T^*_q\Sigma(z)} \left(p^T M^{-1} p_1\right)\left(p^T M^{-1} p_2\right)\kappa^{M^{-1}}_{T^*_q\Sigma(z)}(dp) \\ &= \int_{T^*_q\Sigma(z)} \left(p^T M^{-1} P_M(q) p_1\right)\left(p^T M^{-1} P_M(q) p_2\right)\kappa^{M^{-1}}_{T^*_q\Sigma(z)}(dp). \end{aligned}$$

Denoting the scalar product $\langle p_1, p_2 \rangle_{M^{-1}} = p_1^T M^{-1} p_2$, this yields

$$\begin{aligned} & p_1^T M^{-1} \Sigma M^{-1} p_2 \\ & = \int_{T_q^* \Sigma(z)} \langle p, P_M(q) p_1 \rangle_{M^{-1}} \langle p, P_M(q) p_2 \rangle_{M^{-1}} \frac{\mathrm{e}^{-\frac{\beta}{2}\langle p,p \rangle_{M^{-1}}}}{\left(\frac{2\pi}{\beta}\right)^{(3N-m)/2}} \sigma_{T_q^* \Sigma(z)}^{M^{-1}}(dp), \\ & = \beta^{-1} \langle P_M(q) p_1, P_M(q) p_2 \rangle_{M^{-1}}, \end{aligned}$$

so that eventually:

$$\Sigma = \beta^{-1} P_M(q) M. \tag{3.177}$$

This gives:

$$\int_{T_q^* \Sigma(z)} \mathrm{Hess}_q(\xi)(M^{-1}p, M^{-1}p)\, \kappa_{T_q^* \Sigma(z)}^{M^{-1}}(dp) = \beta^{-1} \mathrm{Hess}_q(\xi) : \left(M^{-1} P_M(q)\right).$$

Averaging (3.103) over momenta thus yields the desired result. □

Using Lemma 3.65, free energy derivatives can be computed by averaging $\overline{f}_{\mathrm{rgd}}^M(q)$ with respect to the distribution $\mu_{T^*\Sigma(z)}(dq\,dp)$. The function $\overline{f}_{\mathrm{rgd}}^M(q)$ may thus be called rigid local mean force. Note that using the averaged version $\overline{f}_{\mathrm{rgd}}^M$ instead of the original f_{rgd}^M in the estimator of $\nabla_z F_{\mathrm{rgd}}(z)$ in (3.173) may decrease the variance, or at least, should not increase it. Indeed, the average over momenta has been taken into account analytically in the expression for $\overline{f}_{\mathrm{rgd}}^M$.

3.3.6.4 *Free energy from Lagrange multipliers*

Free energy derivatives can be computed using the Lagrange multipliers of a Langevin constrained process, similarly to the overdamped case in (3.89). This technique avoids the possibly cumbersome computation of second order derivatives $\mathrm{Hess}_q(\xi)$ of the reaction coordinate which appear in the expression of f_{rgd}^M or $\overline{f}_{\mathrm{rgd}}^M$. It relies on the following convergence result:

Theorem 3.66. *Consider the rigidly constrained Langevin process solution of (3.148) with associated Lagrange multipliers λ_t. Assume that $\nabla\xi(q)$, $G_M^{-1}(q)$, and $\sigma(q)$ are bounded function of $q \in \Sigma(z)$. The following almost sure convergence holds:*

$$\lim_{T \to +\infty} \frac{1}{T} \int_0^T d\lambda_t = \nabla_z F_{\mathrm{rgd}}^M(z) \qquad \text{a.s.} \tag{3.178}$$

Proof. Recall the expression of the Lagrange multipliers, which can be decomposed as the sum of the constraining force, a dissipation term and a martingale (fluctuation) term:

$$\begin{aligned} d\lambda_t =& - G_M^{-1}\mathrm{Hess}_{q_t}(\xi)(M^{-1}p_t, M^{-1}p_t)dt \\ &+ G_M^{-1}\nabla\xi(q_t)^T M^{-1}\left(\nabla V(q_t)dt + \gamma(q_t)M^{-1}p_t\, dt - \sigma(q_t)dW_t\right) \\ =& f_{\mathrm{rgd}}^M(q_t,p_t)\, dt + G_M^{-1}\nabla\xi(q_t)^T M^{-1}\left(\gamma(q_t)M^{-1}p_t\, dt - \sigma(q_t)dW_t\right). \end{aligned} \tag{3.179}$$

The result follows from three elementary facts. First, the process is ergodic with respect to the equilibrium distribution $\mu_{T^*\Sigma(z)}(dp\, dq)$ (see Proposition 3.53) and the average of f_{rgd}^M with respect to this measure gives the rigid free energy derivative according to Proposition 3.64. Second, the Gaussian distribution of $\mu_{T^*\Sigma(z)}(dq\, dp)$ with respect to momenta variables is centered, which yields:

$$\int_{T^*\Sigma(z)} G_M^{-1}(q)\nabla\xi(q)^T M^{-1}\gamma(q)M^{-1}\, p\, \mu_{T^*\Sigma(z)}(dq\, dp) = 0.$$

Third, the following variance estimate holds:

$$\begin{aligned} &\mathbb{E}\left|\frac{1}{\sqrt{T}}\int_0^T G_M^{-1}(q_t)\nabla\xi^T(q_t)M^{-1}\sigma(q_t)dW_t\right|^2 \\ &\le \left\|\mathrm{Tr}(G_M^{-1}\nabla\xi^T M^{-1}\sigma\sigma^T M^{-1}\nabla\xi G_M^{-1})\right\|_\infty. \end{aligned}$$

The boundedness of the left-hand side when $T \to +\infty$ implies the almost sure convergence of $\displaystyle\frac{1}{T}\int_0^T G_M^{-1}(q_t)\nabla\xi^T(q_t)M^{-1}\sigma(q_t)dW_t$ to 0 (see for example Theorem 1.3.15 in [Duflo (1997)]). □

To avoid unnecessary sources of variance in computations, the free energy may rather be computed equivalently only with the "Hamiltonian part" of the Lagrange multipliers, which amounts to computing:

$$\begin{aligned} d\lambda_t^{\mathrm{ham}} &= d\lambda_t + G_M^{-1}(\nabla\xi)^T(q_t)M^{-1}\left(-\gamma(q_t)M^{-1}p_t\, dt + \sigma(q_t)dW_t\right) \\ &= f_{\mathrm{rgd}}^M(q_t,p_t)\, dt, \end{aligned} \tag{3.180}$$

for which a convergence result similar to (3.178) holds:

$$\lim_{T\to+\infty}\frac{1}{T}\int_0^T d\lambda_t^{\mathrm{ham}} = \nabla_z F_{\mathrm{rgd}}^M(z).$$

After time discretization, the following approximation formula may be used:

$$\nabla_z F_{\mathrm{rgd}}^M(z) \simeq \frac{1}{N\Delta t}\sum_{n=0}^{N-1}(\lambda^{n+1/2} + \lambda^{n+3/4}) \tag{3.181}$$

where $(\lambda^{n+1/2}, \lambda^{n+3/4})$ are the Lagrange multipliers in the velocity Verlet part of the discretization of Langevin processes (3.160). The consistency of the scheme is given by the following proposition.

Proposition 3.67 (Consistency). *The approximation formula* (3.181) *is consistent when the numerical scheme* (3.160) *is used. More precisely, the Lagrange multipliers* $(\lambda^{n+1/2}, \lambda^{n+3/4})$ *in* (3.160) *are both equivalent when* $\Delta t \to 0$ *to the constraining force defined in* (3.103)*:*

$$\begin{cases} \lambda^{n+1/2} = f_{\mathrm{rgd}}^M(q^n, p^{n+1/2}) \dfrac{\Delta t}{2} + \mathrm{O}(\Delta t^2), \\ \lambda^{n+3/4} = f_{\mathrm{rgd}}^M(q^{n+1}, p^{n+1/2}) \dfrac{\Delta t}{2} + \mathrm{O}(\Delta t^2). \end{cases}$$

Moreover, the following second order consistency holds for the sum of the Lagrange multipliers:

$$\lambda^{n+1/2} + \lambda^{n+3/4} = f_{\mathrm{rgd}}^M(q^n, p^{n+1/2}) \frac{\Delta t}{2} + f_{\mathrm{rgd}}^M(q^{n+1}, p^{n+1/2}) \frac{\Delta t}{2} + \mathrm{O}(\Delta t^3),$$

together with the variant:

$$\lambda^{n+1/2} + \lambda^{n+3/4} = f_{\mathrm{rgd}}^M(q^n, p^{n+1/4}) \frac{\Delta t}{2} + f_{\mathrm{rgd}}^M(q^{n+1}, p^{n+3/4}) \frac{\Delta t}{2} + \mathrm{O}(\Delta t^3). \tag{3.182}$$

The variant (3.182) will be used in (3.184) below to estimate the time-step error in the thermodynamic integration method.

Proof. For a sufficiently small time-step Δt, the implicit function theorem ensures that the projection steps in (3.160) are well defined. A Taylor expansion with respect to Δt of the position constraints gives

$$\begin{aligned} z &= \xi(q^{n+1}) \\ &= \xi(q^n + \Delta t M^{-1} p^{n+1/2}) \\ &= \xi(q^n) + \Delta t \nabla \xi^T(q^n) M^{-1} p^{n+1/2} \\ &\quad + \frac{\Delta t^2}{2} \mathrm{Hess}_{q^n}(\xi)(M^{-1}p^{n+1/2}, M^{-1}p^{n+1/2}) \\ &\quad + \frac{\Delta t^3}{6} \mathrm{D}^3_{q^n}(\xi)(M^{-1}p^{n+1/2}, M^{-1}p^{n+1/2}, M^{-1}p^{n+1/2}) + \mathrm{O}(\Delta t^4), \end{aligned}$$

where $D^3_q(\xi)(x, y, z) \in \mathbb{R}^m$ denotes the order 3 differential of ξ computed at q and evaluated with the vectors $x, y, z \in \mathbb{R}^{3N}$. We denote

$$\alpha^{n+1/2}(q) := G_M^{-1}(q) \mathrm{D}^3_q(\xi)(M^{-1}p^{n+1/2}, M^{-1}p^{n+1/2}, M^{-1}p^{n+1/2});$$

then the fact that $z = \xi(q^{n+1}) = \xi(q^n)$ and the identity

$$\nabla\xi^T(q^n)M^{-1}p^{n+1/2} = -\frac{\Delta t}{2}\nabla\xi^T(q^n)M^{-1}\nabla V(q^n) + G_M(q^n)\lambda^{n+1/2}$$

yields the following expansion of $\lambda^{n+1/2}$ in terms of $(q^n, p^{n+1/2})$:

$$\lambda^{n+1/2} = f_{\text{rgd}}^M(q^n, p^{n+1/2})\frac{\Delta t}{2} - \frac{\Delta t^2}{6}\alpha^{n+1/2}(q^n) + \mathrm{O}(\Delta t^3).$$

By time symmetry, the same computation holds starting from $(q^{n+1}, p^{n+3/4})$ for $\lambda^{n+3/4}$ and by formally replacing $\Delta t \to -\Delta t$. This can be double checked by Taylor expanding with respect to Δt in the position constraints:

$$\begin{aligned}
z &= \xi(q^n) \\
&= \xi(q^{n+1} - \Delta t M^{-1}p^{n+1/2}) \\
&= \xi(q^{n+1}) - \Delta t \nabla\xi^T(q^{n+1})M^{-1}p^{n+1/2} \\
&\quad + \frac{\Delta t^2}{2}\mathrm{Hess}_{q^{n+1}}(\xi)(M^{-1}p^{n+1/2}, M^{-1}p^{n+1/2}) \\
&\quad - \frac{\Delta t^3}{6}\mathrm{D}^3{}_{q^{n+1}}(\xi)(M^{-1}p^{n+1/2}, M^{-1}p^{n+1/2}, M^{-1}p^{n+1/2}) + \mathrm{O}(\Delta t^4).
\end{aligned}$$

Then, the fact that $z = \xi(q^{n+1}) = \xi(q^n)$ and the identity

$$\nabla\xi^T(q^{n+1})M^{-1}p^{n+1/2} = \frac{\Delta t}{2}\nabla\xi^T(q^{n+1})M^{-1}\nabla V(q^{n+1}) - G_M(q^{n+1})\lambda^{n+3/4}$$

yields the following expansion of $\lambda^{n+3/4}$ in terms of $(q^{n+1}, p^{n+1/2})$:

$$\lambda^{n+3/4} = f_{\text{rgd}}^M(q^{n+1}, p^{n+1/2})\frac{\Delta t}{2} + \frac{\Delta t^2}{6}\alpha^{n+1/2}(q^{n+1}) + \mathrm{O}(\Delta t^3).$$

The sum of the multipliers yields

$$\begin{aligned}
&\lambda^{n+1/2} + \lambda^{n+3/4} - f_{\text{rgd}}^M(q^n, p^{n+1/2})\frac{\Delta t}{2} - f_{\text{rgd}}^M(q^{n+1}, p^{n+1/2})\frac{\Delta t}{2} \\
&\quad = \frac{\Delta t^2}{6}\left(\alpha^{n+1/2}(q^{n+1}) - \alpha^{n+1/2}(q^n)\right) + \mathrm{O}(\Delta t^3) \\
&\quad = \mathrm{O}(\Delta t^3).
\end{aligned}$$

Now using the previous calculations, we remark that:

$$\begin{cases}
p^{n+1/2} = p^{n+1/4} - \dfrac{\Delta t}{2}\nabla V(q^n) + \dfrac{\Delta t}{2}\nabla\xi(q^n) f_{\text{rgd}}^M(q^n, p^{n+1/2}) + \mathrm{O}(\Delta t^2), \\
p^{n+1/2} = p^{n+3/4} + \dfrac{\Delta t}{2}\nabla V(q^{n+1}) - \dfrac{\Delta t}{2}\nabla\xi(q^{n+1}) f_{\text{rgd}}^M(q^{n+1}, p^{n+1/2}) + \mathrm{O}(\Delta t^2).
\end{cases}$$

Thus, it holds

$$
\begin{aligned}
& f_{\mathrm{rgd}}^M(q^n, p^{n+1/2}) + f_{\mathrm{rgd}}^M(q^{n+1}, p^{n+1/2}) \\
& = f_{\mathrm{rgd}}^M(q^n, p^{n+1/4}) + f_{\mathrm{rgd}}^M(q^{n+1}, p^{n+3/4}) \\
& + \nabla_p f_{\mathrm{rgd}}^M(q^n, p^{n+1/4}) \left(-\frac{\Delta t}{2} \nabla V(q^n) + \frac{\Delta t}{2} \nabla \xi(q^n) f_{\mathrm{rgd}}^M(q^n, p^{n+1/2}) \right) \\
& - \nabla_p f_{\mathrm{rgd}}^M(q^{n+1}, p^{n+3/4}) \left(-\frac{\Delta t}{2} \nabla V(q^{n+1}) + \frac{\Delta t}{2} \nabla \xi(q^{n+1}) f_{\mathrm{rgd}}^M(q^{n+1}, p^{n+1/2}) \right) \\
& + \mathrm{O}(\Delta t^2) \\
& = f_{\mathrm{rgd}}^M(q^n, p^{n+1/4}) + f_{\mathrm{rgd}}^M(q^{n+1}, p^{n+3/4}) + \mathrm{O}(\Delta t^2).
\end{aligned}
$$

This gives the claimed second order consistency of the sum of the Lagrange multipliers. □

Let us now discuss the convergence of the numerical schemes. Consider the constrained splitting scheme (3.160) or (3.161) as a discretization of Langevin processes (3.148). Suppose $\Delta t \to 0$ with $N_{\Delta t} \Delta t \to T$, then the following convergence in probability distribution (law) is expected:

$$
\lim_{\Delta t \to 0} \mathrm{Law} \left(\frac{1}{N_{\Delta t} \Delta t} \sum_{n=0}^{N_{\Delta t}-1} (\lambda^{n+1/2} + \lambda^{n+3/4}) \right) = \mathrm{Law} \left(\frac{1}{T} \int_0^T d\lambda_t^{\mathrm{ham}} \right), \tag{3.183}
$$

where λ_t^{ham} is defined by (3.180). The rigorous justification of this convergence follows the arguments of Remark 3.55.

The limit (3.183) still probably holds when considering first the long time limit $N \to +\infty$, and then $\Delta t \to 0$. A rigorous justification requires the proof of the Δt-consistency of the invariant measure of the numerical scheme, typically by the introduction of some cut-off on velocities, and by assuming exponential convergence towards equilibrium, in the spirit of [Faou and Lelièvre (2009)].

When the scheme is used with a Metropolis step, the long time limit $N \to +\infty$ followed by $\Delta t \to 0$ can be proven directly, up to assuming irreducibility. Indeed, consider the GHMC scheme in Algorithm 3.56, and assume (i) the irreducibility of the associated Markov chain, and (ii) that appropriate rejections outside the set $D_{\Delta t} \subset \mathbb{R}^{6N}$ are made in the different steps of the algorithm, so that the projection steps associated with the nonlinear constraints in step (2) of Algorithm 3.56 are well defined. Then, by ergodicity, the following longtime averaging holds:

$$\lim_{N\to+\infty} \frac{1}{\Delta t N} \sum_{n=0}^{N-1} (\lambda^{n+1/2} + \lambda^{n+3/4})$$
$$= \frac{\displaystyle\int_{D_{\Delta t}} f_{\mathrm{rgd}}^{M}(q,p)\mu_{T^*\Sigma(z)}(dq\,dp)}{\displaystyle\int_{D_{\Delta t}} \mu_{T^*\Sigma(z)}(dq\,dp)} + \mathrm{O}(\Delta t^2) \quad \text{a.s.,} \tag{3.184}$$

where we have used the estimate (3.182) on the Lagrange multipliers in Proposition 3.67. The limit $\Delta t \to 0$ is obtained by a dominated convergence argument:

$$\lim_{\Delta t\to 0} \lim_{N\to+\infty} \frac{1}{\Delta t N} \sum_{n=0}^{N-1} (\lambda^{n+1/2} + \lambda^{n+3/4}) = \nabla_z F_{\mathrm{rgd}}^{M}(z) \quad \text{a.s.}$$

In conclusion, note that when using a sampling method with a Metropolis correction (Algorithm 3.56 or Algorithm 3.58), the time-step error in the sampling of the invariant measure is removed. The only remaining time-step error is of order 2 and arises from the Lagrange multipliers, when the latter are used to compute the free energy. If the analytic expression of the local rigid mean force $\overline{f}_{\mathrm{rgd}}^{M}$ given in (3.176) is computed and averaged, using the convergence

$$\lim_{N\to+\infty} \frac{1}{N} \sum_{n=0}^{N-1} \overline{f}_{\mathrm{rgd}}^{M}(q^n) = \nabla_z F_{\mathrm{rgd}}^{M}(z) \quad \text{a.s.,}$$

then the free energy is computed without time-step error.

Finally, it should be emphasized that computing free energy derivatives by (3.181) with the splitting scheme (3.160) can be performed within the overdamped Langevin framework, using the method of Proposition 3.57, *i.e.* by chosing the following relation in (3.160):

$$\frac{\Delta t}{4}\gamma = M \propto \mathrm{Id}.$$

This leads to original schemes, which can be seen as variants of the overdamped case in (3.89).

3.3.6.5 *Numerical illustration*

We consider the dimer model described in Section 1.3.2.4, with the same parameters as in Section 2.5.2.3. For this system, $M = \mathrm{Id}$, $|\nabla\xi|$ is constant, and the rigid free energy $F^M_{\rm rgd}(z)$ is therefore equal to the free energy $F(z)$. Likewise, $\overline{f}^M_{\rm rgd} = f$ where f is the local mean force defined in (3.82).

The mean force is estimated at the values $z_i = z_{\min} + i\Delta z$, with $z_{\min} = -0.1$ and $\Delta z = 0.012$, by ergodic averages obtained with the projected dynamics with Metropolis correction (Algorithm 3.56, where in the simple case considered here the fluctuation-dissipation part may be integrated exactly), integrated on a time $T = 2 \times 10^3$ with a step size $\Delta t = 0.02$ and a friction $\gamma = 1$. The time-step chosen here is much larger than the time-step chosen for the numerical implementation of the projected overdamped dynamics in Section 3.2.5.4. In both cases, we chose a time-step as large as possible while still having a numerically stable dynamics.

The resulting mean force profile, obtained by the time averaging of the local mean force f, is presented in Figure 3.6, together with the associated potential of mean force. Note that these profiles are close to the ones obtained in Section 2.5.2.3 for the same model. Figure 3.7 compares the

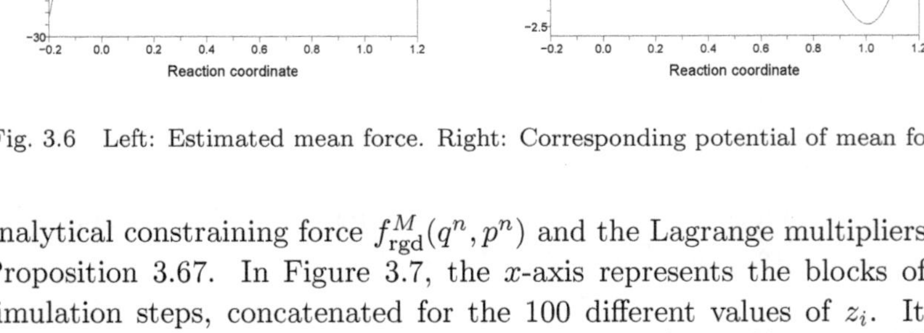

Fig. 3.6 Left: Estimated mean force. Right: Corresponding potential of mean force.

analytical constraining force $f^M_{\rm rgd}(q^n, p^n)$ and the Lagrange multipliers, see Proposition 3.67. In Figure 3.7, the x-axis represents the blocks of 10^5 simulation steps, concatenated for the 100 different values of z_i. It can be checked numerically that the difference $|\lambda^{n+1/2} - f^M_{\rm rgd}(q^n, p^{n+1/2})\frac{\Delta t}{2}|$ and $|\lambda^{n+3/4} - f^M_{\rm rgd}(q^{n+1}, p^{n+1/2})\frac{\Delta t}{2}|$ are indeed of order Δt^2, and that the difference $\left|\lambda^{n+1/2} + \lambda^{n+3/4} - f^M_{\rm rgd}(q^n, p^{n+1/2})\frac{\Delta t}{2} - f^M_{\rm rgd}(q^{n+1}, p^{n+1/2})\frac{\Delta t}{2}\right|$ is indeed of order Δt^3 (by computing the average of these elementary differences for various step sizes). The Lagrange multipliers are in any case very good approximations of the constraining force $f^M_{\rm rgd}$.

Let us finally discuss the efficiency of the different estimators of the

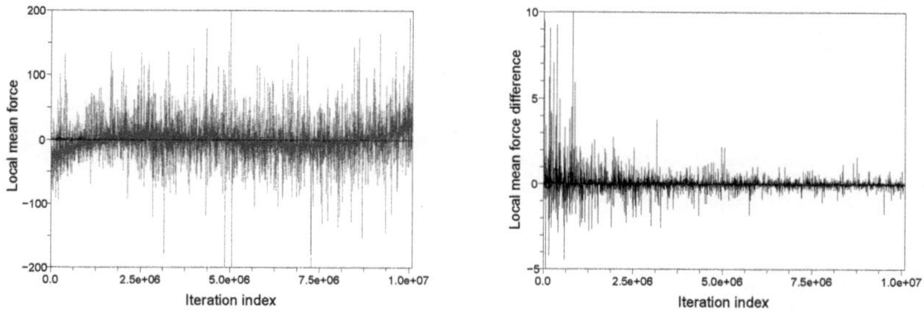

Fig. 3.7 Left: The constraining force $f_{\rm rgd}^M(q^n, p^n)$ (pale line); and the difference between the constraining force and its estimate from the Lagrange multipliers (dark line). Right: Zoom on the difference between the constraining force and the Lagrange multipliers (dark line). Note the difference of scale for the y-axis.

mean force, in terms of their variances. They can all be written as the empirical average of a random sequence (X_n), with the following choices for X_n:

$$f_{\rm rgd}^M(q^n, p^n), \qquad \overline{f}_{\rm rgd}^M(q^n), \qquad \frac{\lambda^{n+1/2} + \lambda^{n+3/4}}{\Delta t},$$

where $q^n, p^n, \lambda^{n+1/2}, \lambda^{n+3/4}$ are given by the numerical scheme (Algorithm 3.56 with the modification mentioned above). Since the correlations between the iterates are very similar for the three methods, we simply computed the variance of the sample points considered as independent. Table 3.1 compares the so-obtained standard errors over 10^5 time-steps with $\Delta t = 0.02$ (simulation time $T = 2{,}000$ for each value of the reaction coordinate). The results show that the different estimators are more or less equivalent, except for the lowest value of the reaction coordinate where the estimator based on the averaged local mean force $\overline{f}_{\rm rgd}^M(q^n)$ leads to some noticeable reduction.

Table 3.1 Standard error of the mean force estimator (with correlations neglected), for different values z of the reaction coordinate.

z	$f^M_{\rm rgd}(q^n, p^n)$	$\dfrac{\lambda^{n+1/2} + \lambda^{n+3/4}}{\Delta t}$	$\overline{f}^M_{\rm rgd}(q^n)$
-0.2	22.1	21.9	14.7
0.0	16.0	15.5	15.4
0.2	23.1	22.5	22.9
0.4	21.1	20.4	21.0
0.6	21.4	20.7	21.3
0.8	21.6	20.9	21.5
1.0	21.4	20.6	21.4
1.2	21.0	20.3	20.9

Chapter 4

Nonequilibrium methods

Nonequilibrium methods for free energy computations are quite recent compared to the more traditional techniques presented in Chapter 2 (free energy perturbation and histogram methods) and Chapter 3 (thermodynamic integration and computation of the mean force using constrained processes). The methods presented in this chapter are termed "nonequilibrium" since the transition from one value of the reaction coordinate (or alchemical parameter) to another one is imposed *a priori*, with a given deterministic schedule, and may be arbitrarily fast. Therefore, even if the system starts at equilibrium, it does not remain at equilibrium. In the limit of infinitely fast switchings, free energy perturbation is recovered. It is therefore no surprise that nonequilibrium methods share some similarities with free energy perturbation, especially in the study of the statistical error of the free energy estimators (which have the same structure, upon replacing energy-difference distributions in free energy perturbation, by work distributions).

The practical use of nonequilibrium equalities was triggered off by [Jarzynski (1997b)]. We present the method in the alchemical case in Section 4.1, in order to highlight some important features which are also valid in the reaction coordinate case — though they may be obscured by geometrical issues when presented in the most general setting. In Section 4.2, a generalized detailed balance condition leading to Jarzynski-Crooks nonequilibrium equalities is presented. The latter can be understood as the extension of the notion of reversibility at equilibrium (see Section 2.2) to the nonequilibrium context. In Section 4.3, the reaction coordinate case is studied: Some reaction coordinates of the system are constrained and then switched according to a predefined path in reaction coordinate space. This can be done both for overdamped Langevin dynamics, or Langevin dynamics in phase space (see respectively Sections 4.3.1

and 4.3.2). Finally, we show in Section 4.4 how nonequilibrium equalities, once restated as equalities on path ensembles, can be approximated numerically by sampling ensembles of nonequilibrium switching paths with some Metropolis-Hastings procedure on paths.

4.1 The Jarzynski equality in the alchemical case

The alchemical setting is presented in Section 1.3.2. It consists in considering transitions where the Hamiltonian function varies. These variations are indexed by some external parameter $\lambda \in [0,1]$. The following notation is used in this section. The variable x can represent the whole degrees of freedom (q,p) of the system, or only the configurational part q. Depending on the context, the invariant measure for a given value λ of the alchemical parameter is therefore the canonical measure

$$\mu_\lambda(dq\,dp) = \frac{1}{Z_{\mu,\lambda}} \mathrm{e}^{-\beta H_\lambda(q,p)}\,dq\,dp, \quad Z_{\mu,\lambda} = \int_{T^*\mathcal{D}} \mathrm{e}^{-\beta H_\lambda(q,p)}\,dq\,dp, \tag{4.1}$$

or its marginal with respect to the momenta, which reads

$$\nu_\lambda(dq) = \frac{1}{Z_{\nu,\lambda}} \mathrm{e}^{-\beta V_\lambda(q)}\,dq, \quad Z_{\nu,\lambda} = \int_{\mathcal{D}} \mathrm{e}^{-\beta V_\lambda(q)}\,dq.$$

When we do not wish to precise further the dynamics, we simply call

$$x \in \mathcal{S} := \mathcal{D} \text{ or } T^*\mathcal{D}$$

the configuration of the system, and $E_\lambda(x) := E(x;\lambda)$ its energy. The associated free energy is

$$F(\lambda) = -\frac{1}{\beta} \ln \int_{\mathcal{S}} \mathrm{e}^{-\beta E_\lambda(x)}\,dx.$$

We start the section by proving an equality for nonequilibrium switching dynamics (Sections 4.1.1 and 4.1.2), where the free energy is recovered by exponentially averaging nonequilibrium works. We then turn to the practical implementation of the method in Section 4.1.3. An important numerical issue is the degeneracy of the quantities entering the free energy estimator, which we highlight in Section 4.1.4. This degeneracy can be quantified with some error analysis, see Section 4.1.5.

4.1.1 *Markovian nonequilibrium simulations*

The usual way to achieve a nonequilibrium switching is to first select a switching time $T > 0$ and a sufficiently smooth[1] transition schedule

[1] In this section, $\Lambda \in \mathrm{C}^1([0,T],\mathbb{R})$ is sufficient.

$t \mapsto \Lambda(t)$ verifying $\Lambda(0) = 0$ and $\Lambda(T) = 1$, and then to perform a time-inhomogeneous Markovian dynamics $t \mapsto X_t$ $(t \in [0, T])$, *starting from equilibrium*:

$$X_0 \sim Z_0^{-1}\, \mathrm{e}^{-\beta E_0(x)}\, dx.$$

There are no other *a priori* constraints on the switching schedule: Λ need not be increasing, and could even have values outside of $[0, 1]$. The choice of the schedule is an important question for numerical purposes.

On the other hand, there are some constraints on the possible dynamics, whose transition kernels actually depend on the time-varying alchemical parameter; in particular the canonical measure should be invariant when the value λ of the alchemical parameter is fixed.

More precisely, the dynamics used is such that the associated *time-homogeneous* probability transitions for a *fixed* time value $t \in [0, T]$ leaves the Boltzmann-Gibbs distribution

$$\pi_t = Z_t^{-1}\, \mathrm{e}^{-\beta E_{\Lambda(t)}(x)}\, dx, \qquad Z_t = \int_S \mathrm{e}^{-\beta E_{\Lambda(t)}(x)}\, dx,$$

invariant. For example, Langevin dynamics (2.39) or their overdamped limits (2.34), associated with H_λ or V_λ respectively, may be considered. For example, when the transition indexes changes only in the potential energy, a possible time-inhomogeneous Langevin process with scalar friction $\gamma > 0$ is, for a given schedule Λ,

$$\begin{cases} dq_t = M^{-1} p_t\, dt, \\ dp_t = -\nabla V_{\Lambda(t)}\,(q_t)\, dt - \gamma\, M^{-1} p_t\, dt + \sqrt{\dfrac{2\gamma}{\beta}}\, dW_t, \end{cases} \tag{4.2}$$

with associated generator (see (2.43))

$$\mathcal{L}_t = p^T M^{-1} \nabla_q - \nabla V_{\Lambda(t)} \cdot \nabla_p - \gamma p^T M^{-1} \nabla_p + \frac{\gamma}{\beta} \Delta_p. \tag{4.3}$$

The overdamped version reads

$$dq_t = -\nabla V_{\Lambda(t)}\,(q_t)\; dt + \sqrt{\frac{2}{\beta}}\, dW_t, \tag{4.4}$$

with generator (see (2.35))

$$\mathcal{L}_t = -\nabla_q V_{\Lambda(t)} \cdot \nabla_q + \frac{1}{\beta} \Delta_q. \tag{4.5}$$

In both cases, the stationarity of the canonical measure is indeed satisfied since

$$\forall t \in [0, T],\ \forall \varphi, \qquad \int_S \mathcal{L}_t(\varphi)\, d\pi_t = 0. \tag{4.6}$$

where $\mathcal{L}_t$ is the infinitesimal generator of the dynamics at time t.

4.1.2 *Importance weights of nonequilibrium simulations*

We wish here to derive the Jarzynski equality through a Feynman-Kac formula, as was done in [Hummer and Szabo (2001)]. For a given nonequilibrium run $\{X_t\}_{0 \le t \le T}$, we denote by

$$\mathcal{W}_t\big(\{X_s\}_{0 \le s \le t}\big) = \int_0^t \frac{\partial E_{\Lambda(s)}}{\partial \lambda}(X_s)\dot{\Lambda}(s)\, ds \tag{4.7}$$

the virtual work induced on the system by the modification of the parameter λ during the time interval $[0, t]$. In this expression, $\partial_\lambda E_{\Lambda(s)}$ denotes the partial derivative $\partial_\lambda E_\lambda$ at $\lambda = \Lambda(s)$. Typically, $X_t = q_t$ solution of (4.4) and $E_\lambda = V_\lambda$, or $X_t = (q_t, p_t)$ solution of (4.2) and $E_\lambda = H_\lambda$.

The work is therefore a function of the trajectory. With a slight abuse of notation, we will denote in the sequel by $\mathcal{W}_t$ both the path functional defined in (4.7) and the random variable $\mathcal{W}_t\big(\{X_s\}_{0 \le s \le t}\big)$ giving the work value when applied to a given realization of the path.

The quantity $\mathcal{W}_t$ is used to define the importance weight of nonequilibrium simulations with respect to the target equilibrium distribution. Note also that this definition is consistent with the definition used for Hamiltonian dynamics (see (1.73)).

Proposition 4.1 (Jarzynski nonequilibrium equality). *Consider a switching schedule $\Lambda \in \mathrm{C}^1([0, T], \mathbb{R})$ such that $\Lambda(0) = 0$ and $\Lambda(T) = 1$, and a stochastic process $t \mapsto X_t$ starting from equilibrium $X_0 \sim \pi_0$, and evolving according to a dynamics whose generator $\mathcal{L}_t$ is such that*

$$\forall t \in [0, T], \ \forall \varphi, \qquad \int_{\mathcal{S}} \mathcal{L}_t(\varphi)\, d\pi_t = 0. \tag{4.8}$$

Then, for all test functions φ,

$$\frac{Z_t}{Z_0} \int_{\mathcal{S}} \varphi \, d\pi_t = \mathbb{E}\Big(\varphi(X_t)\, \mathrm{e}^{-\beta \mathcal{W}_t}\Big), \tag{4.9}$$

where the expectation is an expectation both over the initial conditions and over all the possible realizations of the switching dynamics for a given initial condition.

In particular, the choice $\varphi = 1$ leads to the so-called Jarzynski equality

$$\mathbb{E}\Big[\mathrm{e}^{-\beta \mathcal{W}_t}\Big] = \frac{Z_t}{Z_0} = \mathrm{e}^{-\beta \Delta F(\Lambda(t))}, \tag{4.10}$$

where $\Delta F(\Lambda(t)) = F(\Lambda(t)) - F(0)$. Note also that the free energy perturbation equality (2.88) is recovered in the limit when the final time T goes to 0.

Proof. Consider the transition operator $P^w_{s,t}$ $(0 \le s \le t \le T)$ associated with the inhomogeneous dynamics where the paths $(X_s)_{0 \le s \le t}$ are weighted by the factor $\exp(-\beta \mathcal{W}_t)$: for all test functions φ,

$$P^w_{s,t}\varphi(x) = \mathbb{E}\left(\varphi(X_t)\, \mathrm{e}^{-\beta(\mathcal{W}_t - \mathcal{W}_s)} \,\middle|\, X_s = x \right), \tag{4.11}$$

where the expectation is taken over all the possible realizations starting from $X_s = x$. This operator satisfies the following backward Kolmogorov evolution:

$$\partial_s P^w_{s,t} = -\mathcal{L}_s P^w_{s,t} + \beta \frac{\partial E_{\Lambda(s)}}{\partial \lambda}\, \dot{\Lambda}(s)\, P^w_{s,t}. \tag{4.12}$$

Indeed, consider the following partial differential equation:

$$\partial_s \Phi(s,x) = -\mathcal{L}_s \Phi(s,x) + \beta \dot{\Lambda}(s)\, \partial_\lambda E_{\Lambda(s)}(x)\, \Phi(s,x), \tag{4.13}$$

with the final condition $\Phi(t,x) = \varphi(x)$. We want to show that $\Phi(s,x) = P^w_{s,t}\varphi(x)$. First, denoting by $X^{s,x}_r$ the solution at time r of the stochastic process starting from x at time s, we compute the following variation using Itô calculus:

$$\begin{aligned} d&\left[\Phi(r, X^{s,x}_r) \exp\left(-\beta \int_s^r \dot{\Lambda}(\tau)\, \partial_\lambda E_{\Lambda(\tau)}(X^{s,x}_\tau)\, d\tau \right) \right] \\ &= \left[\partial_r + \mathcal{L}_r - \beta \dot{\Lambda}(r)\, \partial_\lambda E_{\Lambda(r)}(X^{s,x}_r) \right] (\Phi)(r, X^{s,x}_r)\, \mathrm{e}^{-\beta(\mathcal{W}_r - \mathcal{W}_s)}\, dr \\ &\quad + dM_r, \end{aligned} \tag{4.14}$$

where M_r is a martingale. The precise form of dM_r depends on the dynamics at hand. It vanishes for Hamiltonian dynamics, and is equal to

$$dM_r = \sqrt{\frac{2}{\beta}}\, \nabla \Phi(r, X^{s,x}_r) \cdot dW_r$$

for overdamped Langevin dynamics. In any case, $\mathbb{E}(M_r) = 0$, and the first term on the right-hand side of (4.14) vanishes in view of (4.13). Therefore, taking expectations in (4.14) and integrating the equality on $[s,t]$,

$$\begin{aligned} \Phi(s,x) &= \mathbb{E}\left[\Phi(t, X^{s,x}_t) \exp\left(-\beta \int_s^t \dot{\Lambda}(\tau)\, \partial_\lambda E_{\Lambda(\tau)}(X^{s,x}_\tau)\, d\tau \right) \right] \\ &= \mathbb{E}\left[\varphi(X^{s,x}_t) \exp\left(-\beta(\mathcal{W}_t - \mathcal{W}_s) \right) \right] \end{aligned}$$

in view of the final condition $\Phi(t,x) = \varphi(x)$. The comparison with (4.11) shows that $\Phi(s,x) = P^w_{s,t}\varphi(x)$, as claimed. Finally, (4.12) follows from (4.13).

Now, with the identity (4.12), it is straightforward to check that, for any smooth test function φ,

$$\frac{d}{ds}\left(\int_{\mathcal{S}} P^w_{s,t}\varphi(x)\,\mathrm{e}^{-\beta E_{\Lambda(s)}(x)}\,dx\right) = 0. \tag{4.15}$$

Indeed,

$$\begin{aligned}
&\frac{d}{ds}\left(\int_{\mathcal{S}} P^w_{s,t}\varphi(x)\,\mathrm{e}^{-\beta E_{\Lambda(s)}(x)}\,dx\right) \\
&\quad = \int_{\mathcal{S}}\left(\partial_s P^w_{s,t}\varphi(x) - \beta\dot{\Lambda}(s)\,\frac{\partial E_{\Lambda(s)}(x)}{\partial\lambda}\,P^w_{s,t}\varphi(x)\right)\mathrm{e}^{-\beta E_{\Lambda(s)}(x)}\,dx \\
&\quad = -Z_s\int_{\mathcal{S}}\mathcal{L}_s\left(P^w_{s,t}\varphi\right)d\pi_s = 0,
\end{aligned}$$

where the last step is a consequence of the invariance of the canonical measure expressed by (4.8). Therefore, integrating (4.15) on $[0,t]$,

$$Z_0\int_{\mathcal{S}} P^w_{0,t}\varphi\,d\pi_0 = Z_t\int_{\mathcal{S}} P^w_{t,t}\varphi\,d\pi_t = Z_t\int_{\mathcal{S}}\varphi\,d\pi_t.$$

Since

$$\int_{\mathcal{S}} P^w_{0,t}\varphi\,d\pi_0 = \int_{\mathcal{S}}\mathbb{E}\left(\varphi(X_t)\,\mathrm{e}^{-\beta\mathcal{W}_t}\;\middle|\;X_0 = x\right)\pi_0(dx),$$

the fundamental Feynman-Kac fluctuation equality is finally recovered:

$$\frac{Z_t}{Z_0}\int_{\mathcal{S}}\varphi\,d\pi_t = \mathbb{E}\left(\varphi(X_t)\,\mathrm{e}^{-\beta\mathcal{W}_t}\right),$$

which concludes the proof. □

Remark 4.2 (Initial conditions). *It may be worth emphasizing again that the correct ratio of partition functions is obtained if and only if the nonequilibrium dynamics is started at equilibrium. The Feynman-Kac equality is precisely aimed at reweighting the nonequilibrium evolution in order to transform the current nonequilibrium distribution into some equilibrium distribution.*

Remark 4.3 (Hamiltonian dynamics). *The above derivation still makes sense for a Hamiltonian switching (which corresponds to the case* $\gamma = 0$ *of the nonequilibrium Langevin dynamics* (4.2)*). Indeed, we do not make use of any (hypo)ellipticity property, and simply relied on the local (*i.e. *at fixed* λ*) invariance of the canonical measure* (4.8).

Remark 4.4 (A free energy inequality). *Jensen's inequality applied to* (4.10) *reads, for a given convex function f,*

$$f\left(\mathrm{e}^{-\beta\Delta F(\Lambda(t))}\right) = f\left[\mathbb{E}\left(\mathrm{e}^{-\beta\mathcal{W}_t}\right)\right] \leq \mathbb{E}\left[f\left(\mathrm{e}^{-\beta\mathcal{W}_t}\right)\right].$$

The choice $f(x) = -\beta^{-1}\ln x$ *leads to*

$$\mathbb{E}(\mathcal{W}_t) \geq \Delta F(\Lambda(t)). \tag{4.16}$$

It is expected that the difference $\mathbb{E}(\mathcal{W}_t) - \Delta F(\Lambda(t))$, *sometimes called* dissipated work, *decreases when the switching is slower, and increases when the switching is faster.*

The inequality (4.16) *can be interpreted as a microscopic analogue of the second law of thermodynamics, which states that the average work exerted on the system, at equilibrium in the initial and final stages of the transformation, is always larger than the free energy difference. This is in accordance with the macroscopic laws of thermodynamics, where the entropy creation involved in a thermodynamic transformation is always nonnegative. More precisely, consider a switching function such that* $\Lambda(0) = 0$, *and* $\Lambda(t) = 1$ *for* $t \geq t_0 > 0$. *When the switching is performed in a time* $T \geq t_0$, *the process can be decomposed into two stages:* (i) *an actual switching of the parameter from 0 to 1, in a time* t_0*;* (ii) *an equilibration step with the value of the parameter fixed to 1, for a time* $T - t_0$. *In accordance with the macroscopic laws of thermodynamics, let us* define *the entropy variation of the system during* $[0, T]$ *as*

$$\Delta S := \beta\left(\Delta U - \Delta F\right),$$

where the energy variation ΔU *is*

$$\Delta U := \int_{\mathcal{S}} E_{\Lambda(T)}\, d\pi_T - \int_{\mathcal{S}} E_0\, d\pi_0,$$

and $\Delta F = F(\Lambda(T)) - F(0)$. *This entropy variation can be decomposed as*

$$\Delta S = \delta S_{\mathrm{exch}}(T) + \delta S_{\mathrm{creation}}(T).$$

The entropy variation $\delta S_{\mathrm{exch}}(T)$ *arising from heat exchanges is, by definition,*

$$\delta S_{\mathrm{exch}}(T) = \beta\,\mathbb{E}(\mathcal{Q}_T),$$

where the exchanged heat is given by the energy conservation principle:

$$\mathcal{Q}_T := E_{\Lambda(T)}(X_T) - E_0(X_0) - \mathcal{W}_T.$$

In the limit $T \to +\infty$, and for a fixed t_0, the equilibration procedure on the time interval $[t_0, T]$ and the fact that the system initially starts at equilibrium imply that

$$\lim_{T \to +\infty} \mathbb{E}(\mathcal{Q}_T + \mathcal{W}_T) = \Delta U.$$

The inequality (4.16) *reads, for any $T \geq t_0$ (remember that the actual switching is performed on $[0, t_0]$):*

$$\mathbb{E}(\mathcal{W}_T) \geq \Delta F,$$

so that

$$\begin{aligned}
\delta S_{\text{creation}}(T = +\infty) &= \Delta S - \delta S_{\text{exch}}(T = +\infty) \\
&= \beta(\Delta U - \Delta F) - \beta \mathbb{E}(\mathcal{Q}_{T=+\infty}) \\
&= \beta\Big(\mathbb{E}(\mathcal{W}_{T=+\infty}) - \Delta F\Big) \geq 0,
\end{aligned}$$

which can be interpreted as the non-negativity of the entropy creation.

The inequality (4.16) *is an equality if and only if the transformation is quasi-static (infinitely slow), which amounts to considering the limiting regime where $t_0 \to +\infty$. In this case, there is no entropy creation.*

4.1.3 *Practical implementation*

The nonequilibrium switching dynamics requires a switching schedule $\Lambda(t)$ for $0 \leq t \leq T$, and a dynamics. Given a time-step Δt, and denoting by N the number of time-steps required to reach the time T (so that $N\Delta t = T$), a possible implementation of the nonequilibrium switching dynamics is to set $\lambda^n = \Lambda(n\Delta t)$ for $1 \leq n \leq N$, and obtain X^n from X^{n-1} using a numerical discretization of a continuous dynamics leaving $\exp(-\beta E_{\lambda^n})$ invariant. The work at time $n\Delta t$ for a single realization of the switching process is then defined as

$$\mathcal{W}^n = \sum_{i=0}^{n-1} \Big(E_{\lambda^{i+1}}(X^i) - E_{\lambda^i}(X^i) \Big). \tag{4.17}$$

For instance, a possible implementation for overdamped Langevin dynamics is

$$q^n = q^{n-1} - \Delta t \, \nabla V_{\lambda^n}(q^n) + \sqrt{\frac{2\Delta t}{\beta}} \, G^n, \tag{4.18}$$

where $(G^n)_n$ are i.i.d. centered Gaussian random vectors with covariance Id_{3N}, while for Hamiltonian dynamics the Verlet scheme can be used,

$$\begin{cases} p^{n-1/2} = p^{n-1} - \dfrac{\Delta t}{2} \nabla V_{\lambda^n}(q^{n-1}), \\ q^n = q^{n-1} + \Delta t \, M^{-1} p^{n-1/2}, \\ p^n = p^{n-1/2} - \dfrac{\Delta t}{2} \nabla V_{\lambda^n}(q^n). \end{cases} \tag{4.19}$$

The average in (4.10) can be approximated by an empirical average over M independent realizations of the switching process. To this end, M independent initial conditions should first be sampled according to the canonical measure $Z_0^{-1} \exp(-\beta E_0(x))\,dx$, and then the switching is performed for each of them, using independent random numbers for each trajectory of the Monte Carlo or stochastic dynamics. Labelling the so-obtained works by an additional upper index $1 \le m \le M$, an estimator of the free energy difference $\Delta F = F(1) - F(0)$ is

$$\widehat{\Delta F}_M = -\beta^{-1} \ln \left(\frac{1}{M} \sum_{m=1}^{M} \mathrm{e}^{-\beta \mathcal{W}^{N,m}} \right). \tag{4.20}$$

It is clear that the estimation of ΔF by (4.20) is straightforward to parallelize since the works $(\mathcal{W}^{N,m})_{1 \le m \le M}$ are independent. More general estimators are presented in Section 4.1.5.

The estimator (4.20) is a consistent estimator of the free energy difference in the limits $\Delta t \to 0$ (hence $N \to +\infty$) and $M \to +\infty$. This is a consequence of the fact that the empirical average over independent random variables

$$\frac{1}{M} \sum_{m=1}^{M} \mathrm{e}^{-\beta \mathcal{W}^{N,m}} \tag{4.21}$$

is a consistent and unbiased estimator of $\mathrm{e}^{-\beta \Delta F}$ in the limit $\Delta t \to 0$ when consistent discretizations of SDEs are used. A consequence of the unbiasedness of the estimator (4.21) is that (4.20) is a *biased* estimator of the free energy difference (due to the nonlinearity of the log function). The study of statistical properties of such estimators (variance, finite sampling bias) is performed in Section 4.1.5.

Remark 4.5 (Time-discrete version of the Jarzynski equality). *An equality of the form* (4.10) *can also be obtained for discrete time dynamics, for a given discrete schedule* $(\lambda^0, \lambda^1, \ldots, \lambda^N)$. *More precisely, consider a sequence of configurations* $X^n \sim P_{\lambda^n}(X^{n-1}, dy)$, *where the transition* *probability* $P_{\lambda^n}(x, dy)$ *leaves* π_{λ^n} *invariant, and* X^0 *is distributed according to* π_{λ^0}. *For instance, it is possible to use a Metropolis scheme where the current configuration* X^{n-1} *is modified in a symmetric way by some random displacement, and then accepted or rejected according to the Metropolis ratio* $\pi_{\lambda^n}(X^n)/\pi_{\lambda^n}(X^{n-1})$. *The work is still computed as* (4.17). *In this case,*

$$\mathbb{E}\left(\mathrm{e}^{-\beta\mathcal{W}^N}\right) = \frac{Z_{\lambda^N}}{Z_{\lambda^0}}.$$

The proof of the unbiasedness of the estimator (4.21) *relies on the observation that the probability to observe a given discrete path* $(X^0, \ldots, X^N)$ *is (see also Section* 4.4.1.1*)*

$$\Pi(dX) = \pi_{\lambda^0}(dX^0)\prod_{i=1}^{N} P_{\lambda^i}(X^{i-1}, dX^i),$$

so that

$$\begin{aligned}
\mathbb{E}\left(\mathrm{e}^{-\beta\mathcal{W}^N}\right) &= \int_{\mathcal{S}^N}\prod_{i=0}^{N-1}\exp\left[-\beta\Big(E_{\lambda^{i+1}}(X^i) - E_{\lambda^i}(X^i)\Big)\right]\Pi(dX) \\
&= \frac{1}{Z_{\lambda^0}}\int_{\mathcal{S}^N}\mathrm{e}^{-\beta E_{\lambda^1}(X^0)}\prod_{i=1}^{N-1}\exp\left[-\beta\Big(E_{\lambda^{i+1}}(X^i) - E_{\lambda^i}(X^i)\Big)\right] \\
&\quad\times\prod_{i=1}^{N} P_{\lambda^i}(X^{i-1}, dX^i)\, dX^0 \\
&= \frac{Z_{\lambda^1}}{Z_{\lambda^0}}\int_{\mathcal{S}^{N-1}}\prod_{i=1}^{N-1}\exp\left[-\beta\Big(E_{\lambda^{i+1}}(X^i) - E_{\lambda^i}(X^i)\Big)\right] \\
&\quad\times\pi_{\lambda^1}(dX^1)\prod_{i=2}^{N} P_{\lambda^i}(X^{i-1}, dX^i)
\end{aligned}$$

since, by the invariance of π_{λ^1} *through* P_{λ^1},

$$\int_{\mathcal{S}}\mathrm{e}^{-\beta E_{\lambda^1}(X^0)}\, P_{\lambda^1}(X^0, dX^1)\, dX^0 = Z_{\lambda^1}\pi_{\lambda^1}(dX^1).$$

A repeated use of the above manipulation leads to

$$\mathbb{E}\left(\mathrm{e}^{-\beta\mathcal{W}^N}\right) = \frac{Z_{\lambda^N}}{Z_{\lambda^0}},$$

which is the desired equality.

This may be seen as a way to eliminate the time-step error in the estimate (4.21) *of the free energy difference upon resorting to a time-inhomogeneous Metropolis-Hastings dynamics, instead of plain discretizations of the underlying diffusion process.*

4.1.4 *Degeneracy of weights*

4.1.4.1 *Work distributions*

As elegant as the Jarzynski equality may be, it is often the case in practice, unless the switching is very slow, that the work distribution is very spread out. In this case, only a few work values (the smallest ones, which are also typically the most unlikely ones) are really important in the nonlinear average (4.20). The work distribution $P_t(dW)$ at time t, for a switching performed with a given schedule, is defined through the following equality: for any test function g,

$$\int_{\mathbb{R}} g(W)\, P_t(dW) = \mathbb{E}\Big[g\big(\mathcal{W}_t(\{X_s\}_{0\le s\le t})\big)\Big], \tag{4.22}$$

where the work function is defined in (4.7). We will often assume for simplicity that the work distribution is absolutely continuous with respect to the Lebesgue measure, and simply write it $P_t(W)\,dW$ with a slight abuse of notation. Then, the equality (4.10) can be rewritten as

$$\exp\Big(-\beta\Delta F(\Lambda(t))\Big) = \int_{\mathbb{R}} \mathrm{e}^{-\beta W}\, P_t(W)\, dW. \tag{4.23}$$

Some examples of work distributions are displayed below in Figure 4.1 (see Section 4.1.5.3) and Figure 4.3 (Section 4.2.3).

Considering (4.23), it is clear that the most negative values of W are the most important in the integral. On the other hand, these values typically have very small probabilities. In other words, in a discretization using Monte Carlo methods such as (4.21), only a few realizations are important in the evaluation of the empirical mean. Many realizations have small weights and only a few have large weights. This is called the *degeneracy of weights.*

In essence, the ideal work distribution is the Dirac mass $P_t(dW) = \delta_{\Delta F(\Lambda(t))}(dW)$, whereas typical work distributions have finite widths. An idea of the width of the work distribution is obtained by computing the dissipated work given by (4.16):

$$\mathbb{E}(\mathcal{W}_t) - \Delta F(\Lambda(t)) = \int_{\mathbb{R}} \Big(W - \Delta F(\Lambda(t))\Big) P_t(W)\, dW \ge 0.$$

When the work distribution is very peaked, the dissipated work is positive but small. It is believed that a large dissipated work is equivalent to a large variance of the estimator (4.21). The latter variance reads

$$\mathrm{Var}\left(\widehat{\Delta F}_M\right) = \frac{\mathrm{Var}\Big(\exp\Big[-\beta\big(\mathcal{W}_t - \Delta F(\Lambda(t))\big)\Big]\Big)}{\beta^2 M},$$

see (4.32) below. This can be shown rigorously for Gaussian work distributions (see below), but we are not aware of a general result in this direction.

Let us detail the case when the work distribution at the final time T is Gaussian with mean ω and variance Ω^2, namely

$$P_T(W) = \frac{1}{\sqrt{2\pi\Omega^2}} \exp\left(-\frac{(W-\omega)^2}{2\Omega^2}\right).$$

In this case,

$$\begin{aligned}\mathbb{E}\left(e^{-\beta\mathcal{W}_T}\right) &= \int_{\mathbb{R}} e^{-\beta W}\, P_T(W)\, dW \\ &= \exp\left(\frac{\beta^2\Omega^2}{2} - \beta\omega\right) \frac{1}{\sqrt{2\pi\Omega^2}} \int_{\mathbb{R}} \exp\left(-\frac{(W-\omega+\beta\Omega^2)^2}{2\Omega^2}\right) dW \quad (4.24)\\ &= \exp\left(\frac{\beta^2\Omega^2}{2} - \beta\omega\right),\end{aligned}$$

so that

$$\Delta F(\Lambda(T)) = -\frac{1}{\beta}\ln \mathbb{E}\left(e^{-\beta\mathcal{W}_T}\right) = \omega - \frac{\beta\Omega^2}{2}. \quad (4.25)$$

The dissipated work

$$\mathbb{E}(\mathcal{W}_T) - \Delta F(\Lambda(T)) = \frac{\beta\Omega^2}{2}$$

is proportional to the variance of the work distribution. Besides, it is clear from (4.24) that the work values which count most in the expectation are the ones around the maxima of the Gaussian function to be integrated, *i.e.* work values around $\omega - \beta\Omega^2$. These values are difficult to sample when Ω^2 is large since the typical values of the sampled works fall in the vicinity of the average work ω.

Gaussian work distributions are encountered in practice for slow switchings, or for toy models such as the one presented in Section 4.1.4.2. In general, the work distribution is not Gaussian, and strategies to reduce the variance of the estimator (4.20) by diminishing the width of the work distribution should be employed, see Section 4.1.4.3.

Remark 4.6 (Typical vs dominant work values). *A classification of work values into "dominant" and "typical" values is proposed in [Jarzynski (2006)]. The dominant values are the ones in the lower tail of the work distribution which are the most important in the exponential average* (4.10) *(i.e. the maximizers of $P_t(W)\exp(-W)$), while the typical ones are in* *the vicinity of maximizers of $P_t(W)$. Dominant work values are actually (the opposite of) typical work values when the switching is performed using the time-reversed schedule $\Lambda(T-t)$, starting at equilibrium from the final canonical distribution π_T. This is the essence of the Crooks-Jarzynski equality, see Section 4.2 for further precision.*

4.1.4.2 *An analytical example*

The above heuristic observations on the weight degeneracy can be made rigorous in some situations, where analytical computations can be performed. For instance, a pulled harmonic oscillator can be considered, the dynamics being either an overdamped Langevin dynamics [Mazonka and Jarzynski (1999)], or the Hamiltonian dynamics [Hummer (2007)]. We present here the former case.

The transformation is described by the family of potentials

$$V_\lambda(q) = \frac{1}{2}K(q-\lambda L)^2,$$

and the schedule is assumed to be linear: $\Lambda(t) = t/T$. Note that $\Delta F(\Lambda(t)) = 0$ at all times since $\int_{\mathbb{R}} \exp(-\beta V_\lambda(q))\,dq$ does not depend on λ.

A linear switching in a time T amounts to displacing the equilibrium position of the harmonic oscillator at a constant velocity $v = L/T$. The dynamics of the system is

$$dq_t = -K(q_t - vt)\,dt + \sqrt{\frac{2}{\beta}}\,dW_t, \qquad q_0 \sim \mathcal{N}\left(0, \frac{1}{\beta K}\right),$$

since the dynamics is started at equilibrium. It is convenient to restate the dynamics in the frame centered on the instantaneous equilibrium position of the harmonic oscillator, by introducing $Q_t = q_t - vt$. Then, the system evolves as

$$dQ_t = -(KQ_t + v)\,dt + \sqrt{\frac{2}{\beta}}\,dW_t, \qquad Q_0 = q_0, \tag{4.26}$$

while the work associated with a trajectory $(Q_s)_{0\leq s\leq t}$ is

$$\mathcal{W}_t = -Kv\int_0^t Q_s\,ds. \tag{4.27}$$

By linearity of the expressions, since q_0 is distributed initially according to a Gaussian law, the law of the random variable Q_t, as well as the law of the work $\mathcal{W}_t$, are Gaussian functions at all times, and $(Q_t, \mathcal{W}_t)$ is a

Gaussian vector. Thus, the laws of these random variables are completely characterized by their averages

$$A_t = \mathbb{E}(Q_t), \qquad B_t = \mathbb{E}(\mathcal{W}_t),$$

variances

$$C_t = \mathbb{E}\left[\left(Q_t - A_t\right)^2\right], \qquad D_t = \mathbb{E}\left[\left(\mathcal{W}_t - B_t\right)^2\right],$$

and the covariance

$$E_t = \mathbb{E}\left[\left(Q_t - A_t\right)\left(\mathcal{W}_t - B_t\right)\right].$$

The initial values are $A_0 = B_0 = D_0 = E_0 = 0$ and $C_0 = (\beta K)^{-1}$. The expressions for these quantities can be obtained from (4.26)-(4.27) using Itô calculus. More precisely, (4.26) implies that

$$dA_t = -(KA_t + v)\,dt, \qquad dC_t = -2K\left(C_t - \frac{1}{\beta K}\right)dt,$$

while (4.27) leads to

$$B_t = -Kv\int_0^t A_s\,ds, \qquad dD_t = -2Kv\,E_t\,dt.$$

The evolution of the conditional covariance is

$$dE_t = -K[E_t + v\,C_t]\,dt.$$

These equations are easily integrated as

$$A_t = -\frac{v}{K}\left(1 - \mathrm{e}^{-Kt}\right), \quad B_t = v^2\left(t - \frac{1 - \mathrm{e}^{-Kt}}{K}\right), \tag{4.28}$$

$$C_t = \frac{1}{\beta K}, \quad D_t = \frac{2}{\beta}B_t, \quad E_t = \frac{1}{\beta}A_t. \tag{4.29}$$

The quantity A_t is the average distance at which particles lag behind the equilibrium position of the translated harmonic oscillator. At the endtime, and for slow switchings, the lag is negative but small since

$$\mathbb{E}(q_T) = L + A_T = L\left(1 - \frac{1 - \mathrm{e}^{-KT}}{KT}\right) \sim L\left(1 - \frac{1}{KT}\right)$$

as $T \to +\infty$; while $\mathbb{E}(q_T) \to 0$ as $T \to 0$ since the particle does not have time to move.

Note also the remarkable equality

$$\mathbb{E}\left[\left(\mathcal{W}_t - \mathbb{E}(\mathcal{W}_t)\right)^2\right] = D_t = \frac{2}{\beta}\,B_t = \frac{2}{\beta}\mathbb{E}(\mathcal{W}_t). \tag{4.30}$$

It can easily be checked that the Jarzynski equality (4.10) is valid using (4.25) and (4.30).

The relation (4.24) shows that the most important work values in the nonlinear average (4.20) (the dominant ones with the terminology introduced in Remark 4.6) are around $B_t - \beta D_t = -B_t = -\mathbb{E}(\mathcal{W}_t)$.

These important values may be quite unlikely when the switching is fast. In the limit $T \to 0$, a work distribution of finite mean and variance

$$\mathbb{E}(\mathcal{W}_T) \sim KL^2, \qquad \mathrm{Var}(\mathcal{W}_T) \sim \frac{2KL^2}{\beta}$$

is found (actually identical to the distribution of energy differences found by free energy perturbation, see Section 2.4.1). It is difficult to sample values around $-\mathbb{E}(\mathcal{W}_T)$ when KL^2 or β are large. Since $\Delta F(\Lambda(t)) = 0$ at all times, it is clear on this example that

$$\forall t > 0, \qquad B_t = \mathbb{E}(\mathcal{W}_t) > 0 = \Delta F(\Lambda(t)).$$

In view of (4.28), the difference $\mathbb{E}(\mathcal{W}_t) - \Delta F(\Lambda(t)) = B_t$ is an increasing function of the time t and the piston velocity v.

In the limit $T \to +\infty$, the work distribution at the endtime has a vanishing mean and variance:

$$\mathbb{E}(\mathcal{W}_T) \sim \frac{L^2}{T} = Lv, \qquad \mathrm{Var}(\mathcal{W}_T) \sim \frac{2L^2}{\beta T},$$

and the dissipated work is thus of order $\mathrm{O}(T^{-1})$ in this case.

4.1.4.3 *Reducing the width of work distributions*

Selection strategies. To avoid the degeneracy of weights, especially when the switching is not slow, a possible strategy is to use selection mechanisms on replicas of the system simulated in parallel in order to focus the computational resources on the lower tail of the work distribution. This selection uses an interacting system of particles, a strategy inspired by resampling techniques (see the literature on sequential Monte Carlo algorithms, in particular [Doucet *et al.* (2001)] and the review paper [Doucet *et al.* (2006)]). We refer to Section 6.1 for further precision.

Importance sampling in nonequilibrium switching path space. Another strategy, proposed in [Vaikuntanathan and Jarzynski (2008)], consists in adding a drift term proportional to $\dot{\Lambda}(t)$ in the evolution equation of the system in order to reduce the so-called dynamical lag of the system, *i.e.* to "escort" the system closer to equilibrium. With the notation of Section 4.1.2, the generator of the dynamics is transformed into

$$\widetilde{\mathcal{L}}_t = \mathcal{L}_t + \dot{\Lambda}(t)\, u(x, \Lambda(t)) \cdot \nabla_x,$$

where $\mathcal{L}_t$ corresponds to the switching dynamics used for the transition (without escorting), and $u(x, \lambda)$ is the additional drift term. A formula similar to (4.9) can be derived, following the proof of Proposition 4.1. To this end, define the density kernel associated with the inhomogeneous dynamics where the paths $(\widetilde{X}_s)_{0 \le s \le t}$ (obtained from a dynamics with generator $\widetilde{\mathcal{L}}_t$) are weighted by a factor $\exp(-\beta \widetilde{\mathcal{W}}_t)$ which should be chosen correctly in order to have an equality similar to (4.9). Define the transition operators $\widetilde{P}^w_{s,t}$ as: for any test function φ,

$$\widetilde{P}^w_{s,t}\varphi(x) = \mathbb{E}\left(\varphi(\widetilde{X}_t)\, \mathrm{e}^{-\beta(\widetilde{\mathcal{W}}_t - \widetilde{\mathcal{W}}_s)} \,\middle|\, \widetilde{X}_s = x \right).$$

When the modified works are sought for under the form

$$\widetilde{\mathcal{W}}_t = \int_0^t \omega_s(\widetilde{X}_s)\, \dot{\Lambda}(s)\, ds,$$

the operators $\widetilde{P}^w_{s,t}$ are characterized by the following backward Kolmogorov evolution, generalizing (4.12) to the case when $u \neq 0$:

$$\partial_s \widetilde{P}^w_{s,t} = -\widetilde{\mathcal{L}}_s \widetilde{P}^w_{s,t} + \beta \omega_s\, \dot{\Lambda}(s)\, \widetilde{P}^w_{s,t}.$$

The work increment ω_s should be chosen such that

$$\frac{d}{ds}\left(\int_{\mathcal{S}} \widetilde{P}^w_{s,t}\varphi(x)\, \mathrm{e}^{-\beta E_{\Lambda(s)}(x)}\, dx \right) = 0.$$

Some straightforward computations show that this is the case for

$$\omega_s(x) = \frac{\partial E_{\Lambda(s)}(x)}{\partial \lambda} + u(x, \Lambda(s)) \cdot \nabla_x E_{\Lambda(s)}(x) - \frac{1}{\beta} \nabla_x \cdot u(x, \Lambda(s)).$$

Finally, the fundamental Feynman-Kac fluctuation equality is recovered (by a proof similar to the end of the proof of Proposition 4.1):

$$\frac{Z_t}{Z_0} \int_{\mathcal{S}} \varphi\, d\pi_t = \mathbb{E}\left(\varphi(\widetilde{X}_t)\, \mathrm{e}^{-\beta \widetilde{\mathcal{W}}_t} \right),$$

where $\widetilde{X}_0 \sim \pi_0$ and $\widetilde{X}_s$ evolves with a dynamics biased by a drift term u.

Now, the additional drift u should be chosen so that the resulting work distribution is as narrow as possible. The optimal choice corresponds of course to the case when all the works are precisely equal to $\Delta F(\Lambda(t))$.

This is the case when $u = u^*$, where the optimal drift term u^* satisfies the following equation:

$$\frac{\partial E_\lambda}{\partial \lambda}(x) + u^*(x,\lambda) \cdot \nabla_x E_\lambda(x) - \frac{1}{\beta}\nabla_x \cdot u^*(x,\lambda) = F'(\lambda),$$

which is equivalent to

$$-\nabla_x \cdot \left(u^*(x,\lambda)\,\mathrm{e}^{-\beta E_\lambda}(x)\right) = \beta\left(F'(\lambda) - \frac{\partial E_\lambda(x)}{\partial \lambda}\right)\mathrm{e}^{-\beta E_\lambda(x)}.$$

In practice however, the optimal drift is unknown since the right-hand side of the above equation involves the free energy, and should therefore be approximated. A more practical idea is to design it in order to reduce the dynamical lag, *i.e.* somehow push the system in the direction where the most important modes of the equilibrium distribution are moving, see [Vaikuntanathan and Jarzynski (2008)] for a simple one-dimensional example.

4.1.5 *Error analysis*

We present here some elements on the statistical error of free energy estimators for nonequilibrium dynamics (bias and variance). We leave out the errors arising from the numerical integration of the dynamics, related to the finiteness of the time-step Δt.

For simplicity, we do not write the time index in the works in this section, but the values $\mathcal{W}^m$ should be thought of as work values $\mathcal{W}_t^m$ (resp. $\mathcal{W}^{m,N}$) obtained after a switching on the time interval $[0,t]$ (resp. after N steps of the numerical integration). Similarly, the free energy difference to be estimated will always be denoted by ΔF in this section, and $\mathrm{e}^{-\beta\Delta F} = \mathbb{E}(\mathrm{e}^{-\beta\mathcal{W}})$.

4.1.5.1 *One-sided averages*

We first consider the straightforward estimator of the free energy difference obtained in the case of a switching from an initial state described by the energy E_0 to a final state described by $E_{\Lambda(T)} = E_1$. More precisely, for M independent initial conditions sampled according to the canonical measure associated with E_0, and independent switching realizations,

$$\widehat{\Delta F}_M = -\beta^{-1}\ln\left(\frac{1}{M}\sum_{m=1}^{M}\mathrm{e}^{-\beta\mathcal{W}^m}\right), \tag{4.31}$$

where the works $(\mathcal{W}^m)_{m=1,\dots,M}$ computed from (4.17) are i.i.d. The fact that the work values are independent is an ideal situation that is usually not encountered in practice: Initial points are often not strictly uncorrelated since they are generated from a (subsampled) MCMC trajectory. Besides, time-step or discretization errors in the integration of the underlying dynamics may have an influence on the work distribution. We will however overlook these issues.

We present in this section some results on the statistical properties of the estimator (4.31) (possibly only in some limiting regime), considering successively the finite sampling bias $\mathbb{E}\left(\widehat{\Delta F}_M\right) - \Delta F$ and the variance $\mathbb{E}\left(\widehat{\Delta F}_M^2\right) - \left[\mathbb{E}(\widehat{\Delta F}_M)\right]^2$.

Monotonicity of the finite sampling bias. Even for independently distributed works, the estimator $\widehat{\Delta F}_M$ is a biased estimator for a fixed number $M < +\infty$ or realizations of the switching process. Indeed, $\exp(-\beta\widehat{\Delta F}_M)$ is an unbiased estimator of $\exp(-\beta\Delta F)$:

$$\mathbb{E}\left(\mathrm{e}^{-\beta\widehat{\Delta F}_M}\right) = \mathrm{e}^{-\beta\Delta F}.$$

Therefore, the following inequality is obtained as in Remark 4.4, using the concavity of the logarithm and Jensen's inequality:

$$\mathbb{E}\left(\widehat{\Delta F}_M\right) \geq \Delta F.$$

In fact, it can be shown that the finite sampling bias of the free energy estimator (4.31) decreases with the sample size (see [Zuckerman and Woolf (2004)]).

Lemma 4.7. *The bias of estimator $\mathbb{E}(\widehat{\Delta F}_M)$ of the free energy difference is a decreasing function of M:*

$$\Delta F \leq \cdots \leq \mathbb{E}\left(\widehat{\Delta F}_{M+1}\right) \leq \mathbb{E}\left(\widehat{\Delta F}_M\right) \leq \cdots \leq \mathbb{E}\left(\widehat{\Delta F}_1\right) = \mathbb{E}(\mathcal{W}).$$

Proof. Consider the random variable

$$G_n = g^{-1}\left(\frac{1}{n}\sum_{i=1}^{n} g(Z_i)\right),$$

where the smooth function $z \in \mathbb{R} \mapsto g(z)$ is strictly decreasing and convex, and the random variables $(Z_i)_{i\geq 1}$ with values in $\mathbb{R}$ are i.i.d. according to the measure $\rho(dz)$. Note that the inverse function g^{-1} is well defined (since g is strictly decreasing), and is in fact also strictly decreasing, and convex.[2]

[2]The latter property arises from the fact that $(g^{-1})' = 1/(g' \circ g^{-1})$ is increasing (recall that $g' \circ g^{-1}$ is decreasing since g' is increasing in view of the convexity of g, while g^{-1} is decreasing because g is).

The estimator (4.31) is recovered for $g(z) = \exp(-\beta z)$ and $Z_i = \mathcal{W}^i$. Since

$$G_n = g^{-1}\left(\frac{1}{n}\sum_{i=1}^{n} g(Z_i)\right) = g^{-1}\left[\frac{1}{n}\sum_{i=1}^{n}\left(\frac{1}{n-1}\sum_{j\neq i} g(Z_j)\right)\right],$$

the convexity of g^{-1} implies

$$G_n \leq \frac{1}{n}\sum_{i=1}^{n} g^{-1}\left(\frac{1}{n-1}\sum_{j\neq i} g(Z_j)\right).$$

Denoting by $\mathbb{E}_n$ the expectation with respect to $Z_1, \dots, Z_n$, *i.e.*

$$\mathbb{E}_n\left[f(Z_1, \dots, Z_n)\right] = \int_{\mathbb{R}^n} f(z_1, \dots, z_n)\, \rho(dz_1) \dots \rho(dz_n),$$

and noticing that

$$\forall i = 1, \dots, n, \qquad \mathbb{E}_n\left[g^{-1}\left(\frac{1}{n-1}\sum_{j\neq i} g(Z_j)\right)\right] = \mathbb{E}_{n-1}(G_{n-1}),$$

it follows

$$\mathbb{E}_n(G_n) \leq \frac{1}{n}\sum_{i=1}^{n} \mathbb{E}_{n-1}(G_{n-1}) = \mathbb{E}_{n-1}(G_{n-1}),$$

which gives the result. □

Variance and finite sampling bias in the large M limit. The computations presented in this paragraph can be made rigorous using Lemma 2.24, or techniques similar to the ones used in its proof. In the large sample limit, a central limit theorem may be applied, and the expectation of an empirical mean over i.i.d. samples may be approximated as

$$\frac{1}{M}\sum_{i=1}^{M} Z_i \simeq \mathbb{E}(Z) + \frac{1}{\sqrt{M}}\sqrt{\mathrm{Var}(Z)}\, G_M^Z,$$

where $Z, Z_1, \dots, Z_M$ have the same law, $Z_1, \dots, Z_M$ being independent, and

$$G_M^Z = \frac{\sqrt{M}}{\mathrm{Var}(Z)}\left(\frac{1}{M}\sum_{i=1}^{M} Z_i - \mathbb{E}(Z)\right),$$

converges in law to a standard random Gaussian variable. Note that $\mathbb{E}(G_M^Z) = 0$. The image of the average by a (nonlinear) function is then

$$f\left(\frac{1}{M}\sum_{i=1}^{M} Z_i\right) \simeq f\Big(\mathbb{E}(Z)\Big) + \frac{1}{\sqrt{M}} f'\Big(\mathbb{E}(Z)\Big)\sqrt{\mathrm{Var}(Z)}\, G_M^Z + \frac{1}{2M} f''\Big(\mathbb{E}(Z)\Big)\,\mathrm{Var}(Z)\,(G_M^Z)^2 + \dots$$

From this formula, it is readily seen that the estimator

$$\widehat{f}_M = f\left(\frac{1}{M}\sum_{i=1}^{M} Z_i\right)$$

of $f(\mathbb{E}(Z))$ has a bias

$$\mathbb{E}\Big(\widehat{f}_M - f(\mathbb{E}(Z))\Big) \simeq \frac{1}{2M} f''\Big(\mathbb{E}(Z)\Big)\,\mathrm{Var}(Z),$$

while its variance is

$$\mathbb{E}\left[\Big(\widehat{f}_M - (\mathbb{E}(\widehat{f}_M))\Big)^2\right] \simeq \frac{1}{M} f'\Big(\mathbb{E}(Z)\Big)^2\,\mathrm{Var}(Z).$$

Corrections to the systematic bias and estimates of the variance of the estimator (4.31) can therefore be obtained upon considering $f(z) = -\beta^{-1}\ln z$ and $Z_i = \mathcal{W}^i$ (see [Zuckerman and Woolf (2002, 2004); Gore *et al.* (2003)]). At first orders in $1/M$, the estimator $\widehat{\Delta F}_M$ given by (4.31) is such that

$$\begin{aligned}\mathbb{E}\left(\widehat{\Delta F}_M\right) - \Delta F &= \frac{\mathrm{Var}(\mathrm{e}^{-\beta\mathcal{W}})}{2\beta M \mathrm{e}^{-2\beta\Delta F}} + \mathrm{O}\left(\frac{1}{M^2}\right)\\ &= \frac{\mathrm{Var}(\mathrm{e}^{-\beta\mathcal{W}_{\mathrm{diss}}})}{2\beta M} + \mathrm{O}\left(\frac{1}{M^2}\right),\end{aligned}$$

where the realization-dependent dissipated work is the random variable $\mathcal{W}_{\mathrm{diss}} = \mathcal{W} - \Delta F$, while

$$\mathrm{Var}\left(\widehat{\Delta F}_M\right) = \frac{\mathrm{Var}(\mathrm{e}^{-\beta\mathcal{W}_{\mathrm{diss}}})}{\beta^2 M} + \mathrm{O}\left(\frac{1}{M^2}\right). \qquad (4.32)$$

This shows that the asymptotic sampling error (related to the standard deviation of the estimator, *i.e.* the square root of the variance) is expected to scale as $M^{-1/2}$ in the large sample limit, and therefore dominates the finite sampling bias, which scales as M^{-1}. This situation is standard for Monte Carlo methods.

Scaling with the dimension. For nonequilibrium methods, and actually also for the free energy perturbation method described in Section 2.4.1, it is expected that the asymptotic sampling error (when considering a large number of replicas M in (4.20)) grows exponentially with the system size, or at least with the number of degrees of freedom directly affected by the alchemical parameter, as is the case for free energy perturbation (and in contrast with thermodynamic integration since in the latter case canonical averages are computed). See for instance the remark after Eq. (34) in [Jarzynski (2006)].

These heuristic intuitions can be made precise using again the analytical example studied in Section 4.1.4.2. Consider a system consisting of d independent harmonic oscillators, described by the potential energy function

$$V_\lambda(q_1, \dots, q_d) = \sum_{i=1}^{d} v_\lambda(q_i), \qquad v_\lambda(q_i) = \frac{1}{2}K(q_i - \lambda L)^2.$$

In the thermodynamic integration method, an approximation of the mean force is obtained for instance with a numerical discretization of the overdamped dynamics:

$$q^{n+1} = q^n - \Delta t\, \nabla V_\lambda(q^n) + \sqrt{\frac{2}{\beta}}\, G^n,$$

and then approximating the mean force as

$$\frac{\displaystyle\int_{\mathbb{R}^d} \partial_\lambda V_\lambda\, \mathrm{e}^{-\beta V_\lambda}}{\displaystyle\int_{\mathbb{R}^d} \mathrm{e}^{-\beta V_\lambda}} \simeq \sum_{i=1}^{d} A_i^N, \qquad A_i^N = \frac{1}{N}\sum_{n=1}^{N} \partial_\lambda v_\lambda(q_i^n)$$

where the random variables A_i^N are i.i.d. since the variables $(q_i^n)_{i=1,\dots,d}$ are i.i.d. for all $n \geq 1$. This shows that the variance of the estimator of the mean force is d times larger than the variance of the mean force in the case of a single harmonic oscillator.

For nonequilibrium switchings, the total work is the sum of individual i.i.d. works:

$$\mathcal{W}_t = \int_0^t \partial_\lambda V_{\Lambda(s)}(q_{1,s}, \dots, q_{d,s})\, \dot{\Lambda}(s)\, ds = \sum_{i=1}^{d} B_{i,t},$$

with

$$B_{i,t} = \int_0^t \partial_\lambda v_{\Lambda(s)}(q_{i,s})\, \dot{\Lambda}(s)\, ds.$$

The i.i.d. random variables $B_{i,t}$ follow a Gaussian distribution, with a mean denoted by ω_t, and a variance $2\omega_t/\beta$, see (4.28)-(4.29). Therefore, the total work $\mathcal{W}_t$ follows a Gaussian law with mean $d\omega_t$ and variance $2d\omega_t/\beta$, so that the realization-dependent dissipated work $\mathcal{W}_t^{\mathrm{diss}} = \mathcal{W}_t - d\,\omega_t$ follows a centered Gaussian law of variance $2d\omega_t/\beta$. In view of (4.32), this shows that the variance of the free energy estimator

$$-\frac{1}{\beta}\ln\left(\frac{1}{M}\sum_{m=1}^{M}\mathrm{e}^{-\beta\mathcal{W}_t^m}\right)$$

is equal, at first order in $1/M$ and up to a factor $\beta^{-2}M^{-1}$, to $\exp(4\beta d\omega_t) - \exp(2\beta d\omega_t)$, and therefore asymptotically scales exponentially with the dimension d.

4.1.5.2 *Double-sided averages*

A very important assessment of the quality of the result can be obtained by performing backward switchings. Backward switchings are related to Crooks-Jarzynski equalities, which are presented below. Here, we give only the outline of the method, and refer to Section 4.2 for more details. For a backward switching, the free energy is estimated starting at equilibrium for the canonical distribution at $\lambda = 1$, and using a reverse schedule

$$t \mapsto \Lambda^{\mathrm{b}}(t) = \Lambda(T-t),$$

thus switching the alchemical parameter backward from 1 to 0. The nonequilibrium equality (4.10) then becomes:

$$\mathbb{E}\left[\mathrm{e}^{-\beta\mathcal{W}_T^{\mathrm{b}}}\right] = \mathrm{e}^{\beta\Delta F},$$

where $\mathcal{W}_t^{\mathrm{b}}$ is the exchanged work during the backward switching, and ΔF is still the free energy difference $F(\Lambda(T)) - F(\Lambda(0)) = -[F(\Lambda^{\mathrm{b}}(T)) - F(\Lambda^{\mathrm{b}}(0))]$. In practice, this can be done only when it is possible to sample the canonical distribution at $\lambda = 1$.

Bridge estimators in the spirit of the Bennett acceptance method (see Section 2.5) can then be used to tentatively improve the quality of the estimates, see Section 4.2.3. In any case, a rough comparison of forward and reverse work distributions should be attempted. If the distributions do not overlap, it is likely that the switching process is so out of equilibrium that the values predicted by the estimators cannot be trusted.

Remark 4.8 (Reverse or forward switching?). *If only one work distribution can be sampled, it seems that only the distribution with the larger* *dissipated work*[3] *should be sampled [Jarzynski (2006)] – although some authors suggest to sample the distribution for which* $e^{-\beta \mathcal{W}}$ *has the lowest variance (recall that a larger dissipated work is often associated with a broader work distribution and so, a larger variance* $\mathrm{Var}(e^{-\beta \mathcal{W}})$*).*

4.1.5.3 *Influence of the parameters*

Influence of the switching schedule. In general, the accuracy of the results strongly depends on the chosen schedule $t \mapsto \Lambda(t)$. The heuristic idea is that the switching should be sufficiently slow in regions where the free energy varies rapidly. In [Atilgan and Sun (2004)], the authors suggest to average works over generalized ensembles where both nonequilibrium trajectories and switching schedules are sampled. Some recent studies suggest to optimize the switching path in order to minimize the dissipated work [Schmiedl and Seifert (2007)]. However, we are not aware of any work studying carefully the influence of the switching schedule on the accuracy of the free energy estimator.

Influence of the dynamics. A point that is often overlooked in nonequilibrium switching simulations is how the results depend on the dynamics used to generate the switching paths (Langevin dynamics, Hamiltonian dynamics, overdamped Langevin dynamics, etc.). This possibly biases the efficiency comparisons between results obtained from nonequilibrium switching and those obtained with other methods.

We believe that it is difficult to give a general answer to the question. Let us nevertheless study one specific example, namely the 2D potential (1.66) at inverse temperature $\beta = 5$, with a linear switching schedule such that $\Lambda(0)$ and $\Lambda(T)$ correspond respectively to $x = -1.2$ and $x = 0$. The switching time is $T = 2$, and trajectories are integrated with a time-step $\Delta t = 0.001$. The initial conditions are generated from an undersampled Langevin dynamics, where configurations are retained at times separated by $T_{\mathrm{sample}} = 1$. The results are presented in Table 4.1 for different sample sizes M, and the variance of the results is estimated by running N_{runs} independent realizations of the whole procedure.

The results show that the overdamped Langevin dynamics performs better than Langevin and Hamiltonian dynamics since the free energy estimates display a lower variability (the variance over the runs is noticeably

[3]The question being, of course, which one has the largest dissipated work when only one of the distributions can be sampled?

Table 4.1 Comparison of the free energy estimators for different dynamics. The bias and the standard deviation of the estimators are computed using N_{runs} independent runs (in particular, different initial conditions are sampled each time). For each run, M trajectories are generated, with a switching time $T = 2$ and a time-step $\Delta t = 0.001$. The results are presented as "average (standard deviation)". The reference free energy difference has been computed using standard numerical integration techniques: $\Delta F = 0.9186$.

M	N_{runs}	Overdamped Langevin	Langevin ($\gamma = 1$)	Hamiltonian
10^3	10^4	0.923 (0.044)	0.968 (0.120)	0.936 (0.080)
10^4	10^3	0.918 (0.016)	0.930 (0.066)	0.921 (0.034)
10^5	10^3	0.917 (0.006)	0.920 (0.028)	0.917 (0.012)
10^6	10^2	0.917 (0.002)	0.919 (0.010)	0.917 (0.005)

smaller), and are closer to the analytical results for small sample sizes.

This is confirmed by the typical work distributions obtained from these switchings (see Figure 4.1): In the overdamped case, the dissipated work is lower than in the Hamiltonian case, and the peak of the work distribution is closer to the reference value of the free energy. The explanation of the different behaviors is understood by looking at typical switching trajectories displayed in Figure 4.2: The inertia of the Langevin and Hamiltonian dynamics may have dramatic consequences for unfortunate choices of the initial velocity, the system being sent in those cases to high energy regions.

Let us however emphasize that these results hold only for the parameters and the system considered here. Other studies, such as [Athènes (2004)], find the Langevin dynamics more efficient than the overdamped dynamics for the problems considered. In any case, the hierarchy may be different for longer switching times T.

4.1.5.4 *Numerical results for Widom insertion*

We now present some results for Widom insertion, where the nonequilibrium switching is performed with Langevin dynamics. The parameters of the system and of the dynamics are the same as in Section 2.4.1.1 (in particular, $\Delta t = 0.005$ and $\gamma = 1$). The initial conditions are obtained by subsampling a Langevin trajectory at $\lambda = 0$, the time spacing being $T_{\mathrm{sample}} = 1$.

In order to determine the switching schedule to use, we perform some preliminary computations for a given switching time $T = 1$ and a given number of replicas $M = 1000$. The proposed switchings have the functional shape $\Lambda(t) = (t/T)^{\alpha}$. The results are presented in Table 4.2. A comparison

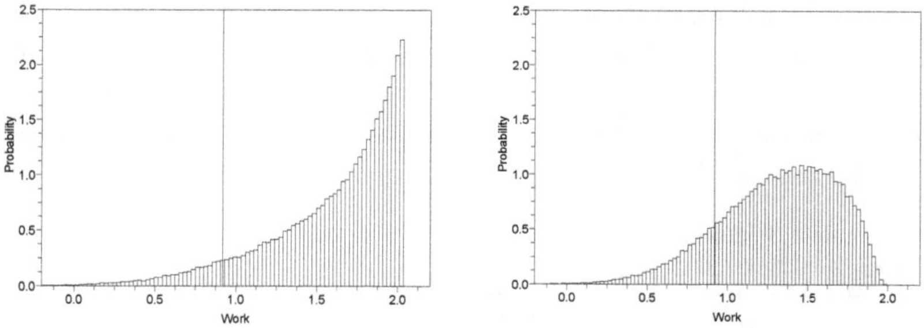

Fig. 4.1 Distributions of work for a single realization of the switching procedure ($T = 2$, $\Delta t = 0.001$) with $M = 10^5$ trajectories. Left: Hamiltonian dynamics. Right: Overdamped Langevin dynamics. The vertical line gives the reference value of the free energy difference in both cases.

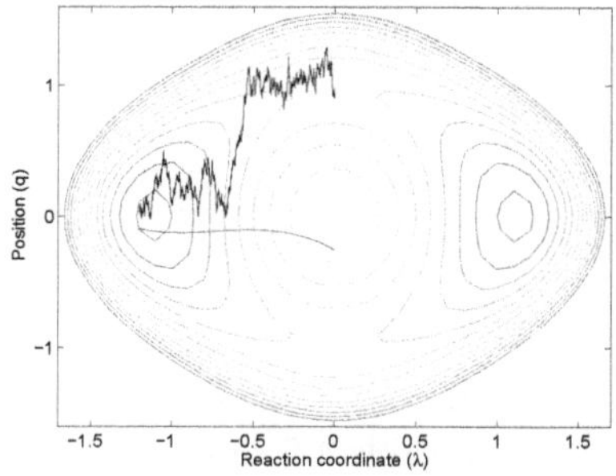

Fig. 4.2 Plot of a single trajectory for a switching performed with the overdamped Langevin dynamics (upper path) and Hamiltonian dynamics (lower path).

Table 4.2 Estimates of $\mu^{\rm ex}$ for different switching schedules $\Lambda(t) = (t/T)^\alpha$. The reference value is $\mu^{\rm ex} = 1.317$.

Parameter α	1	2	3	4
Average	1.36	1.33	1.38	1.35
Standard deviation	0.20	0.19	0.18	0.19

with the reference value given in Section 2.4.1.1, shows that the choice $\Lambda(t) = (t/T)^2$ is better than the plain linear switching schedule for which the interaction of the inserted particle is switched on too abruptly. On the other hand, slowing down further the initial transition ($\alpha = 3$ or 4) is not a good idea since the computed average estimate of the free energy difference departs from the reference value.

We then study the influence of the switching time T and number of replicas M for the schedule $\Lambda(t) = (t/T)^2$, see Tables 4.3 and 4.4. As expected, the finite sampling bias and the variance of the estimator decrease as M or T increases. In order to discuss the trade-off between an increase

Table 4.3 Estimates of μ^{ex} for $T = 1$. The reference value is $\mu^{ex} = 1.317$.

M	100	300	600	1000	3,000	10,000
Average	1.46	1.38	1.36	1.33	1.33	1.33
Standard deviation	0.64	0.35	0.24	0.19	0.10	0.06

Table 4.4 Estimates of μ^{ex} for $M = 1000$. The reference value is $\mu^{ex} = 1.317$.

Time T	0.1	0.3	0.6	1	3	10
Average	1.44	1.39	1.40	1.33	1.33	1.32
Standard deviation	0.38	0.29	0.26	0.19	0.11	0.06

in the number of replicas and longer switching times, we also performed numerical experiments at fixed computational cost, *i.e.* MT fixed in the present context. The results of Table 4.5 show that the accuracies are quite comparable.

Table 4.5 Estimates of μ^{ex} for a fixed computational cost $MT = 1000$. The reference value is $\mu^{ex} = 1.317$.

Time T	0.1	0.25	0.5	1	2	5	10
M	10000	4000	2000	1000	500	200	100
Average	1.38	1.36	1.35	1.33	1.38	1.33	1.35
Standard deviation	0.12	0.15	0.19	0.19	0.19	0.18	0.19

4.2 Generalized Jarzynski-Crooks fluctuation identity

We prove in this section a fluctuation identity more general than (4.9), called Crooks identity. This equality was proved for Monte Carlo dynamics of Metropolis-Hastings type in [Crooks (1998, 1999)], and we show here how it can be extended to more general nonequilibrium (deterministic or stochastic) processes (see [Lelièvre *et al.* (2010)]). The Crooks equality relates the free energy difference and work values from a switching process and its backward version (a notion to be defined below) when these two processes start at equilibrium. The proof relies on the fundamental balance condition (4.43) stated below. In order to be more pedagogical, we consider an alchemical transition as a running example. We emphasize however that the results given in this section hold in a very general framework, and will be very useful for nonequilibrium switching dynamics in the reaction coordinate case (see Section 4.3).

4.2.1 *Derivation of the identity*

Evolving measures. Consider a family $t \mapsto \Pi_t$ of non-normalized Boltzmann-Gibbs distributions on $\mathcal{S} = \mathcal{D}$ (when only positions q are considered) or $\mathcal{S} = T^*\mathcal{D}$ (when a phase space dynamics is considered); and denote the associated normalization by

$$Z_t = \int_{\mathcal{S}} d\Pi_t.$$

We use the notation of Section 4.1 for the alchemical case, except that we now consider the non-normalized Boltzmann-Gibbs distributions

$$\Pi_t(dx) = \mathrm{e}^{-\beta E_{\Lambda(t)}(x)}\, dx.$$

For phase space processes, $\Pi_t(dx) = \mathrm{e}^{-\beta H_{\Lambda(t)}(q,p)}\, dq\, dp$, while, for processes on the position q only, $\Pi_t(dx) = \mathrm{e}^{-\beta V_{\Lambda(t)}(q)}\, dq$. Note that, for any test function φ,

$$\frac{d}{dt}\left(\int_{\mathcal{S}} \varphi(x)\, \Pi_t(dx)\right) = -\beta \dot{\Lambda}(t) \int_{\mathcal{S}} \varphi\, \partial_\lambda E_{\Lambda(t)}\, d\Pi_t. \tag{4.33}$$

Definition of the dynamics. Consider two (deterministic or stochastic) Markov processes X_t^{f} and $X_{t'}^{\mathrm{b}}$, with associated generators $\mathcal{L}_t^{\mathrm{f}}$ and $\mathcal{L}_{t'}^{\mathrm{b}}$, evolving in $\mathcal{S}$, and starting at equilibrium (a notion which will be precised below). These processes are both considered over the time-interval $[0, T]$, and the time t' in $X_{t'}^{\mathrm{b}}$ is related to the time t of the process X_t^{f} as

$$t' = T - t.$$

The process $X_{t'}^{\mathrm{b}}$ is related to the time-reversed process X_{T-t}^{f}, and will therefore be referred to as the *backward* process in the sequel.

In the alchemical case, a first example of forward and backward dynamics can be given for the overdamped Langevin dynamics: the forward process $X_t^{\rm f} = q_t^{\rm f}$ is characterized by $q_0^{\rm f} \sim Z_0^{-1}\nu_0$ and

$$dq_t^{\rm f} = -\nabla V_{\Lambda(t)}\left(q_t^{\rm f}\right) dt + \sqrt{\frac{2}{\beta}}\, dW_t^{\rm f}, \tag{4.34}$$

and the associated backward dynamics is obtained for a switching function $t' \mapsto \Lambda(T-t')$:

$$dq_{t'}^{\rm b} = -\nabla V_{\Lambda(T-t')}\left(q_{t'}^{\rm b}\right) dt' + \sqrt{\frac{2}{\beta}}\, dW_{t'}^{\rm b}, \tag{4.35}$$

with initial condition $q_0^{\rm b} \sim Z_T^{-1}\nu_T$. The generator $\mathcal{L}_{t'}^{\rm b}$ of the backward dynamics is in this case related to the generator of the forward dynamics as

$$\mathcal{L}_{t'}^{\rm b} = \mathcal{L}_{T-t'}^{\rm f}.$$

Another example is the Langevin dynamics: the forward evolution $X_t^{\rm f} = (q_t^{\rm f}, p_t^{\rm f})$, whose generator is denoted by $\mathcal{L}_t^{\rm f}$ (see Section 2.2.3.1), is characterized by $(q_0^{\rm f}, p_0^{\rm f}) \sim Z_0^{-1}\mu_0$, and

$$\begin{cases} dq_t^{\rm f} = M^{-1} p_t^{\rm f}\, dt, \\ dp_t^{\rm f} = -\nabla V_{\Lambda(t)}\left(q_t^{\rm f}\right) dt - \gamma\, M^{-1} p_t^{\rm f}\, dt + \sigma\, dW_t^{\rm f}, \end{cases} \tag{4.36}$$

with $\sigma\sigma^T = 2\,\beta^{-1}\gamma$. The associated backward dynamics is, for a given initial condition $(q_0^{\rm b}, p_0^{\rm b}) \sim Z_T^{-1}\mu_T$,

$$\begin{cases} dq_{t'}^{\rm b} = -M^{-1} p_{t'}^{\rm b}\, dt', \\ dp_{t'}^{\rm b} = \nabla V_{\Lambda(T-t')}\left(q_{t'}^{\rm b}\right) dt' - \gamma\, M^{-1} p_{t'}^{\rm b}\, dt' + \sigma\, dW_{t'}^{\rm b}. \end{cases} \tag{4.37}$$

This amounts to saying that $(q_{t'}^{\rm b}, -p_{t'}^{\rm b})$ is a solution (in law) of the forward evolution (4.36) with the schedule $t' \mapsto \Lambda(T-t')$. The backward dynamics may also be understood as a dynamics of generator $\mathcal{L}_{T-t'}^{\rm f}$ (a forward Langevin dynamics with a time-reversed switching function $t' \mapsto \Lambda(T-t')$) but where momenta are first reversed in the initial condition, and then are reversed back after the time evolution. The generator $\mathcal{L}_{t'}^{\rm b}$ of the backward dynamics is thus related to the generator of the forward dynamics as

$$\mathcal{L}_{t'}^{\rm b} = \mathcal{R}\, \mathcal{L}_{T-t'}^{\rm f}\, \mathcal{R},$$

where $\mathcal{R}$ denotes the momentum flip operator $\mathcal{R} : \phi \mapsto \phi \circ S$ with $S(q,p) = (q, -p)$.

Definition of the works. We associate work functions to each of the processes X_t^{f} and $X_{t'}^{\mathrm{b}}$, which are integrals of some instantaneous work function. More precisely, consider two instantaneous work functions w^{f} and w^{b}, from $[0,T] \times \mathcal{S}$ to $\mathbb{R}$, related through the identity:

$$w^{\mathrm{b}}(t',\cdot) = -w^{\mathrm{f}}(T-t',\cdot). \tag{4.38}$$

The work associated with the forward process $\{X_t^{\mathrm{f}}\}_{0\le t\le T}$ is

$$\mathcal{W}_{t,T}^{\mathrm{f}} = \int_t^T w^{\mathrm{f}}(s, X_s^{\mathrm{f}})\, ds,$$

while the work associated with the backward process $\{X_{t'}^{\mathrm{b}}\}_{0\le t'\le T}$ is

$$\mathcal{W}_{t',T}^{\mathrm{b}} = \int_{t'}^T w^{\mathrm{b}}(s', X_{s'}^{\mathrm{b}})\, ds'.$$

For example, in the alchemical case, considering the time-reversed switching $\Lambda^{\mathrm{b}}(t') = \Lambda(T-t')$, the works are defined as

$$\mathcal{W}_{t,T}^{\mathrm{f}}\Big(\{X_s^{\mathrm{f}}\}_{0\le s\le T}\Big) = \int_t^T \partial_\lambda E_{\Lambda(s)}(X_s^{\mathrm{f}})\, \dot{\Lambda}(s)\, ds, \tag{4.39}$$

and

$$\mathcal{W}_{t',T}^{\mathrm{b}}\Big(\{X_{s'}^{\mathrm{b}}\}_{0\le s'\le T}\Big) = \int_{t'}^T \partial_\lambda E_{\Lambda^{\mathrm{b}}(s')}(X_{s'}^{\mathrm{b}})\, \dot{\Lambda}^{\mathrm{b}}(T-s')\, ds', \tag{4.40}$$

$$= -\int_{t'}^T \partial_\lambda E_{\Lambda(T-s')}(X_{s'}^{\mathrm{b}})\, \dot{\Lambda}(T-s')\, ds', \tag{4.41}$$

and they correspond to

$$w^{\mathrm{f}}(s,x) = \partial_\lambda E_{\Lambda(s)}(x)\, \dot{\Lambda}(s), \tag{4.42}$$

$$w^{\mathrm{b}}(s',x) = -\partial_\lambda E_{\Lambda(T-s')}(x)\, \dot{\Lambda}(T-s').$$

The relation (4.38) is indeed satisfied.

Generalized detailed balance condition. The Crooks-Jarzynski fluctuation identity (see (4.46) below) relies on a generalized detailed balance condition, which is a nonequilibrium version of the usual detailed balance property (2.30): for any smooth test functions φ_1 and φ_2, and for any $t \in [0,T]$,

$$\begin{aligned}&\int_{\mathcal{S}} \varphi_2\, \mathcal{L}_t^{\mathrm{f}}(\varphi_1)\, d\Pi_t - \int_{\mathcal{S}} \varphi_1\, \mathcal{L}_{T-t}^{\mathrm{b}}(\varphi_2)\, d\Pi_t \\ &\quad = \frac{d}{dt}\left(\int_{\mathcal{S}} \varphi_1\, \varphi_2\, d\Pi_t\right) + \beta \int_{\mathcal{S}} \varphi_1\, \varphi_2\, w^{\mathrm{f}}(t,\cdot)\, d\Pi_t.\end{aligned} \tag{4.43}$$

Remark 4.9 (Equivalent reformulation). *In view of* (4.38), *the generalized detailed balance condition* (4.43) *is equivalent to the symmetric backward version:*

$$\int_{\mathcal{S}} \varphi_2 \, \mathcal{L}^{\mathrm{b}}_{t'}(\varphi_1) \, d\Pi_{T-t'} - \int_{\mathcal{S}} \varphi_1 \, \mathcal{L}^{\mathrm{f}}_{T-t'}(\varphi_2) \, d\Pi_{T-t'}$$
$$= \frac{d}{dt'} \left(\int_{\mathcal{S}} \varphi_1 \, \varphi_2 \, d\Pi_{T-t'} \right) + \beta \int_{\mathcal{S}} \varphi_1 \, \varphi_2 \, w^{\mathrm{b}}(t', \cdot) \, d\Pi_{T-t'}.$$

The equality (4.43) indeed holds for the running example of alchemical transition, *i.e.* for the dynamics (4.34)-(4.35) or (4.36)-(4.37). Actually, in this case, the left-hand side in (4.43) vanishes since the times in the forward and backward evolutions are matched, and because of the equilibrium detailed balance condition expressing the reversibility of the dynamics (possibly up to momentum reversal), see Section 2.2.2.1 for overdamped Langevin dynamics, and Section 2.2.3.1 for Langevin dynamics. In view of the expression (4.33) of the time derivative of the normalization, and the definition of the forward work (4.42), the right-hand side of (4.43) vanishes as well. The fact that the right-hand side and the left-hand side of (4.43) are both equal to 0 is however not general (see the dynamics in Section 4.3).

Generalized work fluctuation identity. We are now in position to state the Crooks-Jarzynski identity, which involves path functionals. A path functional is an application from the set of continuous functions on $[0, T]$ (equipped with the supremum norm) with values in $\mathbb{R}$.

It is convenient, in the proof, to introduce the weighted transition operators

$$P^{\mathrm{f}}_{t,T}(\varphi)(x) = \mathbb{E}\Big(\varphi(X^{\mathrm{f}}_T) \, \mathrm{e}^{-\beta \mathcal{W}^{\mathrm{f}}_{t,T}} \,\Big|\, X^{\mathrm{f}}_t = x \Big), \tag{4.44}$$

$$P^{\mathrm{b}}_{t',T}(\varphi)(x) = \mathbb{E}\Big(\varphi(X^{\mathrm{b}}_T) \,\Big|\, X^{\mathrm{b}}_{t'} = x \Big). \tag{4.45}$$

We assume in the sequel that, for all $t \in [0, T]$, and all bounded smooth test function φ, the transition semi-group of the forward dynamics $P^{f}_{t,T}(\varphi)$ is smooth and well defined in an open neighborhood of the support of Π_t, and, for all $t' \in [0, T]$, the transition semi-group of the backward dynamics $P^{b}_{t',T}(\varphi)$ is smooth and well defined in an open neighborhood of the support of $\Pi_{t'}$.

Theorem 4.10. *Consider a forward process* X^{f}_t, *starting from* X^{f}_0 *distributed according to the equilibrium probability measure* $Z_0^{-1} \, d\Pi_0$, *and a backward process* $X^{\mathrm{b}}_{t'}$, *starting from* X^{b}_0 *distributed according to* $Z_T^{-1} \, d\Pi_T$.

Assume that for any test functions φ_1, φ_2, and all $t \in [0,T]$, the generalized detailed balance condition (4.43) is satisfied.

Then, the following Crooks-Jarzynski equality holds for any $\theta \in [0,1]$:

$$\frac{Z_T}{Z_0}\,\mathbb{E}\left(\varphi^{\mathrm{r}}_{[0,T]}(X^{\mathrm{b}})\,\mathrm{e}^{-\beta\theta\mathcal{W}^{\mathrm{b}}_{0,T}}\right) = \mathbb{E}\left(\varphi_{[0,T]}(X^{\mathrm{f}})\,\mathrm{e}^{-\beta(1-\theta)\mathcal{W}^{\mathrm{f}}_{0,T}}\right), \qquad (4.46)$$

where $\varphi_{[0,T]}$ is a bounded measurable path functional, and $\varphi^{\mathrm{r}}_{[0,T]}$ denotes the composition of $\varphi_{[0,T]}$ with the operation of time-reversal on paths:

$$\varphi^{\mathrm{r}}_{[0,T]}\Big(\{X_{t'}\}_{0\le t'\le T}\Big) = \varphi_{[0,T]}\Big(\{X_{T-t}\}_{0\le t\le T}\Big). \qquad (4.47)$$

As a consequence, the Jarzynski equality (4.10) is recovered with the choice $\varphi_{[0,T]} = 1$ and $\theta = 0$, while the fluctuation identity (4.9) corresponds to $\varphi_{[0,T]}(X) = \phi(X_T)$ and $\theta = 0$.

Remark 4.11 (On the parameter θ in (4.46)). *The result for an arbitrary parameter $\theta \in [0,1]$ can be deduced from (4.46) in the case $\theta = 0$ upon considering the path functional $\widetilde{\varphi} = \varphi_{[0,T]}\,\exp(\theta\beta\mathcal{W}^{\mathrm{f}}_{0,T})$. Indeed, using (4.38),*

$$\begin{aligned}\left(\mathcal{W}^{\mathrm{f}}_{0,T}\right)^{\mathrm{r}}(X) &= \int_0^T w^{\mathrm{f}}(s, X_{T-s})\,ds = \int_0^T w^{\mathrm{f}}(T-t, X_t)\,dt \\ &= -\int_0^T w^{\mathrm{b}}(t, X_t)\,dt = -\mathcal{W}^{\mathrm{b}}_{0,T}(X),\end{aligned}$$

so that the time-reversed path functional is $\widetilde{\varphi}^{\mathrm{r}} = \varphi^{\mathrm{r}}_{[0,T]}\,\exp(-\theta\beta\mathcal{W}^{\mathrm{b}}_{0,T})$.

Proof. By Remark 4.11, it is enough to show the result for $\theta = 0$. The transition operators (4.44)-(4.45) satisfy the following backward Kolmogorov evolution equations (see the proof of Proposition 4.1):

$$\begin{cases} \partial_t P^{\mathrm{f}}_{t,T} = -\mathcal{L}^{\mathrm{f}}_t P^{\mathrm{f}}_{t,T} + \beta\, w^{\mathrm{f}}(t,\cdot)\, P^{\mathrm{f}}_{t,T}, \\ P^{\mathrm{f}}_{T,T} = \mathrm{Id}, \end{cases} \qquad \begin{cases} \partial_{t'} P^{\mathrm{b}}_{t',T} = -\mathcal{L}^{\mathrm{b}}_{t'} P^{\mathrm{b}}_{t',T}, \\ P^{\mathrm{b}}_{T,T} = \mathrm{Id}. \end{cases}$$

When deterministic processes are considered, the initial condition completely determines the future evolution, and no expectation has to be taken in (4.44)-(4.45). The evolution equations are however unchanged.

Consider two test functions φ_0 and φ_T, from $\mathcal{S}$ to $\mathbb{R}$. The balance condition (4.43) implies

$$\frac{d}{dt}\left(\int_{\mathcal{S}} P^{\mathrm{f}}_{t,T}(\varphi_T)\, P^{\mathrm{b}}_{T-t,T}(\varphi_0)\, d\Pi_t\right) = 0.$$

Integrating this equality on $[0, T]$ yields

$$\int_{\mathcal{S}} P^{\mathrm{f}}_{0,T}(\varphi_T)\, \varphi_0 \, d\Pi_0 = \int_{\mathcal{S}} \varphi_T \, P^{\mathrm{b}}_{0,T}(\varphi_0)\, d\Pi_T, \tag{4.48}$$

which is the Crooks identity (4.46) for $\theta = 0$ and path functionals of the form

$$\varphi_{[0,T]}(X) = \varphi_0(X_0)\, \varphi_T(X_T).$$

Indeed,

$$\begin{aligned}\int_{\mathcal{S}} P^{\mathrm{f}}_{0,T}(\varphi_T)\, \varphi_0 \, d\Pi_0 &= Z_0 \, \mathbb{E}\Big[\varphi_0(X^{\mathrm{f}}_0)\, \varphi_T(X^{\mathrm{f}}_T)\, \mathrm{e}^{-\beta \mathcal{W}^{\mathrm{f}}_{0,T}}\Big] \\ &= Z_0 \, \mathbb{E}\Big[\varphi_{[0,T]}(X^{\mathrm{f}})\, \mathrm{e}^{-\beta \mathcal{W}^{\mathrm{f}}_{0,T}}\Big],\end{aligned}$$

while

$$\int_{\mathcal{S}} \varphi_T \, P^{\mathrm{b}}_{0,T}(\varphi_0)\, d\Pi_T = Z_T \, \mathbb{E}\Big[\varphi_T(X^{\mathrm{b}}_0)\, \varphi_0(X^{\mathrm{b}}_T)\Big] = Z_T \, \mathbb{E}\Big[\varphi^{\mathrm{r}}_{[0,T]}(X^{\mathrm{b}})\Big].$$

Similar computations show that (4.48) can be extended to any subinterval $[t_k, t_{k+1}] \subset [0, T]$ for path functionals of the form $\varphi_{[t_k,t_{k+1}]}(X) = \varphi_k(X_{t_k})\, \varphi_{k+1}(X_{t_{k+1}})$. On this subinterval, $t \in [t_k, t_{k+1}]$ while $t' = T - t \in [T - t_{k+1}, T - t_k]$. This suggests to consider the following quantity:

$$\frac{d}{dt}\left(\int_{\mathcal{S}} P^{\mathrm{f}}_{t,t_{k+1}}(\varphi_{k+1})\, P^{\mathrm{b}}_{T-t,T-t_k}(\varphi_k)\, d\Pi_t\right) = 0$$

where we again used the balance condition (4.43). An integration over the time interval $[t_k, t_{k+1}]$ gives

$$\int_{\mathcal{S}} P^{\mathrm{f}}_{t_k,t_{k+1}}(\varphi_{k+1})\, \varphi_k \, d\Pi_{t_k} = \int_{\mathcal{S}} \varphi_{k+1} \, P^{\mathrm{b}}_{T-t_{k+1},T-t_k}(\varphi_k)\, d\Pi_{t_{k+1}}. \tag{4.49}$$

Then, (4.48) can be extended to finite-dimensional path functionals of the form:

$$\varphi_{[0,T]}(X) = \varphi_0(X_0) \ldots \varphi_k(X_{t_k}) \ldots \varphi_K(X_T) \tag{4.50}$$

with $0 = t_0 < t_1 \cdots < t_K = T$ by repeatedly using (4.49) on time subintervals $[t_k, t_{k+1}]$. This allows us to conclude since finite dimensional time marginal laws characterize the distribution on continuous paths, see for instance [Ethier and Kurtz (1986)].

Let us detail the proof of the extension to functionals (4.50) in the case

$$\varphi_{[0,T]}(X) = \varphi_0(X_0)\varphi_\tau(X_\tau)\varphi_T(X_T)$$

with $0 < \tau < T$. First, note that, for $t_1 < t_2$,

$$\varphi_1(x)\, P^{\mathrm{f}}_{t_1,t_2}\varphi_2(x) = \mathbb{E}\left[\varphi_1(X^{\mathrm{f}}_{t_1})\, \varphi_2(X^{\mathrm{f}}_{t_2})\, \mathrm{e}^{-\beta \mathcal{W}^{\mathrm{f}}_{t_1,t_2}} \,\middle|\, X^{\mathrm{f}}_{t_1} = x\right]$$

so that the expectation over forward trajectories can be rewritten as

$$\begin{aligned}
\mathbb{E}\left[\varphi_{[0,T]}(X^{\mathrm{f}})\,\mathrm{e}^{-\beta\mathcal{W}^{\mathrm{f}}_{0,T}}\right] &= \mathbb{E}\left[\varphi_0(X^{\mathrm{f}}_0)\varphi_\tau(X^{\mathrm{f}}_\tau)\varphi_T(X^{\mathrm{f}}_T)\,\mathrm{e}^{-\beta\mathcal{W}^{\mathrm{f}}_{0,T}}\right] \\
&= \mathbb{E}\left[\varphi_0(X^{\mathrm{f}}_0)\,\mathrm{e}^{-\beta\mathcal{W}^{\mathrm{f}}_{0,\tau}}\,\varphi_\tau(X^{\mathrm{f}}_\tau)\mathbb{E}\left(\varphi_T(X^{\mathrm{f}}_T)\,\mathrm{e}^{-\beta\mathcal{W}^{\mathrm{f}}_{\tau,T}}\,\middle|\,X^{\mathrm{f}}_\tau\right)\right] \\
&= \mathbb{E}\left[\varphi_0(X^{\mathrm{f}}_0)\,\mathrm{e}^{-\beta\mathcal{W}^{\mathrm{f}}_{0,\tau}}\varphi_\tau(X^{\mathrm{f}}_\tau)P^{\mathrm{f}}_{\tau,T}\,\varphi_T(X^{\mathrm{f}}_\tau)\right] \\
&= \mathbb{E}\left[\varphi_0(X^{\mathrm{f}}_0)P^{\mathrm{f}}_{0,\tau}\left(\varphi_\tau P^{\mathrm{f}}_{\tau,T}\,\varphi_T\right)(X^{\mathrm{f}}_0)\right] \\
&= \frac{1}{Z_0}\int_{\mathcal{S}}\varphi_0(x)P^{\mathrm{f}}_{0,\tau}\left(\varphi_\tau P^{\mathrm{f}}_{\tau,T}\,\varphi_T\right)(x)\,\Pi_0(dx).
\end{aligned}\tag{4.51}$$

Now, using (4.49) twice,

$$\begin{aligned}
\int_{\mathcal{S}}\varphi_0\,P^{\mathrm{f}}_{0,\tau}\left(\varphi_\tau P^{\mathrm{f}}_{\tau,T}\,\varphi_T\right)d\Pi_0 &= \int_{\mathcal{S}}\left(\varphi_\tau P^{\mathrm{f}}_{\tau,T}\,\varphi_T\right)P^{\mathrm{b}}_{T-\tau,T}\varphi_0\,d\Pi_\tau \\
&= \int_{\mathcal{S}}\left(P^{\mathrm{f}}_{\tau,T}\,\varphi_T\right)\left(\varphi_\tau P^{\mathrm{b}}_{T-\tau,T}\varphi_0\right)d\Pi_\tau \\
&= \int_{\mathcal{S}}\varphi_T\,P^{\mathrm{b}}_{0,T-\tau}\left(\varphi_\tau P^{\mathrm{b}}_{T-\tau,T}\varphi_0\right)d\Pi_T,
\end{aligned}$$

and the last expression is equal to $Z_T\,\mathbb{E}\left(\varphi^{\mathrm{r}}_{[0,T]}(X^{\mathrm{b}})\right)$ by computations similar to the ones performed above to rewrite $\mathbb{E}\left[\varphi_{[0,T]}(X^{\mathrm{f}})\,\mathrm{e}^{-\beta\mathcal{W}^{\mathrm{f}}_{0,T}}\right]$ as (4.51). More generally, for $0 \le t_0 < t_1 \cdots < t_K \le T$,

$$\int_{\mathcal{S}}\varphi_0 P^{\mathrm{f}}_{t_0,t_1}\left[\varphi_1\cdots P^{\mathrm{f}}_{t_k,t_{k+1}}\left(\varphi_{k+1}\cdots P^{\mathrm{f}}_{t_{K-1},t_K}(\varphi_K)\right)\right]d\Pi_{t_0} =$$

$$\int_{\mathcal{S}}\varphi_K P^{\mathrm{b}}_{T-t_K,T-t_{K-1}}\left[\varphi_{K-1}\cdots P^{\mathrm{b}}_{T-t_k,T-t_{k-1}}\left(\varphi_{k-1}\cdots P^{\mathrm{b}}_{T-t_1,T-t_0}(\varphi_0)\right)\right]d\Pi_{t_K},$$

which proves the result once the integrals have been rewritten as expectations of path functionals. □

4.2.2 *Relationship with standard equalities in the physics and chemistry literature*

It is instructive to consider applications of the general work fluctuation identity (4.46) in the alchemical case. First, the standard Jarzynski fluctuation identity (4.9) is recovered in the case when $\theta = 0$ in (4.46) (the weight is completely borne by the forward switching process) and for a path functional $\varphi_{[0,T]}(X) = \phi(X_T)$ (see also the comment after Theorem 4.10).

Besides, the standard Crooks relation is often stated in terms of work distributions in the physics literature. Consider path functionals depending only on the work (4.39):

$$\varphi_{[0,T]}(X) = g\left(\mathcal{W}^{\mathrm{f}}_{0,T}(X)\right) = g\left(\int_0^T \partial_\lambda E_{\Lambda(s)}(X_s)\,\dot{\Lambda}(s)\,ds\right) \tag{4.52}$$

for a given function g. As already mentioned in Remark 4.11, $(\mathcal{W}^{\mathrm{f}}_{0,T})^{\mathrm{r}} = -\mathcal{W}^{\mathrm{b}}_{0,T}$, so that

$$\left(g\left(\mathcal{W}^{\mathrm{f}}_{0,T}\right)\right)^{\mathrm{r}} = g\left(-\mathcal{W}^{\mathrm{b}}_{0,T}\right).$$

Denote by $p^{\mathrm{f}}_T(dW)$ the law of the random variable $\mathcal{W}^{\mathrm{f}}_{0,T}(X^{\mathrm{f}})$ and by $p^{\mathrm{b}}_T(dW)$ the law of the random variable $\mathcal{W}^{\mathrm{b}}_{0,T}(X^{\mathrm{b}})$. The equality (4.46) for the path functional (4.52):

$$\frac{Z_T}{Z_0}\,\mathbb{E}\left[\left(g\left(\mathcal{W}^{\mathrm{f}}_{0,T}\right)\right)^{\mathrm{r}}(X^{\mathrm{b}})\,\mathrm{e}^{-\beta\theta\mathcal{W}^{\mathrm{b}}_{0,T}}\right] = \mathbb{E}\left[g\left(\mathcal{W}^{\mathrm{f}}_{0,T}(X^{\mathrm{f}})\right)\,\mathrm{e}^{-\beta(1-\theta)\mathcal{W}^{\mathrm{f}}_{0,T}}\right]$$

can now be rewritten as

$$\mathrm{e}^{-\beta\Delta F(T)}\int_{\mathbb{R}} g(-W)\,\mathrm{e}^{-\beta\theta W}\,p^{\mathrm{b}}_T(dW) = \int_{\mathbb{R}} g(W)\,\mathrm{e}^{-\beta(1-\theta)W}\,p^{\mathrm{f}}_T(dW)$$

with $\exp(-\beta\Delta F(T)) = Z_T/Z_0$. Assuming that the work distributions p^{f}_T and p^{b}_T are absolutely continuous with respect to the Lebesgue measure dW, and denoting (with an abuse of notation) by $p^{\mathrm{f}}_T(W)$ and $p^{\mathrm{b}}_T(W)$ their densities, it finally holds

$$\mathrm{e}^{-\beta\Delta F(T)}\,p^{\mathrm{b}}_T(-W) = \mathrm{e}^{-\beta W}\,p^{\mathrm{f}}_T(W). \tag{4.53}$$

Again, the θ variable plays no role (as expected from Remark 4.11). The remarkable equality (4.53) has several consequences. First, it gives two free energy estimators depending on whether only forward or backward paths are sampled. These estimators are obtained from the following equalities, which are derived from (4.53) by integration:

$$\mathrm{e}^{-\beta\Delta F(T)} = \int_{\mathbb{R}} \mathrm{e}^{-\beta W}\,p^{\mathrm{f}}_T(W)\,dW \tag{4.54}$$

and

$$\mathrm{e}^{\beta\Delta F(T)} = \int_{\mathbb{R}} \mathrm{e}^{\beta W}\,p^{\mathrm{b}}_T(-W)\,dW = \int_{\mathbb{R}} \mathrm{e}^{-\beta W}\,p^{\mathrm{b}}_T(W)\,dW. \tag{4.55}$$

More interestingly,

$$\mathrm{e}^{-\beta\Delta F(T)} = \mathrm{e}^{-\beta W}\,\frac{p^{\mathrm{f}}_T(W)}{p^{\mathrm{b}}_T(-W)}. \tag{4.56}$$

If both work distributions can be sampled, estimates of the free energy difference can be obtained on a whole range of work values by (4.56). This is sometimes used to obtain estimates more accurate than those based on (4.54) or (4.55) (see [Hummer (2007)] for instance). Besides, there is a *single* value W^* for which $p_T^{\mathrm{f}}(W^*) = p_T^{\mathrm{b}}(-W^*)$, and this value is such that $W^* = \Delta F(T)$. This equality has a nice graphical illustration, see Figure 4.3 below.

4.2.3 *Numerical strategies*

Free energy estimators based on Crooks equality. For simplicity, and when there is no ambiguity, we drop in this section the mention of the time arguments in the works and the free energy difference

$$\Delta F = -\frac{1}{\beta} \ln \frac{Z_T}{Z_0}$$

to be estimated. Estimators for the free energy difference can be obtained from Crooks equality with the path functional $\varphi_{[0,T]} = 1$. The usual one-sided estimator (4.31)

$$\widehat{\Delta F}^{\mathrm{f}}_M = -\frac{1}{\beta} \ln \left(\frac{1}{M} \sum_{i=1}^{M} \mathrm{e}^{-\beta W_i^{\mathrm{f}}} \right),$$

for i.i.d. work values sampled from $p^{\mathrm{f}}(dW)$, is based on the the forward dynamics only (which amounts to setting $\theta = 0$ in (4.46)). Similarly, an estimator based on the backward dynamics only is obtained for the choice $\theta = 1$:

$$\widehat{\Delta F}^{\mathrm{b}}_M = \frac{1}{\beta} \ln \left(\frac{1}{M} \sum_{i=1}^{M} \mathrm{e}^{-\beta W_i^{\mathrm{b}}} \right), \tag{4.57}$$

for i.i.d. work values sampled from $p^{\mathrm{b}}(dW)$.

The equality (4.53) for work distributions however suggests to resort to some bridge sampling estimator (see Section 2.4.2). More precisely, an estimator of the free energy difference is

$$\widehat{\Delta F}^{\mathrm{bridge},g}_{M_{\mathrm{f}},M_{\mathrm{b}}} = -\frac{1}{\beta} \ln \frac{\displaystyle \frac{1}{M_{\mathrm{f}}} \sum_{i=1}^{M_{\mathrm{f}}} g(W_i^{\mathrm{f}})\, \mathrm{e}^{-\beta W_i^{\mathrm{f}}}}{\displaystyle \frac{1}{M_{\mathrm{b}}} \sum_{i=1}^{M_{\mathrm{b}}} g(-W_i^{\mathrm{b}})},$$

which is indeed a consistent estimator of the free energy difference ΔF for any function g when M_{f} independent values of forward works $W_i^{\mathrm{f}} \sim p^{\mathrm{f}}(dW)$

and $M_{\rm b}$ independent values of backward works $W_i^{\rm b} \sim p^{\rm b}(dW)$ are sampled. This corresponds to the framework of Section 2.4.2 with the choice $f_2(W) = Z_0\, p^{\rm f}(W)$ and $f_1(W) = Z_T\, p^{\rm b}(-W)$, with the important remark that the ratio of these densities can be computed analytically from (4.53):

$$\frac{f_2(W)}{f_1(W)} = \mathrm{e}^{-\beta W}.$$

Besides, the function α in (2.108) is sought here under the form $\alpha = g/f_2$. By the results of Section 2.4.2, minimizing the variance of the estimator $\widehat{\Delta F}^{\rm bridge,\mathit{g}}_{M_{\rm f},M_{\rm b}}(T)$ with respect to the function g leads to the bridge sampling estimator $\widehat{\Delta F}^{\rm bridge}_{M_{\rm f},M_{\rm b}}(T)$, solution of the nonlinear equation (2.113) which writes here:

$$\begin{aligned}
&\sum_{i=1}^{M_{\rm b}} \frac{M_{\rm f}}{M_{\rm f} + M_{\rm b} \exp\left[\beta\left(\widehat{\Delta F}^{\rm bridge}_{M_{\rm f},M_{\rm b}}(T) - W_i^{\rm b}\right)\right]} \\
&\quad = \sum_{i=1}^{M_{\rm f}} \frac{M_{\rm b}}{M_{\rm f} \exp\left[\beta\left(W_i^{\rm f} - \widehat{\Delta F}^{\rm bridge}_{M_{\rm f},M_{\rm b}}(T)\right)\right] + M_{\rm b}}.
\end{aligned} \tag{4.58}$$

In the limit $M_{\rm b}/M_{\rm f} \to 0$ the forward one-sided estimator (4.31) is recovered, while (4.57) is obtained in the limit $M_{\rm b}/M_{\rm f} \to +\infty$.

Let us finally mention that, once the free energy difference between the initial and the final state is computed, efficient estimates of the free energy profile can be obtained with the Crooks identity (4.46), following [Minh and Adib (2008)].

Application to Widom insertion. We now present some numerical results for Widom insertion. The data have been obtained with the schedule $\Lambda(t) = (t/T)^2$, using $M_{\rm f} = M_{\rm b} = M = 10^5$ for $T = 3$ and $M_{\rm f} = M_{\rm b} = M = 10^4$ replicas when $T = 10$, the other parameters being the same as in Section 4.1.5.4. The so-obtained work distributions $p_T^{\rm f}(W)$ and $p_T^{\rm b}(-W)$ are displayed in Figure 4.3. Note that the forward and backward work distributions indeed intersect at the value $W^* = \Delta F$, as predicted by (4.53).

Resorting to the bridge estimator (4.58) may be a good idea when work distributions overlap sufficiently, *i.e.* when the switching is slow enough, see Table 4.6. The bridge sampling estimate has been obtained by the following simple fixed-point iteration:

$$r^{k+1} = r^k - \alpha\left(\sum_{i=1}^{M_{\rm b}} \frac{M_{\rm f}}{M_{\rm f} + M_{\rm b}\,(r^k)^{-1}\,\mathrm{e}^{-\beta W_i^{\rm b}}} - \sum_{i=1}^{M_{\rm f}} \frac{M_{\rm b}}{M_{\rm f}\, r^k\, \mathrm{e}^{\beta W_i^{\rm f}} + M_{\rm b}}\right)$$

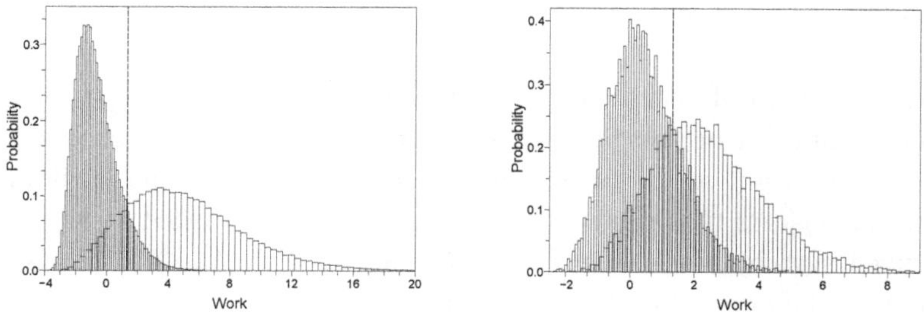

Fig. 4.3 Work distributions $p_T^{\rm f}(W)$ and $p_T^{\rm b}(-W)$ for $T = 3$ (left) and $T = 10$ (right). The reference free energy difference is given by the vertical dashed line. Note that the work distributions have a single intersection point, for W^* equal to the free energy difference.

Table 4.6 Comparison of one-sided and bridge estimators. The standard error of the estimator, evaluated by performing 100 independent realizations of the estimation process, is mentioned in brackets. The reference free energy difference obtained in Section 2.4.1.1 is 1.317 ± 0.001.

Number of replicas	Switching time	$\widehat{\Delta F}_M^{\rm fwd}$	$\widehat{\Delta F}_M^{\rm bck}$	$\widehat{\Delta F}_{M,M}^{\rm bridge}$
$M = 100$	$T = 1$	1.51 (0.68)	0.09 (0.88)	1.40 (0.31)
$M = 1000$	$T = 1$	1.29 (0.18)	0.66 (0.90)	1.42 (0.10)
$M = 10,000$	$T = 1$	1.29 (0.06)	1.02 (0.61)	1.44 (0.03)
$M = 100$	$T = 10$	1.34 (0.17)	1.26 (0.24)	1.32 (0.09)
$M = 1000$	$T = 10$	1.31 (0.05)	1.29 (0.10)	1.31 (0.03)

with $\alpha = 0.001$, where r^k is an estimate of $\mathrm{e}^{-\beta\Delta F}$. The initial value r^0 is for instance the average of the foward and backward estimators of $\mathrm{e}^{-\beta\Delta F}$:

$$r^0 = \frac{1}{2}\left(\frac{1}{M_{\rm f}}\sum_{i=1}^{M_{\rm f}} \mathrm{e}^{-\beta W_i^{\rm f}} + \left(\frac{1}{M_{\rm b}}\sum_{i=1}^{M_{\rm b}} \mathrm{e}^{\beta W_i^{\rm b}}\right)^{-1}\right).$$

In all cases, 100 independent realizations of the estimation procedure have been performed. In the case $T = 10$, the bridge estimator has a reduced variance compared to the forward one-sided estimator, while the backward one is really off the result (which is not surprising since it is well known in the physics and chemistry communities that Widom insertion is to be preferred to Widom deletion, see the discussion in Section 2.4.1.4). In the case $T = 1$, the simple forward estimator performs better than the bridge estimator. This is related to the fact that work distributions do not overlap sufficiently.

4.3 Nonequilibrium stochastic methods in the reaction coordinate case

When free energy differences are indexed by a reaction coordinate, nonequilibrium dynamics can still be considered, but some care has to be taken in the computation of the work since there is some additional switching force which ensures that the constraint $\xi(q_t) = z(t)$ is satisfied at all times for a given switching schedule $z : [0,T] \to \mathcal{M}$, with $\mathcal{M} \subset \mathbb{R}^m$. As for thermodynamic integration (see Chapter 3), two prototypical processes are considered: overdamped Langevin dynamics (see Section 4.3.1, which summarizes [Lelièvre *et al.* (2007a)]) and phase space dynamics of Langevin type (Section 4.3.2 and [Lelièvre *et al.* (2010)]).

4.3.1 *Overdamped nonequilibrium dynamics*

This section follows the notation of Section 3.2.1 where constrained overdamped processes are considered. Let us first briefly recall some notation. The ambient space is $\mathcal{D} = \mathbb{R}^{3N}$ (though other spaces could be considered, see Remark 1.1). First, a reaction coordinate $\xi : \mathbb{R}^{3N} \to \mathbb{R}^m$ is introduced. Its Gram tensor

$$G_{\alpha,\gamma}(q) = \nabla\xi_\alpha(q) \cdot \nabla\xi_\gamma(q)$$

is assumed to be invertible in a neighborhood of the submanifold

$$\Sigma(z) = \left\{ q \,\middle|\, \xi(q) = z \right\}$$

defined by the reaction coordinate, for all values z visited by the switching schedule (defined in Section 4.3.1.1). The canonical measure reads

$$\nu(dq) = Z_\nu^{-1} \exp(-\beta V(q))\, dq, \qquad Z_\nu = \int_{\mathcal{D}} \exp(-\beta V(q))\, dq.$$

Conditional averages at a fixed value z of the reaction coordinate are defined as averages with respect to the following measure with support on $\Sigma(z)$:

$$\begin{aligned} \nu^\xi(dq \,|\, z) &= Z_z^{-1} \exp\left(-\beta V(q)\right) |\det G(q)|^{-1/2}\, \sigma_{\Sigma(z)}(dq) \\ &= Z_z^{-1} \exp\left(-\beta V^\xi(q)\right)\, \sigma_{\Sigma(z)}(dq), \end{aligned}$$

where the modified potential

$$V^\xi = V + \frac{1}{\beta} \ln\left((\det G)^{1/2} \right)$$

is defined in (3.32), and

$$Z_z = \int_{\Sigma(z)} \mathrm{e}^{-\beta V^\xi}\, d\sigma_{\Sigma(z)}.$$

The associated free energy F reads (see (3.22)):

$$F(z) = -\frac{1}{\beta} \ln \left(\int_{\Sigma(z)} Z_\nu^{-1} \mathrm{e}^{-\beta V(q)} \, |\det G(q)|^{-1/2} \, \sigma_{\Sigma(z)}(dq) \right). \tag{4.59}$$

The mean force ∇F can be expressed as the average of the local mean force f with respect to the conditional probability measures $\nu^\xi(\cdot \,|\, z)$ (see Lemma 3.9):

$$\nabla F(z) = \int_{\Sigma(z)} f(q) \, \nu^\xi(dq \,|\, z), \tag{4.60}$$

where the components of the local mean force $f = (f_1, \ldots, f_m)$ are, for $\alpha \in \{1, \ldots, m\}$,

$$f_\alpha = \sum_{\gamma=1}^{m} G_{\alpha,\gamma}^{-1} \nabla \xi_\gamma \cdot \nabla V - \beta^{-1} \mathrm{div} \left(\sum_{\gamma=1}^{m} G_{\alpha,\gamma}^{-1} \nabla \xi_\gamma \right). \tag{4.61}$$

The orthogonal projector onto the normal space of $\Sigma(\xi(q))$ at a given point q is

$$P^\perp(q) = \sum_{\alpha,\gamma=1}^{m} G_{\alpha,\gamma}^{-1}(q) \nabla \xi_\alpha(q) \otimes \nabla \xi_\gamma(q),$$

and the orthogonal projection operator onto the tangent space of $\Sigma(\xi(q))$ reads

$$P(q) = \mathrm{Id} - P^\perp(q).$$

Remark 4.12 (Free energy and geometry). *As for thermodynamic integration (see Section* 3.2.5.3*), if the system is initially described by a potential V, the classical definition of free energy is given by* (4.59) *above. The associated projected dynamics (see* (4.62) *below) however requires the computation of the gradient of the modified potential V^ξ, which involves some (possibly cumbersome) second order derivatives of the constraints ξ. In practice, as suggested in Section* 3.2.5.3, *it is possible instead to use ∇V as a drift term in the projected dynamics, thereby computing the rigid free energy*

$$F_{\mathrm{rgd}}(z) = -\frac{1}{\beta} \ln \left(\int_{\Sigma(z)} Z_\nu^{-1} \mathrm{e}^{-\beta V(q)} \, \sigma_{\Sigma(z)}(dq) \right).$$

The usual free energy (4.59) *is then recovered as*

$$F(z) - F_{\mathrm{rgd}}(z) = -\beta^{-1} \ln \left(\frac{\displaystyle\int_{\Sigma(z)} (\det G)^{-1/2} \, \mathrm{e}^{-\beta V} \, d\sigma_{\Sigma(z)}}{\displaystyle\int_{\Sigma(z)} \mathrm{e}^{-\beta V} \, d\sigma_{\Sigma(z)}} \right).$$

The right-hand side is a conditional canonical average, which can be evaluated by reweighting nonequilibrium trajectories using the nonequilibrium works, see Remark 4.15 *below.*

4.3.1.1 *Definition of the switched dynamics*

Let us first define the nonequilibrium overdamped process associated with a smooth switching schedule $z = (z_1, \dots, z_m) : [0,T] \to \mathbb{R}^m$. As in the alchemical case, the dynamics starts at equilibrium: $q_0 \sim \nu^\xi(\cdot \,|\, z(0))$, but is driven out of equilibrium by the time evolution of the constraint.

Proposition 4.13. *There is a unique process* $t \mapsto q_t$ *satisfying for some Lagrange multipliers* $t \mapsto (\lambda_{1,t}, \dots, \lambda_{m,t})_{t\in[0,T]} \in \mathbb{R}^m$*:*

$$\begin{cases} q_0 \sim \nu^\xi(\cdot \,|\, z(0)), \\ dq_t = -\nabla V^\xi(q_t)\,dt + \sqrt{\frac{2}{\beta}}\,dW_t + \sum_{\alpha=1}^m \nabla\xi_\alpha(q_t)\,d\lambda_{\alpha,t}, \\ \xi(q_t) = z(t), \end{cases} \tag{4.62}$$

where W_t *is a standard* $3N$*-dimensional Wiener process, and* $t \mapsto \lambda_t$ *is an adapted*[4] *Itô process. Moreover, for all* $1 \le \alpha \le m$*, the process* $(\lambda_{\alpha,t})_{t\in[0,T]}$ *can be decomposed as*

$$\lambda_{\alpha,t} = \lambda^{\mathrm{m}}_{\alpha,t} + \lambda^{\mathrm{mf}}_{\alpha,t} + \lambda^{\mathrm{ext}}_{\alpha,t}, \tag{4.63}$$

with the martingale part

$$d\lambda^{\mathrm{m}}_{\alpha,t} = -\sqrt{\frac{2}{\beta}} \sum_{\gamma=1}^m G^{-1}_{\alpha,\gamma}(q_t)\nabla\xi_\gamma(q_t)\cdot dW_t,$$

the local mean force part (see (4.61) *for the definition of the local mean force)*

$$d\lambda^{\mathrm{mf}}_{\alpha,t} = f_\alpha(q_t)\,dt,$$

and the external forcing (or switching) term

$$d\lambda^{\mathrm{ext}}_{\alpha,t} = \sum_{\gamma=1}^m G^{-1}_{\alpha,\gamma}(q_t)\,\dot z_\gamma(t)\,dt. \tag{4.64}$$

[4] That is to say, continuous in time, and such that λ_t depends only on the past values of $s \mapsto W_s$ up to time t.

Proof. Assume that (4.63) holds, so that

$$\begin{aligned}\nabla\xi_\alpha(q_t)\,d\lambda_{\alpha,t} &= Y_{\alpha,t}\nabla\xi_\alpha(q_t)\,dt - \sqrt{\frac{2}{\beta}}\sum_{\gamma=1}^{m}\Big(G^{-1}_{\alpha,\gamma}(q_t)\nabla\xi_\alpha(q_t)\otimes\nabla\xi_\gamma(q_t)\Big)\,dW_t \\ &= Y_{\alpha,t}\nabla\xi_\alpha(q_t)\,dt - \sqrt{\frac{2}{\beta}}P^{\perp}(q_t)\,dW_t\end{aligned}$$

with

$$Y_{\alpha,t} = f_\alpha(q_t) + \sum_{\gamma=1}^{m} G^{-1}_{\alpha,\gamma}(q_t)\,\dot z_\gamma(t). \tag{4.65}$$

Then,

$$dq_t = \left(-\nabla V^{\xi}(q_t) + \sum_{\alpha=1}^{m} Y_{\alpha,t}\nabla\xi_\alpha(q_t)\right)dt + \sqrt{\frac{2}{\beta}}\,P(q_t)\,dW_t,$$

and, by Itô's calculus,

$$\begin{aligned}d\big[\xi_\alpha(q_t)\big] &= \nabla\xi_\alpha(q_t)\cdot dq_t + \frac{1}{\beta}P:\nabla^2\xi_\alpha(q_t)\,dt \\ &= \nabla\xi_\alpha(q_t)\cdot\left(-\nabla V^{\xi}(q_t) + \sum_{\gamma=1}^{m} Y_{\gamma,t}\nabla\xi_\gamma(q_t)\right)dt + \frac{1}{\beta}P:\nabla^2\xi_\alpha(q_t)\,dt\end{aligned}$$

since $\nabla\xi_\alpha(q_t)\cdot(P(q_t)\,dW_t) = (P(q_t)\nabla\xi_\alpha(q_t))\cdot dW_t = 0$. Using the equality $P:\nabla^2\xi_\gamma = -\nabla\xi_\gamma\cdot\mathcal{H}$ (see (3.46)) and the expression (3.82) of the local mean force, it follows that

$$\begin{aligned}&\sum_{\gamma=1}^{m} G^{-1}_{\alpha,\gamma}(q_t)\,d\big[\xi_\gamma(q_t)\big] \\ &= \sum_{\gamma=1}^{m} G^{-1}_{\alpha,\gamma}(q_t)\left[-\nabla\xi_\gamma(q_t)\cdot\Big(\nabla V^{\xi}(q_t) + \beta^{-1}\mathcal{H}\Big) + \sum_{\delta=1}^{m} G_{\gamma,\delta}(q_t)Y_{\delta,t}\right]dt \\ &= \Big(-f_\alpha(q_t) + Y_{\alpha,t}\Big)\,dt \\ &= \sum_{\gamma=1}^{m} G^{-1}_{\alpha,\gamma}(q_t)\,\dot z_\gamma(t)\,dt\end{aligned}$$

by the definition (4.65) of $Y_{\alpha,t}$. Since we assumed $G(q_t)$ to be invertible,

$$d\big[\xi_\alpha(q_t)\big] = \dot z_\alpha(t)\,dt.$$

Besides, $\xi_\alpha(q_0) = z(0)$, so that the constraints are indeed satisfied: $\xi(q(t)) = z(t)$ for all $t \in [0,T]$.

The uniqueness of the process (4.62) is a consequence of the fact that the stochastic process $t \mapsto q_t$ can actually be characterized by the following stochastic differential equation:

$$\begin{cases} q_0 \sim \nu^\xi(\cdot \mid z(0)), \\ dq_t = -P(q_t)\nabla V^\xi(q_t)\,dt + \sqrt{\dfrac{2}{\beta}}P(q_t) \circ dW_t + \displaystyle\sum_{\alpha=1}^m \nabla\xi_\alpha(q_t)\,d\lambda_{\alpha,t}^{\mathrm{ext}}, \\ d\lambda_{\alpha,t}^{\mathrm{ext}} = \displaystyle\sum_{\gamma=1}^m G_{\alpha,\gamma}^{-1}(q_t)\,\dot{z}_\gamma(t)\,dt, \qquad \forall\, 1 \le \alpha \le m, \end{cases} \tag{4.66}$$

which is well posed when the potential V and the reaction coordinate ξ are smooth. □

The expression of the generator of the dynamics at a fixed value of the reaction coordinate (which corresponds in this context to $d\lambda_{\alpha,t}^{\mathrm{ext}} = 0$), is given in Remark 3.23. The infinitesimal generator of (4.62) is obtained by taking into account the additional drift term corresponding to $d\lambda_{\alpha,t}^{\mathrm{ext}}$ defined by (4.64): for all test functions φ,

$$\mathcal{L}_t^{\mathrm{f}}\varphi = \frac{1}{\beta}\,\mathrm{e}^{\beta V^\xi}\mathrm{div}_{\Sigma(z(t))}\left(\mathrm{e}^{-\beta V^\xi}P\nabla\varphi\right) + \sum_{\alpha,\gamma=1}^m \dot{z}_\gamma(t)\,G_{\alpha,\gamma}^{-1}\nabla\xi_\alpha\cdot\nabla\varphi, \tag{4.67}$$

where the divergence on the submanifold $\Sigma(z(t))$ is defined by $\mathrm{div}_{\Sigma(z(t))}(\varphi) = \mathrm{Tr}(P\nabla\varphi)$ for positions q such that $\xi(q) = z(t)$. The superscript "f" in (4.67) reminds the reader that the associated dynamics will be considered as the forward switching dynamics, see Section 4.2 for the terminology.

4.3.1.2 *Jarzynski-Crooks identity*

A notion of work and a backward switching dynamics should now be defined in order to state the Jarzynski-Crooks identity. The work mentioned in this section should always be understood as the forward work with the terminology of Section 4.2, and we therefore drop the superscript "f" in the corresponding notation.

It turns out that the nonequilibrium work exerted on the diffusion q_t solution to (4.66) is appropriately defined by the integral of the local mean force as

$$\mathcal{W}_{0,t}\left(\{q_s\}_{0\le s\le t}\right) = \sum_{\alpha=1}^m \int_0^t f_\alpha(q_s)\,\dot{z}_\alpha(s)\,ds = \sum_{\alpha=1}^m \int_0^t \dot{z}_\alpha(s)\,d\lambda_{\alpha,s}^{\mathrm{mf}}. \tag{4.68}$$

Indeed, this definition enables us to rewrite equilibrium expectation as weighted averages of nonequilibrium trajectories, see (4.71) below. It may be motivated heuristically by observing that the so-defined work is the integral of the constraining force multiplied by the variation of the constraint.

The backward switching dynamics is obtained by time-reversing the switching schedule as (see Section 4.2)

$$\begin{cases} q_0^{\mathrm{b}} \sim \nu^\xi(\cdot\,|\,z(T)), \\ dq_{t'}^{\mathrm{b}} = -\nabla V^\xi(q_{t'}^{\mathrm{b}})\,dt' + \sqrt{\dfrac{2}{\beta}}\,dW_{t'}^{\mathrm{b}} + \displaystyle\sum_{\alpha=1}^{m} \nabla\xi_\alpha(q_{t'}^{\mathrm{b}})\,d\lambda_{\alpha,t'}, \\ \xi(q_{t'}^{\mathrm{b}}) = z(T-t'), \end{cases} \tag{4.69}$$

the associated generator being: for all test functions φ,

$$\mathcal{L}_{t'}^{\mathrm{b}}\varphi = \frac{1}{\beta}\,\mathrm{e}^{\beta V^\xi}\mathrm{div}_{\Sigma(z(T-t'))}\left(\mathrm{e}^{-\beta V^\xi}P\nabla\varphi\right) - \sum_{\alpha,\gamma=1}^{m} \dot{z}_\gamma(T-t')\,G_{\alpha,\gamma}^{-1}\nabla\xi_\alpha\cdot\nabla\varphi. \tag{4.70}$$

With these definitions, the following result holds.

Theorem 4.14 (Jarzynski-Crooks fluctuation identity). *Denote by q^{f} the forward switched dynamics (4.66) and by q^{b} the backward switched dynamics (4.69). Then, for any bounded measurable path functional $\varphi_{[0,T]}$,*

$$\frac{Z_{z(T)}}{Z_{z(0)}}\,\mathbb{E}\left(\varphi_{[0,T]}^{\mathrm{r}}(q^{\mathrm{b}})\right) = \mathbb{E}\left(\varphi_{[0,T]}(q^{\mathrm{f}})\,\mathrm{e}^{-\beta\mathcal{W}_{0,T}(q^{\mathrm{f}})}\right), \tag{4.71}$$

where $\mathcal{W}_{0,T}(q^{\mathrm{f}})$ is defined in (4.68), and $\varphi_{[0,T]}^{\mathrm{r}}$ denotes the composition of $\varphi_{[0,T]}$ with the operation of time-reversal on paths:

$$\varphi_{[0,T]}^{\mathrm{r}}\left(\{q_t\}_{0\le t\le T}\right) = \varphi_{[0,T]}\left(\{q_{T-t'}\}_{0\le t'\le T}\right).$$

In particular, the choice $\varphi_{[0,T]} = 1$ leads to the following work fluctuation identity for all $t \in [0,T]$:

$$\Delta F(z(t)) = F(z(t)) - F(z(0)) = -\frac{1}{\beta}\ln\left[\mathbb{E}\left(\mathrm{e}^{-\beta\mathcal{W}_{0,t}(q^{\mathrm{f}})}\right)\right], \tag{4.72}$$

which allows to estimate free energy differences.

Note also that an equality similar to (4.71) can be obtained with works from backward switchings, see (4.46) and Remark 4.11.

Remark 4.15 (Recovering canonical conditional averages). *Conditional canonical averages can be computed by reweighting nonequilibrium trajectories. For a given observable A, an application of (4.71) for* *the path functionals* $\varphi_{[0,T]}(\{q_t\}_{0\le t\le T}) = A(q_T)$ *and* $\varphi_{[0,T]}(\{q_t\}_{0\le t\le T}) = 1$ *indeed leads to*

$$\int_{\Sigma(z(T))} A\, d\nu^{\xi}(\cdot \,|\, z(T)) = \frac{\displaystyle\int_{\Sigma(z(T))} A(q)\, \mathrm{e}^{-\beta V^{\xi}(q)}\, \sigma_{\Sigma(z(T))}(dq)}{\displaystyle\int_{\Sigma(z(T))} \mathrm{e}^{-\beta V^{\xi}(q)}\, \sigma_{\Sigma(z(T))}(dq)}$$

$$= \frac{\mathbb{E}\left(A(q^{\mathrm{f}}_T)\, \mathrm{e}^{-\beta \mathcal{W}_{0,T}(q^{\mathrm{f}})}\right)}{\mathbb{E}\left(\mathrm{e}^{-\beta \mathcal{W}_{0,T}(q^{\mathrm{f}})}\right)}.$$

A similar result holds for any time $0 \le t \le T$. *Canonical conditional averages with respect to the measure* $\widetilde{Z}_z^{-1}\, \mathrm{e}^{-\beta V}\, d\sigma_{\Sigma(z)}$ *instead of* $Z_z^{-1}\, \mathrm{e}^{-\beta V^{\xi}}\, d\sigma_{\Sigma(z)}$ *are obtained for switching dynamics* (4.62) *where the force term is* $-\nabla V$ *instead of* $-\nabla V^{\xi}$*. This remark is important to compute the correction to the rigid free energy, as explained in Remark* 4.12.

Proof. The proof follows by an application of Theorem 4.10. However, we first need to define weighted forward transition operators in the neighborhood of $\Sigma(z(t))$ to this end. For any $t \in [0,T]$ and x in a neighborhood of $\Sigma(z(t))$, let us introduce the forward process $(q_s^{\mathrm{f},t,x})_{s\in[t,T]}$ satisfying the SDE (4.66), starting from x at time t:

$$\begin{cases} q_t^{\mathrm{f},t,x} = x, \\ dq_s^{\mathrm{f},t,x} = -P(q_s^{\mathrm{f},t,x})\nabla V^{\xi}(q_s^{\mathrm{f},t,x})\, ds + \sqrt{\dfrac{2}{\beta}}\, P(q_s^{\mathrm{f},t,x}) \circ dW_s^{\mathrm{f}} \\ \qquad\qquad + \displaystyle\sum_{\alpha=1}^{m} \nabla\xi_\alpha(q_s^{\mathrm{f},t,x})\, d\lambda_{\alpha,s}^{\mathrm{ext}}, \\ d\lambda_{\alpha,s}^{\mathrm{ext}} = \displaystyle\sum_{\gamma=1}^{m} G_{\alpha,\gamma}^{-1}(q_s^{\mathrm{f},t,x})\, \dot z_\gamma(s)\, ds, \qquad \forall 1 \le \alpha \le m. \end{cases} \tag{4.73}$$

This process satisfies $d\xi(q_s^{\mathrm{f},t,x}) = \dot z(s)\, ds$, so that

$$\xi(q_s^{\mathrm{f},t,x}) = \xi(x) + z(s) - z(t).$$

For any $t \in [0,T]$, there is therefore an open neighborhood $(t^-, t^+) \times \mathcal{D}_t$ of $(t, \Sigma(z(t)))$ such that the diffusion $q_s^{\mathrm{f},t,x} \in \mathcal{D}_s$ almost surely. This gives usual regularity assumptions sufficient to obtain the transition semi-group defined in (4.44):

$$P_{t,T}^{\mathrm{f}}\varphi(x) = \mathbb{E}\left[\varphi(q_T^{\mathrm{f},t,x}) \exp\left(-\beta \int_t^T \sum_{\alpha=1}^{m} f_\alpha(q_s^{\mathrm{f},t,x})\, \dot z_\alpha(s)\, ds\right)\right],$$

satisfying the following partial differential equation on $(t^-, t^+) \times \mathcal{D}_t$:

$$\partial_t \left(P^{\mathrm{f}}_{t,T} \varphi \right) = \left(-\mathcal{L}^{\mathrm{f}}_t + \beta \sum_{\alpha=1}^m \dot{z}_\alpha(t) f_\alpha \right) P^{\mathrm{f}}_{t,T} \varphi,$$

where $\mathcal{L}^{\mathrm{f}}_t$ is given in (4.67).

The cornerstone of the proof is now to show that the balance condition (4.43) holds with $\Pi_t(dq) = \mathrm{e}^{-\beta V^\xi(q)}\, \sigma_{\Sigma(z(t))}(dq)$. Using the reversibility of the canonical measure (see Remark 3.23), and the expressions (4.67) and (4.70) of the generators of the forward and backward dynamics respectively,

$$\begin{aligned}
&\int_{\Sigma(z(t))} \left(\varphi_1\, \mathcal{L}^{\mathrm{f}}_t \varphi_2 - \varphi_2\, \mathcal{L}^{\mathrm{b}}_{T-t} \varphi_1 \right) \mathrm{e}^{-\beta V^\xi}\, d\sigma_{\Sigma(z(t))} \\
&= \sum_{\alpha,\gamma=1}^m \dot{z}_\gamma(t) \int_{\Sigma(z(t))} G^{-1}_{\alpha,\gamma} \left(\varphi_1\, \nabla \xi_\alpha \cdot \nabla \varphi_2 + \varphi_2\, \nabla \xi_\alpha \cdot \nabla \varphi_1 \right) \mathrm{e}^{-\beta V^\xi}\, d\sigma_{\Sigma(z(t))} \\
&= \sum_{\alpha,\gamma=1}^m \dot{z}_\gamma(t) \int_{\Sigma(z(t))} G^{-1}_{\alpha,\gamma} \nabla \xi_\alpha \cdot \nabla(\varphi_1 \varphi_2)\, \mathrm{e}^{-\beta V^\xi}\, d\sigma_{\Sigma(z(t))}.
\end{aligned}$$

On the other hand,

$$\begin{aligned}
&\frac{d}{dt} \left(\int_{\Sigma(z(t))} \varphi_1 \varphi_2\, \mathrm{e}^{-\beta V^\xi}\, d\sigma_{\Sigma(z(t))} \right) \\
&= \sum_{\alpha=1}^m \dot{z}_\alpha(t)\, \nabla_{z_\alpha} \left(\int_{\Sigma(z)} \varphi_1 \varphi_2\, \mathrm{e}^{-\beta V}\, (\det\, G)^{-1/2}\, d\sigma_{\Sigma(z)} \right) \Bigg|_{z=z(t)} \\
&= \sum_{\alpha,\gamma=1}^m \dot{z}_\alpha(t) \int_{\Sigma(z(t))} \Big[G^{-1}_{\alpha,\gamma} \nabla \xi_\gamma \cdot \nabla \left(\varphi_1 \varphi_2\, \mathrm{e}^{-\beta V} \right) \\
&\qquad\qquad + \mathrm{div} \left(G^{-1}_{\alpha,\gamma} \nabla \xi_\gamma \right) \varphi_1 \varphi_2\, \mathrm{e}^{-\beta V} \Big] (\det\, G)^{-1/2} d\sigma_{\Sigma(z(t))},
\end{aligned}$$

where we have used Lemma 3.10 with $\psi = \varphi_1 \varphi_2 \exp(-\beta V)$. Therefore, in view of the expression (4.61) of the local mean force,

$$\begin{aligned}
&\frac{d}{dt} \left(\int_{\Sigma(z(t))} \varphi_1 \varphi_2\, \mathrm{e}^{-\beta V^\xi}\, d\sigma_{\Sigma(z(t))} \right) \\
&= \sum_{\alpha=1}^m \dot{z}_\alpha(t) \int_{\Sigma(z(t))} \left(-\beta f_\alpha \varphi_1 \varphi_2 + \sum_{\gamma=1}^m G^{-1}_{\alpha,\gamma} \nabla \xi_\gamma \cdot \nabla(\varphi_1 \varphi_2) \right) \mathrm{e}^{-\beta V^\xi}\, d\sigma_{\Sigma(z(t))}.
\end{aligned}$$

This shows that the balance condition (4.43) indeed holds (with $w^{\mathrm{f}}(t,q) = \sum_{\alpha=1}^m \dot{z}_\alpha(t) f_\alpha(q)$). The remainder of the proof follows the lines of the proof of Theorem 4.10. $\square$

4.3.1.3 *Numerical implementation*

In order to compute a numerical approximation of the free energy difference, a first task is to discretize the dynamics (4.62). To this end, the schedule is replaced by a discrete schedule $\{z(0), \ldots, z(t_{N_T})\}$ where N_T is the number of time-steps, and $t_{N_T} = T$. For example, equal time increments can be used, in which case $\Delta t = \frac{T}{N_T}$ and $t_n = n\Delta t$.

Possible generalizations of the schemes (3.66) and (3.67) for constrained equilibrium sampling to the nonequilibrium case are respectively:

$$\left\{ \begin{aligned} & q^{n+1} = q^n - \Delta t \, \nabla V^\xi(q^n) + \sqrt{\frac{2\Delta t}{\beta}} \, G^n + \sum_{\alpha=1}^m \lambda_\alpha^{n+1} \, \nabla \xi_\alpha(q^{n+1}), \\ & \quad \text{where } (\lambda_\alpha^{n+1})_{1 \le \alpha \le m} \text{ is such that } \xi(q^{n+1}) = z(t_{n+1}), \end{aligned} \right. \tag{4.74}$$

and

$$\left\{ \begin{aligned} & q^{n+1} = q^n - \Delta t \, \nabla V^\xi(q^n) + \sqrt{\frac{2\Delta t}{\beta}} \, G^n + \sum_{\alpha=1}^m \lambda_\alpha^{n+1} \, \nabla \xi_\alpha(q^n), \\ & \quad \text{where } (\lambda_\alpha^{n+1})_{1 \le \alpha \le m} \text{ is such that } \xi(q^{n+1}) = z(t_{n+1}), \end{aligned} \right. \tag{4.75}$$

where $(G^n)_{n \ge 0}$ are $3N$-dimensional Gaussian random vectors with mean 0 and covariance Id. Let us recall that, when the computation of ∇V^ξ is cumbersome, it is possible to integrate a dynamics with ∇V^ξ replaced by ∇V upon correcting the bias as explained in Remark 4.12. It is also possible to compute approximations of ∇V^ξ using finite differences, see Remark 3.30.

The simplest strategy to compute the works is to discretize the time integral (4.68). The following approximation $\mathcal{W}^n$ of $\mathcal{W}_{0,t_n}$ may be considered:

$$\left\{ \begin{aligned} & \mathcal{W}^0 = 0, \\ & \mathcal{W}^{n+1} = \mathcal{W}^n + \sum_{\alpha=1}^m \Big(z_\alpha(t_{n+1}) - z_\alpha(t_n) \Big) f_\alpha(q^n). \end{aligned} \right.$$

Estimators of the free energy difference, based on (4.72) can then be constructed as in the alchemical case, see Sections 4.1.3, 4.1.4 and 4.1.5 for further precision.

Alternatively, the local mean force $f_\alpha(q^n)$ can be replaced by an approximation of the local mean force part of the Lagrange multiplier $\lambda^{\mathrm{mf}}_{\alpha,t_{n+1}}$ (see (4.63)), for instance

$$\lambda_\alpha^{\mathrm{mf},n+1} = \lambda_\alpha^{n+1} - \sum_{\gamma=1}^m G^{-1}_{\alpha,\gamma}(q^n) \left(z_\gamma(t_{n+1}) - z_\gamma(t_n) - \sqrt{\frac{2\Delta t}{\beta}} \, \nabla \xi_\gamma(q^n) \cdot G^n \right).$$

According to (4.68), the work is then updated as

$$\mathcal{W}^{n+1} = \mathcal{W}^n + \sum_{\alpha=1}^m \frac{z_\alpha(t_{n+1}) - z_\alpha(t_n)}{t_{n+1} - t_n} \lambda_\alpha^{\mathrm{mf},n+1}.$$

A generalization of the variance reduction method using antithetic variables (see (3.91) and [Lelièvre *et al.* (2007a)]) is also possible. Consider the following dynamics with a locally time-reversed constrained evolution (written here for the scheme (4.74)):

$$q^{\mathrm{R},n+1} = q^n - \Delta t\, \nabla V^\xi(q^n) - \sqrt{\frac{2\Delta t}{\beta}}\, G^n + \sum_{\alpha=1}^m \lambda_\alpha^{\mathrm{R},n+1}\, \nabla\xi_\alpha(q^{\mathrm{R},n+1}),$$

with $\lambda^{\mathrm{R},n+1}$ such that

$$\frac{1}{2}\Big(\xi(q^{\mathrm{R},n+1}) + \xi(q^{n+1})\Big) = \xi(q^n).$$

The position $q^{\mathrm{R},n+1}$ is therefore computed as q^{n+1} in (4.74), but with a negative Brownian increment $-\sqrt{\Delta t}\, G^n$, and a projection on $\Sigma(2\xi(q^n) - \xi(q^{n+1}))$ instead of $\Sigma(\xi(q^{n+1}))$. Note that in the case of a linear schedule and if the constraints are exactly satisfied at the previous time steps, $2\xi(q^n) - \xi(q^{n+1}) = z(t_{n-1})$. The force part $\lambda_\alpha^{\mathrm{mf},n+1}$ is then obtained through

$$\lambda_\alpha^{\mathrm{mf},n+1} = \frac{1}{2}(\lambda_\alpha^{n+1} + \lambda_\alpha^{\mathrm{R},n+1}),$$

which can be shown to be a consistent time discretization of $d\lambda^{\mathrm{mf}}_{\alpha,t_{n+1}}$.

4.3.2 *Hamiltonian and Langevin nonequilibrium dynamics*

This section presents nonequilibrium dynamics for Hamiltonian and Langevin dynamics with time-evolving constraints, see [Lelièvre *et al.* (2010)]. The notation of Section 3.3 for Hamiltonian dynamics and Langevin processes with constraints are used. The smooth switching schedule is a C^2 function $z : [0,T] \to \mathbb{R}^m$ between two extremal values of the reaction coordinates. The out-of-equilibrium Langevin process is a modification of the constrained Langevin equation (3.148):

$$\begin{cases} dq_t = M^{-1} p_t\, dt, \\ dp_t = -\nabla V(q_t)\, dt - \gamma(q_t) M^{-1} p_t\, dt + \sigma(q_t)\, dW_t + \nabla\xi(q_t)\, d\lambda_t, \\ \xi(q_t) = z(t), \qquad (C_q(t)) \end{cases} \tag{4.76}$$

where the Lagrange multiplier $t \mapsto \lambda_t \in \mathbb{R}^m$ is an adapted process enforcing the constraints $(C_q(t))$, and the usual fluctuation-dissipation identity holds:

$$\sigma\sigma^T = \frac{2\gamma}{\beta}.$$

4.3.2.1 *Generalized free energy and notation*

We recall in this section some notation from Sections 3.3 and 3.3.6 used in the remainder of Section 4.3.2. The appropriate variant of free energy which arises naturally in this context and allows to state the Crooks-Jarzynski equality (4.106) below was already introduced in (3.169).

In view of the time-evolution of the constraints,

$$\dot{z}(t) = \frac{d\xi(q_t)}{dt} = \nabla\xi(q_t)^T M^{-1} p_t,$$

the system belongs at time t to the state space:

$$\Sigma_{\xi,v_\xi}(z(t),\dot{z}(t)) = \left\{(q,p) \,\middle|\, \xi(q) = z(t),\ \nabla\xi(q)^T M^{-1} p = \dot{z}(t)\right\},$$

whose associated phase space measure (induced by the symplectic structure) is denoted by

$$\sigma_{\Sigma_{\xi,v_\xi}(z(t),\dot{z}(t))}(dq\,dp).$$

The effective velocity of the reaction coordinates ξ, defined in (3.107) as $v_\xi(q,p) := \nabla\xi(q)^T M^{-1} p$, allows to rewrite the constraints satisfied by the elements of $\Sigma_{\xi,v_\xi}(z(t),\dot{z}(t))$ as

$$\Xi(q,p) = \zeta(t),$$

where

$$\begin{cases} \Xi(q,p) := \begin{pmatrix} \xi(q) \\ v_\xi(q,p) \end{pmatrix} := \begin{pmatrix} \xi(q) \\ \nabla\xi(q)^T M^{-1} p \end{pmatrix}, \\ \zeta(t) := \begin{pmatrix} z(t) \\ \dot{z}(t) \end{pmatrix}. \end{cases} \tag{4.77}$$

The Gram tensor (the inverse of the effective mass of the reaction coordinates ξ) is

$$G_M(q) = \nabla\xi(q)^T M^{-1} \nabla\xi(q),$$

and the orthogonal projector on the cotangent space $T_q^*\Sigma(z)$ is

$$P_M(q) = \mathrm{Id} - \nabla\xi(q) G_M^{-1}(q) \nabla\xi(q)^T M^{-1}.$$

The projected fluctuation-dissipation matrices defined in (3.151) read

$$(\sigma_P, \gamma_P) := (P_M(q)\,\sigma, P_M(q)\,\gamma\,P_M(q)^T), \tag{4.78}$$

and the constraining force $f_{\mathrm{rgd}}^M : T^*\Sigma(z) \to \mathbb{R}^m$ introduced in (3.103) reads

$$f_{\mathrm{rgd}}^M(q,p) = G_M^{-1}(q) \nabla\xi(q)^T M^{-1} \nabla V(q) - G_M^{-1}(q) \mathrm{Hess}_q(\xi)(M^{-1}p, M^{-1}p). \tag{4.79}$$

The symplectic Gram tensor Γ defined componentwise as

$$\Gamma_{a,b} := \{\Xi_a, \Xi_b\} \qquad a, b = 1, \ldots, 2m,$$

reads in this context:

$$\Gamma = \begin{pmatrix} 0 & G_M \\ -G_M & \{v_\xi, v_\xi\} \end{pmatrix}.$$

Therefore

$$\Gamma^{-1} = \begin{pmatrix} G_M^{-1}\{v_\xi, v_\xi\} G_M^{-1} & -G_M^{-1} \\ G_M^{-1} & 0 \end{pmatrix}. \tag{4.80}$$

The associated Poisson bracket with constraints of two test functions (φ_1, φ_2) then reads

$$\{\varphi_1, \varphi_2\}_\Xi = \{\varphi_1, \varphi_2\} - \{\varphi_1, \Xi\} \Gamma^{-1} \{\Xi, \varphi_2\}. \tag{4.81}$$

As will be made precise below, the free energy difference actually computed in this section by the Jarzynski relation (4.106) is in fact a generalized version of the rigid free energy $F_{\rm rgd}$ defined in (3.169), see Section 3.3.6.1. In the present case, this generalized rigid free energy is associated with the effective velocity:

$$F_{\rm rgd}^{\xi, v_\xi}(z(t), \dot{z}(t)) = -\frac{1}{\beta} \ln \int_{\Sigma_{\xi, v_\xi}(z(t), \dot{z}(t))} \mathrm{e}^{-\beta H(q,p)} \sigma_{\Sigma_{\xi, v_\xi}(z(t), \dot{z}(t))}(dq\, dp). \tag{4.82}$$

This free energy is proportional to the logarithm of the partition function of the following phase space probability measure:

$$\begin{cases} \mu_{\Sigma_{\xi, v_\xi}(z(t), \dot{z}(t))}(dq\, dp) := \dfrac{1}{Z_{z(t), \dot{z}(t)}} \mathrm{e}^{-\beta H(q,p)} \sigma_{\Sigma_{\xi, v_\xi}(z(t), \dot{z}(t))}(dq\, dp), \\ Z_{z(t), \dot{z}(t)} := \displaystyle\int_{\Sigma_{\xi, v_\xi}(z(t), \dot{z}(t))} \mathrm{e}^{-\beta H}\, d\sigma_{\Sigma_{\xi, v_\xi}(z(t), \dot{z}(t))}. \end{cases} \tag{4.83}$$

The generalized rigid free energy (4.82) can be explicitly related to the usual free energy as follows. First, denoting by

$$\Sigma_{v_\xi(q,\cdot)}(v_z) = \left\{ p \in \mathbb{R}^{3N} \,\middle|\, \nabla\xi(q)^T M^{-1}(p - \nabla\xi(q) G_M^{-1}(q) v_z) = 0 \right\}$$

the set of momenta satisfying the constraints for a given position q, and using the change of variables $p \to p - \nabla\xi(q) G_M^{-1}(q) v_z$, the following equalities hold:

$$\begin{aligned} &\int_{\Sigma_{v_\xi(q,\cdot)}(v_z)} \exp\left(-\frac{\beta}{2} p^T M^{-1} p\right) \sigma^{M^{-1}}_{\Sigma_{v_\xi(q,\cdot)}(v_z)}(dp) \\ &= \exp\left(-\frac{\beta}{2} v_z^T G_M^{-1}(q) v_z\right) \int_{T_q^*\Sigma(z)} \mathrm{e}^{-\frac{\beta}{2} p^T M^{-1} p} \sigma^{M^{-1}}_{T_q^*\Sigma(z)}(dp), \\ &= \exp\left(-\frac{\beta}{2} v_z^T G_M^{-1}(q) v_z\right) \left(2\pi\beta^{-1}\right)^{\frac{3N-m}{2}}. \end{aligned}$$

Note that $\frac{1}{2}v_z^T G_M^{-1}(q)v_z$ can be interpreted as the kinetic energy of the reaction coordinate ξ. Using the decomposition of measures (3.130) and the above calculations, the following expression for the generalized free energy is therefore obtained:

$$F_{\rm rgd}^{\xi,v_\xi}(z,v_z) = -\frac{1}{\beta}\ln\int_{\Sigma(z)} \exp\left[-\beta\left(V(q)+\frac{1}{2}v_z^T G_M^{-1}(q)v_z\right)\right]\sigma_{\Sigma(z)}^M(dq) + \mathrm{C}, \tag{4.84}$$

where C denotes a numerical constant that may vary from line to line. As a consequence, the original free energy, defined for instance in (3.165), can be easily recovered using relations similar to (3.168). Indeed, with (4.84), the difference between the two free energies writes:

$$F(z) - F_{\rm rgd}^{\xi,v_\xi}(z,v_z) = \\ -\frac{1}{\beta}\ln\int_{\Sigma_{\xi,v_\xi}(z,v_z)} (\det G_M(q))^{-1/2}\exp\left(\frac{\beta}{2}v_z^T G_M^{-1}(q)v_z\right)\mu_{\Sigma_{\xi,v_\xi}(z,v_z)}(dq\,dp) \tag{4.85}$$

up to an unimportant additive constant. To recover the actual free energy profile from nonequilibrium computations, it therefore suffices to add a correction term to the work value in the Jarzynski estimator, see (4.109) and (4.110) in Section 4.3.2.4.

4.3.2.2 *Dynamics and generators*

The explicit computation of Lagrange multipliers is obtained similarly to the computation of (3.101) in the unconstrained case, see also Section 3.3.4.1 for constrained processes. Differentiating twice the constraints with respect to time:

$$\ddot{z}(t) = \frac{d^2}{dt^2}\xi(q_t),$$

leads to

$$\begin{aligned} d\lambda_t = &- G_M^{-1}(q_t)\mathrm{Hess}_{q_t}(\xi)(M^{-1}p_t, M^{-1}p_t)\,dt + G_M^{-1}(q_t)\ddot{z}(t)dt \\ &+ G_M^{-1}\nabla\xi(q_t)^T M^{-1}\left(\nabla V(q_t)\,dt + \gamma(q_t)M^{-1}p_t\,dt - \sigma(q_t)\,dW_t\right) \\ = &\, f_{\rm rgd}^M(q_t,p_t)\,dt + G_M^{-1}(q_t)\ddot{z}(t)dt \\ &+ G_M^{-1}\nabla\xi(q_t)^T M^{-1}\left(\gamma(q_t)M^{-1}p_t\,dt - \sigma(q_t)\,dW_t\right). \end{aligned} \tag{4.86}$$

In order to rewrite the momenta evolution more explicitly, we use the decomposition

$$P_M(q_t)p_t = p_t - \nabla\xi(q_t)G_M^{-1}(q_t)\nabla\xi(q_t)^T M^{-1}p_t = p_t - \nabla\xi(q_t)G_M^{-1}(q_t)\dot{z}(t).$$

Multiplying each term by $P_M\gamma M^{-1}$ and using $M^{-1}P_M = P_M^T M^{-1}$,

$$\begin{aligned} &P_M(q_t)\gamma(q_t)\, M^{-1}p_t = \\ &\quad \gamma_P(q_t)M^{-1}p_t + P_M(q_t)\gamma(q_t)\, M^{-1}\nabla\xi(q_t)G_M^{-1}(q_t)\dot{z}(t). \end{aligned}$$

The momenta evolution in (4.76) can then be rewritten as follows:

$$\begin{aligned} dp_t = &-\nabla V(q_t)\, dt + \nabla\xi(q_t) f_{\mathrm{rgd}}^M(q_t, p_t)\, dt \\ &-\gamma_P(q_t)M^{-1}p_t\, dt + \sigma_P(q_t)\, dW_t \\ &-P_M(q_t)\gamma(q_t)\, M^{-1}\nabla\xi(q_t)G_M^{-1}(q_t)\dot{z}(t) + \nabla\xi(q_t)G_M^{-1}(q_t)\ddot{z}(t)\, dt, \end{aligned} \tag{4.87}$$

where γ_P and σ_P are defined in (4.78). If we assume moreover that the matrices γ and σ verify

$$(\gamma, \sigma) = (\gamma_P, \sigma_P), \tag{4.88}$$

the term involving $\dot{z}$ disappears and the momentum evolution (4.87) simplifies to:

$$\begin{aligned} dp_t = &- \nabla V(q_t)\, dt + \nabla\xi(q_t) f_{\mathrm{rgd}}^M(q_t, p_t)\, dt \\ &- \gamma_P(q_t)M^{-1}p_t\, dt + \sigma_P(q_t)\, dW_t \\ &+ \nabla\xi(q_t)G_M^{-1}(q_t)\ddot{z}(t)\, dt. \end{aligned}$$

Assumption (4.88) will be used in the following, in Sections 4.3.2.4 and 4.3.2.5.

Denote by $\mathcal{L}_t^{\mathrm{f}}$ the generator of the forward dynamics $t \mapsto (q_t, p_t)$ defined in (4.76). The latter has a backward switching version (see Section 4.2),

$$t' \mapsto (q_{t'}^{\mathrm{b}}, p_{t'}^{\mathrm{b}}),$$

obtained by using a time-reversed switching, and by first reversing momenta in the initial condition, and then reversing them back after the time evolution. More precisely, it can be defined through its generator

$$\mathcal{L}_{t'}^{\mathrm{b}} = \mathcal{R}\, \mathcal{L}_{T-t'}^{\mathrm{f}}\, \mathcal{R}, \tag{4.89}$$

where $\mathcal{L}_{T-t'}^{\mathrm{f}}$ is the generator of the forward process at time $t = T - t'$, and $\mathcal{R} : \phi \mapsto \phi \circ S$ is the momentum flip operator with $S(q,p) = (q, -p)$. Note that $t' \mapsto (q_{t'}^{\mathrm{b}}, -p_{t'}^{\mathrm{b}})$ is a solution of the forward evolution equation (4.76) with a switching time schedule $t' \mapsto z(T - t')$. The time evolution of the backward dynamics is therefore given by

$$\begin{cases} dq^{\mathrm{b}}_{t'} = -M^{-1} p^{\mathrm{b}}_{t'} dt', \\ dp^{\mathrm{b}}_{t'} = \nabla V(q^{\mathrm{b}}_{t'})\, dt' - \gamma(q^{\mathrm{b}}_{t'}) M^{-1} p^{\mathrm{b}}_{t'}\, dt' + \sigma(q^{\mathrm{b}}_{t'})\, dW^{\mathrm{b}}_{t'} + \nabla\xi(q^{\mathrm{b}}_{t'})\, d\lambda^{\mathrm{b}}_{t'}, \\ \xi(q^{\mathrm{b}}_{t'}) = z(T - t'). \end{cases} \tag{4.90}$$

The following lemma gives the expressions of $\mathcal{L}^{\mathrm{f}}_t$ and $\mathcal{L}^{\mathrm{b}}_{t'}$. Note that a formulation of the generators $\mathcal{L}^{\mathrm{f}}_t$ and $\mathcal{L}^{\mathrm{b}}_t$ without the full constraints notation (Ξ, Γ) is tedious, which justifies the introduction of such notation. We refer the reader to Section 3.3.3 for more detail on the latter notation.

Lemma 4.16. *We denote by $\mathcal{L}^{\mathrm{thm}}_{\Xi}$ the fluctuation-dissipation operator defined in Proposition* 3.52 *by:*

$$\mathcal{L}^{\mathrm{thm}}_{\Xi} = \frac{1}{\beta} \mathrm{e}^{\beta H} \mathrm{div}_p \left(\mathrm{e}^{-\beta H} P_M \gamma P_M^T \nabla_p \cdot \right). \tag{4.91}$$

The generator of the forward process (4.76) *at time $t \in [0, T]$ reads:*

$$\mathcal{L}^{\mathrm{f}}_t = \{\cdot, H\}_{\Xi} + \mathcal{L}^{\mathrm{thm}}_{\Xi} + \{\cdot, \Xi\}\, \Gamma^{-1} \dot{\zeta}(t) - \dot{z}(t)^T G_M^{-1} (\nabla\xi)^T M^{-1} \gamma\, P_M^T \nabla_p.$$

The generator of the backward process (4.90) *at time $t' \in [0, T]$ writes:*

$$\mathcal{L}^{\mathrm{b}}_{t'} = -\{\cdot, H\}_{\Xi} + \mathcal{L}^{\mathrm{thm}}_{\Xi} - \{\cdot, \Xi\}\, \Gamma^{-1} \dot{\zeta}(T - t') + \dot{z}(T - t')^T G_M^{-1} (\nabla\xi)^T M^{-1} \gamma P_M^T \nabla_p.$$

Proof. First, let us consider the terms in (4.76) arising from the Hamiltonian evolution and from the switching (*i.e.* without fluctuation/dissipation, which amounts to setting $\gamma = 0$ and $\sigma = 0$):

$$\begin{cases} dq_t = M^{-1} p_t\, dt \\ dp_t = -\nabla V(q_t)\, dt + \nabla\xi(q_t) f^M_{\mathrm{rgd}}(q_t, p_t)\, dt + \nabla\xi(q_t) G_M^{-1}(q_t) \ddot{z}(t)\, dt. \end{cases} \tag{4.92}$$

Since during this dynamics $v_\xi(q_t, p_t) = \dot{z}(t)$, formula (3.137) implies:

$$\begin{aligned} \{\Xi, H\}(q_t, p_t) &= \begin{pmatrix} \nabla\xi(q_t)^T M^{-1} p_t \\ \mathrm{Hess}_{q_t}(\xi)(M^{-1} p_t, M^{-1} p_t) - \left(\nabla\xi^T M^{-1} \nabla V\right)(q_t) \end{pmatrix} \\ &= \begin{pmatrix} \dot{z}(t) \\ \mathrm{Hess}_{q_t}(\xi)(M^{-1} p_t, M^{-1} p_t) - \left(\nabla\xi^T M^{-1} \nabla V\right)(q_t) \end{pmatrix}. \end{aligned}$$

In view of (4.79) and (4.80),

$$\Gamma^{-1}(q_t, p_t) \Big(\{\Xi, H\}(q_t, p_t) - \dot{\zeta}(t) \Big) = \begin{pmatrix} G_M^{-1}(q_t) \ddot{z}(t) + f^M_{\mathrm{rgd}}(q_t, p_t) \\ 0 \end{pmatrix}. \tag{4.93}$$

Therefore, the identity $\{\varphi, \Xi\} \cdot \begin{pmatrix} a \\ 0 \end{pmatrix} = -a^T \nabla \xi^T \nabla_p \varphi$ for all $a \in \mathbb{R}^m$ implies, for any test function $\varphi : \mathbb{R}^{6N} \to \mathbb{R}$:

$$\begin{aligned} &\{\varphi, \Xi\} \Gamma^{-1}(q_t, p_t) \Big(\dot{\zeta}(t) - \{\Xi, H\}(q_t, p_t) \Big) = \\ &\left(G_M^{-1}(q_t)\ddot{z}(t) + f_{\mathrm{rgd}}^M(q_t, p_t)\right)^T \nabla \xi(q_t)^T \nabla_p \varphi(q_t, p_t). \end{aligned} \tag{4.94}$$

Now the transport terms of the switched dynamics (4.92) can be recognized in (4.94), so that the generator of (4.92) reads, for any smooth test function φ:

$$\begin{aligned} \mathcal{L}_t^{\mathrm{trans}} \varphi &= \Big(\nabla \xi f_{\mathrm{rgd}}^M + \nabla \xi G_M^{-1} \ddot{z}(t) \Big)^T \nabla_p \varphi - (\nabla V)^T \nabla_p \varphi + p^T M^{-1} \nabla_q \varphi \\ &= \{\varphi, \Xi\} \Gamma^{-1} \Big(\dot{\zeta}(t) - \{\Xi, H\} \Big) + \{\varphi, H\} \\ &= \{\varphi, H\}_\Xi + \{\varphi, \Xi\} \Gamma^{-1} \dot{\zeta}(t), \end{aligned} \tag{4.95}$$

recognizing the Poisson bracket (4.81). The expression of the generator $\mathcal{L}_t^{\mathrm{f}}$ is then obtained by adding the terms arising from the fluctuation/dissipation. These terms are obtained directly from the terms involving γ and σ in (4.87).

The generator of the backward switching process given by (4.90) can be obtained from similar computations. First, consider the Hamiltonian part in the dynamics (4.90). By definition of the backward dynamics, the transport part inferred from (4.95) is

$$\begin{aligned} \mathcal{L}_{t'}^{\mathrm{b,trans}}(\varphi)(q,p) &= \mathcal{R} \mathcal{L}_{T-t'}^{\mathrm{trans}} \left(\mathcal{R}(\varphi) \right)(q,p) \\ &= \Big(\nabla \xi(q) f_{\mathrm{rgd}}^M(q,p) + \nabla \xi(q) G_M^{-1}(q) \ddot{z}(T-t') \Big)^T (-\nabla_p \varphi)(q,p) \\ &\quad - \nabla V(q)^T (-\nabla_p \varphi) - p^T M^{-1} \nabla_q \varphi(q,p). \end{aligned}$$

This gives, with (4.95),

$$\begin{aligned} \mathcal{L}_{T-t'}^{\mathrm{b,trans}} \varphi &= - \{\varphi, \Xi\} \Gamma^{-1} \Big(\dot{\zeta}(T-t') - \{\Xi, H\} \Big) - \{\varphi, H\} \\ &= - \{\varphi, H\}_\Xi - \{\varphi, \Xi\} \Gamma^{-1} \dot{\zeta}(T-t'). \end{aligned}$$

On the other hand, the thermostat part in (4.90) and in (4.76) are the same, with a schedule $t' \mapsto z(T-t')$ instead of $t \mapsto z(t)$. As a consequence, the generators of the thermostat part in (4.90) and in (4.76) are the same, upon replacing $\dot{z}(t)$ by:

$$\frac{d}{dt'} \Big(z(T-t') \Big) = -\dot{z}(T-t').$$

This gives the claimed expression for $\mathcal{L}_{t'}^{\mathrm{b}}$. □

4.3.2.3 *Definition of the work*

We define the work $(\mathcal{W}_t)_{t\geq 0}$ associated with the constraining force $\nabla\xi(q_t)\,d\lambda_t$ in (4.76) as the physical displacement multiplied by the force:

$$\begin{aligned} d\mathcal{W}_t &:= \left(\frac{dq_t}{dt}\right)^T \circ \left(\nabla\xi(q_t)\,d\lambda_t\right) &(4.96)\\ &= \left(\frac{dq_t}{dt}\right)^T \nabla\xi(q_t) \circ d\lambda_t \\ &= \dot{z}^T(t) \circ d\lambda_t \\ &= \dot{z}^T(t)\,d\lambda_t. &(4.97)\end{aligned}$$

By convention, $\mathcal{W}_0 = 0$. In the above, we have used successively the fact that $t \mapsto \xi(q_t)$, and then $t \mapsto z(t)$ are differentiable processes, so that Stratonovitch and Itô integrations are equivalent.

Remark 4.17 (Energy and heat exchanges). *The work exerted on the system during the switching can be decomposed as the sum of an energy variation term, and a term interpreted as the heat exchanged with the stochastic thermostat. Indeed, the energy variation of the Langevin process* (4.76) *over time is computed by Stratonovitch stochastic differentiation (see* (2.18) *and Remark* 2.6 *in Section* 2.2.1*):*

$$\begin{aligned} dH(q_t,p_t) &= p_t^T M^{-1} \circ dp_t + p_t^T M^{-1}\nabla V(q_t)\,dt \\ &= p_t^T M^{-1} \circ \left(-\gamma(q_t)\,M^{-1}p_t\,dt + \sigma(q_t)\,dW_t + \nabla\xi(q_t)\,d\lambda_t\right) \\ &= d\mathcal{W}_t + d\mathcal{Q}_t,\end{aligned}$$

where the exchanged heat is defined as

$$d\mathcal{Q}_t := -p_t^T M^{-1}\gamma(q_t)\,M^{-1}p_t\,dt + p_t^T M^{-1} \circ \sigma(q_t)\,dW_t,$$

with again the convention $\mathcal{Q}_0 = 0$. Note that $\mathcal{Q}_t$ vanishes when the coupling with the thermostat is turned off ($\gamma = 0$ and $\sigma = 0$).

The work can further be decomposed into a "Hamiltonian part", interpreted as the work exchange with the system, and a "thermostat" part, interpreted as the work exchange with the thermostat. This is made precise in the following lemma:

Lemma 4.18. *The infinitesimal variations of the work* (4.96) *can be decomposed as:*

$$d\mathcal{W}_t = d\mathcal{W}_t^{\mathrm{ham}} + d\mathcal{W}_t^{\mathrm{thm}}, \qquad (4.98)$$

where

$$d\mathcal{W}_t^{\text{ham}} = \dot{\zeta}(t)^T \Gamma^{-1} \{\Xi, H\} (q_t, p_t)\, dt, \tag{4.99}$$

$$= \dot{z}(t)^T \Big(G_M^{-1}(q_t) \ddot{z}(t)\, dt + f_{\text{rgd}}^M(q_t, p_t) \Big) dt, \tag{4.100}$$

and

$$d\mathcal{W}_t^{\text{thm}} = \dot{z}(t)^T G_M(q_t)^{-1} \nabla\xi(q_t)^T M^{-1} \Big(\gamma(q_t)\, M^{-1} p_t\, dt - \sigma(q_t)\, dW_t \Big).$$

In the above, the term $\mathcal{W}_t^{\text{ham}}$ of the right-hand side corresponds to the Hamiltonian dynamics, while $\mathcal{W}_t^{\text{thm}}$ arises from the fluctuation/dissipation forces, and vanishes when the fluctuation/dissipation matrices are of the form (3.151)*: $(\gamma(q), \sigma(q)) = (\sigma_P(q), \gamma_P(q))$.*

Proof. The expression of the Lagrange multipliers in (4.86) yields:

$$\dot{z}(t)^T d\lambda_t = \dot{z}(t)^T \Big(G_M^{-1}(q_t) \ddot{z}(t)\, dt + f_{\text{rgd}}^M(q_t, p_t)\, dt \Big) + \dot{z}(t)^T G_M^{-1} \nabla\xi(q_t)^T M^{-1} \Big(\gamma(q_t)\, M^{-1} p_t\, dt - \sigma(q_t)\, dW_t \Big).$$

Moreover, (4.93) gives:

$$\begin{aligned} &\dot{z}^T(t) \Big(G_M^{-1}(q_t) \ddot{z}(t) + f_{\text{rgd}}^M(q_t, p_t) \Big) dt \\ &\quad = \dot{\zeta}(t)^T \Gamma^{-1}(q_t, p_t) \Big(\{\Xi, H\} (q_t, p_t) - \dot{\zeta}(t) \Big) \\ &\quad = \dot{\zeta}(t)^T \Gamma^{-1} \{\Xi, H\} (q_t, p_t), \end{aligned}$$

where in the last line we used $\dot{\zeta}(t)^T \Gamma^{-1} \dot{\zeta}(t) = 0$. This gives (4.100). Finally, notice that, by definition, $\nabla\xi(q)^T M^{-1} \gamma_P(q) = 0$ and $\nabla\xi(q)^T M^{-1} \sigma_P(q) = 0$, which shows that $\mathcal{W}_t^{\text{thm}}$ vanishes when considering tangential fluctuation/dissipation matrices $(\gamma(q), \sigma(q))$. □

4.3.2.4 *Jarzynski-Crooks identity*

We are now in a position to state the Jarzynski-Crooks identity for Langevin processes with constraints. As mentioned above, the work (4.98) contains terms from the thermostat part, which require a special care in the theoretical analysis, and lead to unnecessary variance in numerical schemes. Thus for simplicity, it will be assumed that the fluctuation/dissipation terms only apply to the tangential part of the constraints, which amounts to considering position dependent tensors $(\sigma_P(q), \gamma_P(q))$ of the form (4.78).

We therefore consider the following forward dynamics, labelled with the superscript f:

$$\begin{cases} dq_t^{\mathrm{f}} = M^{-1} p_t^{\mathrm{f}}\, dt, \\ dp_t^{\mathrm{f}} = -\nabla V(q_t^{\mathrm{f}})\, dt - \gamma_P(q_t^{\mathrm{f}})\, M^{-1} p_t^{\mathrm{f}}\, dt + \sigma_P(q_t^{\mathrm{f}})\, dW_t^{\mathrm{f}} + \nabla \xi(q_t^{\mathrm{f}})\, d\lambda_t^{\mathrm{f}}, \\ \xi(q_t^{\mathrm{f}}) = z(t). \qquad (C_q(t)) \end{cases} \tag{4.101}$$

The associated backward process writes

$$\begin{cases} dq_{t'}^{\mathrm{b}} = -M^{-1} p_{t'}^{\mathrm{b}}\, dt', \\ dp_{t'}^{\mathrm{b}} = \nabla V(q_{t'}^{\mathrm{b}})\, dt' - \gamma_P(q_{t'}^{\mathrm{b}}) M^{-1} p_{t'}^{\mathrm{b}}\, dt' + \sigma_P(q_{t'}^{\mathrm{b}})\, dW_{t'}^{\mathrm{b}} + \nabla \xi(q_{t'}^{\mathrm{b}})\, d\lambda_{t'}^{\mathrm{b}}, \\ \xi(q_{t'}^{\mathrm{b}}) = z(T - t'). \end{cases} \tag{4.102}$$

Since we consider fluctuation/dissipation matrices $(\sigma_P(q), \gamma_P(q))$ of the form (3.151), the generators of the forward switched and backward switched dynamics simplify respectively to (see Lemma 4.16):

$$\mathcal{L}_t^{\mathrm{f}} = \{\cdot, H\}_\Xi + \mathcal{L}_\Xi^{\mathrm{thm}} + \{\cdot, \Xi\}\, \Gamma^{-1} \dot{\zeta}(t), \tag{4.103}$$

and

$$\mathcal{L}_{t'}^{\mathrm{b}} = -\{\cdot, H\}_\Xi + \mathcal{L}_\Xi^{\mathrm{thm}} - \{\cdot, \Xi\}\, \Gamma^{-1} \dot{\zeta}(T - t'). \tag{4.104}$$

In the same way, the thermostat part in the exchanged work (4.98) vanishes for all $t \in [0, T]$, and the forward work then reads:

$$\mathcal{W}_{0,t}^{\mathrm{f}} = \int_0^t w^{\mathrm{f}}(s, q_s^{\mathrm{f}}, p_s^{\mathrm{f}})\, ds,$$

with

$$w^{\mathrm{f}}(t, q, p) = \dot{\zeta}(t)^T \Gamma^{-1} \{\Xi, H\}(q, p). \tag{4.105}$$

We are now in position to state the main theorem of this section.

Theorem 4.19 (Jarzynski-Crooks). *Denote by* $(q_t^{\mathrm{f}}, p_t^{\mathrm{f}})_{0 \le t \le T}$ *the forward Langevin process solution of* (4.101) *with initial conditions sampled according to:*

$$(q_0, p_0) \sim \mu_{\Sigma_{\xi, v_\xi}(z(0), \dot{z}(0))}(dq\, dp),$$

and by $(q_{t'}^{\mathrm{b}}, p_{t'}^{\mathrm{b}})_{0 \le t' \le T}$ *(for* $t' = T - t \in [0, T]$*) the backward process, solution of* (4.102) *with initial conditions sampled according to:*

$$(q_0^{\mathrm{b}}, p_0^{\mathrm{b}}) \sim \mu_{\Sigma_{\xi, v_\xi}(z(T), \dot{z}(T))}(dq\, dp).$$

Then the Crooks-Jarzynski identity holds on $[0,T]$*: for any bounded path functional* $\varphi_{[0,T]}$,

$$\frac{Z_{z(T),\dot{z}(T)}}{Z_{z(0),\dot{z}(0)}}\mathbb{E}\left(\varphi^{\mathrm{r}}_{[0,T]}(q^{\mathrm{b}},p^{\mathrm{b}})\right) = \mathbb{E}\left(\varphi_{[0,T]}(q^{\mathrm{f}},p^{\mathrm{f}})\,\mathrm{e}^{-\beta \mathcal{W}^{\mathrm{f}}_{0,T}}\right),$$

where $(\,\cdot\,)^{\mathrm{r}}$ *denotes the composition with the operation of time reversal of paths, defined in* (4.47)*, and the distribution* $\mu_{\Sigma_{\xi,v_\xi}(z(t),\dot{z}(t))}$ *and the associated partition function* $Z_{z(t),\dot{z}(t)}$ *are given in* (4.83).

Note that the theorem still holds in the Hamiltonian case, where $\gamma = \sigma = 0$. Besides, the choice $\varphi_{[0,T]} = 1$ leads to the following work fluctuation identity:

$$F^{\xi,v_\xi}_{\mathrm{rgd}}(z(T),\dot{z}(T)) - F^{\xi,v_\xi}_{\mathrm{rgd}}(z(0),\dot{z}(0)) = -\frac{1}{\beta}\ln\left[\mathbb{E}\left(\mathrm{e}^{-\beta \mathcal{W}^{\mathrm{f}}_{0,T}}\right)\right]. \quad (4.106)$$

Estimators of the free energy difference, based on (4.106) can then be constructed as in the alchemical case, see Sections 4.1.3, 4.1.4 and 4.1.5 for further precision. The choice $\varphi_{[0,T]}(q,p) = \phi(q_T,p_T)$ leads to the following representation of the canonical distribution with constraints:

$$\frac{\mathbb{E}\left(\phi(q^{\mathrm{f}}_T,p^{\mathrm{f}}_T)\mathrm{e}^{-\beta \mathcal{W}^{\mathrm{f}}_{0,T}}\right)}{\mathbb{E}\left(\mathrm{e}^{-\beta \mathcal{W}^{\mathrm{f}}_{0,T}}\right)} = \int_{\Sigma_{\xi,v_\xi}(z(T),\dot{z}(T))} \phi(q,p)\,\mu_{\Sigma_{\xi,v_\xi}(z(T),\dot{z}(T))}(dq\,dp). \quad (4.107)$$

The usual free energy profile $z \mapsto F(z)$ can then be computed using the relation (4.85), and by combining (4.106) and (4.107). Introducing the corrector

$$C(t,q) = \frac{1}{2\beta}\ln\left(\det G_M(q)\right) - \frac{1}{2}\dot{z}(t)^T G_M^{-1}(q)\dot{z}(t), \quad (4.108)$$

the following equality holds:

$$F(z(T)) - F^{\xi,v_\xi}_{\mathrm{rgd}}(z(0),\dot{z}(0)) = -\frac{1}{\beta}\ln\left[\mathbb{E}\left(\mathrm{e}^{-\beta\left(\mathcal{W}^{\mathrm{f}}_{0,T}+C(T,q^{\mathrm{f}}_T)\right)}\right)\right]. \quad (4.109)$$

In (4.108), $\frac{1}{2\beta}\ln(\det G_M)$ is the Fixman entropic term due to the geometry of the position constraints (see also Remark 3.51), and $\frac{1}{2}\dot{z}(t)^T G_M^{-1}\dot{z}(t)$ is the kinetic energy term due to the effective velocity of the switched reaction coordinates.

Therefore, free energy differences can be estimated as

$$F(z(T)) - F(z(0)) = -\frac{1}{\beta}\ln\left[\frac{\mathbb{E}\left(\mathrm{e}^{-\beta\left(\mathcal{W}^{\mathrm{f}}_{0,T}+C(T,q^{\mathrm{f}}_T)\right)}\right)}{\mathbb{E}\left(\mathrm{e}^{-\beta C(0,q^{\mathrm{f}}_0)}\right)}\right]. \quad (4.110)$$

Again, estimators of the free energy in (4.110) can then be constructed as in the alchemical case, see Sections 4.1.3-4.1.5 for further precision.

Proof. The proof can be carried out by checking the conditions of the general Jarzynski-Crooks Theorem 4.10. Considering the generator $\mathcal{L}_t^{\mathrm{f}}$ of the forward process (4.103), and the generator $\mathcal{L}_{t'}^{\mathrm{b}}$ of the backward process (4.104), the nonequilibrium balance condition (4.43) to be satisfied reads in this context: for any two smooth test functions φ_1 and φ_2,

$$\begin{aligned} &\int_{\Sigma_{\xi,v_\xi}(z(t),\dot{z}(t))} \Big(\varphi_1 \mathcal{L}_t^{\mathrm{f}}(\varphi_2) - \varphi_2 \mathcal{L}_{T-t}^{\mathrm{b}}(\varphi_1)\Big) \mathrm{e}^{-\beta H} \, d\sigma_{\Sigma_{\xi,v_\xi}(z(t),\dot{z}(t))} \\ &\quad = \int_{\Sigma_{\xi,v_\xi}(z(t),\dot{z}(t))} \beta w^{\mathrm{f}}(t,\cdot)\varphi_1\varphi_2 \mathrm{e}^{-\beta H} \, d\sigma_{\Sigma_{\xi,v_\xi}(z(t),\dot{z}(t))} \\ &\qquad + \frac{d}{dt}\left(\int_{\Sigma_{\xi,v_\xi}(z(t),\dot{z}(t))} \varphi_1\varphi_2 \mathrm{e}^{-\beta H} \, d\sigma_{\Sigma_{\xi,v_\xi}(z(t),\dot{z}(t))} \right). \end{aligned} \tag{4.111}$$

We then need to verify two points:

(i) the existence of an open neighborhood of $\Sigma_{\xi,v_\xi}(z(t),\dot{z}(t))$ (resp. $\Sigma_{\xi,v_\xi}(z(t'),\dot{z}(t')))$ for any $t \in [0,T]$ (resp. $t' \in [0,T]$) where the probability transition semi-groups associated with the generator $\mathcal{L}_t^{\mathrm{f}}$ (resp. $\mathcal{L}_{t'}^{\mathrm{b}}$) is smooth;
(ii) the equality (4.111) itself.

The first point is verified in the overdamped case in the proof of Theorem 4.14. The method is directly adaptable in the present context, and details are therefore omitted.

We now concentrate on point (ii). First, using Lemma 3.63 of Section 3.3.6, the variation of the canonical equilibrium distribution with constraints with respect to the switching is

$$\begin{aligned} &\frac{d}{dt}\left(\int_{\Sigma_{\xi,v_\xi}(z(t),\dot{z}(t))} \varphi_1\varphi_2 \, \mathrm{e}^{-\beta H} \, d\sigma_{\Sigma_{\xi,v_\xi}(z(t),\dot{z}(t))} \right) \\ &\quad = \int_{\Sigma_{\xi,v_\xi}(z(t),\dot{z}(t))} \dot{\zeta}(t)^T \Gamma^{-1} \left\{ \Xi, \varphi_1\varphi_2 \, \mathrm{e}^{-\beta H} \right\} \, d\sigma_{\Sigma_{\xi,v_\xi}(z(t),\dot{z}(t))}. \end{aligned} \tag{4.112}$$

On the other hand, using the expressions (4.103)-(4.104) of the generators of the switched processes, as well as the definition of the forward work increment (4.105) and the expression of the generator of the fluctuation/dissipation part (4.91), we obtain:

$$\begin{aligned}
&\varphi_1 \mathcal{L}^{\mathrm{f}}_t(\varphi_2) - \varphi_2 \mathcal{L}^{\mathrm{b}}_{T-t}(\varphi_1) - \beta w^{\mathrm{f}}(t,\cdot)\varphi_1\varphi_2 \\
&\quad = \{\varphi_1\varphi_2, H\}_\Xi + \mathrm{e}^{\beta H}\left\{\varphi_1\varphi_2 \mathrm{e}^{-\beta H}, \Xi\right\}\Gamma^{-1}\dot{\zeta}(t) \\
&\quad + \varphi_1 \frac{1}{\beta}\mathrm{e}^{\beta H}\mathrm{div}_p\left(\mathrm{e}^{-\beta H} P_M \gamma P_M^T \nabla_p \varphi_2\right) \\
&\quad - \varphi_2 \frac{1}{\beta}\mathrm{e}^{\beta H}\mathrm{div}_p\left(\mathrm{e}^{-\beta H} P_M \gamma P_M^T \nabla_p \varphi_1\right).
\end{aligned} \tag{4.113}$$

Now, (4.111) is verified in two steps. First the last two terms in (4.113) cancel out after integration against $\mathrm{e}^{-\beta H}\, d\sigma_{\Sigma_{\xi,v_\xi}(z(t),\dot{z}(t))}$. To see it, consider the decomposition of measures (3.130) in Proposition 3.40 and the divergence formula on affine spaces, similar to (3.157):

$$\int_{\Sigma_{v_\xi(q,\cdot)}(\dot{z}(t))} \mathrm{div}_p\left(P_M(q)\phi\right)\, \sigma^{M-1}_{\Sigma_{v_\xi(q,\cdot)}(\dot{z}(t))}(dp) = 0. \tag{4.114}$$

The choice $\phi = \varphi_1 \mathrm{e}^{-\beta H}\gamma P_M^T \nabla_p \varphi_2$ shows that,

$$\begin{aligned}
&\frac{1}{\beta}\int_{\Sigma_{\xi,v_\xi}(z(t),\dot{z}(t))} \varphi_1 \mathrm{div}_p\left(\mathrm{e}^{-\beta H} P_M \gamma P_M^T \nabla_p \varphi_2\right) d\sigma_{\Sigma_{\xi,v_\xi}(z(t),\dot{z}(t))} \\
&\quad = -\frac{1}{\beta}\int_{\Sigma_{\xi,v_\xi}(z(t),\dot{z}(t))} \nabla_p \varphi_1^T P_M \gamma P_M^T \nabla_p \varphi_2\, \mathrm{e}^{-\beta H}\, d\sigma_{\Sigma_{\xi,v_\xi}(z(t),\dot{z}(t))}.
\end{aligned}$$

The latter expression is symmetric with respect to (φ_1, φ_2), which yields the claimed cancellation in (4.113). Second, the term $\{\varphi_1\varphi_2, H\}_\Xi$ in (4.113) vanishes after integration against $\mathrm{e}^{-\beta H}\, d\sigma_{\Sigma_{\xi,v_\xi}(z(t),\dot{z}(t))}$ thanks to the divergence theorem (see Proposition (3.46)), and the equality

$$\{\varphi_1\varphi_2, H\}_\Xi\, \mathrm{e}^{-\beta H} = -\frac{1}{\beta}\left\{\varphi_1\varphi_2, \mathrm{e}^{-\beta H}\right\}_\Xi.$$

Finally, an integration of (4.113) against $\mathrm{e}^{-\beta H}\, d\sigma_{\Sigma_{\xi,v_\xi}(z(t),\dot{z}(t))}$ gives, in view of (4.112),

$$\begin{aligned}
&\int_{\Sigma_{\xi,v_\xi}(z(t),\dot{z}(t))} \left(\varphi_1 \mathcal{L}^{\mathrm{f}}_t(\varphi_2) - \varphi_2 \mathcal{L}^{\mathrm{b}}_{T-t}(\varphi_1) - \beta w^{\mathrm{f}}(t,\cdot)\varphi_1\varphi_2\right)\mathrm{e}^{-\beta H}\, d\sigma_{\Sigma_{\xi,v_\xi}(z(t),\dot{z}(t))} \\
&\quad = \int_{\Sigma_{\xi,v_\xi}(z(t),\dot{z}(t))} \left\{\varphi_1\varphi_2 \mathrm{e}^{-\beta H}, \Xi\right\}\Gamma^{-1}\dot{\zeta}(t)\, d\sigma_{\Sigma_{\xi,v_\xi}(z(t),\dot{z}(t))} \\
&\quad = \frac{d}{dt}\left(\int_{\Sigma_{\xi,v_\xi}(z(t),\dot{z}(t))} \varphi_1\varphi_2\, \mathrm{e}^{-\beta H}\, d\sigma_{\Sigma_{\xi,v_\xi}(z(t),\dot{z}(t))}\right),
\end{aligned}$$

which is indeed (4.111). □

4.3.2.5 *Numerical schemes*

We present here a numerical scheme in the special case treated in the previous section, namely for fluctuation/dissipation tensors given by (3.151), that is to say of the form $(\sigma_P, \gamma_P) := (P_M(q)\,\sigma, P_M(q)\,\gamma\, P_M(q)^T)$. For simplicity, notation of the possible dependance with respect to q in (γ, σ) will be omitted. As for Langevin processes with constraints in Section 3.3, a splitting between the Hamiltonian part and the thermostat parts of the dynamics (4.76) gives rise to simple and natural schemes.

The reaction coordinate path is first discretized as $\{z(0), \dots, z(t_{N_T})\}$ where N_T is the number of time-steps. Equal time increments can be used, in which case $\Delta t = \frac{T}{N_T}$ and $t_n = n\Delta t$. The deterministic equations of motion (4.92) with switched position constraints $\xi(q) = z(t)$ can be integrated by a velocity-Verlet algorithm with constraints similar to (3.158). It reads as follows:

$$\begin{cases} p^{n+1/2} = p^n - \dfrac{\Delta t}{2}\nabla V(q^n) + \nabla\xi(q^n)\lambda^{n+1/2}, \\ q^{n+1} = q^n + \Delta t\; M^{-1}p^{n+1/2}, \\ \xi(q^{n+1}) = z(t_{n+1}), & (C_q) \\ p^{n+1} = p^{n+1/2} - \dfrac{\Delta t}{2}\nabla V(q^{n+1}) + \nabla\xi(q^{n+1})\lambda^{n+1}, \\ \nabla\xi(q^{n+1})^T M^{-1} p^{n+1} = \dfrac{z(t_{n+2}) - z(t_{n+1})}{\Delta t}, & (C_p) \end{cases} \tag{4.115}$$

where Δt is the time-step. The Lagrange multipliers $\lambda^{n+1/2}$ are associated with the position constraints (C_q), and the Lagrange multipliers λ^{n+1} are associated with the velocity constraints (C_p). Note that the velocity of the switching at time t_{n+1} is discretized as:

$$\dot{z}(t_{n+1}) \simeq \frac{z(t_{n+2}) - z(t_{n+1})}{\Delta t}.$$

The latter choice is motivated by the following observation: The position after one step of an unconstrained motion, given by

$$\tilde{q}^{n+1} = q^n + \Delta t\; M^{-1}p^n - \frac{\Delta t^2}{2}M^{-1}\nabla V(q^n),$$

already satisfies (C_q) at order 2 with respect to Δt. Indeed, using the constraint (C_p) at the previous step:

$$\begin{aligned} \xi(\tilde{q}^{n+1}) &= \xi(q^n) + \Delta t \nabla\xi(q^n)^T M^{-1} p^n + \mathrm{O}(\Delta t^2) \\ &= z(t_{n+1}) + \mathrm{O}(\Delta t^2). \end{aligned}$$

Such a property was also satisfied by the Verlet scheme with time-independent constraints (RATTLE), see (3.158). It is useful to ensure a rapid convergence of the numerical algorithm solving the nonlinear constraints (C_q).

The fluctuation-dissipation term in the constrained Langevin equation with switching (4.76) can be integrated similarly to the constrained case without switching (3.159), using an Ornstein-Uhlenbeck process on momenta computed by a midpoint Euler scheme. Indeed, let us rewrite the fluctuation-dissipation equation on momenta for a given position q as

$$\begin{cases} dp_t = -\gamma_P(q) M^{-1} p_t \, dt + \sigma_P(q) \, dW_t + \nabla\xi(q) \, d\lambda_t, \\ \nabla\xi(q)^T M^{-1} p_t = \dot{z}(t). \qquad (C_p) \end{cases} \tag{4.116}$$

The latter evolution can be divided in two parts: (i) the evolution of the tangential part of the momenta:

$$\begin{aligned} d\left(P_M(q) p_t\right) &= -\,\gamma_P(q) M^{-1} p_t \, dt + \sigma_P(q) \, dW_t \\ &= -\,P_M(q)\gamma M^{-1}\left(P_M(q) p_t\right) dt + P_M(q)\sigma \, dW_t; \end{aligned}$$

and (ii) the evolution of the orthogonal part of the momenta:

$$d\left(\mathrm{Id} - P_M(q)\right) p_t = \nabla\xi(q)^T G_M^{-1}(q) \dot{z}(t) \, dt.$$

Assume that at step n, the following constraint on momenta holds (as in (4.115)):

$$\nabla\xi^T(q) M^{-1} p^n = \frac{z(t_{n+1}) - z(t_n)}{\Delta t}.$$

Then, the midpoint Euler scheme for the fluctuation-dissipation equation (4.116) reads:

$$\begin{cases} p^{n+1} = p^n - \dfrac{\Delta t}{2}\gamma M^{-1}(p^{n+1} + p^n) + \sqrt{\Delta t}\,\sigma\, G^n \\ \qquad +\Delta t \gamma M^{-1} \nabla\xi(q) G_M(q)^{-1} \dfrac{z(t_{n+1}) - z(t_n)}{\Delta t} + \nabla\xi(q)\lambda^{n+1}, \\ \nabla\xi(q)^T M^{-1} p^{n+1} = \dfrac{z(t_{n+1}) - z(t_n)}{\Delta t}, \end{cases} \tag{4.117}$$

where $(G^n)_{n\geq 0}$ are i.i.d. centered and normalized Gaussian vectors, and λ^{n+1} are the Lagrange multipliers associated with the momenta constraints.

Indeed, upon remarking that (using the definition of P_M)

$$p^{n+1} + p^n = P_M(q)(p^{n+1} + p^n) + 2\nabla\xi(q)G_M(q)^{-1}\frac{z(t_{n+1}) - z(t_n)}{\Delta t},$$

and inserting this expression in the damping term in (4.117), it is possible to compute p^{n+1} by decomposing it as the sum of two contributions: (i) the tangential part

$$P_M(q)p^{n+1} = P_M(q)p^n - \frac{\Delta t}{2}\gamma_P(q)M^{-1}(p^{n+1} + p^n) + \sqrt{\Delta t}\,\sigma_P(q)\,G^n,$$

and (ii) the orthogonal part

$$\nabla\xi(q)^T M^{-1} p^{n+1} = \frac{z(t_{n+1}) - z(t_n)}{\Delta t}. \tag{4.118}$$

This second part is responsible for the Lagrange multiplier λ^{n+1} in (4.117). In practice, (4.117) may be rewritten into the more explicit form:

$$\begin{aligned}\left(\mathrm{Id} + \frac{\Delta t}{2}\gamma M^{-1}\right)p^{n+1} &= \left(\mathrm{Id} - \frac{\Delta t}{2}\gamma M^{-1}\right)p^n \\ &+ \Delta t\gamma M^{-1}\nabla\xi(q)G_M(q)^{-1}\frac{z(t_{n+1}) - z(t_n)}{\Delta t} \\ &+ \sqrt{\Delta t}\,\sigma\,G^n + \nabla\xi(q)\,\lambda^{n+1}.\end{aligned}$$

The Lagrange multiplier λ^n can then be computed from (4.118) by solving a well-posed linear system, see the discussion below the scheme (3.159) for the case of stationary constraints. Note that the latter computation becomes particularly simple when choosing a dissipation matrix of the form $\gamma = \frac{\gamma_0}{m_0}M$, where $m_0, \gamma_0 > 0$ are tunable mass and friction coefficients.

Finally the splitting scheme for the Langevin dynamics with constraints and a switching schedule reads as follows:

$$\begin{cases} p^{n+1/4} = p^n - \dfrac{\Delta t}{4}\gamma M^{-1}(p^{n+1/4} + p^n) + \sqrt{\dfrac{\Delta t}{2}}\sigma G^n \\ \qquad\qquad + \dfrac{\Delta t}{2}\gamma M^{-1}\nabla\xi(q^n)G_M(q^n)^{-1}\dfrac{z(t_{n+1}) - z(t_n)}{\Delta t} + \nabla\xi(q^n)\lambda^{n+1/4}, \\ \nabla\xi(q^n)^T M^{-1} p^{n+1/4} = \dfrac{z(t_{n+1}) - z(t_n)}{\Delta t}, \end{cases}$$

$$\begin{cases} p^{n+1/2} = p^{n+1/4} - \dfrac{\Delta t}{2}\nabla V(q^n) + \nabla\xi(q^n)\lambda^{n+1/2}, \\ q^{n+1} = q^n + \Delta t\, M^{-1} p^{n+1/2}, \\ \xi(q^{n+1}) = z(t_{n+1}), \\ p^{n+3/4} = p^{n+1/2} - \dfrac{\Delta t}{2}\nabla V(q^{n+1}) + \nabla\xi(q^{n+1})\lambda^{n+3/4}, \\ \nabla\xi(q^{n+1})^T M^{-1} p^{n+1} = \dfrac{z(t_{n+2}) - z(t_{n+1})}{\Delta t}, \end{cases}$$

$$\begin{cases} p^{n+1} = p^{n+3/4} - \dfrac{\Delta t}{4}\gamma M^{-1}(p^{n+1} + p^{n+3/4}) + \sqrt{\dfrac{\Delta t}{2}}\sigma G^{n+1/2} \\ \qquad\qquad + \dfrac{\Delta t}{2}\gamma M^{-1}\nabla\xi(q^{n+1})G_M(q^{n+1})^{-1}\dfrac{z(t_{n+2}) - z(t_{n+1})}{\Delta t} + \nabla\xi(q^{n+1})\lambda^{n+1}, \\ \nabla\xi(q^{n+1})^T M^{-1} p^{n+1} = \dfrac{z(t_{n+2}) - z(t_{n+1})}{\Delta t}. \end{cases} \tag{4.119}$$

The approximation of the work $\mathcal{W}_{0,t}(q,p)$ using Lagrange multipliers according to formula (4.105) can be used, for instance with the following discretization formula based on (4.96):

$$\begin{cases} \mathcal{W}^0 = 0, \\ \mathcal{W}^{n+1} = \mathcal{W}^n + \left(\dfrac{z(t_{n+1}) - z(t_n)}{\Delta t}\right)^T \left(\lambda^{n+1/2} + \lambda^{n+3/4}\right), \end{cases} \tag{4.120}$$

for $n = 0 \dots N_T - 1$. Alternatively, $\mathcal{W}^{n+1}$ can be computed using the explicit expression (4.100) of the constraining force, for instance with the discretization formula

$$\mathcal{W}^{n+1} = \mathcal{W}^n + (\dot{z}^n)^T \left(G_M^{-1}(q^n)\ddot{z}^n + f_{\mathrm{rgd}}^M(q^n, p^n)\right)\Delta t, \tag{4.121}$$

with

$$\dot{z}^n = \frac{z(t_{n+1}) - z(t_n)}{\Delta t}, \qquad \ddot{z}^n = \frac{\dot{z}^{n+1} - \dot{z}^n}{\Delta t}.$$

The free energy profile is then estimated as explained in Section 4.1, by using several simulated replicas of the switched system, and by discretizing the fluctuation identity (4.110). The consistency of the scheme is given by the following proposition.

Proposition 4.20 (Consistency). *The approximation formula* (4.120) *is consistent. More precisely, the Lagrange multipliers* $(\lambda^{n+1/2}, \lambda^{n+3/4})$ *in the Verlet scheme with switching* (4.115) *verify*

$$\begin{cases} \lambda^{n+1/2} = f_{\mathrm{rgd}}^M(q^n, p^{n+1/2})\dfrac{\Delta t}{2} + \mathrm{O}(\Delta t^2) \\ \lambda^{n+3/4} = f_{\mathrm{rgd}}^M(q^{n+1}, p^{n+1/2})\dfrac{\Delta t}{2} + G_M^{-1}(q^{n+1})\ddot{z}(t_{n+1})\,\Delta t + \mathrm{O}(\Delta t^2). \end{cases}$$

Moreover, the following second order consistency holds for the sum of the multipliers which appear in (4.120)*:*

$$\begin{aligned} \lambda^{n+1/2} + \lambda^{n+3/4} =& f_{\mathrm{rgd}}^M(q^n, p^{n+1/2})\frac{\Delta t}{2} + f_{\mathrm{rgd}}^M(q^{n+1}, p^{n+1/2})\frac{\Delta t}{2} \\ &+ G_M^{-1}(q^{n+1})\ddot{z}(t_{n+1})\,\Delta t + \mathrm{O}(\Delta t^3). \end{aligned}$$

Proof. The proof is similar to the proof of Proposition 3.67. The computation of $\lambda^{n+1/2}$ follows the lines of the corresponding computation in the proof of Proposition 3.67 by remarking that, with the choice of the discretization of the velocity constraint,

$$\begin{aligned} \xi(q^{n+1}) - \xi(q^n) - \Delta t \nabla\xi(q^n)^T M^{-1} p^n &= z(t_{n+1}) - z(t_n) - \Delta t \frac{z(t_{n+1}) - z(t_n)}{\Delta t} \\ &= 0. \end{aligned}$$

For the computation of $\lambda^{n+3/4}$, the additional term $G_M^{-1}(q^{n+1})\ddot{z}(t_{n+1})$ comes from the following approximation:

$$\begin{aligned} &\xi(q^n) - \xi(q^{n+1}) + \Delta t \nabla\xi(q^{n+1})^T M^{-1} p^{n+1} = \\ &\quad z(t_n) - z(t_{n+1}) + \Delta t \frac{z(t_{n+2}) - z(t_{n+1})}{\Delta t} = \ddot{z}(t_{n+1})\,\Delta t^2 + \mathrm{O}(\Delta t^4). \end{aligned}$$

The summation of the Lagrange multipliers yields the second order consistency, by computations similar to the ones performed in the proof of Proposition 3.67. □

Note that computing the work by (3.181) with the splitting scheme (4.119) can be performed with overdamped velocities, using the method of Proposition 3.57, *i.e.* by choosing the following relation in (4.119):

$$\frac{\Delta t}{4}\gamma = M = \mathrm{Id}.$$

This leads to original schemes that can be seen as variants of (4.74)-(4.75) which were derived directly in the overdamped case.

4.3.3 *Numerical results*

We present some free energy profiles obtained with nonequilibrium switching dynamics for the model system described in Section 1.3.2.4 (dimer in a solvent), with the same parameters as in Section 2.5.2.3.

In the overdamped case, the numerical scheme (4.74) is used, the work being computed with the analytic expression of the mean force, see (4.3.1.3). The time-step is $\Delta t = 2.5 \times 10^{-4}$, and the switching schedule reads

$$z(t) = z_{\min} + (z_{\max} - z_{\min})\frac{t}{T}$$

with $z_{\min} = -0.1$ and $z_{\max} = 1.1$. The initial equilibrium conditions are obtained by subsampling a constrained equilibrium dynamics at $\xi(q) = z_{\min}$, with a time spacing $T_{\text{sample}} = 0.25$.

For Langevin dynamics, we use the same schedule as for the overdamped dynamics. In the specific case at hand, the corrector term (4.108) is constant, and free energies differences are equal to differences of rigid free energies. The dynamics used to integrate the nonequilibrium dynamics is based on a splitting strategy, analogous to (4.119), except that the midpoint integration of the Ornstein-Uhlenbeck part is replaced by an exact integration for the unconstrained dynamics, followed by a projection. This can be done here since we choose a friction matrix of the form $\gamma\,\mathrm{Id}$ (recall also that $M = \mathrm{Id}$). More precisely, the corresponding scheme is obtained by replacing (4.117) with

$$\widetilde{p}^{n+1} = \alpha p^n + \sqrt{\frac{1-\alpha^2}{\beta}}\,G^n,$$

and $p^{n+1} = \widetilde{p}^{n+1} + \lambda^{n+1}\nabla\xi(q)$ with λ^{n+1} chosen such that

$$\nabla\xi(q)^T M^{-1} p^{n+1} = \frac{z(t_{n+1}) - z(t_n)}{\Delta t}.$$

The time-step is $\Delta t = 0.01$, and $\gamma = 1$. The initial conditions are obtained by first subsampling a constrained dynamics with $\xi(q) = z_{\min}$ and $v_\xi(q,p) = 0$, with a time spacing $T_{\text{sample}} = 1$; and then adding the required component $\nabla\xi(q)G_M^{-1}\dot{z}(0)$ to the momentum (with $\dot{z}(0) = (z_{\max} - z_{\min})/T$).

Figure 4.4 presents estimates obtained with M independent realizations of the switching dynamics for different switching times T. In all cases, the product MT is kept constant. The free energy profile becomes closer to the reference curve as T is decreased. Of course, more reliable profiles could be obtained with backward switchings and Crooks relation, see Section 4.2.3 and [Minh and Adib (2008)].

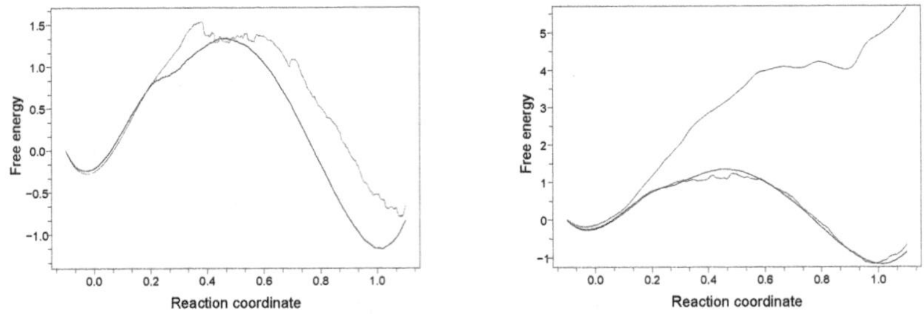

Fig. 4.4 Left: Overdamped case, from top to bottom: $T = 1$ with $M = 10^4$, and $T = 100$ and $M = 100$. Right: Langevin dynamics. The top curve corresponds to $T = 1$ with $= 10^5$, while the two other curves were obtained for $T = 10$ with $M = 10^4$ and $T = 100$ with $M = 10^3$ (smoothest curve).

4.4 Path sampling strategies

The Jarzynski equality can be rewritten as an average over work values, see (4.54) and (4.23):

$$e^{-\beta \Delta F(t)} = \int_{\mathbb{R}} e^{-\beta W} P_t(dW), \tag{4.122}$$

where the work distribution P_t is defined in (4.22). The idea behind path sampling methods is that, instead of sampling initial conditions and performing independent switchings, it is possible instead to sample directly switching paths, by a Metropolis-Hastings procedure where a new path is obtained by a perturbation of a previous one. This may be interesting if the initial conditions are difficult to sample (due to the existence of several metastable states for example), but most importantly because many refinements may be used to efficiently sample paths.

For simplicity, the presentation is restricted to the case of alchemical transitions, but the method can be used in the reaction coordinate case as well. The path ensemble is defined in Section 4.4.1, and some sampling algorithms are then presented in Section 4.4.2.

4.4.1 *The path ensemble*

We define in this section the measure on path space, for dynamics which are *discrete in time* (discretizations of continuous in time dynamics, or Markov chain dynamics in the Metropolis-Hastings fashion, see Section 4.1.3).

4.4.1.1 *Equilibrium paths*

We first consider ensembles of paths obtained from a time-homogeneous dynamics. In the sequel, the variable x represents either the configurational part q of the phase space (when overdamped Langevin dynamics are used for instance), or the full phase space variables (q, p) (for Hamiltonian or Langevin dynamics). In both cases, the accessible space is denoted by $\mathcal{S}$. The energy $E(x)$ is accordingly either $H(q,p) = \frac{1}{2}p^T M^{-1} p + V(q)$ or $V(q)$. Denote by Δt the time-step used to discretize a trajectory of length $T = L\Delta t$ (or directly a number of iteration steps L when discrete time dynamics are considered from the beginning). A discrete trajectory is a sequence

$$x = (x_0, \dots, x_L) \in \mathcal{S}^{L+1} = \mathcal{P}. \tag{4.123}$$

Note that, contrarily to the standard notation in this book, the iteration index along a given trajectory is in subscript (and not in superscript), since the superscript index is used for the numbering of the Monte-Carlo iterations modifying the trajectories.

The measure on the set of all trajectories (4.123) is

$$\pi(dx) = \rho(x_0)\, dx_0 \prod_{i=0}^{L-1} \mathrm{p}(x_i, dx_{i+1}), \tag{4.124}$$

where $\rho(x_0) = Z^{-1} \mathrm{e}^{-\beta E(x_0)}$ is the Boltzmann weight of the initial configuration, and $\mathrm{p}(x, dy)$ is the transition probability of the dynamics. Note that the measures on the space of discretized paths depend of course on the time-step. For the ease of notation, we will however omit this dependence in the sequel.

The transition probability $\mathrm{p}(x, dy)$ depends on the underlying dynamics and its discretization. We will focus in this section on two paradigmatic dynamics: the Hamiltonian dynamics, which is often used in this context, and the overdamped Langevin dynamics, which is representative of stochastic dynamics (for Langevin dynamics, see [Stoltz (2007)]). For Hamiltonian dynamics, the state $x_{i+1} = (q_{i+1}, p_{i+1})$ is a deterministic function of the previous state $x_i = (q_i, p_i)$, obtained by an application of the Verlet integrator $\Phi_{\Delta t}^{\mathrm{Verlet}}$ associated with (1.22). Therefore, the measure describing paths $x = ((q_0, p_0), \dots, (q_L, p_L))$ obtained with the Verlet integrator reads

$$\pi(dx) = Z^{-1}\, \mathrm{e}^{-\beta H(q_0, p_0)}\, dq_0\, dp_0 \prod_{i=1}^{L} \delta_{\Phi_{\Delta t}^{\mathrm{Verlet}}(q_{i-1}, p_{i-1})}(dq_i\, dp_i),$$

where δ_a is the Dirac mass at $a \in T^*\mathcal{D}$. For overdamped Langevin dynamics discretized with the Euler-Maruyama scheme

$$q_{i+1} = q_i - \Delta t \, \nabla V(q_i) + \sqrt{\frac{2\Delta t}{\beta}} \, G_i, \qquad G_i \sim \mathcal{N}(0, \mathrm{Id}_{3N}),$$

the transition probability $\mathrm{p}(q_i, dq_{i+1})$ is

$$\mathrm{p}(q_i, dq_{i+1}) = \left(\frac{\beta}{4\pi\Delta t}\right)^{3N/2} \exp\left(-\frac{\beta}{4\Delta t} \left|q_{i+1} - q_i + \Delta t \, \nabla V(q_i)\right|^2\right) dq_{i+1}.$$

4.4.1.2 *Switching paths*

For nonequilibrium switching processes, the energy function (hence the force, and the transition probabilities), depend on a parameter which changes with time. We denote by E_λ the energies V_λ or H_λ, depending on the context. We consider in the sequel that a sequence of switching parameters

$$\Lambda = (\lambda_0, \ldots, \lambda_L) \in [0,1]^{L+1},$$

with $\lambda_0 = 0$ and $\lambda_L = 1$, is given. Resorting to the scheme presented in Section 4.1.3, the measure on the space of discretized paths (4.123), indexed by the sequence Λ, is

$$\pi(dx; \Lambda) = \rho(x_0) \, dx_0 \prod_{i=0}^{L-1} \mathrm{p}(x_i, dx_{i+1}; \lambda_{i+1}), \qquad (4.125)$$

where $\rho(x_0) = Z_0^{-1} \mathrm{e}^{-\beta E_{\lambda_0}(x_0)}$, and the transition probability $\mathrm{p}(x, dy; \lambda)$ depends on the chosen dynamics.

When Hamiltonian dynamics is used (see (4.19)), the new state is again completely determined by the previous one, so that the measure on switching paths reads

$$\pi(dx; \Lambda) = Z_0^{-1} \exp(-\beta H_{\lambda_0}(q_0, p_0)) \, dq_0 \, dp_0 \prod_{i=1}^{L} \delta_{\Phi_{\Delta t}^{\mathrm{Verlet}}(q_{i-1}, p_{i-1}; \lambda_i)}(dq_i \, dp_i),$$

where $\Phi_{\Delta t}^{\mathrm{Verlet}}(q, p; \lambda)$ is the Verlet integrator associated with the Hamiltonian H_λ. For overdamped Langevin dynamics, with the discretization (4.18), the transition probability reads

$$\mathrm{p}(q_i, dq_{i+1}; \lambda_{i+1}) = \left(\frac{\beta}{4\pi\Delta t}\right)^{3N/2} \exp\left(-\frac{\beta}{4\Delta t} \left|q_{i+1} - q_i + \Delta t \, \nabla V_{\lambda_{i+1}}(q_i)\right|^2\right) dq_{i+1}.$$

The expectation in the Jarzynski equality (4.10) can be approximated (up to a time discretization error that we neglect in this section) by an integral over all possible paths:

$$\mathrm{e}^{-\beta \Delta F} = \int_{\mathcal{P}} \mathrm{e}^{-\beta \mathcal{W}(x;\,\Lambda)}\, \pi(dx;\Lambda), \tag{4.126}$$

since, given the schedule Λ, the work (4.17) is a function of the discrete trajectory:

$$\mathcal{W}(x;\,\Lambda) = \sum_{i=0}^{n-1} E_{\lambda_{i+1}}(x_i) - E_{\lambda_i}(x_i).$$

The time-step error in estimated value of the free energy difference vanishes in the limit $\Delta t \to 0$, or when the switching is performed with a Metropolis-Hastings scheme (see Remark 4.5).

4.4.2 *Sampling switching paths*

After a presentation of the general Metropolis-Hastings algorithm for path sampling in Section 4.4.2.1, we present two proposal moves in Sections 4.4.2.2 and 4.4.2.3, as well as some numerical refinements in Sections 4.4.2.4 and 4.4.2.6. After a brief discussion of the efficiency of the method (Section 4.4.2.5), we end this section with a numerical illustration (Section 4.4.2.7).

4.4.2.1 *General sampling strategy*

The set of all nonequilibrium switching trajectories can be (theoretically) obtained by sampling infinitely many initial conditions according to the canonical measure $Z^{-1} \exp(-\beta E_{\lambda_0}(x))\, dx$, and integrating switching trajectories (considering all the possible realizations starting from a given initial condition when stochastic dynamics are used).

Path sampling methods are other numerical strategies to sample the measure (4.125). The most efficient methods currently available rely on the Metropolis-Hastings algorithm. More precisely, a new path $y \in \mathcal{P}$ can be generated from a previous path $x \in \mathcal{P}$ by first selecting a discrete time index $0 \leq k \leq L$ (called "shooting index") at random along the path, and then proposing a new path y according to a proposal probability $T(x, dy;\, k, \Lambda)$ which depends on the chosen index k. This new path is then accepted or rejected according to the usual Metropolis-Hastings ratio

$$r(x, y;\, k, \Lambda) = \frac{\pi(dy;\Lambda)\, T(y, dx;\, k, \Lambda)}{\pi(dx;\Lambda)\, T(x, dy;\, k, \Lambda)}. \tag{4.127}$$

A path sampling method is therefore completely characterized by the way a proposed path is generated from a given path once a shooting point is fixed. We present in Sections 4.4.2.2 and 4.4.2.3 two types of proposals. Other possible proposals can be found in [Dellago *et al.* (2002)].

The precise algorithm reads as follows.

Algorithm 4.21 (Metropolis-Hastings algorithm for path sampling). *Assume that the Metropolis-Hasting ratio* (4.127) *exists and is positive for* $\pi(dx;\Lambda)\,T(x,dy;\,k,\Lambda)$*-almost all* $(x,y)\in\mathcal{P}^2$ *and all indices* $k=0,\ldots,L$. *Generate an initial path* x^0 *(for instance by a realization of a nonequilibrium switching dynamics), and iterate on* $n\geq 0$*:*

(1) Choose an iteration index $k\in\{0,\ldots,L\}$ *at random, according to a predefined discrete probability distribution;*
(2) Propose a new path y^{n+1} *from* x^n*, with the proposition probability* $T(x^n,dy;\,\Lambda,k)$*;*
(3) Accept the proposition with probability

$$\min\Big(1, r\big(x^n, y^{n+1};\, k,\Lambda\big)\Big),$$

and set in this case $x^{n+1}=y^{n+1}$*; otherwise, set* $x^{n+1}=x^n$.

It is easily seen that this algorithm leaves the path measure $\pi(dx;\Lambda)$ invariant, since the Metropolis-Hastings acceptance/rejection is constructed so that the distribution $\pi(dx;\Lambda)$ remains invariant for each shooting index k. The ergodicity therefore follows provided the irreducibility can be proven. This property has to be checked for each proposal, see below. When ergodicity holds, an estimator of the free energy difference can be constructed from (4.126) as

$$\lim_{N\to+\infty} -\frac{1}{\beta}\ln\left(\frac{1}{N}\sum_{n=1}^{N} \mathrm{e}^{-\beta\mathcal{W}(x^n;\,\Lambda)}\right) = \mathrm{e}^{-\beta\Delta F}. \qquad (4.128)$$

Note that the paths are correlated (so that the work values $\mathcal{W}(x^n;\,\Lambda)$ are also correlated), whereas the realizations of nonequilibrium switching dynamics starting from independent canonical initial conditions are independent. The hope is that the increased correlation is balanced by a more efficient sampling of the lower work values, so that the variance of the estimator (4.128) is lower than for i.i.d. realizations. As usual for Metropolis-Hastings algorithms, the performance of the proposal function can be measured by the decorrelation between successive paths. In particular, rejection rates should not be too large, otherwise the sampled path distribution will be degenerate. This reflects the usual trade-off between acceptance and decorrelation of accepted proposed moves.

4.4.2.2 *Shooting moves*

In the shooting algorithm [Dellago *et al.* (2002)], the configuration x_k^n corresponding to the chosen index k along the path x^n is first modified, and then a new path is generated by integrating the dynamics forward using the underlying dynamics, and proposing some initial part prior to the modified configuration using a convenient integrator. The so-constructed trajectory should be typical enough for the measure on path space, otherwise it is likely to be rejected by the Metropolis-Hastings algorithm. This requires that the construction of the part prior to the modified configuration is done properly.

The modification of a selected configuration x_k^n is usually performed using some symmetric modification, such as adding a random perturbation (for instance, Gaussian) to the momenta and/or to the positions. Such a modification is indeed important when Hamiltonian dynamics are used, in order to modify the energy levels. This is not necessary when stochastic dynamics are used. The new k-th configuration is denoted by y_k^{n+1}, and the associated generation probability is $\mathrm{p}_{\mathrm{gen}}(x_k^n, dy_k^{n+1})$.

Once the configuration has been modified, the remaining part of the trajectory is generated by integrating the dynamics starting from y_k^{n+1}, using the dynamics at hand, thus successively obtaining $y_{k+1}^{n+1}, \dots, y_L^{n+1}$. The generation of the part prior to the modified configuration requires more care. We denote by $\overline{\mathrm{p}}\left(y_{i+1}^{n+1}, dy_i^{n+1}; \lambda_{i+1}\right)$ the transition probability to generate a previous configuration y_i^{n+1} starting from y_{i+1}^{n+1}. The specific form of this transition probability depends on the dynamics under study, see below.

The proposal probability finally writes, for a given shooting index k,

$$T(x, dy;\, \Lambda, k) = \mathrm{p}_{\mathrm{gen}}(x_k, dy_k) \prod_{j=0}^{k-1} \overline{\mathrm{p}}(y_{j+1}, dy_j; \lambda_{j+1}) \prod_{j=k}^{L-1} \mathrm{p}(y_j, dy_{j+1}; \lambda_{j+1}).$$

Using (4.125) and (4.127), the Metropolis-Hastings ratio for a given shooting index k is then

$$\begin{aligned} r(x, y; k, \Lambda) = & \frac{\rho(y_0)\, dy_0}{\rho(x_0)\, dx_0} \frac{\mathrm{p}_{\mathrm{gen}}(y_k, dx_k)}{\mathrm{p}_{\mathrm{gen}}(x_k, dy_k)} \\ & \times \prod_{j=0}^{k-1} \frac{\mathrm{p}(y_j, dy_{j+1}; \lambda_{j+1})}{\overline{\mathrm{p}}(y_{j+1}, dy_j; \lambda_{j+1})} \frac{\overline{\mathrm{p}}(x_{j+1}, dx_j; \lambda_{j+1})}{\mathrm{p}(x_j, dx_{j+1}; \lambda_{j+1})}. \end{aligned} \tag{4.129}$$

Hamiltonian dynamics. In the Hamiltonian case, a natural idea to generate the part of the trajectory prior to the shooting point is to use the integrator

$$\Phi_{-\Delta t}^{\rm Verlet} = S \circ \Phi_{\Delta t}^{\rm Verlet} \circ S,$$

where S is the momentum reversal operator (*i.e.* $Sy_i = (q_i, -p_i)$ when $y_i = (q_i, p_i)$) and $\Phi_{\Delta t}^{\rm Verlet}$ is the Verlet integrator given by (1.22). The part of the trajectory prior to the shooting point is constructed iteratively as

$$\forall i = k-1, \ldots, 0, \qquad y_i = \left(S \circ \Phi_{\Delta t}^{\rm Verlet}(\,\cdot\,; \lambda_{i+1}) \circ S\right) y_{i+1}.$$

The associated transition probability is

$$\overline{\mathrm{p}}\,(y_{i+1}, dy_i; \lambda_{i+1}) = \delta_{\Phi_{-\Delta t}^{\rm Verlet}(y_{i+1}\,;\lambda_{i+1})}(dy_i),$$

so that

$$\overline{\mathrm{p}}\,(y_{i+1}, dy_i; \lambda_{i+1})\; dy_{i+1} = \mathrm{p}\,(y_i, dy_{i+1}; \lambda_{i+1})\; dy_i.$$

The ratio (4.129) then simplifies as

$$r(x, y;\, k, \Lambda) = \frac{\rho(y_0)\; \mathrm{p_{gen}}(y_k, dx_k)\, dy_k}{\rho(x_0)\; \mathrm{p_{gen}}(x_k, dy_k)\, dx_k}.$$

This expression shows that reasonable acceptance rates (*i.e.* $r(x, y;\, k, \Lambda)$ not too small) can be obtained by choosing suitable modifications of the shooting point. When symmetric modifications are considered, the acceptance rate depends only on the initial conditions.

Overdamped Langevin dynamics. A first idea to generate the part of the overdamped Langevin trajectory prior to the shooting point is to use a numerical scheme with negated time increments. This is however difficult to use in practice since gradient dynamics with negative time increments are usually very unstable.

An alternative strategy consists in using the forward dynamics to generate the trajectory prior to the shooting point, as

$$q_i = q_{i+1} - \Delta t\, \nabla V_{\lambda_{i+1}}(q_{i+1}) + \sqrt{\frac{2\Delta t}{\beta}}\, G_i,$$

where the Gaussian vectors G_i are i.i.d. with mean 0 and identity covariance matrix. Then, $\overline{\mathrm{p}}\,(q_{i+1}, dq_i; \lambda_{i+1}) = \mathrm{p}\,(q_{i+1}, dq_i; \lambda_{i+1})$.

Langevin dynamics. Let us now present a discretization of the Langevin dynamics such that ratio (4.129) has a simple expression. The integrator, for a fixed value of λ, is based on the following splitting:

$$\Phi_{\lambda,\Delta t} = \Phi_{\Delta t}^{\mathrm{OU}} \circ \Phi_{\lambda,\Delta t}^{\mathrm{Verlet}}, \tag{4.130}$$

where $\Phi_{\lambda,\Delta t}^{\mathrm{Verlet}}$ is the Verlet integrator (1.22) corresponding to the Hamiltonian H_λ associated with (1.61), while $\Phi_{\Delta t}^{\mathrm{OU}}$ corresponds to an exact integration of the Ornstein-Uhlenbeck process on the momenta:

$$dp_t = -\gamma p_t\, dt + \sqrt{\frac{2\gamma}{\beta}}\, dW_t$$

over a time-step Δt. Such a splitting strategy in the context of path sampling was used in [Adjanor *et al.* (2006)] for instance. A simple computation shows that

$$\Phi_{\Delta t}^{\mathrm{OU}}\left(q^n, p^n\right) = \left(q^n, \alpha p^n + \sqrt{\frac{1-\alpha^2}{\beta}}\, G^n\right),$$

where $\alpha = \exp(-\gamma \Delta t)$, and where G^n are i.i.d. standard Gaussian random vectors.

A possible dynamics to generate configurations prior to the shooting point is

$$\overline{\Phi}_{\lambda,\Delta t} = \Phi_{\lambda,-\Delta t}^{\mathrm{Verlet}} \circ \Phi_{\Delta t}^{\mathrm{OU}} = S \circ \Phi_{\lambda,\Delta t}^{\mathrm{Verlet}} \circ S \circ \Phi_{\Delta t}^{\mathrm{OU}}.$$

Note that the order of the Verlet and fluctuation/dissipation steps has changed. Denoting by $\widetilde{x}_{k+1} = (q_{k+1}, \widetilde{p}_{k+1}) = \Phi_{\lambda,\Delta t}^{\mathrm{Verlet}}(x_k)$ with $x_k = (q_k, p_k)$, and by $x_{k+1} = \Phi_{\Delta t}^{\mathrm{OU}}(\widetilde{x}_{k+1})$, a simple computation shows that

$$\frac{\mathrm{p}(x_k, dx_{k+1};\, \lambda)\, dx_k}{\overline{\mathrm{p}}(x_{k+1}, dx_k;\, \lambda)\, dx_{k+1}} = \exp\left(-\frac{\beta}{2}\left(p_{k+1}^2 - \widetilde{p}_{k+1}^2\right)\right).$$

It is then very easy to compute the Metropolis-Hastings ratio (4.129), when the shooting point is not modified. In the latter case,

$$r(x, y;\, k, \Lambda) = \frac{\rho(y_0)}{\rho(x_0)} \exp\left(-\frac{\beta}{2} \sum_{j=1}^{k} \left(p_j^2 - \widetilde{p}_j^2\right)\right).$$

4.4.2.3 *Brownian tube moves for stochastic dynamics*

The above described shooting moves may be inefficient for stochastic dynamics: There is no way to tune the acceptance/rejection rate for those dynamics since the newly-generated trajectories use entirely different random numbers. The tuning of the acceptance is important in order to have some balance between decorrelation and acceptance, hence to obtain reliable results, see the discussion in Section 2.1.2. For Hamiltonian dynamics, the acceptance rate can be controlled by the magnitude of the perturbations of the shooting point (when there is no modification of the shooting point, the same trajectory is recovered, and the acceptance rate is 1).

A refined strategy for stochastic dynamics consists in integrating the parts of the trajectory before and after the shooting index with random numbers correlated to the ones used to generate the previous trajectory. The amount of correlation then controls the acceptance rate. This strategy is called the *Brownian tube* proposal [Stoltz (2007)]. For simplicity, we present the implementation in the case of overdamped Langevin dynamics, but the method can be extended to Langevin dynamics (by resorting to convenient numerical schemes, such as the splitting scheme (4.130) or the schemes studied in [Stoltz (2007)]).

A path $q = (q_0, \dots, q_L)$ generated with the numerical scheme (4.18) is now described as an initial configuration q_0 and a sequence of random noises $(G_0, \dots, G_{L-1})$ given by

$$G_i = \sqrt{\frac{\beta}{2\Delta t}} \Big(q_{i+1} - q_i + \Delta t \, \nabla V_{\lambda_{i+1}}(q_i) \Big).$$

We denote in the sequel $Q = (q_0, G_0, \dots, G_{L-1})$ such a path. The probability measure (4.125) reads with this formulation

$$\rho(dq_0) \prod_{i=0}^{L-1} N(g_i)\, dg_i, \tag{4.131}$$

with $\rho(dq_0) = Z_0^{-1} \mathrm{e}^{-\beta V_{\lambda_0}(q_0)}\, dq_0$ and

$$N(g) = \left(\frac{1}{2\pi} \right)^{3N/2} \exp\left(-\frac{|g|^2}{2} \right).$$

Consider a given path $Q = (q_0, G_0, \dots, G_{L-1})$, sampled from (4.131). When a shooting index k is chosen, the path is decomposed as a forward part $(q_k, \dots, q_L)$ and a part prior to the shooting index $(q_k, \dots, q_0)$. As described in Section 4.4.2.2, one possible choice to obtain the part prior to the shooting index is to use the same numerical scheme, starting from q_k, and successively obtaining $q_{k-1}, \dots, q_0$. The associated noises are denoted by G_j for the forward part of the paths, and by $\overline{G}_j$ for the part prior to the shooting point. More precisely,

$$G_j = \sqrt{\frac{\beta}{2\Delta t}} \Big(q_{j+1} - q_j + \Delta t \, \nabla V_{\lambda_{j+1}}(q_j) \Big) \qquad (k \le j \le L-1),$$

and

$$\overline{G}_j = \sqrt{\frac{\beta}{2\Delta t}}\Big(q_j - q_{j+1} + \Delta t\,\nabla V_{\lambda_{j+1}}(q_{j+1})\Big) \qquad (0 \le j \le k-1). \quad (4.132)$$

Another equivalent description of the path is therefore the knowledge of the configuration q_k at the shooting index, and the sequence of random noises $(\overline{G}_0, \dots, \overline{G}_{k-1}, G_k, \dots, G_{L-1})$. Note however that, since $q_{j+1} = q_j - \Delta t\,\nabla V_{\lambda_{j+1}}(q_j) + \sqrt{2\beta^{-1}\Delta t}$, the random variable $\overline{G}_j$ obtained from (4.132) is not distributed according to a Gaussian law (because of the finite time-step errors).

Shooting moves described in Section 4.4.2.2 consist in generating a new path described by the configuration q_k at the shooting index, and a sequence of new random noises $(\overline{G}'_0, \dots, \overline{G}'_{k-1}, G'_k, \dots, G'_{L-1})$, independent from the previous ones. A new path with an arbitrary correlation with the previous one may however be generated upon replacing the random noises by new ones, with a controlled level of correlation:

$$G'_j = \alpha_j G_j + \sqrt{1-\alpha_j^2}\,\mathcal{G}_j, \quad j = k, \dots, L-1, \qquad (4.133)$$

$$\overline{G}'_j = \alpha_j \overline{G}_j + \sqrt{1-\alpha_j^2}\,\mathcal{G}_j, \quad j = 1, \dots, k-1, \qquad (4.134)$$

where $(\mathcal{G}_j)_{j=0,\dots,L-1}$ are i.i.d. $3N$-dimensional centered Gaussian vectors with identity covariance matrix. The coefficients $0 \le \alpha_j \le 1$ allow to tune the local correlations of the noises. The probability to generate the path $Q' = (q'_0, G'_0, \dots, G'_{L-1})$ from the path $Q = (q_0, G_0, \dots, G_{L-1})$, given an index $k \in \{0, \dots, L\}$, is therefore

$$T(Q, Q') = \prod_{0 \le j \le k-1} \mathrm{p}_{\alpha_j}(\overline{G}_j, \overline{G}'_j) \prod_{k \le j \le L-1} \mathrm{p}_{\alpha_j}(G_j, G'_j), \qquad (4.135)$$

where

$$\mathrm{p}_\alpha(G, G') = \left(\frac{1}{2\pi(1-\alpha^2)}\right)^{3N/2} \exp\left(-\frac{|G' - \alpha G|^2}{2(1-\alpha^2)}\right).$$

In conclusion, given some path $x = (q_0, \dots, q_L)$, equivalently described by $Q = (q_0, G_0, \dots, G_{L-1})$, the algorithm proceeds as follows: (i) choose a shooting index k, and compute $\overline{G}_1, \dots, \overline{G}_{k-1}$ from the knowledge of x, using (4.132); (ii) propose a modification of the sequence of noises according to (4.133)-(4.134); (iii) integrate the equations of motions with the new noises before and after the shooting point, to obtain a new path $y = (q'_0, \dots, q'_L)$, equivalently described by $Q' = (q'_0, G'_0, \dots, G'_{L-1})$; (iv) compute the transition probabilities $T(Q, Q')$ and $T(Q', Q)$ according to (4.135), and the path probabilities according to (4.131). The Metropolis-Hastings ratio is then known, and the acceptance/rejection step can be performed.

As already hinted at above, shooting moves are recovered in the special case when $\alpha_0 = \cdots = \alpha_{L-1} = 0$. It is therefore not surprising that some choices of local correlation functions α_j may lead to better sampling efficiencies, see [Stoltz (2007)] for a case study (in this case, the correlation coefficients α_j are small around the shooting point, and close to 1 far away from it). In essence, the improvement relies on the fact that an arbitrary acceptance/rejection rate in the Metropolis-Hastings step can be obtained, allowing to increase the acceptance rate in situations when standard shooting moves are often rejected.

4.4.2.4 *Weighted path ensembles*

A straightforward estimator of the free energy difference based on the formula (4.122), such as (4.128):

$$\widehat{\Delta F}_M = -\beta^{-1} \ln \left(\frac{1}{M} \sum_{m=1}^{M} \mathrm{e}^{-\beta \mathcal{W}(x^m;\, \Lambda)} \right),$$

where the works are sampled using a sequence of paths obtained with the Metropolis algorithm 4.21, is likely to have a large variance. This is related to the fact that the largest values of the function $W \mapsto \mathrm{e}^{-\beta W} P(W)$ are obtained for work values for which $P(W)$ is small, see the discussion in Section 4.1.5. Besides, the correlation between paths contributes to an increased statistical error, the hope being however that this increased variance is balanced by the fact that the lower tail of the work distribution is better sampled.

Some smooth interpolation between the distribution $P(W)\,dW$ and the distribution proportional to $\mathrm{e}^{-\beta W} P(W)\,dW$ may help increasing the accuracy of free energy estimators. Consider for instance the image by $\mathcal{W}(\cdot\,;\,\Lambda)$ of the family of (non-normalized) path measures

$$\pi_\alpha(dx;\,\Lambda) = \mathrm{e}^{-\alpha\beta \mathcal{W}(x;\,\Lambda)}\,\pi(dx;\,\Lambda), \qquad 0 \le \alpha \le 1. \tag{4.136}$$

There are two interesting limiting cases for the associated work distributions. The measure $\pi_0(dx;\,\Lambda)$ is the measure (4.125) on forward switching paths, so that the associated work distribution is $p^{\mathrm{f}}_{L\Delta t}(W)\,dW$ with the notation of Section 4.2.2. On the other hand, by the Crooks equality (4.56), the work distribution associated with the measure $\pi_1(dx;\,\Lambda)$ is $p^{\mathrm{b}}_{L\Delta t}(-W)\,dW$.

The free energy difference can be computed from the relation

$$\mathrm{e}^{-\beta\Delta F} = f(1), \qquad f(\alpha) = \int_{\mathcal{P}} \pi_\alpha(dx;\Lambda) = \int_{\mathbb{R}} P_T(W)\,\mathrm{e}^{-\alpha\beta W}\,dW.$$

The free energy difference may therefore be seen as a partition function in path space (rather than regular phase space). Any method to compute alchemical free energy differences may now be used to estimate it, including nonequilibrium methods and adaptive methods.

To illustrate the discussion, consider thermodynamic integration in path space, as first proposed in [Sun (2003)]. This method relies on an explicit expression of the derivative

$$f'(\alpha) = -\beta\,\mathbb{E}_{\pi_\alpha}(\mathcal{W})\,f(\alpha),$$

where

$$\mathbb{E}_{\pi_\alpha}(\mathcal{W}) = \frac{\displaystyle\int_{\mathcal{P}} \mathcal{W}(x;\,\Lambda)\pi_\alpha(dx;\,\Lambda)\,dx}{\displaystyle\int_{\mathcal{P}} \pi_\alpha(dx;\,\Lambda)\,dx},$$

so that

$$f(\alpha) = \exp\left(-\beta\int_0^\alpha \mathbb{E}_{\pi_s}(\mathcal{W})\,ds\right),$$

and finally

$$\Delta F = \int_0^1 \mathbb{E}_{\pi_\alpha}(\mathcal{W})\,d\alpha.$$

The free energy difference is therefore rewritten as an integral of an average quantity (the work here), as for usual thermodynamic integration.

The method requires the computation of averages in a work biased ensemble of switching trajectories (*i.e.* with respect to the measure $\pi_\alpha(dx;\,\Lambda)$), and path sampling techniques are particularly convenient to this end. The generation of proposal paths can be done as described in Section 4.4.1.2, and accepted or rejected after a suitable modification of the acceptance/rejection rate (4.127) in the Metropolis-Hastings algorithm. More precisely, (4.129) should be replaced by

$$\begin{aligned} r(x,y;\,k,\Lambda) = \mathrm{e}^{-\alpha\beta(\mathcal{W}(y;\,\Lambda)-\mathcal{W}(x;\,\Lambda))}\,\frac{\rho(y_0)}{\rho(x_0)}\frac{dy_0}{dx_0}\frac{\mathrm{p}_{\mathrm{gen}}(y_k,dx_k)}{\mathrm{p}_{\mathrm{gen}}(x_k,dy_k)} \\ \times\prod_{j=0}^{k-1}\frac{\mathrm{p}(y_j,dy_{j+1};\lambda_{j+1})}{\overline{\mathrm{p}}(y_{j+1},dy_j;\lambda_{j+1})}\frac{\overline{\mathrm{p}}(x_{j+1},dx_j;\lambda_{j+1})}{\mathrm{p}(x_j,dx_{j+1};\lambda_{j+1})}. \end{aligned} \qquad (4.137)$$

4.4.2.5 *Importance sampling*

Importance sampling techniques may be used to bias the sampling towards paths corresponding to unlikely low values of the work. In this case, some measure $\Pi(x)$ is sampled, and averages over the path ensemble are recovered as

$$\langle A \rangle = \frac{\displaystyle\int_{\mathcal{P}} A(x) \frac{\pi(dx;\,\Lambda)}{\Pi(dx)}\,\Pi(dx)}{\displaystyle\int_{\mathcal{P}} \frac{\pi(dx;\,\Lambda)}{\Pi(dx)}\,\Pi(dx)}.$$

Having in mind the computation of the average of $A(x) = \exp(-\beta\mathcal{W}(x;\,\Lambda))$, biasing measures should be chosen which allow to interpolate between the denominator, namely $\pi(dx;\,\Lambda)$, and the numerator, namely $\pi(dx;\,\Lambda)\exp(-\beta\mathcal{W}(x;\,\Lambda))$. For instance,

$$\Pi(dx) \propto \pi(dx;\,\Lambda)\exp\left(-\frac{\beta}{2}\mathcal{W}(x;\,\Lambda)\right) = \pi_{1/2}(dx;\,\Lambda)$$

(with the notations of the previous section) was proposed in [Ytreberg and Zuckerman (2004); Athènes (2004)]. More generally, path sampling algorithms may be straightforwardly modified to sample importance functions of the form $\Pi(dx) \propto \pi(dx;\,\Lambda)\,g(\mathcal{W}(x;\Lambda))$, since the path sampling algorithms allow to generate paths distributed according to $\pi(dx;\,\Lambda)$, and the weighting term $g(\mathcal{W}(x;\,\Lambda))$ can be taken into account by modifying accordingly the Metropolis-Hastings acceptance rule.

An optimal importance sampling function can be derived as in Section 2.4.1, under similar assumptions (bias of the estimator negligible, truncation of the terms in the variance, i.i.d. samples), see [Oberhofer and Dellago (2008)].

4.4.2.6 *Efficiency of the path sampling approach*

Since free energy perturbation and thermodynamic integration can be seen, at least heuristically, as limiting cases of the nonequilibrium method for respectively infinitely fast and infinitely slow switchings, it seems plausible that there may be some optimum switching rate for which nonequilibrium methods, enhanced by importance sampling on path space, outperform some other standard approach.

No definite conclusion can however be drawn from the computational studies performed so far [Lechner and Dellago (2007); Oberhofer *et al.* (2005); Ytreberg *et al.* (2006)] since the switching dynamics used varies from one study to the other (recall that the switching dynamics, be it Hamiltonian dynamics or Langevin dynamics or overdamped Langevin dynamics, has an influence on the results, see Section 4.1.5.3), the physical systems for which the comparison is performed are different (it is believed that the optimal method depends on the system at hand), and the methods to which path sampling techniques are compared differ from one study to the other (not to mention their implementations).

Besides, techniques based on nonequilibrium simulations are much more recent than conventional techniques, and it can be expected that there is still room for improvement. On the other hand, the error analysis performed in Section 4.1.5 (in particular the scaling of the error with the dimension of the system) casts some doubts on the practical interest of nonequilibrium methods to compute free energy differences.

4.4.2.7 *Application to Widom insertion*

We consider a nonequilibrium switching performed with Langevin dynamics, using a standard shooting algorithm, with the implementation described in Section 4.4.2.2. Work distributions for several ensembles are presented

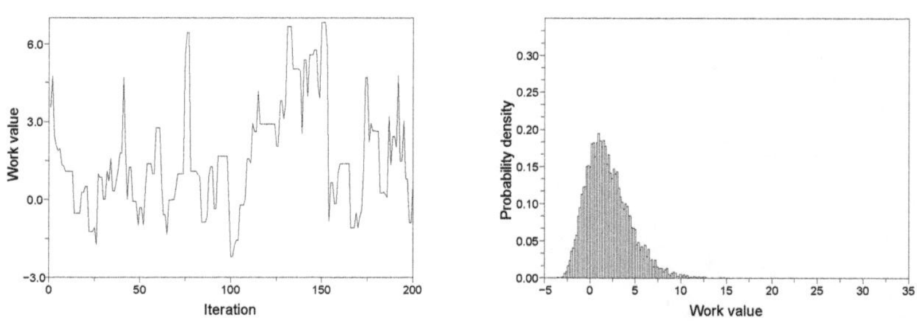

Fig. 4.5 Left: First iterations of a path sampling $P(W)\,e^{-\alpha\beta W}$ with $\alpha = 0.3$ and $T = 4$. Right: Resulting work distribution.

in Figures 4.5 and 4.6, in the case $T = 4$, $\Delta t = 0.005$ and $\gamma = 0.1$. Note first the correlation between successive paths, which is obvious in Figure 4.5 (Left) since there are rejections (in which case two successive work values are equal), and because the sampled work values vary more or less continuously. The work distribution for $\alpha = 0.3$ in (4.136), plotted in Figure 4.5 (Right), is intermediate between the work distributions for $\alpha = 0$ (standard forward switching), and the work distribution for $\alpha = 1$ (which is the work

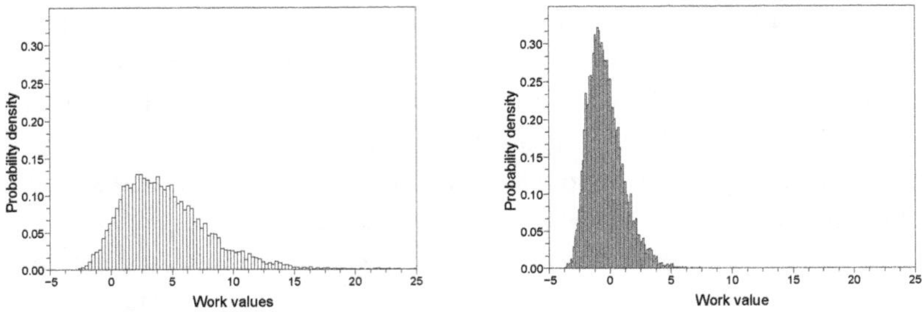

Fig. 4.6 Work distribution for $\alpha = 0$ (Left) and $\alpha = 1$ (Right). These distributions correspond respectively to the work distributions of forward and backward paths (with the terminology of Section 4.2).

distribution of backward paths, up to a sign change, see Section 4.4.2.4), see Figure 4.6. All the work distributions have been plotted using the same axis, for the ease of comparison. Free energy estimates can then be obtained from (4.128) using averages along a sequence of paths, such as the one depicted in Figure 4.6 (left).

Chapter 5

Adaptive methods

Adaptive methods are the most recent techniques which have been designed in order to compute free energy differences. Three typical examples are: the Wang-Landau approach [Wang and Landau (2001a)], the adaptive biasing force method [Darve and Porohille (2001); Hénin and Chipot (2004)] or the nonequilibrium metadynamics [Laio and Parrinello (2002); Bussi *et al.* (2006)]. The principle of adaptive methods is to modify the potential V experienced by the particles during the simulation, in order to remove the metastable features of the dynamics. This modification uses an approximation (also computed during the simulation) of the free energy F associated with some reaction coordinate of interest. The free energy is thus also an output of such computations.

The biasing potential is actually adjusted on-the-fly, depending on the whole trajectory of the system. Adaptive methods can therefore be seen as adaptive importance sampling strategies, which makes these methods appealing for general purpose sampling strategies and problems *a priori* unrelated to free energy computations (remember the introductory Section 1.3.3 on metastability issues). Another important point is that the biasing potential depends only on the reaction coordinate, which is of small dimension, so that the biasing potential can be stored on a low-dimensional grid of the reaction coordinate space (it would be for instance unrealistic to consider an efficient biasing potential depending on the whole configurational variable q, which lives in a high dimensional space).

The bottom line of adaptive dynamics is that, if the free energy F was known, the dynamics driven by the potential $V - F \circ \xi$ (where $\circ$ is the composition operator *i.e.* $(F \circ \xi)(q) = F(\xi(q))$), would be less metastable than the dynamics driven by the potential V. The strategy then consists in (i) replacing the interaction potential V by

$$\mathcal{V}_t = V - F_t \circ \xi$$

and (ii) giving some rule to update the biasing term F_t in such a way that F_t approximates F in the longtime limit. When the method is constructed carefully, a *consistency* result can be proved: *When* F_t converges (which is usually not a trivial result), then it converges to the free energy F, up to an additive constant. Such consistency results are presented in Section 5.1 after a general introduction to adaptive methods in some unified framework. In some cases, a *convergence* result can be shown (see Section 5.2), under some assumptions on the potential V and on the dynamics used. In Section 5.1.5, some numerical examples illustrate the interest of the approach.

5.1 Adaptive algorithms: A general framework

The aim of this section is to present a general mathematical framework for adaptive methods, in which many adaptive methods used in practice may be understood. This framework may also be used to propose new adaptive methods, is suitable for numerical and theoretical analysis of the convergence of the methods (see Section 5.2), and leads to new numerical discretization strategies based on multiple replica techniques (see Section 5.1.3.1). This section is mainly based on [Lelièvre *et al.* (2007a)].

As already mentioned in Sections 1.3.4 and 2.3.2, averages over trajectories of the simple gradient (or overdamped Langevin) dynamics:

$$dq_t = -\nabla V(q_t)\, dt + \sqrt{\frac{2}{\beta}}\, dW_t \tag{5.1}$$

are not efficient to sample the Boltzmann-Gibbs measure

$$\nu(dq) = Z_\nu^{-1} \exp(-\beta V(q))\, dq \tag{5.2}$$

(and *a fortiori* to compute free energy differences) since the potential V typically contains many local minima separated by high barriers, so that the dynamics of the stochastic process (q_t) is metastable. In words, q_t (or at least some of its components) essentially does not move for very long period of times (it is stuck in a metastable state), before hopping to a new metastable state, where it remains again for a very long time. This observation, as well as what is presented in this section, holds for the gradient dynamics (5.1), but also for other dynamics sampling the canonical measure, such as the Langevin dynamics in phase space, or Metropolis-Hastings dynamics.

The information needed to build an adaptive method is a description of the metastable features of the dynamics associated to the potential V through the *reaction coordinate*

$$\xi : \mathcal{D} \to \mathcal{M},$$

where the set $\mathcal{M}$ of admissible values of the reaction coordinate is typically a subset of $\mathbb{R}^m$, or of $\mathbb{T}^m$ (or more generally of $\mathbb{T}^{m-m'} \times \mathbb{R}^{m'}$ for some $m' \in \{0, \dots, m\}$). Adaptive methods (and actually also thermodynamic integration methods or nonequilibrium methods) rely on the fundamental assumption that the essential metastable features of q_t are contained in $\xi(q_t)$. A precise mathematical content to this statement will be given in Section 5.2. Roughly speaking, the idea is that, if the dynamics associated to V is indeed metastable, then the dynamics associated to the free energy biased potential

$$V - F \circ \xi \tag{5.3}$$

where F is the free energy associated with ξ, is not (or, at least, less) metastable. We refer to Section 1.3.3 for some numerical illustrations of this fact for simple two-dimensional toy examples. It may also be helpful to recall at this stage the definitions of free energy and associated mean force obtained in Sections 3.2 and 3.2.2. The free energy F associated with ξ is defined by (see (3.22)):

$$F(z) = -\beta^{-1} \ln \left(\int_{\Sigma(z)} Z_\nu^{-1} \mathrm{e}^{-\beta V} \, (\det G)^{-1/2} \, d\sigma_{\Sigma(z)} \right), \tag{5.4}$$

where $\Sigma(z) = \{q, \ \xi(q) = z\}$ and $G_{\alpha,\beta} = \nabla \xi_\alpha \cdot \nabla \xi_\beta$ for $1 \leq \alpha, \beta \leq m$. The free energy is such that the image of the canonical measure ν by ξ, denoted by ν^ξ, is:

$$\nu^\xi(dz) = \left(\int_{\Sigma(z)} Z_\nu^{-1} \mathrm{e}^{-\beta V} \, (\det G)^{-1/2} \, d\sigma_{\Sigma(z)} \right) dz = \mathrm{e}^{-\beta F(z)} \, dz. \tag{5.5}$$

The mean force ∇F can then be expressed as an average of the local mean force f with respect to the conditional probability measures $\nu^\xi(\cdot \,|\, z)$ (see Lemma 3.9):

$$\nabla F(z) = \int_{\Sigma(z)} f(q) \, \nu^\xi(dq \,|\, z), \tag{5.6}$$

where the components of the local mean force $f = (f_1, \dots, f_m)$ are, for $\alpha \in \{1, \dots, m\}$,

$$f_\alpha = \sum_{\gamma=1}^m G_{\alpha,\gamma}^{-1} \nabla \xi_\gamma \cdot \nabla V - \beta^{-1} \mathrm{div} \left(\sum_{\gamma=1}^m G_{\alpha,\gamma}^{-1} \nabla \xi_\gamma \right), \tag{5.7}$$

and the conditional probability measures $\nu^\xi(\cdot\,|\,z)$ are:

$$\nu^\xi(dq\,|\,z) = \frac{\mathrm{e}^{-\beta V(q)}\,(\det G(q))^{-1/2}\,\sigma_{\Sigma(z)}(dq)}{\displaystyle\int_{\Sigma(z)} \mathrm{e}^{-\beta V}\,(\det G)^{-1/2}\,d\sigma_{\Sigma(z)}}. \tag{5.8}$$

Of course, the free energy F is not known in practice, so that it is not possible to use the potential (5.3) as a biasing potential. The idea of adaptive methods is to use a *time-dependent potential*

$$\mathcal{V}_t = V - F_t \circ \xi \tag{5.9}$$

where F_t is an *approximation* of the free energy F at time t, in view of the regions of the conformational space visited so far. It may be helpful to have the following prototypical biased gradient dynamics in mind:

$$\begin{aligned} dq_t &= -\nabla(V - F_t \circ \xi)(q_t)\,dt + \sqrt{\frac{2}{\beta}}\,dW_t \\ &= \left(-\nabla V(q_t) + \sum_{\alpha=1}^{m} \nabla_\alpha F_t(\xi(q_t))\nabla\xi_\alpha(q_t)\right) dt + \sqrt{\frac{2}{\beta}}\,dW_t, \end{aligned} \tag{5.10}$$

although all the following considerations also apply to other dynamics consistent with the canonical ensemble. Once such a biased dynamics is chosen, the question is: How to compute F_t in such a way that F_t converges to F, up to an additive constant (and thus that the process (q_t) does not remain trapped in some metastable states)?

There are actually two types of adaptive methods, depending on the way the bias is updated. Adaptive Biasing Potential (ABP) methods update the potential F_t; whereas Adaptive Biasing Force (ABF) methods update the vector field ∇F_t, which will be denoted Γ_t in the following. In general, for $m \geq 2$, Γ_t does not need to be a gradient. For ABF methods, the problem is then how to update Γ_t in such a way that it converges to ∇F, and a typical dynamics to have in mind is

$$dq_t = \left(-\nabla V(q_t) + \sum_{\alpha=1}^{m} \left[\Gamma_t(\xi(q_t))\right]_\alpha \nabla\xi_\alpha(q_t)\right) dt + \sqrt{\frac{2}{\beta}}\,dW_t. \tag{5.11}$$

Remark 5.1 (Open question: about an optimal bias).
From a mathematical viewpoint, adaptive methods can be seen as adaptive importance sampling techniques, where the importance function is adapted on-the-fly. An interesting open question is the following: Given the fact that we want to bias the dynamics by a function of ξ, what is the optimal

choice of biasing function? Is it indeed the free energy? For the question to be well posed, some cost function or criterion to optimize should of course be precised, such as the rate of convergence to equilibrium of the biased dynamics (related to the spectral gap of the associated operator), or some (asymptotic) variance for a class of test functions or observables.

5.1.1 *Updating formulas*

We proceed in two steps. First, we motivate the definitions of the observed free energy and mean force, which are somehow the instantaneous or current versions of the (equilibrium) free energy and mean force. We then propose an update of the bias using these observed quantities, in such a way that the biasing potential F_t (or the biasing force Γ_t) converges to the free energy (or the mean force) in the longtime limit.

5.1.1.1 *Observed free energy and mean force*

Let us denote by $\psi(t,\cdot)$ the probability density function of q_t where the process (q_t) is a solution to the adaptive dynamics (5.10) or (5.11), depending on whether adaptive potentials or forces are considered. Define, for a given time t,

- the *observed free energy*

$$F_{\mathrm{obs}}(t,z) = -\beta^{-1} \ln \psi^{\xi}(t,z), \tag{5.12}$$

  where

$$\psi^{\xi}(t,z) = \int_{\Sigma(z)} \psi(t,q)\,|\det G(q)|^{-1/2}\,\sigma_{\Sigma(z)}(dq) \tag{5.13}$$

  is the image of the measure $\nu(t,dq) = \psi(t,q)\,dq$ by ξ (denoted $\nu^{\xi}(t,dz) = \psi^{\xi}(t,z)\,dz$ in the following). Note that (5.12) is analogous to (5.4), with the canonical probability density function $Z_{\nu}^{-1}\mathrm{e}^{-\beta V(q)}\,dq$ replaced by $\psi(t,q)$;
- the *observed mean force*

$$\Gamma_{\mathrm{obs}}(t,z) = \int_{\Sigma(z)} f(q)\,\nu^{\xi}(t,dq\,|\,z), \tag{5.14}$$

  which is a vector in $\mathcal{M}$. This definition is motivated by the definition (5.6) of the mean force, with the conditional measure $\nu^{\xi}(\cdot\,|\,z)$ at a fixed value $\xi(q) = z$ of the reaction coordinate (obtained from the canonical measure) replaced by the conditional measure

$\nu^\xi(t,\cdot\,|\,z)$ obtained from the instantaneous law $\psi(t,q)\,dq$ of the process:

$$\nu^\xi(t,dq\,|\,z) = \frac{\psi(t,q)\,|\det G(q)|^{-1/2}\,\sigma_{\Sigma(z)}(dq)}{\displaystyle\int_{\Sigma(z)} \psi\,|\det G|^{-1/2}\,d\sigma_{\Sigma(z)}}. \tag{5.15}$$

Note that the definition of the observed free energy (5.12) involves the marginal distribution ψ^ξ, which is a probability density function (with respect to the Lebesgue measure on $\mathcal{M}$), while the definition of the observed mean force (5.14) involves the conditional measure $\nu^\xi(t,\cdot|z)$, which is a probability measure with support $\Sigma(z)$. Formulas (5.15) and (5.13) are direct consequences of the co-area formula, see Corollary 3.3.

Of course, and as already mentioned, formulas (5.4) and (5.6) are particular cases of (respectively) (5.12) and (5.14) in the case $\psi(t,\cdot) = Z_\nu^{-1}\exp(-\beta V)$. Note also that, for a general function ψ, the vector field $\Gamma_{\rm obs}$ is not a gradient, and is therefore different from $\nabla F_{\rm obs}$.

5.1.1.2 *Updating the bias with the observed quantities*

Let us now go back to our original question: Using the observed quantities $F_{\rm obs}$ and $\Gamma_{\rm obs}$, how to update F_t (resp. Γ_t) in order for them to converge to F, up to an additive constant (resp. to ∇F)? We proceed in two steps, assuming first that the biased dynamics is instantaneously at equilibrium with respect to the biased potential, and then using this analysis as a guideline to derive appropriate updating formulas.

Instantaneous equilibrium. Let us start by assuming that the process (q_t) is instantaneously at equilibrium with respect to the biased potential $\mathcal{V}_t$, *i.e.* that

$$\psi(t,q) \equiv \psi^{\rm eq}(t,q) = Z_t^{-1}\mathrm{e}^{-\beta[V(q)-F_t(\xi(q))]},$$

where $Z_t = \displaystyle\int_{\mathcal{D}} \exp(-\beta(V - F_t\circ\xi))$ is a normalizing constant. Note that, for ABF methods, this requires the additional assumption that there exists a function F_t such that $\Gamma_t = \nabla F_t$ (which is of course satisfied if $m=1$, but is false in general in higher dimensions).

It is then easy to check that

$$\begin{aligned} F^{\rm eq}_{\rm obs}(t,z) &= -\beta^{-1}\ln\left(\int_{\Sigma(z)} \psi^{\rm eq}(t,\cdot)\,(\det G)^{-1/2}\,d\sigma_{\Sigma(z)}\right) \\ &= F(z) - F_t(z) + \beta^{-1}\ln Z_t, \end{aligned} \tag{5.16}$$

and

$$\Gamma_{\rm obs}^{\rm eq}(t,z) = \frac{\displaystyle\int_{\Sigma(z)} f(q)\psi^{\rm eq}(t,q)\,(\det G(q))^{-1/2}\,\sigma_{\Sigma(z)}(dq)}{\displaystyle\int_{\Sigma(z)} \psi^{\rm eq}(t,q)\,(\det G(q))^{-1/2}\,\sigma_{\Sigma(z)}(dq)} = \nabla F(z). \quad (5.17)$$

In view of these formulas, it seems that a typical convenient updating law is:

- For Adaptive Biasing Potential methods, (5.16) suggests that
$$\frac{dF_t(z)}{dt} = F_{\rm obs}^{\rm eq}(t,z).$$
This would indeed ensure that F_t converges exponentially fast to F (up to an additive constant);
- For Adaptive Biasing Force methods, (5.17) suggests
$$\Gamma_t(z) = \Gamma_{\rm obs}^{\rm eq}(t,z),$$
which would ensure $\Gamma_t = \nabla F$ instantaneously!

The general case. Now, of course, the process (q_t) is not instantaneously at equilibrium with respect to the biased potential $\mathcal{V}_t$. We nonetheless use formulas (5.16) and (5.17) as guidelines to propose updating procedures which should ensure convergence to the free energy or the mean force:

- A typical Adaptive Biasing Potential (ABP) dynamics is:
$$\begin{cases} dq_t = -\nabla(V - F_t\circ\xi)(q_t)\,dt + \sqrt{\dfrac{2}{\beta}}\,dW_t, \\ \dfrac{dF_t(z)}{dt} = -\beta^{-1}\ln\left(\displaystyle\int_{\Sigma(z)} \psi(t,\cdot)\,(\det G)^{-1/2}\,d\sigma_{\Sigma(z)}\right). \end{cases} \quad (5.18)$$
The ABP dynamics has an appealing intuitive interpretation: It penalizes the regions visited so far in the reaction coordinate space by increasing the potential in these regions, therefore enforcing the exploration of unexplored regions.
- A typical Adaptive Biasing Force (ABF) dynamics is:
$$\begin{cases} dq_t = \left(-\nabla V(q_t) + \displaystyle\sum_{\alpha=1}^{m} \left[\Gamma_t(\xi(q_t))\right]_\alpha \nabla\xi_\alpha(q_t)\right) dt + \sqrt{\dfrac{2}{\beta}}\,dW_t, \\ \Gamma_t(z) = \displaystyle\int_{\Sigma(z)} f(q)\,\nu^\xi(t,dq\,|\,z). \end{cases} \quad (5.19)$$
An ABF dynamics makes an approximation of the mean force in the regions visited so far, and subtracts this approximated average force from ∇V.

Following the idea that, if the process was at equilibrium ($\psi = \psi^{\rm eq}$), the convergence of F_t to F (up to an additive constant) or the convergence of Γ_t to ∇F should be ensured, more general updating formulas may be proposed, relying on the following observation: If $(h_t)_{t\geq 0}$ is a family of strictly increasing and continuous functions (with $h_\infty = \lim_{t\to+\infty} h_t$ strictly increasing and continuous), and $t \mapsto y(t)$, solution of the equation

$$y'(t) = h_t(y_0 - y(t)),$$

converges to some value y_∞ as $t \to +\infty$, then $y'(t) \to 0$ and $y_\infty = y_0 - c$ as $t \to +\infty$, where $c = h_\infty^{-1}(0)$. For ABP methods, given a family of strictly increasing function $\mathcal{F}_t : \mathbb{R} \to \mathbb{R}$, the following update can then be considered:

$$\frac{dF_t(z)}{dt} = \mathcal{F}_t\left(-\beta^{-1}\ln\left[\int_{\Sigma(z)} \psi(t,\cdot)\,(\det G)^{-1/2}\,d\sigma_{\Sigma(z)}\right]\right). \tag{5.20}$$

Note that (5.18) is obtained for the specific choice $\mathcal{F}_t(x) = x$. Likewise, for ABF methods, a good candidate is an updating law of the form:

$$\frac{d\Gamma_t(z)}{dt} = \mathcal{G}_t\left(\int_{\Sigma(z)} f\,d\nu^\xi(t,\cdot\,|\,z) - \Gamma_t(z)\right), \tag{5.21}$$

where the updating function can be written as

$$\mathcal{G}_t(x_1,\ldots,x_m) = (\mathcal{G}_t^1(x_1),\ldots,\mathcal{G}_t^m(x_m)) \,:\, \mathbb{R}^m \to \mathbb{R}^m,$$

with strictly increasing functions $\mathcal{G}_t^\alpha : \mathbb{R} \to \mathbb{R}$ $(1 \leq \alpha \leq m)$ such that $\mathcal{G}_t^\alpha(0) = 0$. Equation (5.19) corresponds formally to the choice $\mathcal{G}_t(x) = x/\tau$, in the limit $\tau \to \infty$.

We would like to emphasize again that ABP and ABF methods are different in nature. In particular, in ABP methods, the dynamics is biased by ∇F_t, the gradient of a potential, while this is not the case for ABF methods since Γ_t is not a gradient in general (for $m \geq 2$).

5.1.1.3 *Consistency of adaptive methods.*

We discuss here the consistency of adaptive methods, namely that if some stationary state is reached, then it is the expected one. Assume to this end that some equilibrium is reached in the limit $t \to +\infty$, more precisely that

$$\psi(t,\cdot) \to \psi_\infty, \qquad \partial_t\psi(t,\cdot) \to 0,$$

and

$$F_t \to F_\infty \text{ (for ABP methods)},$$

or

$$\Gamma_t \to \Gamma_\infty \text{ and } \Gamma_\infty \text{ is a gradient (for ABF methods).}$$

Note again the assumption that Γ_∞ is a gradient is always satisfied in dimension $m = 1$.

Under these assumptions, necessarily, F_∞ is the free energy (up to an additive constant) and Γ_∞ is the mean force. Indeed, for the ABP methods, if an equilibrium is reached, $\psi_\infty = \exp(-\beta(V - F_\infty \circ \xi))$ (up to a multiplicative normalizing constant) and then, $\mathcal{F}_\infty(F(z) - F_\infty(z)) = 0$ which indeed implies that $F_\infty = F + \mathcal{F}_\infty^{-1}(0)$, and the free energy is recovered up to an additive constant. For the ABF methods, if Γ_∞ can be written as $\Gamma_\infty = \nabla F_\infty$ for some potential F_∞, then, by a similar reasoning, $\psi_\infty = \exp(-\beta(V - F_\infty \circ \xi))$ (up to a multiplicative normalizing constant) and $\mathcal{G}_\infty(\nabla F(z) - \Gamma_\infty(z)) = 0$, which again implies $\Gamma_\infty = \nabla F$.

Of course, this is not a proof of convergence (see Section 5.2 to this end), but an indication of the consistency of the method.

Remark 5.2 (ABF dynamics with a gradient bias). *In ABF methods, as already mentioned, the force Γ_t is in general not a gradient. An interesting open question is the following: How does the method perform with the* additional constraint *that Γ_t is a gradient? Such a requirement may simply be satisfied by projecting the observed mean force $\int_{\Sigma(z)} f d\nu^\xi(t, \cdot|z)$ onto a gradient field, for instance by solving the following Poisson problem: find $F_t : \mathcal{M} \to \mathbb{R}$ such that*

$$-\Delta F_t(z) = -\text{div}\left(\int_{\Sigma(z)} f d\nu^\xi(t, \cdot|z)\right), \qquad \forall z \in \mathcal{M},$$

and then setting $\Gamma_t(z) = \nabla F_t(z)$. The Poisson problem is supplemented with boundary conditions of the form

$$\nabla F_t(z) \cdot n(z) = \left(\int_{\Sigma(z)} f d\nu^\xi(t, \cdot|z)\right) \cdot n(z), \qquad \forall z \in \partial\mathcal{M},$$

where $n(z)$ denotes the outward unit normal to $\mathcal{M}$. Periodic boundary conditions may replace the Neumann conditions if the reaction coordinate is periodic (like for angles for example). Many efficient numerical methods can be used to discretize such a problem (see also Remark 3.14*).*

This method is consistent, since the longtime limit of Γ_t (if it exists) should indeed be a gradient. Besides, it has the advantage of reducing the noise of the vector field since it is a linear projection of an m-dimensional *vector-field onto a one-dimensional function F_t, so that the variance is typically divided by m.*

On the other hand, we have no experience of the interest of such an approach in practice, neither theoretical results to assess it.

5.1.1.4 *Generalized adaptive importance sampling strategies*

As mentioned above (see Remark 5.1), the algorithms presented in this chapter may be seen as adaptive importance sampling methods (where the importance function is computed on-the-fly) rather than free energy calculation methods. We refer to Section 2.4.1.4 for a general introduction to importance sampling methods. This viewpoint may be helpful to devise more general adaptive importance sampling techniques. We give here two natural extensions.

Erasing the Jacobian term. In the usual ABF approach, the function f may be difficult to compute because of the so-called Jacobian terms involving second derivatives of ξ. An interesting question is the following: If a simplified local mean force, for example (for $m = 1$)

$$\widetilde{f} = \frac{\nabla V \cdot \nabla \xi}{|\nabla \xi|^2},$$

is used in an ABF strategy instead of the local mean force

$$f = \nabla V \cdot \nabla \xi |\nabla \xi|^{-2} - \beta^{-1} \mathrm{div}(\nabla \xi |\nabla \xi|^{-2}),$$

does the algorithm converge?

If the algorithm indeed converges, it is then possible to unbias the distribution (once the bias has converged), and compute quantities at equilibrium, such as free energy differences. More precisely, in the case $m = 1$ it is straightforward to unbias the stationary distribution (by integrating F'). In the case $m > 1$, we would recommend to impose the bias to be a gradient (as explained in Remark 5.2) in order to be able to easily unbias the stationary distribution.

Using many reaction coordinates. Another interesting question is to treat the case of many reaction coordinates. The classical approaches presented above are limited to a small number of reaction coordinates (typically $m \leq 3$) because the number of degrees of freedom to represent the biasing function grows typically exponentially with respect to m (see Section 5.1.3.3), which means that the storage and the computation of such a function is impossible in high dimensions.

A natural generalization of the usual ABF strategy would then be to assume some separated representation of the bias (the total bias being the sum of low-dimensional functions), in order to compute low-dimensional biasing functions, while using many reaction coordinates. As an example, the bias may be assumed to be proportional to

$$\sum_{\alpha=1}^{m} A_t^\alpha \circ \xi_\alpha, \tag{5.22}$$

where A_t^α is updated in such a way that it converges to the mean force associated to ξ_α. Other expressions could be used, like a sum of functions involving two or three reaction coordinates. Such a method could in principle enable the use of many reaction coordinates since only low-dimensional functions have to be computed.

Note that in the case when "the directions ξ_α are independent" (namely if the images of the measure μ by ξ_α, $1 \leq \alpha \leq m$ lead to independent probability measures), the free energy associated with ξ can indeed be written as a sum of functions of ξ_1, ξ_2, ..., ξ_m, which justifies the separated form (5.22). On the other side of the spectrum, if, say, the directions ξ_1 and ξ_2 are strongly correlated, namely if ξ_2 is actually a function of ξ_1, then it is natural to bias the dynamics by a function of ξ_1 only. In both cases (strongly correlated reaction coordinates or independent reaction coordinates), it seems sensible to bias the dynamics by a sum of one-dimensional functions.

We would like to emphasize that we have no definite answer about the practical interest of such approaches, nor theoretical results to support their effectiveness. This is the subject of ongoing researches [Chipot *et al.* (2010)].

5.1.1.5 *Adaptive biasing force or adaptive biasing potential?*

At this point, we can compare the two approaches: ABP and ABF adaptive dynamics.

The main difficulty for ABP dynamics is that no stationary state can be reached "spontaneously": The biasing potential F_t is permanently evolving because new samples are always added in visited regions. This is related to the fact that the computed quantity (namely F_t) is only defined up to an additive constant. Such methods therefore require some vanishing adaptation parameter in practice, which artificially goes to zero as time evolves, in order for dF_t/dt to go to zero in the longtime limit. With the notation used here, this means that the general updating function $\mathcal{F}_t(z)$ has to go to zero in the longtime limit, for all $z \in \mathcal{M}$. It is often not so easy to choose *a priori* the rate of decay to zero of these parameters (see for example the Wang-Landau algorithm in Section 5.1.4.2). Such a difficulty is not raised by ABF dynamics for which conditional expectations approximations used in practice can indeed reach a stationary state. However, there exist recent ABP-like methods (which do not enter exactly the framework presented above) without such vanishing adaption parameters, see [Marsili *et al.* (2006); Dickson *et al.* (2009)] and Section 5.1.4.5.

Another argument in favor of ABF methods is that it is generally better numerically, in terms of accuracy, to approximate a gradient and then to integrate it, rather than approximating a function and then differentiating it. Indeed, it is observed numerically that Γ_t is typically less smooth (as a function of z) than the free energy computed from Γ_t. On the contrary, for ABP methods, the bias F_t is typically rather rough, and its derivative, which is even more noisy, is then used to bias the dynamics.

On the other hand, one advantage of ABP dynamics is that it can easily handle discrete reaction coordinates (such as the magnetization in a spin system, or the number of particles in a Lennard-Jones fluid), while ABF dynamics requires a continuous reaction coordinate for the mean force to be defined.

5.1.2 *Extended dynamics*

The adaptive biasing force method may be cumbersome to use in practice since the computation of the local mean force f may be difficult, because of the terms involving second order derivatives of ξ. There is a way to avoid the technicalities related to the fact that $\Sigma(z)$ is not a flat submanifold. Consider an extended problem set in the space $\mathcal{D} \times \mathcal{M}$, with an extended potential

$$V_\eta^{\mathrm{meta}}(q, z) = V(q) + \frac{1}{2\eta}\|z - \xi(q)\|^2, \tag{5.23}$$

and the following reaction coordinate on the extended space:

$$\xi^{\mathrm{meta}}(q, z) = z. \tag{5.24}$$

The associated free energy is

$$F_\eta^{\mathrm{meta}}(z) = -\beta^{-1} \ln \left(\frac{\displaystyle\int_{\mathcal{D}} \exp(-\beta V_\eta^{\mathrm{meta}}(q, z))\, dq}{\displaystyle\int_{\mathcal{M}} \int_{\mathcal{D}} \exp(-\beta V_\eta^{\mathrm{meta}}(q, z))\, dq\, dz} \right). \tag{5.25}$$

It is easy to check that $F_\eta^{\rm meta}$ is an approximation of the free energy F when η is small enough.

Lemma 5.3 (Convergence of $F_\eta^{\rm meta}$). *The free energy $F_\eta^{\rm meta}$ (for the extended problem) converges to the free energy F (defined by (5.4)) of the original problem in the limit $\eta \to 0$.*

Proof. For $z \in \mathcal{M}$, and using the co-area formula (3.12),

$$\begin{aligned} \mathrm{e}^{-\beta F_\eta^{\rm meta}(z)} &= \frac{\displaystyle\int_{\mathcal{D}} \exp(-\beta V_\eta^{\rm meta}(q,z))\,dq}{\displaystyle\int_{\mathcal{D}}\int_{\mathcal{M}} \exp(-\beta V_\eta^{\rm meta}(q,z))\,dq\,dz} \\ &= \frac{(2\pi\eta\beta^{-1})^{m/2}}{Z_\nu}\int_{\mathcal{D}} \exp(-\beta V(q))\exp\left(-\frac{\beta}{2\eta}\|z-\xi(q)\|^2\right)dq \\ &= \frac{1}{Z_\nu}\int_{\mathcal{M}}\int_{\Sigma(z')} \exp(-\beta V)(\det G)^{-1/2} d\sigma_{\Sigma(z')}\,\chi_\eta(z-z')\,dz' \qquad (5.26) \\ &\xrightarrow[\eta\to 0]{} \frac{1}{Z_\nu}\int_{\Sigma(z)} \exp(-\beta V)(\det G)^{-1/2}\,d\sigma_{\Sigma(z)} = \mathrm{e}^{-\beta F(z)} \end{aligned}$$

since

$$\chi_\eta(z)\,dz = (2\pi\eta\beta^{-1})^{-m/2}\exp\left(-\frac{\beta}{2\eta}\|z\|^2\right)dz \xrightarrow[\eta\to 0]{} \delta_0(dz)$$

in the sense of distributions on $\mathcal{M}$. □

Note that the reaction coordinate $\xi^{\rm meta}$ is very simple, and its derivatives are easily computed. In particular, the expression for the mean force in this case reads

$$\nabla F_\eta^{\rm meta}(z) = \int_{\mathcal{D}} f_\eta^{\rm meta}(q,z)\,\nu_\eta^{\xi^{\rm meta}}(dq\,|\,z), \qquad (5.27)$$

with the local mean force $f_\eta^{\rm meta}$

$$f_\eta^{\rm meta}(q,z) = \nabla_z V_\eta^{\rm meta}(q,z) = \frac{1}{\eta}(z-\xi(q)), \qquad (5.28)$$

and the conditional measure $\nu_\eta^{\xi^{\rm meta}}(\cdot|z)$

$$\nu_\eta^{\xi^{\rm meta}}(dq\,|\,z) = \frac{\exp(-\beta V_\eta^{\rm meta}(q,z))\,dq}{\displaystyle\int_{\mathcal{D}} \exp(-\beta V_\eta^{\rm meta}(q,z))\,dq}. \qquad (5.29)$$

Of course, the difficulty of such an approach is to determine an adequate value for η, sufficiently small for $F_\eta^{\rm meta}$ to be close to F (up to an additive constant), but not too small unless the associated dynamics in extended space becomes stiff, and very small time-steps have to be used in practice.

Combining the idea of extended dynamics and adaptive methods, a typical extended ABP method writes (compare with (5.18)):

$$\left\{\begin{aligned} dq_t &= \left(-\nabla V(q_t) + \frac{1}{\eta}\sum_{\alpha=1}^{m} ((z_t)_\alpha - \xi_\alpha(q_t))\nabla\xi_\alpha(q_t)\right) dt + \sqrt{\frac{2}{\beta}}\, dW_t^q, \\ dz_t &= \left(-\frac{1}{\eta}(z_t - \xi(q_t)) + \nabla F_t(z_t)\right) dt + \sqrt{\frac{2}{\beta}}\, dW_t^z, \\ \frac{dF_t(z)}{dt} &= -\beta^{-1} \ln\left(\int_{\mathcal{D}} \psi(t,q,z)\, dq\right). \end{aligned}\right. \tag{5.30}$$

Extended dynamics for free energy computations were originally proposed in the ABP framework, see [Laio and Parrinello (2002)]. In the same vein, a typical extended ABF dynamics writes (compare with (5.19)):

$$\left\{\begin{aligned} dq_t &= \left(-\nabla V(q_t) + \frac{1}{\eta}\sum_{\alpha=1}^{m} ((z_t)_\alpha - \xi_\alpha(q_t))\nabla\xi_\alpha(q_t)\right) dt + \sqrt{\frac{2}{\beta}}\, dW_t^q, \\ dz_t &= \frac{1}{\eta}(\xi(q_t) - G_t(z_t))\, dt + \sqrt{\frac{2}{\beta}}\, dW_t^z, \\ G_t(z) &= \frac{\displaystyle\int_{\mathcal{D}} \xi(q)\, \psi(t,q,z)\, dq}{\displaystyle\int_{\mathcal{D}} \psi(t,q,z)\, dq}, \end{aligned}\right. \tag{5.31}$$

and the instantaneous approximation to the mean force

$$\Gamma_t(z) = \frac{\displaystyle\int_{\mathcal{D}} f_\eta^{\mathrm{meta}}(q)\, \psi(t,q,z)\, dq}{\displaystyle\int_{\mathcal{D}} \psi(t,q,z)\, dq} = \frac{1}{\eta}(z - G_t(z))$$

is expected to converge to the extended mean force $\nabla F_\eta^{\mathrm{meta}}(z)$. It is of course possible to use more general updating methods (in the spirit of (5.20) or (5.21)) in this context.

Remark 5.4 (Temperature accelerated molecular dynamics). *We would like to mention here a free energy calculation method which is based on the extended potential V_η^{meta}, even though it is not an adaptive dynamics, but rather a sampling method. The temperature accelerated* *method [Maragliano and Vanden-Eijnden (2006)] consists in considering the dynamics:*

$$\begin{cases} dq_t = -\dfrac{1}{\gamma}\left(\nabla V(q_t) + \dfrac{1}{\eta}(\xi(q_t) - z_t)\cdot\nabla\xi(q_t)\right) dt + \sqrt{\dfrac{2}{\beta\gamma}}dW_t^q, \\ dz_t = \dfrac{1}{\overline{\gamma}\eta}(z_t - \xi(q_t))\, dt + \sqrt{\dfrac{2}{\overline{\beta}\,\overline{\gamma}}}dW_t^z. \end{cases}$$

The bottom line of the method is that in the regime

$$\overline{\gamma} \gg \gamma \text{ and } \beta \gg \eta,$$

the dynamics on z_t converges to an effective dynamics of the form:

$$dz_t = -\frac{1}{\overline{\gamma}}\nabla F(z_t)\, dt + \sqrt{\frac{2}{\overline{\beta}\,\overline{\gamma}}}dW_t^z, \tag{5.32}$$

so that the probability measure sampled by z_t is (in this limiting regime) proportional to $\exp(-\overline{\beta}F(z))$ (where F is the free energy associated to the reaction coordinate ξ). The artificial inverse temperature $\overline{\beta}$ is then chosen so that the effective dynamics (5.32) is not too much metastable. Of course, the difficulty of such a method is to choose the numerical parameters $(\gamma, \overline{\gamma}, \eta, \overline{\beta})$ for the method to be efficient in practice.

5.1.3 *Discretization methods*

We present in this section typical methods used in practice to discretize an ABP dynamics (5.18) or an ABF dynamics (5.19). What is needed is:

- An approximation of the probability measures involved in the calculation of Γ_t or F_t: Section 5.1.3.1 is devoted to discretizations using averages over many replicas, while Section 5.1.3.4 is devoted to approximations based on trajectorial averages. Section 5.1.3.2 discusses kernel density estimates which may be used to properly define smooth approximations of the probability measures involved in the calculation of Γ_t or F_t.
- A discretization procedure to store the functions Γ_t or F_t: Section 5.1.3.3 is devoted to the discretization of functions of the reaction coordinate variable (typically, finite element like discretization).

Note that we do not discuss the time discretization, since it is handled with standard finite difference schemes.

5.1.3.1 *Approximation of probability measures*

The probability density function $\psi(t,\cdot)$ of q_t is needed in (5.18) or in (5.19), through its marginal or its conditional measures. In the ABP method, the update of the bias (see (5.18)) may be written formally as:

$$\frac{dF_t(z)}{dt}\,dz = -\beta^{-1}\ln \mathbb{P}\Big(\xi(q_t) \in (z, z+dz)\Big).$$

The formal notation

$$\mathbb{P}\Big(\xi(q_t) \in (z, z+dz)\Big) \simeq \psi^{\xi}(t,z)\,dz, \qquad (5.33)$$

where

$$\psi^{\xi}(t,z) = \int_{\Sigma(z)} \psi(t,\cdot)\,(\det G)^{-1/2}\,d\sigma_{\Sigma(z)}$$

denotes the marginal of ψ along ξ, already gives the idea of the discretization method: compute some (empirical) approximation of the density of $\xi(q_t)$.

In the same vein, the update of the biasing force (see (5.19)) may be written formally as

$$\Gamma_t(z) = \mathbb{E}\Big(f(q_t)\,\Big|\,\xi(q_t) \in (z, z+dz)\Big) = \frac{\mathbb{E}\Big(f(q_t)1_{\{\xi(q_t)\in(z,z+dz)\}}\Big)}{\mathbb{P}\Big(\xi(q_t) \in (z, z+dz)\Big)}. \qquad (5.34)$$

In this case, the conditional expectations have to be approximated by some empirical conditional expectation, based on empirical averages computed at a fixed value of the reaction coordinate.

There are basically two ideas to approximate these quantities: Either perform empirical averages over many replicas, or use trajectorial averages over a single long trajectory. The two strategies may be combined of course, by considering trajectorial averages over many replicas. For the moment, we focus on implementations using many replicas, and we will come back to trajectorial averages in Section 5.1.3.4, but all the considerations below hold for both approaches.

A natural idea to approximate ψ is to introduce many replicas $(q_t^{k,K})_{1\le k\le K}$ of the dynamics, driven by independent Brownian motions $(W_t^k)_{k\ge1}$, but contributing to the same biasing potential. The law of the random variable q_t is then approximated by its empirical counterpart

$$\frac{1}{K}\sum_{k=1}^{K} \delta_{q_t^{k,K}}(dq), \qquad (5.35)$$

where $\delta_{q_t^{k,K}}(dq)$ is a Dirac mass in $\mathbb{R}^n$ at point $q_t^{k,K}$. Likewise, the distribution of the random variable $\xi(q_t)$, is approximated by

$$\frac{1}{K}\sum_{k=1}^{K}\delta_{\xi\left(q_t^{k,K}\right)}(dz),$$

where $\delta_{\xi\left(q_t^{k,K}\right)}(dz)$ is a Dirac mass in $\mathbb{R}^m$ at point $\xi\left(q_t^{k,K}\right)$. Thus, very formally, for an ABP method, one obtains an approximation of F_t satisfying

$$\frac{dF_t^K(z)}{dt}\,dz = -\beta^{-1}\ln\left(\frac{1}{K}\sum_{k=1}^{K}\delta_{\xi\left(q_t^{k,K}\right)}(dz)\right). \tag{5.36}$$

Likewise, for an ABF method and again very formally, the following approximation of Γ_t is obtained:

$$\Gamma_t^K(z) = \frac{\sum_{k=1}^{K} f\left(q_t^{k,K}\right)\delta_{\xi\left(q_t^{k,K}\right)}(dz)}{\sum_{k=1}^{K}\delta_{\xi\left(q_t^{k,K}\right)}(dz)}. \tag{5.37}$$

However, the latter two expressions do not make sense: The Monte Carlo discretization (empirical distribution over many replicas) cannot be considered independently from a discretization (or a regularization) in the z variable, to give a proper meaning to these equations. Indeed, an additional approximation step is needed in order to work with sufficiently smooth functions of z (and not only measures in the z variable). This is necessary both in the ABP framework (to be able to define the log of the marginal density along ξ in (5.36)) and in the ABF framework (to properly define conditional expectations of the empirical distribution (5.35) at a fixed value of the reaction coordinate, see (5.37)). There are basically two routes: Either use a regularization procedure to properly define smooth approximations (see Section 5.1.3.2 for a presentation of kernel density estimates); or combine the approximations by empirical distributions presented above with a discretization of the values taken by ξ, namely a discretization of the functions of the z variable (see Section 5.1.3.3). The latter choice is the more commonly used approach in the molecular dynamics community. In particular, histogram representations fall into this category.

5.1.3.2 *Approximations based on kernel density estimation: regularization and mathematical results*

In order to regularize the empirical distributions, a natural approach is to replace the Dirac masses by approximations of the identity, in (5.36) and (5.37). Thus, the approximated marginal density $\psi^{\xi}_{K,\varepsilon}$ is defined by

$$\psi^{\xi}_{K,\varepsilon}(t,z) = \frac{1}{K}\sum_{k=1}^{K} \delta^{\varepsilon}\left(\xi(q_t^{k,K}) - z\right), \tag{5.38}$$

where

$$\delta^{\varepsilon}(z) = \frac{1}{\varepsilon^m}\chi\left(\frac{z}{\varepsilon}\right),$$

where the smooth function χ satisfies

$$\chi > 0, \qquad \int_{\mathcal{M}} \chi = 1.$$

For instance, χ may be the Gaussian density. In the same vein, the approximated conditional density $\nu^{\xi}_{K,\varepsilon}(t, dq \,|\, z)$ can be defined by duality against smooth test functions: for any test function A,

$$\int_{\mathcal{D}} A(q)\, \nu^{\xi}_{K,\varepsilon}(t, dq \,|\, z) = \frac{\displaystyle\sum_{k=1}^{K} A(q_t^{k,K})\, \delta^{\varepsilon}\left(\xi(q_t^{k,K}) - z\right)}{\displaystyle\sum_{k=1}^{K} \delta^{\varepsilon}\left(\xi(q_t^{k,K}) - z\right)}. \tag{5.39}$$

The smoothing parameter ε plays the same role as the size of the bins in a histogram representation (see Section 5.1.3.3).

Using this smoothing procedure, the multiple replicas discretization of the ABP dynamics is obtained by approximating the current marginal density in (5.18) with the smoothed empirical marginal density (5.38):

$$\begin{cases} dq_t^{k,K} = -\nabla\left(V - F_t^{K,\varepsilon}\circ\xi\right)\left(q_t^{k,K}\right)\, dt + \sqrt{\dfrac{2}{\beta}}\, dW_t^k, \\ \dfrac{dF_t^{K,\varepsilon}(z)}{dt} = -\beta^{-1}\ln\left(\dfrac{1}{K}\displaystyle\sum_{k=1}^{K}\delta^{\varepsilon}\left(\xi\left(q_t^{k,K}\right) - z\right)\right), \end{cases} \tag{5.40}$$

where W_t^k $(1 \le k \le K)$ are independent standard Brownian motions. The multiple replica discretization of the ABF dynamics is obtained by approximating the current mean force in (5.19) with the smoothed empirical conditional measure (5.39):

$$\left\{\begin{aligned} dq_t^{k,K} &= \left(-\nabla V\left(q_t^{k,K}\right) + \sum_{\alpha=1}^{m}\left[\Gamma_t^{K,\varepsilon}\left(\xi\left(q_t^{k,K}\right)\right)\right]_\alpha \nabla\xi_\alpha\left(q_t^{k,K}\right)\right) dt \\ &\quad + \sqrt{\frac{2}{\beta}}\, dW_t^k, \\ \Gamma_t^{K,\varepsilon}(z) &= \frac{\displaystyle\sum_{k=1}^{K} f\left(q_t^{k,K}\right)\delta^\varepsilon\left(\xi\left(q_t^{k,K}\right)-z\right)}{\displaystyle\sum_{k=1}^{K}\delta^\varepsilon\left(\xi\left(q_t^{k,K}\right)-z\right)}, \end{aligned}\right. \tag{5.41}$$

where W_t^k $(1 \leq k \leq K)$ are again independent standard Brownian motions.

The analysis of the convergence of such a multiple replica discretization typically involves two parameters: the smoothing parameter ε and the number K of replicas. The convergence of the method is hopefully obtained in the limit $K \to \infty$ and then $\varepsilon \to 0$ (in this order). For a given $\varepsilon > 0$, a sufficient number of replicas is needed (K sufficiently large) in order to have enough replicas contributing to the approximation of Γ_t or F_t at a fixed reaction coordinate value z. If $m = 1$, it is expected that the statistical error scales as $(\varepsilon K)^{-1/2}$ (since for a given z, only εK replicas typically contribute to the Monte Carlo mean to compute $F_t(z)$ or $\Gamma_t(z)$).

In [Jourdain *et al.* (2009)], it is proved that in the specific (but representative) situation where $\xi(q_1, \ldots, q_N) = q_1$, and the configurational space is $\mathbb{T} \times \mathbb{R}^{n-1}$ or $\mathbb{T}^n$ ($n = dN$, d being the physical dimension of the underlying space), the multiple replicas discretization of the ABF dynamics (5.41) indeed converges to the ABF dynamics when $\varepsilon \to 0$ and $K \to \infty$. More precisely, it can be shown that (see Theorem 1.4 with $\alpha = \sqrt{\varepsilon}$ in [Jourdain *et al.* (2009)]), for all $T > 0$,

$$\mathbb{E}\left(\int_0^T \left\|\Gamma_t^{K,\varepsilon} - F_t'\right\|_{L^\infty(\mathbb{T})} dt\right) = \mathrm{O}\left(\sqrt{\varepsilon} + \frac{1}{\sqrt{K}}\exp(C(T)\varepsilon^{-5/2})\right),$$

where $C(T)$ is a constant depending on time T (but not on ε and K). The exponential factor is certainly not optimal but appears for technical reasons.

5.1.3.3 *Discretization of functions defined on the reaction coordinate space*

In practice, to store the free energy approximation and the mean force approximation on a computer, a suitable discretization in the reaction coordinate variable is needed. Moreover, we will see that this discretization allows to give a proper sense to the equations (5.36) and (5.37), without necessarily using a regularization step as presented in the former section.

Note that in the case when $\mathcal{M}$ is unbounded, the free energy or the mean force is only computed on a bounded subset $\widetilde{\mathcal{M}} \subset \mathcal{M}$ (in particular for convergence reasons, see Section 5.2).

Finite element-like discretization methods. A first approach is to decompose F_t or Γ_t on a finite element space, such as piecewise constant functions ($\mathbb{P}_0$ finite element), or continuous piecewise affine functions ($\mathbb{P}_1$ finite element), for a given mesh of $\widetilde{\mathcal{M}}$. This also requires the definition of a suitable projection operator of the empirical distributions onto the finite element space. We refer for example to Chapter 1 in [Ern and Guermond (2004)].

Let us start with the simple case when a regularization procedure has been used to properly define a smooth approximation of F_t and Γ_t (see Section 5.1.3.2 above). In this case, the Monte Carlo approximations of F_t or Γ_t are continuous functions of z (which are thus defined pointwise) and classical Lagrangian projection operators can be used. This amounts to storing only the values of (the approximations of) F_t or Γ_t at a finite number of points of $\widetilde{\mathcal{M}}$ and then using interpolation methods to give a value to F_t or Γ_t at any point of $\widetilde{\mathcal{M}}$. For kernel density estimates such as those presented in Section 5.1.3.2 for example, the values of $F_t^{K,\varepsilon}$ in (5.40) or $\Gamma_t^{K,\varepsilon}$ in (5.41) would be retained at the nodes of a finite element discretization of functions on $\widetilde{\mathcal{M}}$, associated with a mesh of $\widetilde{\mathcal{M}}$. For example, a continuous piecewise affine discretization ($\mathbb{P}_1$ finite elements) would consist in retaining the values of the potential $F_t^{K,\varepsilon}$ or the force $\Gamma_t^{K,\varepsilon}$ at the mesh nodes, and a global function would be reconstructed by affine interpolations between these nodes. Of course, the typical size Δz of the mesh cells and ε should have the same order of magnitude to represent accurately enough the functions $F_t^{K,\varepsilon}$ or $\Gamma_t^{K,\varepsilon}$ without storing too much useless information.

From now on, we concentrate on the case when the Monte Carlo approximations of the distributions are not defined pointwise (think of formulas (5.36) and (5.37)). Then, other projection operators are needed. For simplicity, consider the one-dimensional case with bins of equal sizes Δz, so that

$$\widetilde{\mathcal{M}} = [z_{\min}, z_{\max}] = \bigcup_{i=0}^{N_z-1} [z_i, z_{i+1}], \quad z_i = z_{\min} + i\Delta z, \quad \Delta z = \frac{z_{\max} - z_{\min}}{N_z}.$$

If $\mathbb{P}_0$ discretization is retained and S denotes any function of z, a natural projection $\pi_0(S)$ of the function S is the following ($\lfloor \cdot \rfloor$ denoting the integer part):

$$\pi_0(S) : z \mapsto S^{\Delta z}\left(\left\lfloor \frac{z - z_{\min}}{\Delta z} \right\rfloor\right), \qquad S^{\Delta z}(n) = \frac{1}{\Delta z} \int_{z_n}^{z_{n+1}} S(z)\, dz. \tag{5.42}$$

Note that this projection operator makes sense even if S is a measure (like a sum of Dirac masses), replacing in the integral $S(z)\, dz$ by $S(dz)$. Generalization to a higher dimensional space $\widetilde{\mathcal{M}}$ is straightforward.

Histograms and bins. In fact, the choice of $\mathbb{P}_0$ finite element discretization and projection operator π_0 (see (5.42)) is the most common choice in practice and corresponds to a so-called histogram method. A histogram method consists in a binning of the reaction coordinate values, and an approximation of functions of the reaction coordinate by piecewise constant functions.

In this method, the marginal probability density $\psi^\xi(t, z)$ (see (5.33)) is approximated as

$$\psi^\xi(t, z) \simeq \psi_t^{\xi,\Delta z}\left(\left\lfloor \frac{z - z_{\min}}{\Delta z} \right\rfloor\right), \qquad \psi_t^{\xi,\Delta z}(n) = \frac{1}{\Delta z} \int_{z_n}^{z_{n+1}} \psi^\xi(t, z)\, dz. \tag{5.43}$$

Note that

$$\psi_t^{\xi,\Delta z}(n) = \frac{1}{\Delta z} \int_{\Sigma^{\Delta z}(n)} \psi(t, q)\, dq,$$

where

$$\Sigma^{\Delta z}(n) = \left\{ q \in \mathcal{D} \;\middle|\; z_n \le \xi(q) \le z_{n+1} \right\}. \tag{5.44}$$

In the same fashion, by applying the projector π_0 at the numerator and denominator of (5.34), the biasing force is approximated as

$$\Gamma_t(z) \simeq \Gamma_t^{\Delta z}\left(\left\lfloor \frac{z - z_{\min}}{\Delta z} \right\rfloor\right), \qquad \Gamma_t^{\Delta z}(n) = \frac{\displaystyle\int_{\Sigma^{\Delta z}(n)} f(q)\, \psi(t, q)\, dq}{\displaystyle\int_{\Sigma^{\Delta z}(n)} \psi(t, q)\, dq}. \tag{5.45}$$

In a multiple replicas implementation, $\psi(t,q)\,dq$ is replaced by its empirical approximation (5.35) in the formulas (5.43)–(5.45), which still make sense. This yields the following updating formulas:

$$\frac{dF_t^{K,\Delta z}(z)}{dt} = -\beta^{-1}\ln\left(\psi_t^{\xi,K,\Delta z}\left(\left\lfloor\frac{z-z_{\min}}{\Delta z}\right\rfloor\right)\right), \tag{5.46}$$

where

$$\psi_t^{\xi,K,\Delta z}(n) = \frac{1}{K\Delta z}\sum_{k=1}^{K} 1_{\Sigma^{\Delta z}(n)}\left(q_t^{k,K}\right).$$

For an ABF method, one obtains the following approximation of Γ_t:

$$\Gamma_t(z) \simeq \Gamma_t^{K,\Delta z}\left(\left\lfloor\frac{z-z_{\min}}{\Delta z}\right\rfloor\right), \quad \Gamma_t^{K,\Delta z}(n) = \frac{\sum_{k=1}^{K} f(q_t^{k,K}) 1_{\Sigma^{\Delta z}(n)}\left(q_t^{k,K}\right)}{\sum_{k=1}^{K} 1_{\Sigma^{\Delta z}(n)}\left(q_t^{k,K}\right)}, \tag{5.47}$$

with the convention, which holds henceforth, that $\frac{0}{0}=0$.

Higher order discretization spaces. Let us emphasize that the use of higher order approximation spaces (like $\mathbb{P}_1$, $\mathbb{P}_2$, or spline functions) should be more advocated, since the computed functions are expected to be rather smooth. One way to use a $\mathbb{P}_1$ discretization (continuous piecewise affine functions), for example, would be to consider the projection operator

$$\pi_1(S) : z \mapsto \sum_{i=1}^{N_z} S_i\,\phi_i(z), \tag{5.48}$$

where $(\phi_i)_{1\le i\le N_z}$ are the hat functions which form a basis of the $\mathbb{P}_1$ finite element space, and

$$S_i = \sum_{j=1}^{N_z} A_{i,j}^{-1} \int_{\widetilde{\mathcal{M}}} S(z)\phi_j(z)\,dz$$

where $A_{i,j}^{-1}$ is the (i,j)-component of the inverse of the so-called mass matrix with coefficients

$$A_{i,j} = \int_{\widetilde{\mathcal{M}}} \phi_i(z)\,\phi_j(z)\,dz.$$

Again, these formulas make sense even if S is a measure (like a sum of Dirac masses), replacing in the integral $S(z)\,dz$ by $S(dz)$. For an ABP method, this yields the updating formula (compare with (5.46))

$$\frac{dF_t^{K,\Delta z}(z)}{dt} = -\beta^{-1}\ln\left(\pi_1\left(\frac{1}{K}\sum_{k=1}^{K}\delta_{\xi(q_t^{k,K})}(dz)\right)\right).$$

And for an ABF methods, Γ_t is approximated by (compare with (5.47)):

$$\Gamma_t^{K,\Delta z}(z) = \frac{\pi_1\left(\sum_{k=1}^{K} f(q_t^{k,K})\delta_{\xi(q_t^{k,K})}(dz)\right)}{\pi_1\left(\sum_{k=1}^{K} \delta_{\xi(q_t^{k,K})}(dz)\right)}.$$

Let us conclude this paragraph by noting that in these approaches, some spectral element basis (like for example the Fourier modes on the interval $\widetilde{\mathcal{M}} = [z_{\min}, z_{\max}]$) could be used instead of a finite element basis, which may help to improve the convergence if the free energy or the mean force are smooth functions. This consists in implementing the previous formulas, with a basis of functions $(\phi_i)_{1\le i\le N_z}$ which are orthogonal for a well chosen scalar product, see for example [Canuto *et al.* (2006)] for a general introduction.

A natural projection operator for the ABF method. In the ABF method, another natural projection method is to use the definition of conditional expectations as L^2-projections (see for example Section 9.3 in [Williams (1991)]).

Let us make this precise. Consider a basis of differentiable functions $(\phi_i)_{1\le i\le N_z}$ with $\phi_i : \widetilde{\mathcal{M}} \to \mathbb{R}$ (one could think for example of the Fourier modes on $\widetilde{\mathcal{M}}$). The objective is to write Γ_t (or more precisely its Monte Carlo approximation through empirical means over K replicas) as a linear combination of the functions $\nabla\phi_i$. Note that, in this case, the biasing force automatically derives from a potential.

It is natural to consider for the coefficients of the linear combination

$$(\alpha_1, \ldots, \alpha_N) = \arg\min_{\gamma_1,\ldots,\gamma_N} \frac{1}{K}\sum_{k=1}^{K}\left| f(q_t^{k,K}) - \sum_{i=1}^{N_z} \gamma_i \nabla\phi_i(\xi(q_t^{k,K}))\right|^2.$$

This amounts to saying that $(\alpha_1, \ldots, \alpha_N)$ is a solution to the linear system: $\forall i \in \{1, \ldots, N_z\}$,

$$\sum_{j=1}^{N_z} C_{i,j}\alpha_j = b_i,$$

with the co-variance matrix

$$C_{i,j} = \frac{1}{K}\sum_{k=1}^{K} \nabla\phi_i \cdot \nabla\phi_j\left(\xi\left(q_t^{k,K}\right)\right)$$

and the right-hand side

$$b_i = \frac{1}{K} \sum_{k=1}^{K} f\left(q_t^{k,K}\right) \cdot \nabla \phi_i \left(\xi\left(q_t^{k,K}\right)\right).$$

This is a small symmetric linear system which has to be solved. In case the matrix C is closed to singular, QR or singular value decomposition should be used (see for example [Demmel (1997); Trefethen and Bau (1997)]).

Complexity. In terms of the number of replicas K, the complexity of all the methods mentioned above are equivalent: they scale linearly with respect to K. In terms of the number of degrees of freedom N_z, zeroth order methods scale linearly with respect to N_z, whereas higher order methods typically scale like $(N_z)^3$ since they involve an inversion of an $N_z \times N_z$ matrix.

5.1.3.4 *Approximation of the law based on trajectorial averages*

In molecular dynamics simulations, most practitioners use trajectorial averages over one single trajectory, rather than empirical means over many replicas. Trajectorial averages, and possibly other averaging procedures, are then used to approximate density functions and to build the bias in adaptive dynamics.

For the ABP dynamics (5.18), the update of the bias is typically replaced by

$$\frac{dF_t(z)}{dt}\, dz \simeq -\beta^{-1} \ln \left(\frac{1}{\theta(t)} \int_{t-\theta(t)}^{t} 1_{\left\{\xi(q_s)\in(z,z+dz)\right\}}\, ds \right), \tag{5.49}$$

where $0 < \theta(t) \leq t$ is a time window on which the local trajectorial average is performed.

For the ABF dynamics (5.19),

$$\Gamma_t(z) \simeq \frac{\displaystyle\int_{t-\theta(t)}^{t} f(q_s) 1_{\left\{\xi(q_s)\in(z,z+dz)\right\}}\, ds}{\displaystyle\int_{t-\theta(t)}^{t} 1_{\left\{\xi(q_s)\in(z,z+dz)\right\}}\, ds}. \tag{5.50}$$

There are basically two choices for $\theta(t)$: Either an average over the whole trajectory is used ($\theta(t) = t$), or an average over a sliding time window of fixed length is performed (θ is a constant function).

Using an average over the whole trajectory ($\theta(t) = t$) is actually quite common. This is the standard implementation of ABF methods, see

Section 5.1.4.3. Such a procedure is also used in the ABP case in the so-called self-healing umbrella sampling method [Marsili *et al.* (2006)], see Section 5.1.4.5.

The implementation of an average over a fixed length time window is easier done by considering exponential memory function:

$$h_t(s) = \exp(-(t-s)/\theta)\, 1_{[0,t]}(s)$$

(rather than the kernel $\theta^{-1} 1_{[t-\theta,t]}(s)$ as mentioned above), where $\theta > 0$ is a fixed positive parameter. For an ABF dynamics, for instance, this leads to

$$\Gamma_t(z) \simeq \frac{\displaystyle\int_0^t f(q_s) h_t(s)\, 1_{\{\xi(q_s)\in(z,z+dz)\}}\, ds}{\displaystyle\int_0^t h_t(s)\, 1_{\{\xi(q_s)\in(z,z+dz)\}}\, ds}.$$

The interest of the exponential time kernel is that, for both the numerator and the denominator, only the current values have to be retained to update the integrals, whereas the whole trajectory over the time window should be kept in memory for the kernel $\theta^{-1} 1_{[t-\theta,t]}(s)$.

In practice, it is recommended to modify the bias only after a certain time, and not at all time-steps, in order to have enough samples in each bin (to ensure that the statistical error is not too large). One way is, for example, to choose a (fixed) time θ, and to update the bias at times $n\theta$ ($n \geq 1$), using the approximation of the marginal distribution by trajectorial averages over the time interval $[n\theta, (n+1)\theta]$. More precisely, in the ABP case (where we recall $\lfloor \cdot \rfloor$ denotes the integer part),

$$\begin{cases} dq_t = -\nabla \left(V - F_t^\theta \circ \xi\right)(q_t)\, dt + \sqrt{\dfrac{2}{\beta}}\, dW_t, \\ F_t^\theta(z) = \overline{F}^{\lfloor t/\theta \rfloor}(z), \\ \overline{F}^{n+1}(z) = \overline{F}^{n}(z) - \beta^{-1} \ln\left(\dfrac{1}{\theta}\displaystyle\int_{n\theta}^{(n+1)\theta} 1_{\{\xi(q_s)\in(z,z+dz)\}}\, ds\right). \end{cases} \tag{5.51}$$

In the ABF case,

$$\begin{cases} dq_t = \left(-\nabla V(q_t) + \displaystyle\sum_{\alpha=1}^m \left[\Gamma_t^\theta(\xi(q_t))\right]_\alpha \nabla \xi_\alpha(q_t)\right) dt + \sqrt{\dfrac{2}{\beta}}\, dW_t, \\ \Gamma_t^\theta(z) = \overline{\Gamma}^{\lfloor t/\theta \rfloor}(z), \\ \overline{\Gamma}^{n}(z) = \dfrac{\displaystyle\int_{n\theta}^{(n+1)\theta} f(q_s) 1_{\{\xi(q_s)\in(z,z+dz)\}}\, ds}{\displaystyle\int_{n\theta}^{(n+1)\theta} 1_{\{\xi(q_s)\in(z,z+dz)\}}\, ds}. \end{cases}$$

(5.52)

Dynamics of this kind were considered in the ABP case in [Zheng and Zhang (2008)], and in the first works on the ABF dynamics [Hénin and Chipot (2004)]. It is also common to apply the bias in a given bin only if the total number of samples in this bin is above a given threshold (again to ensure that the statistical error is not too large).

All the discretization methods (and regularization) in the z-variable discussed in the last two sections also apply to trajectorial averages, by simply replacing the K replicas above by K configurations along the trajectory. Let us emphasize again that the two possible approximation methods (empirical averages over many replicas and trajectorial means) can be used together, by simply considering the ensemble of configurations along the trajectory for all the replicas (see [Minoukadeh *et al.* (2010)]).

From a mathematical viewpoint, discretizations by trajectorial averages are difficult to analyze. As mathematicians, we advocate the use of empirical means over many replicas for the following practical reasons (in addition to the fact that such discretizations can be analyzed mathematically, see [Jourdain *et al.* (2009); Lelièvre *et al.* (2008)] for error estimates):

(i) The parallelization is straightforward, and the efficiency of the parallelization is very good (since the only shared information is the biasing potential or the biasing force). This therefore yields efficient massively parallel algorithms to compute free energy differences, well adapted to the architecture of high-performing computers.

(ii) It is possible to add a selection mechanism to enhance the efficiency of such a discretization method, consisting in duplicating effective replicas while deleting poor ones, according to a fitness function to be chosen. An example of such a fitness function, which favors rapid exploration of the reaction coordinates values, will be provided in Section 6.2.

(iii) It can be shown that the implementation relying upon many replicas is interesting when the reaction coordinate does not describe well all the metastable features of the system, which unfortunately constitutes a generic situation. This is typically the case, for instance, of the so-called bi-channel scenario (namely the free energy landscape features two parallel valleys, which are orthogonal to the reaction coordinate isocontours) or, more generally, when several transition mechanisms are associated to a single reaction coordinate, which is, therefore, not sufficient to parametrize fully the transformation. The underlying idea is that when many replicas are involved, they can visit more efficiently

Table 5.1 The four classes of adaptive methods, with typical examples for each class.

Configurational space	Adaptive Biasing Potential	Adaptive Biasing Force
$q_t \in \mathbb{R}^n$	Wang-Landau or SHUS Sections 5.1.4.2 or 5.1.4.5	ABF Section 5.1.4.3
$(q_t, z_t) \in \mathbb{R}^{n+m}$	Metadynamics Section 5.1.4.1	Section 5.1.4.4

in parallel all the valleys in the direction of the reaction coordinate (see [Minoukadeh *et al.* (2010); Lelièvre and Minoukadeh (2010)]).

Remark 5.5 (Convergence criterium for adaptive methods). *To decide when the longtime convergence of an adaptive method is reached, a simple criterium is to assess whether the free energy (or mean force) approximation F_t (or Γ_t) does not evolve significantly in time anymore. Another criterium (which appears to be more demanding) is to stop the simulation when the marginal probability (in the reaction coordinate) is uniform.*

*Note that, in any case, it is always possible to decide arbitrarily at some point that the exploration is sufficient (*i.e.* that the biasing potential is roughly converged), and to perform an additional equilibrium simulation with the frozen bias. The free energy is then determined from this biased equilibrium simulation, for example by approximating the marginal distribution in ξ by the empirical distribution (for example by a histogram method, see also Section* 2.5*).*

5.1.4 *Classical examples of adaptive methods*

In this section, we show how classical adaptive methods can be recast in the general framework presented above. Two classes of methods have been identified: ABP and ABF methods. Moreover, as shown in Section 5.1.2, it is possible to work either in a configurational space isomorphic to $\mathbb{R}^n$ or in the extended space isomorphic to $\mathbb{R}^{n+m}$. There are thus four possible classes of adaptive methods, and many adaptive methods used in practice enter this classification, see Table 5.1.

5.1.4.1 *Metadynamics*

Metadynamics [Laio and Parrinello (2002); Bussi *et al.* (2006)] consists in using the ABP method in the extended configurational space. The reference dynamics is thus (5.30), with the general updating function in (5.20):

$$\mathcal{F}_t(x) = -\gamma(t)\exp(-\beta x),$$

where $\gamma(t)$ is a parameter which goes to zero in the longtime limit. This yields the following dynamics (starting with $F_0 = 0$):

$$\begin{cases} dq_t = \left(-\nabla V(q_t) + \dfrac{1}{\eta}\displaystyle\sum_{\alpha=1}^{m}((z_t)_\alpha - \xi_\alpha(q_t))\nabla\xi_\alpha(q_t)\right) dt + \sqrt{\dfrac{2}{\beta}}\, dW_t^q, \\ dz_t = \left(-\dfrac{1}{\eta}(z_t - \xi(q_t)) + \nabla F_t(z_t)\right) dt + \sqrt{\dfrac{2}{\beta}}\, dW_t^z, \\ F_t(z) = -\displaystyle\int_0^t \gamma(s) \int_{\mathcal{D}} \psi(s,q,z)\, dq\, ds. \end{cases}$$

The probability density function $\psi(s,q,z)$ is then approximated by a regularized Dirac mass at point (q_s, z_s) (as in (5.40) with only one replica: $K = 1$; or as in (5.49) in the limit $\theta \to 0$), leading to the following dynamics:

$$\begin{cases} dq_t = \left(-\nabla V(q_t) + \dfrac{1}{\eta}\displaystyle\sum_{\alpha=1}^{m}((z_t)_\alpha - \xi_\alpha(q_t))\nabla\xi_\alpha(q_t)\right) dt + \sqrt{\dfrac{2}{\beta}}\, dW_t^q, \\ dz_t = \left(-\dfrac{1}{\eta}(z_t - \xi(q_t)) - \left(\nabla_z \displaystyle\int_0^t \gamma(s)\delta^\varepsilon(z_s - z)\, ds\right)\Big|_{z=z_t}\right) dt \\ \qquad + \sqrt{\dfrac{2}{\beta}}\, dW_t^z, \end{cases} \tag{5.53}$$

and

$$F_t(z) = \int_0^t \gamma(s)\, \delta^\varepsilon(z_s - z)\, ds,$$

where δ^ε is an approximation of the Dirac mass (see Section 5.1.3.2). In practice, one could either store the whole trajectory to compute the bias, or project this quantity on an appropriate discrete functions space (see Section 5.1.3.3), in which case only this projection has to be stored and updated as time evolves. One difficulty with such an approach is that, in order for F_t to converge, $\gamma(t)$ should decrease in time, and eventually converge to 0 (unless F_t keeps on being modified in the visited regions). It is *a priori* difficult to decide at which rate γ has to converge to 0. See however the newly-introduced well-tempered metadynamics [Barducci *et al.* (2008)].

We also refer to [Babin *et al.* (2008)] for a method in the same spirit.

5.1.4.2 *Wang-Landau*

Another famous instance of an ABP dynamics, (originally used for discrete reaction coordinate and for the case when the reaction coordinate is the energy) is the Wang-Landau algorithm [Wang and Landau (2001a)]. As for metadynamics, the updating function is

$$\mathcal{F}_t(x) = -\gamma(t) \exp(-\beta x),$$

where $\gamma(t)$ is a parameter which goes to zero. Typically, a histogram method is used to compute the biasing potential, using only one replica (see Section 5.1.3.3, with $K = 1$). In the case $m = 1$ and $z \in [z_{\min}, z_{\max}]$, the dynamics therefore writes:

$$\begin{cases} dq_t = -\nabla(V - F_t^{\Delta z} \circ \xi)(q_t)\, dt + \sqrt{\dfrac{2}{\beta}}\, dW_t, \\ F_t^{\Delta z}(z) = -\displaystyle\int_0^t \gamma(s)\, 1_{\Sigma^{\Delta z}(\lfloor (z - z_{\min})/\Delta z \rfloor)}(q_s)\, ds, \end{cases} \tag{5.54}$$

where $\Sigma^{\Delta z}(n)$ is defined by (5.44).

This method was originally designed for discrete state spaces (as is the case for spin systems). In this case, the dynamics is some Markov chain (such as a Metropolis Hastings algorithm), and the time integral defining the biasing potential is replaced by a sum.

In the paper [Wang and Landau (2001a)] where the method was first introduced, the Wang-Landau algorithm was proposed with the energy as a reaction coordinate: $\xi = V$. The associated free energy then writes (up to an additive constant) $F(z) = z - \beta^{-1} \ln\left(\int \delta_{V(q)-z}(dq)\right)$, and the probability density proportional to $\exp(-\beta F(z))$ is called the *density of states*.

Again, as for metadynamics, the rate of convergence to zero of the parameter γ is difficult to determine *a priori*. If $\gamma(t)$ goes to 0 slowly enough, it is possible to study (see [Atchade and Liu (2010)]) the convergence of the dynamics, the rate of convergence of $\gamma(t)$ being controlled by the non-uniformity of the histogram of the occupation time in each cell of the reaction coordinate mesh.

5.1.4.3 *The Adaptive Biasing Force method*

The ABF method [Darve and Porohille (2001); Hénin and Chipot (2004)] is based on (5.19), an approximation by trajectorial averages (5.50) (with $\theta(t) = t$), and a $\mathbb{P}_0$ discretization using only one replica (see Section 5.1.3.3, with $K = 1$). This yields the dynamics, written again for simplicity in the case $m = 1$ and $z \in [z_{\min}, z_{\max}]$:

$$\begin{cases} dq_t = \Big(-\nabla V(q_t) + \Gamma_t(\xi(q_t))\nabla\xi(q_t)\Big)\,dt + \sqrt{\frac{2}{\beta}}\,dW_t, \\ \Gamma_t(z) = \dfrac{\displaystyle\int_0^t f(q_s)\,1_{\Sigma^{\Delta z}(\lfloor(z-z_{\min})/\Delta z\rfloor)}(q_s)\,ds}{\displaystyle\int_0^t 1_{\Sigma^{\Delta z}(\lfloor(z-z_{\min})/\Delta z\rfloor)}(q_s)\,ds}, \end{cases} \tag{5.55}$$

where $\Sigma^{\Delta z}(n)$ is defined by (5.44).

In practice, the update is not done at every time-step, but after a certain equilibration time, very much in the spirit of (5.52).

5.1.4.4 *An ABF method in extended space*

The classification of Figure 5.1 may be used to derive new adaptive methods. For example, an ABF method in extended space with trajectorial averages to approximate the conditional expectations, and a regularization of the Dirac mass (as in Sections 5.1.3.2 and 5.1.4.1), would be (see (5.31)):

$$\begin{cases} dq_t = \left(-\nabla V(q_t) + \dfrac{1}{\eta}\displaystyle\sum_{\alpha=1}^m ((z_t)_\alpha - \xi_\alpha(q_t))\nabla\xi_\alpha(q_t)\right)\,dt + \sqrt{\frac{2}{\beta}}\,dW_t^q, \\ dz_t = \dfrac{1}{\eta}\,(\xi(q_t) - G_t(z_t))\,dt + \sqrt{\frac{2}{\beta}}\,dW_t^z, \\ G_t(z) = \dfrac{\displaystyle\int_0^t \xi(q_s)\,\delta^\varepsilon(z_s - z)\,ds}{\displaystyle\int_0^t \delta^\varepsilon(z_s - z)}, \end{cases}$$

and

$$\Gamma_t(z) = \frac{1}{\eta}(z - G_t(z))$$

should converge to the extended mean force $\nabla F_\eta^{\mathrm{meta}}(z)$.

5.1.4.5 *The self-healing umbrella sampling method*

The self-healing umbrella sampling method [Marsili *et al.* (2006); Dickson *et al.* (2009)] is an ABP method which does not require any vanishing adaption parameter. The adaptive dynamics writes:

$$\begin{cases} dq_t = -\nabla(V - V_t^{\text{bias}} \circ \xi)(q_t)\,dt + \sqrt{\dfrac{2}{\beta}}\,dW_t, \\ V_t^{\text{bias}}(z) = -F_t(z) + c_t, \\ \exp(-\beta F_t(z)) = \dfrac{1}{Z_t}\left(1 + \displaystyle\int_0^t \delta^\varepsilon(\xi(q_s) - z)\exp\left(\beta V_s^{\text{bias}}(\xi(q_s))\right) ds\right), \end{cases} \tag{5.56}$$

where δ^ε is an approximation of the Dirac mass (see Section 5.1.3.2). Compared to the classical ABP approach described above (see (5.18)), the idea is to unbias on-the-fly the estimator to directly obtain an approximation F_t of the free energy, and not an approximation of its derivative with respect to time. Thus, a stationary state can be obtained without externally imposed vanishing adaption (as time goes), as required in metadynamics or in the Wang-Landau method.

The basic idea behind the method, as explained in Section 5.1.1.2, is that if the process q_t was instantaneously at equilibrium with respect to the biased potential, namely if q_t had density

$$\psi^{\text{eq}}(t,q) \propto \exp\Big[-\beta\big(V(q) - V_t^{\text{bias}}(\xi(q))\big)\Big]$$

then the average of $\delta^\varepsilon(\xi(q_t) - z)\exp\left(\beta V_t^{\text{bias}}(\xi(q_t))\right)$ would be $\exp(-\beta F(z))$ (in the limit of small ε), where F is the free energy. However, even if the method is based on this instantaneous equilibration idea used above, (5.56) is not exactly of the classical form (5.18) for ABP methods.

The parameter c_t in the definition of V_t^{bias} is very important for the efficiency of the whole procedure, see [Dickson *et al.* (2009)]. In the limit c_t goes to $-\infty$, F_t is not updated, and thus V_t^{bias} does not evolve. In the limit c_t goes to $+\infty$, F_t (and thus V_t^{bias}) varies a lot in time, which may induce numerical instabilities. In the original method [Marsili *et al.* (2006)], c_t is chosen to be zero which does not seem to be the best choice numerically [Dickson *et al.* (2009)]. The thorough mathematical analysis of the whole procedure needed to better understand this point is still missing. It would be interesting to investigate the properties of an implementation using an average over many replicas (rather than a trajectorial average).

5.1.5 *Numerical illustration*

We present some numerical results for the test case of the dimer in a WCA solvent (see Section 1.3.2.4). The physical parameters of the system are the same as in Section 2.5.2.3. The ABF method is used with $N_{\text{replicas}} = 1000$

replicas of the system evolved with an overdamped Langevin dynamics using $\Delta t = 0.00025$. The truncated reaction coordinate space $[\xi_{\min}, \xi_{\max}] = [-0.2, 1.2]$ is separated into $N_z = 100$ bins of equal sizes. The mean force is estimated in each bin by a combination of plain trajectorial averages and averages over replicas. Figures 5.1-5.3 present the evolution of the system in time. Initially, almost all replicas are in the compact state $z = 0$, and

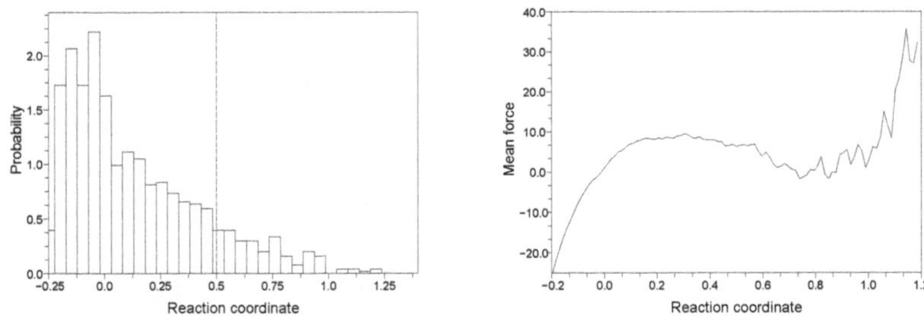

Fig. 5.1 Distribution of reaction coordinates (Left) and current estimate of the mean force (Right) at time $t = 0.05$. Most replicas are in the compact state.

only a rough estimate of the mean force is obtained. After some time,

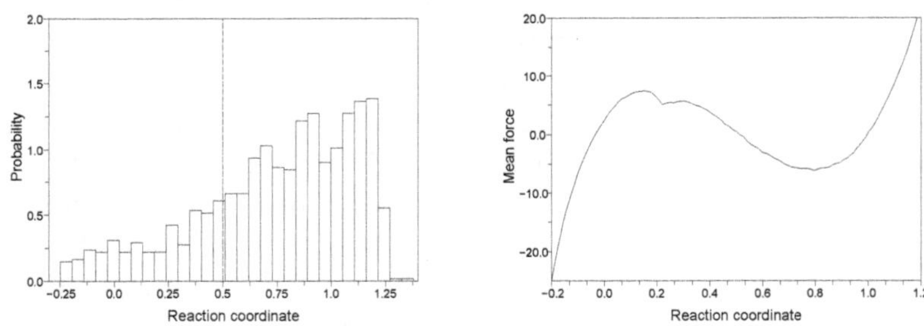

Fig. 5.2 Distribution of reaction coordinates (Left) and current estimate of the mean force (Right) at time $t = 0.625$. Most replicas are in the stretched state.

the biasing term is almost converged in the initial metastable region, and many replicas can therefore escape from this local free energy minimum. They are then stuck in the new free energy minimum corresponding to the compact state ($z = 1$), and the bias has now to be learned in this new region. Once the bias is converged in both metastable regions, the particles can move from one region to the other, and the distribution is uniform in

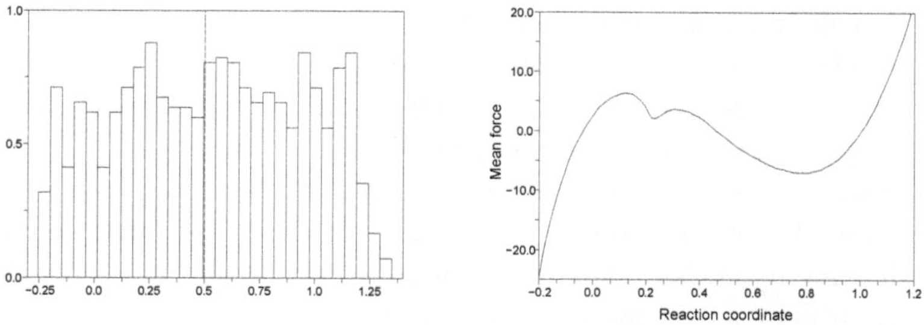

Fig. 5.3 Distribution of reaction coordinates (Left) and current estimate of the mean force (Right) at time $t = 2.5$. The probabilities to be in the compact state or the stretched states are the same.

the reaction coordinate space.

When using the ABF method on this numerical example, three phases are observed in the simulation (see Figure 5.4):

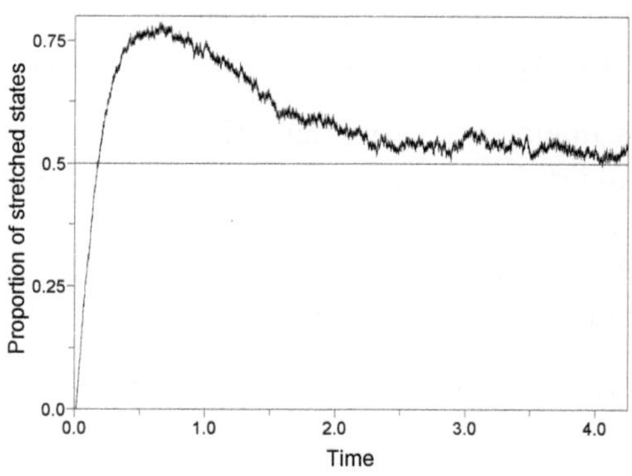

Fig. 5.4 Fraction of replica in the stretched state as a function of time. Initially, all replica start in the compact state, and an even distribution is eventually reached. Note the existence of some intermediate regime where most replica are in the stretched state.

- *Phase 1: All the replicas are in the compact state.* All the replicas start from the compact state, and thus the free energy is first approximated in this region. When the free energy in the compact state is well approximated, the potential $\mathcal{V}_t$ seen by the particles is flat in the compact state region, so that all the replicas go to the stretched state region.

- *Phase 2: Most replicas are in the stretched state.* The replicas remain in the stretched state, since it is a local minimum of $\mathcal{V}_t$ and the free energy is now approximated in this region. When the free energy in the stretched state is well approximated, the potential $\mathcal{V}_t$ seen by the particles is completely flat.
- *Phase 3: The replicas freely go from stretched state to compact state.* The biasing force is converged, and thus $\mathcal{V}_t$ is a flat constant in time potential. Thus the replicas freely visit the whole reaction coordinate space, from compact to stretched state.

Heuristically, adaptive methods penalize the already visited reaction coordinate regions, in order to force the system to visit all the reaction coordinate space.

5.2 Convergence of the adaptive biasing force method

The aim of this section is to propose a mathematical study of an Adaptive Biasing Force method, in particular to give a rigorous formulation and proofs of the following statements (which are the main arguments of practitioners of the field to advocate the use of adaptive methods):

- The ABF technique helps to remove the metastable features of the simple dynamics (5.1), and thus enables efficient exploration of the configuration space.
- With the ABF technique, the free energy F is obtained in the longtime limit, and the convergence is exponentially fast in time.

This section is mainly based on [Lelièvre *et al.* (2008)].

5.2.1 *Presentation of the studied ABF dynamics*

5.2.1.1 *Notation and definitions*

The setting is the following. We consider that the configuration space is $\mathcal{D} \subset \mathbb{R}^n$, and that the reaction coordinate is one-dimensional ($m = 1$):

$$\xi : \mathcal{D} \to \mathcal{M}$$

with $\mathcal{M} = \mathbb{R}$ or $\mathcal{M} = \mathbb{T}$. As before, we suppose that

Assumption 5.6. The function ξ is a smooth function, and $|\nabla \xi(q)| > 0$ for all $q \in \mathcal{D}$.

Thus, the subsets $\Sigma(z) = \{q \in \mathcal{D},\ \xi(q) = z\}$ of $\mathcal{D}$ are smooth submanifolds of co-dimension one which define a partition of $\mathcal{D}$:

$$\mathcal{D} = \bigcup_{z \in \mathcal{M}} \Sigma(z) \text{ and } \Sigma(z) \cap \Sigma(z') = \emptyset \text{ for } z \neq z'.$$

In the case $m = 1$, the formulas (5.4) and (5.6) for the free energy and the mean force have more explicit expressions:

$$F(z) = -\beta^{-1} \ln Z_{\Sigma(z)}, \qquad Z_{\Sigma(z)} = \int_{\Sigma(z)} |\nabla\xi|^{-1} \mathrm{e}^{-\beta V} \, d\sigma_{\Sigma(z)}, \tag{5.57}$$

while

$$F'(z) = \int_{\Sigma(z)} f \, d\nu^{\xi}(\cdot|z),$$

where the local mean force f is

$$f = \frac{\nabla V \cdot \nabla \xi}{|\nabla \xi|^2} - \beta^{-1} \mathrm{div}\left(\frac{\nabla \xi}{|\nabla \xi|^2}\right),$$

and $\nu^{\xi}(\cdot|z)$ is the conditional probability measure

$$d\nu^{\xi}(\cdot \,|\, z) = Z_{\Sigma(z)}^{-1} |\nabla \xi|^{-1} \mathrm{e}^{-\beta V} \, d\sigma_{\Sigma(z)}.$$

The marginal distribution (5.13), the conditional distribution (5.15), and the observed mean force (5.14) also have simpler expressions in terms of the law $\psi(t, \cdot)$ of the stochastic process at time t. The law of $\xi(q_t)$ is the marginal distribution $\psi^{\xi}(t, z)\, dz$ with

$$\psi^{\xi}(t, z) = \int_{\Sigma(z)} |\nabla \xi(q)|^{-1} \, \psi(t, q) \, \sigma_{\Sigma(z)}(dq), \tag{5.58}$$

and the law of q_t conditional to $\{\xi(q_t) = z\}$ is $\nu^{\xi}(t, \cdot \,|\, z)$ defined by

$$\nu^{\xi}(t, dq \,|\, z) = \frac{\psi(t, q) |\nabla \xi(q)|^{-1} \sigma_{\Sigma(z)}(dq)}{\psi^{\xi}(t, z)}. \tag{5.59}$$

The observed mean force is

$$\Gamma_{\text{obs}}(t, z) \equiv F_t'(z) = \int_{\Sigma(z)} f(q) \, d\nu^{\xi}(t, dq \,|\, z).$$

Note that here and in the following, the notation $'$ denotes a derivative with respect to the reaction coordinate values z. Sometimes, the notation $F_t'(z) = \mathbb{E}(f(q_t) \,|\, \xi(q_t) = z)$ is used.

5.2.1.2 *The ABF dynamics*

The ABF dynamics we propose to study is the following nonlinear stochastic differential equation:

$$\begin{aligned} dq_t &= -\nabla\Big(V - F_t \circ \xi + W \circ \xi - \beta^{-1}\ln(|\nabla\xi|^{-2})\Big)(q_t)\,|\nabla\xi|^{-2}(q_t)\,dt \\ &\quad + \sqrt{2\beta^{-1}}|\nabla\xi|^{-1}(q_t)\,dW_t, \qquad (5.60) \\ F_t'(z) &= \mathbb{E}\left(f(q_t)\,\Big|\xi(q_t) = z\right), \qquad (5.61) \end{aligned}$$

where W is an additional well-chosen potential that we will define below. Compared to the simple overdamped Langevin dynamics

$$dq_t = -\nabla V(q_t)\,dt + \sqrt{\frac{2}{\beta}}\,dW_t,$$

three modifications have been made to obtain (5.60):

(1) First and foremost, the potential V has been changed to the biasing potential $V - F_t \circ \xi$, and F_t' is updated as an observed mean force. This is the bottom line of the adaptive strategy, as explained in Section 5.1.1 (see in particular Eq. (5.19));
(2) Second, a potential $W \circ \xi$ has been added. This is actually needed only in the case when $\mathcal{M}$ is an unbounded domain (we recall that $\mathcal{M}$ is the domain where the reaction coordinate lives). In these cases, W is a confining potential chosen so that the law of $\xi(q_t)$ converges exponentially fast to its longtime limit (more precisely, the Fisher information associated with this law converges exponentially fast to zero, see Assumption 5.14 below for a more detailed statement). Besides, from a numerical point of view, such a potential is indeed used in practice in order to separately sample some parts of the reaction coordinate space $\mathcal{M}$ (as in stratified sampling strategies);
(3) Third, some terms depending on $|\nabla\xi|$ have been introduced. This modification is made in order to obtain a simple diffusive behavior for the law of $\xi(q_t)$ (see Proposition 5.11 below). It is expected that the longtime convergence of F_t' towards F' still holds without this modification, when simply considering the gradient dynamics (see Eq. (5.19))
$$dq_t = -\nabla(V - F_t \circ \xi + W \circ \xi)(q_t)\,dt + \sqrt{\frac{2}{\beta}}\,dW_t, \qquad (5.62)$$
with the same definition (5.61) for F_t'. However, we are only able to prove a weaker convergence result in this case, see Sections 5.2.2.5 and 5.2.3.4. Note that if $|\nabla\xi|$ is constant (for example if ξ is a length), a simple change of time relates (5.62) with (5.60).

Remark 5.7 (Computation of the biasing force). *From a practical point of view, with the additional terms mentioned in item 3 above, it is possible to compute the biasing force $F_t'(z)$ without explicitly evaluating f, which may be cumbersome, since (by Itô's calculus on q_t that satisfies* (5.60)*, and assuming $W = 0$ for simplicity)*

$$f(q_t)\,dt = d\xi(q_t) + F_t'(\xi(q_t))\,dt - \sqrt{2\beta^{-1}}\frac{\nabla\xi}{|\nabla\xi|}(q_t)\cdot dW_t. \tag{5.63}$$

By a simple finite difference scheme, we thus have the following approximation

$$\begin{aligned} f(q_{t_{n+1}}) &\simeq F_{t_n}'(\xi(q_{t_n})) \\ &\quad + \frac{1}{\Delta t}\left(\xi(q_{t_{n+1}}) - \xi(q_{t_n}) - \sqrt{2\beta^{-1}}\frac{\nabla\xi}{|\nabla\xi|}(q_{t_n})\cdot(W_{t_{n+1}} - W_{t_n})\right). \end{aligned}$$

Besides, the additional term $\nabla\big(\ln|\nabla\xi|^{-2}\big)$ in the drift term of (5.60) *can be computed using finite differences with a strategy similar to the one proposed in Remark* 3.30. *This strategy to compute F_t' comes at a price in terms of variance, as explained in Remark* 3.34.

We would like to emphasize that our arguments below hold under the following assumption of existence of regular solutions: We assume that we are given a process (q_t) and a function F_t' which satisfy (5.60)–(5.61), and such that the law of q_t at time t has a smooth density $\psi(t,\cdot)$ with respect to the Lebesgue measure on $\mathcal{D}$. We suppose that this density is sufficiently regular so that the computations below are valid. In particular, we assume that the potential V is such that either the stochastic process (q_t) lives in $\mathcal{D}$ and thus that its density $\psi(t,\cdot)$ vanishes sufficiently fast on $\partial\mathcal{D}$; or the stochastic process (q_t) has some reflecting behavior on $\partial\mathcal{D}$ and thus that its density $\psi(t,\cdot)$ has zero normal derivatives on $\partial\mathcal{D}$. In both cases, no boundary terms appear in the integrations by parts we perform to derive the entropy estimates. We refer for example to [Arnold *et al.* (2001)] for an appropriate functional framework in which such entropy estimates hold, and to [Jourdain *et al.* (2009)] for existence and uniqueness results for the solution of ABF dynamics.

5.2.1.3 *Reformulation as a nonlinear partial differential equation*

Since only the law of the process (q_t) at a fixed time t is used in (5.61), it is possible to recast the dynamics in the following nonlinear partial differential equation (Fokker-Planck equation, see Section 2.2.1.1) on the density $\psi(t,\cdot)$ of q_t:

$$\partial_t \psi = \operatorname{div}\left(\frac{\nabla\Big(V - F_t \circ \xi + W \circ \xi - \beta^{-1} \ln\left(|\nabla\xi|^{-2}\right)\Big)}{|\nabla\xi|^2} \psi \right) + \beta^{-1} \Delta\left(\frac{\psi}{|\nabla\xi|^2}\right).$$

Besides, the last term can be rewritten as

$$\begin{aligned}\Delta\left(\frac{\psi}{|\nabla\xi|^2}\right) &= \operatorname{div}\left(\frac{1}{|\nabla\xi|^2}\nabla\psi\right) + \operatorname{div}\left[\left(\nabla(|\nabla\xi|^{-2})\right)\psi\right] \\ &= \operatorname{div}\left(\frac{1}{|\nabla\xi|^2}\nabla\psi\right) + \operatorname{div}\left[\left(\nabla\left(\ln|\nabla\xi|^{-2}\right)\right)\frac{\psi}{|\nabla\xi|^2}\right],\end{aligned}$$

so that finally,

$$\begin{cases} \partial_t \psi = \operatorname{div}\left(\dfrac{\nabla(V - F_t \circ \xi + W \circ \xi)\psi + \beta^{-1}\nabla\psi}{|\nabla\xi|^2} \right), \\ F_t'(z) = \dfrac{\displaystyle\int_{\Sigma(z)} f\, |\nabla\xi|^{-1} \psi(t,\cdot)\, d\sigma_{\Sigma(z)}}{\displaystyle\int_{\Sigma(z)} |\nabla\xi|^{-1} \psi(t,\cdot)\, d\sigma_{\Sigma(z)}}. \end{cases} \tag{5.64}$$

We recall (see the consistency proof in Section 5.1.1.3) that if the potential F_t and the law of q_t reach a stationary state, then this stationary law is proportional to $\exp(-\beta(V(q) - F_\infty \circ \xi(q) + W \circ \xi(q)))\, dq$, where F_∞ denotes the stationary state for F_t and then, from the definition (5.61) of the biasing force, we obtain that, necessarily, $F'_\infty = F'$. This proves the uniqueness of the stationary state for this dynamics. We can thus expect that F_t' converges to the mean force F' in the longtime limit.

The interest of the dynamics (5.60)–(5.61) is actually twofold. First, as expected from the formal argument above, in the longtime limit, F_t' converges to the mean force F' (see the convergence result (5.71) below). Second, using the ABF method, the law of $\xi(q_t)$ has a simple diffusive behavior (see Eq. (5.67) below). The metastable feature of the simple dynamics (5.1) along ξ is thus corrected by the addition of the adaptive potential F_t, and the convergence of the adaptive dynamics to equilibrium is then faster than the convergence of the simple gradient dynamics (see Section 2.3.2).

The aim of the remainder of this section is to give a precise statement for these two assertions. The proof of the longtime convergence relies on entropy techniques (see Sections 2.3.2 and 3.2.6), and requires appropriate assumptions on the potentials V, W and the reaction coordinate ξ.

We prove that under suitable assumptions, the convergence of F'_t to F' is exponentially fast, with a rate of convergence limited, at the macroscopic level, by the rate of convergence of the law of $\xi(q_t)$ to its longtime limit, and, at the microscopic level, by the rate of convergence to the equilibrium conditioned probability measures $\nu^\xi(\cdot\,|\,z)$, for all values z of the reaction coordinate.

All these results are more precisely stated in Section 5.2.2, and the proofs are given in Section 5.2.3. We would like to mention that the main arguments of the proof are presented in a very simple case in Section 5.2.3.1 and that a result of convergence for the dynamics (5.62)–(5.61) is also given in Section 5.2.2.5.

5.2.2 *Precise statements of the convergence results*

In Section 5.2.2.1, we recall some well-known results on entropy and introduce the main notation used in the following to state the convergence result. Section 5.2.2.2 is devoted to the presentation of the convergence result for the dynamics (5.60)–(5.61). Finally, we give in Section 5.2.2.5 a (weaker) convergence result for the dynamics (5.62)–(5.61).

5.2.2.1 *Decomposition of the entropy*

Let us consider ψ and F'_t which satisfy (5.64) and let us introduce the longtime limits of ψ, ψ^ξ (defined by (5.58)) and $\nu^\xi(t,\cdot|z)$ (defined by (5.59)):

$$\psi_\infty = Z_W^{-1}\exp(-\beta(V - F\circ\xi + W\circ\xi)),$$

$$\psi^\xi_\infty(z) = Z_W^{-1}\exp(-\beta W(z)),$$

$$d\nu^\xi(\infty,\cdot\,|\,z) = d\nu^\xi(\cdot\,|\,z) = Z_{\Sigma(z)}^{-1}\exp(-\beta V)|\nabla\xi|^{-1}d\sigma_{\Sigma(z)},$$

where $Z_{\Sigma(z)}$ is given by (5.57) and

$$Z_W = \int_{\mathcal{M}} \exp(-\beta W(z))\,dz.$$

Note that

$$\int_{\mathcal{D}} \psi_\infty = 1, \quad \int_{\mathcal{M}} \psi^\xi_\infty = 1, \quad \int_{\Sigma(z)} d\nu^\xi(\infty,\cdot\,|\,z) = 1,$$

and that the probability measure $\psi^\xi_\infty(z)\,dz$ is the image of the probability measure $\psi_\infty(q)\,dq$ by ξ.

In order to state the results, we also need to introduce the following projection operators (see also the projection operator (3.37) introduced for projected stochastic processes on manifolds). For any $q \in \mathcal{D}$, we denote by

$$P(q) = \mathrm{Id} - \frac{\nabla\xi \otimes \nabla\xi}{|\nabla\xi|^2}(q) \tag{5.65}$$

the orthogonal projection operator onto the tangent space $T_q\Sigma(\xi(q))$ to $\Sigma(\xi(q))$ at point q, and by

$$Q(q) = \frac{\nabla\xi \otimes \nabla\xi}{|\nabla\xi|^2}(q)$$

the orthogonal projection operator onto the normal space $N_q\Sigma(\xi(q))$ to $\Sigma(\xi(q))$ at point q. We recall that $\otimes$ denotes the tensor product: For two vectors $u, v \in \mathcal{D}$, $u \otimes v$ is an $n \times n$ matrix with components $(u \otimes v)_{i,j} = u_i v_j$. We recall the definition of the surface gradient on $\Sigma(z)$ (see (3.60)):

$$\nabla_{\Sigma(z)} = P\nabla, \tag{5.66}$$

where P is the projection operator (5.65).

The "distance" between ψ (respectively ψ^ξ) and ψ_∞ (respectively ψ^ξ_∞) is measured using the relative entropy $H(\psi|\psi_\infty)$ (respectively $H(\psi^\xi|\psi^\xi_\infty)$), that we recall (see Definition 2.19):

$$H(\psi|\psi_\infty) = \int_{\mathcal{D}} \ln\left(\frac{\psi}{\psi_\infty}\right)\psi, \qquad H(\psi^\xi|\psi^\xi_\infty) = \int_{\mathcal{M}} \ln\left(\frac{\psi^\xi}{\psi^\xi_\infty}\right)\psi^\xi.$$

We denote the *total entropy* by

$$E(t) = H\big(\psi(t,\cdot)\,\big|\,\psi_\infty\big),$$

the *macroscopic entropy* by

$$E_M(t) = H\big(\psi^\xi(t,\cdot)\,\big|\,\psi^\xi_\infty\big),$$

and the *microscopic entropy* by

$$E_m(t) = \int_{\mathcal{M}} e_m(t,z)\,\psi^\xi(t,z)\,dz.$$

In this last expression, the "local entropy" at a fixed value z of the reaction coordinate is

$$\begin{aligned} e_m(t,z) &= H\Big(\nu^\xi(t,\cdot\,|\,z)\;\Big|\;\nu^\xi(\infty,\cdot\,|\,z)\Big) \\ &= \int_{\Sigma(z)} \ln\left(\frac{\psi(t,\cdot)}{\psi^\xi(t,z)}\Big/\frac{\psi_\infty}{\psi^\xi_\infty(z)}\right)\frac{\psi(t,\cdot)|\nabla\xi|^{-1}d\sigma_{\Sigma(z)}}{\psi^\xi(t,z)}. \end{aligned}$$

The following result, which can be seen as the extensivity of the entropy, can be obtained by a simple computation:

Lemma 5.8 (Extensivity of entropy). *It holds*

$$E(t) = E_M(t) + E_m(t).$$

We will need in the following another way to compare two probability measures, namely the Wasserstein distance with quadratic cost: For two probability measures π_1 and π_2 defined on a Riemannian manifold Σ,

$$W(\pi_1, \pi_2) = \sqrt{\inf_{\pi \in \Pi(\pi_1,\pi_2)} \int_{\Sigma \times \Sigma} d_\Sigma(x, y)^2 \, \pi(dx, dy)}.$$

In this expression, d_Σ denotes the geodesic distance on Σ: $\forall x, y \in \Sigma$,

$$d_\Sigma(x, y) = \inf \left\{ \sqrt{\int_0^1 |\dot{w}(t)|^2 \, dt} \;\middle|\; w \in \mathcal{C}^1([0, 1], \Sigma), \, w(0) = x, \, w(1) = y \right\}.$$

The set $\Pi(\pi_1, \pi_2)$ denotes the set of coupling probability measures, namely probability measures on $\Sigma \times \Sigma$ such that their marginals are π_1 and π_2: for any $\pi \in \Pi(\pi_1, \pi_2)$,

$$\int_{\Sigma \times \Sigma} \phi(x) \, \pi(dx, dy) = \int_\Sigma \phi \, d\pi_1, \quad \int_{\Sigma \times \Sigma} \psi(y) \, \pi(dx, dy) = \int_\Sigma \psi \, d\pi_2.$$

The following definition will be useful in the sequel:

Definition 5.9 (Talagrand inequality). *The probability measure π_2 is said to satisfy a Talagrand inequality with constant $\rho > 0$ (in short: $T(\rho)$) if for all probability measures π_1 such that $\pi_1 \ll \pi_2$,*

$$W(\pi_1, \pi_2) \leq \sqrt{\frac{2}{\rho} H(\pi_1|\pi_2)}.$$

In the latter definition, it is implicitly assumed that the probability measures have finite moments of order 2. This will always be the case for all the probability measures considered in this section. The following important result will be needed (see Theorem 1 in [Otto and Villani (2000)] and [Bobkov and Götze (1999)]).

Lemma 5.10. *If π_2 satisfies a logarithmic Sobolev inequality with constant ρ (in short LSI(ρ)), then π_2 satisfies $T(\rho)$.*

We recall (see Definition 2.20) that π_2 satisfies LSI(ρ) if, for all probability measures π_1 such that $\pi_1 \ll \pi_2$,

$$H(\pi_1|\pi_2) \leq \frac{1}{2\rho} I(\pi_1|\pi_2),$$

where $I(\pi_1|\pi_2)$ is the Fisher information (see Definition 2.19):

$$I(\pi_1|\pi_2) = \int \left| \nabla \ln \left(\frac{d\pi_1}{d\pi_2} \right) \right|^2 d\pi_1.$$

5.2.2.2 *Convergence of the adaptive dynamics* (5.60)–(5.61)

We are now in a position to state our main results. Concerning the dynamics on the law of $\xi(q_t)$, we have:

Proposition 5.11 (Equation satisfied by the marginal density ψ^ξ). *Let (ψ, F_t') be a smooth solution to* (5.64) *and suppose that Assumption* 5.6 *holds. Then ψ^ξ satisfies the following diffusion equation:*

$$\partial_t \psi^\xi = \partial_z \left(W' \psi^\xi + \beta^{-1} \partial_z \psi^\xi \right) \quad \text{on } \mathcal{M}. \tag{5.67}$$

Equation (5.67) is a simple diffusion equation (which reduces to the heat equation if $W = 0$). Note that even if ψ^ξ satisfies a closed partial differential equation, $\xi(q_t)$ does not satisfy a closed stochastic differential equation (see Eq. (5.63) above).

The fundamental assumptions we need to prove longtime convergence of the dynamics are the following.

Assumption 5.12. The potential V and the reaction coordinate ξ are sufficiently differentiable functions such that

$$\sup_{q \in \mathcal{D}} |\nabla \xi| \leq m < \infty, \qquad \sup_{q \in \mathcal{D}} \left| \nabla_{\Sigma(\xi(q))} f(q) \right| \leq M < \infty.$$

The requirement on f can be seen as a boundedness condition on the coupling between the conditional measures $\nu^\xi(\infty, \cdot \,|\, z)$ and the corresponding marginal ψ^ξ_∞, since it involves the mixed derivatives $P\nabla(Q\nabla V)$, one derivative being along the tangential space and the other one along the normal space of the submanifold $\Sigma(z)$ (see [Otto and Reznikoff (2007); Grunewald *et al.* (2009); Lelièvre (2009)] and Section 5.2.2.3 below).

Assumption 5.13. There exists $\rho > 0$ such that the conditional measure $\nu^\xi(\cdot \,|\, z)$ satisfies LSI(ρ) for all $z \in \mathcal{M}$.

This assumption is an assumption on the potential V and the reaction coordinate ξ. It ensures that, if for a fixed value z of the reaction coordinate, the conditioned probability measure $\nu^\xi(\infty, \cdot \,|\, z)$ were to be sampled by a simple constrained gradient dynamics (such as (3.52), see Section 3.2.6 and [Ciccotti *et al.* (2008)]), the convergence to equilibrium would be exponentially fast with rate ρ. The quantity ρ will therefore be referred to as the *microscopic rate of convergence* in the sequel.

We refer to Section 5.2.3.1 for an explicit framework where Assumptions 5.12 and 5.13 are satisfied, and to Section 5.2.2.4 below for alternative assumptions on V and ξ.

Let us now introduce the needed requirements on W.

Assumption 5.14. The potential W and the initial marginal distribution $\psi^\xi(0,\cdot)$ are such that there exist $I_0 > 0$ and $r > 0$ such that, for all $t \geq 0$,

$$I(\psi^\xi(t,\cdot)|\psi^\xi_\infty) \leq I_0 \exp(-2\beta^{-1} r\, t).$$

This assumption is indeed an assumption on W and $\psi^\xi(0,\cdot)$ because ψ^ξ satisfies the partial differential equation (5.67) where only W appears. We will see below (see Corollary 5.16) some sufficient explicit conditions on W for Assumption 5.14 to be satisfied. Of course, the constant I_0 typically depends on the initial datum $\psi^\xi(0,\cdot)$, while r can be chosen independently of $\psi^\xi(0,\cdot)$.

Assumption 5.14 ensures that the law of $\xi(q_t)$ converges to equilibrium exponentially fast with rate r, which will be referred to as the *macroscopic rate of convergence* in the sequel.

Theorem 5.15 (Exponential convergence of the entropy to zero). *Suppose that Assumptions* 5.6, 5.12, 5.13 *and* 5.14 *are satisfied. Then,*

(i) The microscopic entropy E_m converges exponentially fast to zero: There exist $C > 0$ and $\lambda > 0$ such that, for all $t \geq 0$,

$$\sqrt{E_m(t)} \leq C \exp(-\lambda t). \tag{5.68}$$

More precisely, if $\rho m^{-2} \neq r$, then

$$C = 2 \max\left(\sqrt{E_m(0)}, \frac{M}{\beta^{-1}|\rho m^{-2} - r|}\sqrt{\frac{I_0}{2\rho}}\right)$$

and

$$\lambda = \beta^{-1} \min(\rho m^{-2}, r). \tag{5.69}$$

In the special case $\rho m^{-2} = r$, for all $\lambda < \beta^{-1}\min(\rho m^{-2}, r)$, there exists a positive constant C such that (5.68) *is satisfied.*

(ii) The square root of the total entropy $\sqrt{E}$ and $\|\psi(t,\cdot) - \psi_\infty\|_{L^1(\mathcal{D})}$ both converge exponentially fast to zero with rate λ.

(iii) The biasing force F'_t converges to the mean force F' in the following sense: for all $t \geq 0$,

$$\int_{\mathcal{M}} |F'_t - F'|^2(z)\, \psi^\xi(t,z)\, dz \leq \frac{2M^2}{\rho} E_m(t). \tag{5.70}$$

In the following corollary, we concentrate on two prototypical settings for which Assumption 5.14 is satisfied, and the convergence of F'_t to F' is made precise in these settings.

Corollary 5.16 (Convergence of the biasing force). *Suppose that Assumption* 5.6 *holds.*

(1) If $\mathcal{M} = \mathbb{T}$ *and* $W = 0$, *then Assumption* 5.14 *is satisfied with* $I_0 = I(\psi^\xi(0, \cdot) \,|\, \psi^\xi_\infty)$ *and* $r = 4\pi^2$.
(2) If $\mathcal{M} = \mathbb{R}$ *and* W *is a potential such that* W'' *is bounded from below and there exists* $\overline{r} > 0$ *such that* $Z_W^{-1} \exp(-\beta W)$ *satisfies LSI(*$\overline{r}$*), then Assumption* 5.14 *is satisfied with* $r = \overline{r} - \varepsilon$ *for any* $\varepsilon \in (0, \overline{r})$.

Suppose that Assumptions 5.12 *and* 5.13 *are also satisfied. Then, for both settings given above* (1) *and* (2), F'_t *converges exponentially fast to* F' *in the following sense: For all compact set* $K \subset \mathcal{M}$, *there exist* $\overline{C} > 0$ *and* $t^* > 0$ *such that for all* $t \geq t^*$,

$$\int_K |F'_t - F'|(z) \, \psi^\xi_\infty(z) \, dz \leq \overline{C} \exp(-\lambda t), \tag{5.71}$$

where λ *is the rate of convergence introduced in Eq.* (5.68) *of Theorem* 5.15.

Note that in the case $\mathcal{M} = \mathbb{R}$, the sufficient conditions stated in Corollary 5.16 on W to satisfy Assumption 5.14 (see (2)) are valid for an α-convex potential (namely if $W'' \geq \alpha$ for a positive α), and then it is possible to choose $r = \alpha$ in Assumption 5.14 (see Lemma 5.30 below). We refer to Section 5.2.2.4 below for alternative sufficient conditions on W to satisfy Assumption 5.14.

These results therefore show that F'_t converges exponentially fast to F' (in $L^1(\psi^\xi_\infty(z)\, dz)$-norm) at a rate $\lambda = \beta^{-1} \min(\rho m^{-2}, r)$, which should be compared with (2.87). The limitations on the rate λ are related to the rate of convergence r at the macroscopic level, for Eq. (5.67) to be satisfied by ψ^ξ, and the rate of convergence at the microscopic level, which depends on the constant ρ of the logarithmic Sobolev inequalities (LSI) satisfied by the conditional measures $\nu^\xi(\infty, \cdot \,|\, z)$. There are many ways to improve the macroscopic rate of convergence r (see Section 5.2.2.3), so that the essential limitation in the convergence is the microscopic rate of convergence ρ. This constant ρ of course depends on the choice of the reaction coordinate. In our framework, we could state that *a "good reaction coordinate" is such that* ρ *is as large as possible.*

The proofs of Theorem 5.15 and Corollary 5.16 are given in Sections 5.2.3.1, 5.2.3.2 and 5.2.3.3 below.

5.2.2.3 *Interpretation and discussion of the rate of convergence*

The aim of this section is to discuss the rate of convergence $\lambda = \beta^{-1}\min(\rho m^{-2}, r)$ of the adaptive dynamics.

Comparison with the rate of convergence for the unbiased gradient dynamics. To measure the interest of adaptive dynamics compared to the simple gradient dynamics (5.1), a natural question is: Under the assumptions of Theorem 5.15, what can be said about the constant R for the logarithmic Sobolev inequality satisfied by the measure ν, which controls the rate of convergence to equilibrium for the gradient dynamics (see Section 2.3.2)? Under the following assumptions:

(i) the marginal of ν along ξ satisfies a LSI($\overline{r}$);
(ii) the conditional measures $\nu^{\xi}(\cdot|z)$ satisfy a LSI(ρ) (which is Assumption 5.13 in Theorem 5.15);
(iii) V and ξ are such that $\sup_{q\in\mathcal{D}} |\nabla_{\Sigma(\xi(q))} f(q)| \leq M < \infty$ and $|\nabla\xi|^2 \geq m > 0$ (which is related to Assumption 5.12 in Theorem 5.15),

it is shown in [Lelièvre (2009)] that ν satisfies a LSI with an optimal constant R which satisfies

$$R \geq \frac{1}{2}\left(\overline{r}m + \frac{M^2 m}{\rho} + \rho - \sqrt{\left(\overline{r}m + \frac{M^2 m}{\rho} + \rho\right)^2 - 4\overline{r}m\rho}\right). \qquad (5.72)$$

We also refer to [Grunewald *et al.* (2009)] where such so-called "two-scale criteria" have been introduced in the case of a linear function ξ. Equality holds in (5.72) for normal random variables and a linear function ξ, so that the right-hand side may be considered as a good estimate of R, at least in some particular cases. Note that the right-hand side is bounded from above by $\min(\overline{r}m, \rho)$, which corresponds to the case of zero coupling ($M = 0$). Assumption (iii) is called a bounded coupling condition since, in the particular case of a normal random variable and a linear function ξ, $\nabla_{\Sigma(\xi(q))} f(q)$ measures the covariance between macroscopic components (in $\mathcal{M}$) and microscopic components (in $\Sigma(z)$).

Let us go back to the question asked at the beginning of this paragraph. Two important assumptions required in Theorem 5.15 are that the conditional measures $\nu^{\xi}(\cdot|z)$ satisfy a LSI(ρ) (see Assumption 5.13), and that the marginal distribution ψ^{ξ} converges exponentially fast with rate $\beta^{-1}r$ to equilibrium (see Assumption 5.14), which roughly amounts to saying that ψ^{ξ}_{∞} satisfies a LSI(r). Assuming that the right-hand side in (5.72) is a good estimate of the optimal LSI constant R for the Boltzmann-Gibbs measure ν, it is easy to check that R is typically much smaller than $\min(\rho m^{-2}, r)$ since the LSI constant for the marginal of ν along ξ (denoted $\overline{r}$ above) is typically much smaller than the LSI constant for the marginal of ψ_∞ along ξ (denoted r above):

$$\begin{aligned} R &\simeq \frac{1}{2}\left(\overline{r}m + \frac{M^2 m}{\rho} + \rho - \sqrt{\left(\overline{r}m + \frac{M^2 m}{\rho} + \rho\right)^2 - 4\overline{r}m\rho}\right) \\ &\leq \min(\overline{r}m, \rho) \\ &\ll \min(\rho m^{-2}, r). \end{aligned}$$

A typical example is the simple example (1.68) of Section 1.3.3, for which $\overline{r}$ and R are very small for a high potential barrier in the q_1 variable, compared to r and ρ.

Comparison with thermodynamic integration. As noted in Section 3.2.6, the efficiency of the thermodynamic integration method is also typically limited by the LSI constant of the conditioned probability measures $\nu^\xi(\cdot \mid z)$, so that both methods (adaptive method and thermodynamic integration with constrained sampling) are expected to have similar efficiencies. A natural question is then: What is the interest of an adaptive method compared to thermodynamic integration?

The essential difference is that the number of samples in each submanifold $\Sigma(z)$ is not chosen *a priori* in adaptive methods: The time spent around a submanifold essentially depends on the time needed to obtain an approximation of the mean force which is accurate enough in order to escape from this region.

Moreover, the implementation of adaptive methods using many replicas interacting only through the common bias they are constructing leads to very efficient parallel algorithms, which can be further enhanced using a selection procedure (see Section 6.2).

Adaptive methods thus seem more flexible and easier to implement than thermodynamic integration methods, for which constrained sampling algorithms have to be used (see Chapter 3).

Enhancing the macroscopic rate of convergence. Let us consider the case $\mathcal{M} = \mathbb{R}$ for simplicity. For an α-convex potential W, Corollary 5.16 states that F'_t converges towards F' exponentially fast, with a rate $\lambda = \beta^{-1}\min(\rho m^{-2}, \alpha)$. This may seem surprising since for α large enough, the rate of convergence is no more limited by α. However, in practice, if α is very large, ψ^ξ_∞ is very peaked and some parts of $\mathcal{M}$ are poorly sampled, the variance of the result is large in these areas (which cannot be seen in our convergence result). A good method to enhance the rate of convergence at the macroscopic level while keeping a good sampling and thus low variance, is to use a particle system with many replicas and a selection mechanism. We refer to Section 6.2 and [Lelièvre *et al.* (2007b)] for more detail.

5.2.2.4 *Discussion and extension of Assumptions* 5.12–5.14

Other possible assumptions on V and ξ. We would like to mention other possible Assumptions on V and ξ than Assumptions 5.12 and 5.13 for which the results of Theorem 5.15 still hold.

(i) First, in Assumption 5.12, it is possible to change the assumption $\sup_{q\in\mathcal{D}} |\nabla_{\Sigma(\xi(q))} f(q)| \leq M < \infty$ to

$$\sup_{q\in\mathcal{D}} |f(q)| \leq M < \infty.$$

Indeed, this simply changes the estimate (5.83) in Lemma 5.27 below to the following

$$\begin{aligned} |F_t'(z) - F'(z)| &\leq \sup_{q\in\mathcal{D}} |f(q)| \, \|\nu^\xi(t,\cdot\,|\,z) - \nu^\xi(\infty,\cdot\,|\,z)\|_{TV} \\ &\leq M\sqrt{2H(\nu^\xi(t,\cdot\,|\,z)\,|\,\nu^\xi(\infty,\cdot\,|\,z))}, \end{aligned}$$

using the Csiszár-Kullback inequality (2.84). The rest of the proof remains exactly the same. However, we prefer to concentrate on Assumption 5.12 since it is more realistic from a practical point of view.

(ii) Second, it is possible to obtain a similar result of convergence under slightly different assumptions than Assumptions 5.12 and 5.13 by introducing another Riemannian structure on the submanifolds $\Sigma(z)$. We refer to Appendix B in [Lelièvre *et al.* (2008)].

Reflection boundary conditions and Assumption 5.14. From Lemmas 5.29 and 5.30 below (used to prove Corollary 5.16), it will become clear that Assumption 5.14 is actually satisfied with $W = 0$ as soon as $\mathcal{M}$ is a bounded domain. If $\mathcal{M}$ is an unbounded domain, then a confining potential W with properties such as those stated in Corollary 5.16 is needed.

Another possibility to satisfy Assumption 5.14 is to add reflection boundary conditions to confine the dynamics in a domain $\bigcup_{z\in\mathcal{N}} \Sigma(z)$, where $\mathcal{N}$ is a bounded subset of $\mathcal{M}$. Roughly speaking, this amounts to choosing W which is zero in $\mathcal{N}$ and infinite in $\mathcal{M}\setminus\mathcal{N}$. Let us make this precise. Suppose for example we are interested in the values of $F'(z)$ for $z \in \mathcal{N} = (0,1)$.

The dynamics is confined in the domain $\mathcal{O} = \bigcup_{0<z<1} \Sigma(z)$. The ABF dynamics is

$$\begin{cases} \partial_t \psi = \operatorname{div}\Big(|\nabla \xi|^{-2} \left(\nabla (V - F_t \circ \xi) \psi + \beta^{-1} \nabla \psi \right) \Big), & \text{on } \mathcal{O}, \\ \left(\nabla (V - F_t \circ \xi) \psi + \beta^{-1} \nabla \psi \right) \cdot \nabla \xi = 0, & \text{on } \Sigma(0) \cup \Sigma(1), \\ F_t'(z) = \dfrac{\displaystyle\int_{\Sigma(z)} f \, |\nabla \xi|^{-1} \psi(t,\cdot) d\sigma_{\Sigma(z)}}{\displaystyle\int_{\Sigma(z)} |\nabla \xi|^{-1} \psi(t,\cdot) d\sigma_{\Sigma(z)}}, & \text{for } z \in (0,1). \end{cases}$$

From the point of view of the stochastic process (q_t), the boundary condition translates into a normal reflection on the two submanifolds $\Sigma(0)$ and $\Sigma(1)$. Moreover, it can be checked (using (5.81)) that the boundary condition on ψ translates into a zero Neumann boundary condition on ψ^ξ: $\partial_z \psi^\xi(0) = \partial_z \psi^\xi(1) = 0$. A proof similar to that of Lemma 5.29 then shows that $I(\psi^\xi | \psi^\xi_\infty)$ converges exponentially fast to 0, so that Assumption 5.14 holds. The arguments we use to prove Theorem 5.15 and Corollary 5.16 then show that $\|F_t' - F'\|_{L^2(0,1)}$ goes to 0 exponentially fast.

Vectorial reaction coordinate. From the beginning of Section 5.2, we assume that the reaction coordinate ξ has values in a one-dimensional space, $\mathbb{T}$ or $\mathbb{R}$. The dynamics (5.64) (corresponding to (5.60)–(5.61)) and the results of convergence presented in this section can be extended to the case when $\xi = (\xi_1, \ldots, \xi_m)$ has values in $\mathbb{T}^m$ or $\mathbb{R}^m$, by considering the dynamics:

$$\begin{cases} \partial_t \psi = \operatorname{div}\Bigg[\displaystyle\sum_{\gamma,\delta=1}^m G^{-2}_{\gamma,\delta} \nabla \xi_\delta \, \nabla \xi_\gamma \cdot \Bigg(\Bigg(\nabla (V + W \circ \xi) - \sum_{\alpha=1}^m \Gamma_\alpha \circ \xi \, \nabla \xi_\alpha \Bigg) \psi \\ \qquad\qquad + \beta^{-1} \nabla \psi \Bigg) \Bigg] + \operatorname{div}\left[\kappa P (\nabla V \psi + \beta^{-1} \nabla \psi) \right], \\ \Gamma_\alpha(z) = \dfrac{\displaystyle\int_{\Sigma(z)} f_\alpha \psi (\det G)^{-1/2} d\sigma_{\Sigma(z)}}{\displaystyle\int_{\Sigma(z)} \psi (\det G)^{-1/2} d\sigma_{\Sigma(z)}}, \end{cases} \tag{5.73}$$

where κ is any scalar function bounded from below by a positive constant, $G^{-2}_{\gamma,\delta}$ denotes the components of the matrix G^{-2} and the components $(f_\alpha)_{1 \leq \alpha \leq m}$ of the local mean force are defined by (5.7). The diffusion matrix $\sum_{\gamma,\delta=1}^m G^{-2}_{\gamma,\delta} \nabla \xi_\delta \, \nabla \xi_\gamma$ plays the same role as the term $|\nabla \xi|^{-2}$ in (5.64). The first term on the right-hand side of the partial differential equation in (5.73) is a diffusion along the reaction coordinate, and the second term is a diffusion in the submanifold Σ_z. In particular, it can be checked that the marginal density $\psi^\xi(t,z) = \int_{\Sigma(z)} \psi(\det G)^{-1/2} d\sigma_{\Sigma(z)}$ satisfies the simple diffusion equation

$$\partial_t \psi^\xi = \mathrm{div}_z \left(\nabla_z W \psi^\xi + \beta^{-1} \nabla_z \psi^\xi \right).$$

Then a similar result of convergence as Theorem 5.15 can be proven, following the same proof. However, implementing the dynamics (5.64) may be difficult in practice, since it requires some complicated analytical computations involving derivatives of the reaction coordinate. Note however that in the particular case of orthogonal reaction coordinate:

$$\nabla \xi_\alpha \cdot \nabla \xi_\gamma = \delta_{\alpha,\gamma},$$

where $\delta_{\alpha,\gamma}$ denotes the Kronecker symbol, the partial differential equation in (5.73) simply writes (for $\kappa = 1$):

$$\partial_t \psi = \mathrm{div} \left[\left(\left(\nabla (V + W \circ \xi) - \sum_{\alpha=1}^{m} \Gamma_\alpha \circ \xi \, \nabla \xi_\alpha \right) \psi + \beta^{-1} \nabla \psi \right) \right],$$

and the implementation is straightforward.

When the reaction coordinates are not orthogonal, it is possible to resort to an ABF dynamics in extended space, see the dynamics (5.31):

$$\begin{cases} dq_t = \left(-\nabla V(q_t) + \dfrac{1}{\eta}(Z_t - \xi(q_t)) \nabla \xi(q_t) \right) dt + \sqrt{2\beta^{-1}}\, dW_t^q, \\ dz_t = \dfrac{1}{\eta} \Big(\xi(q_t) - \mathbb{E}(\xi(q_t) \,|\, z_t) \Big) dt + \sqrt{2\beta^{-1}}\, dW_t^z. \end{cases}$$

Alternatively, the dynamics (5.62)–(5.61) (and the result of convergence of Section 5.2.2.5 for this dynamics) can straightforwardly be generalized to a vectorial reaction coordinate, without any orthogonality assumption.

On Assumption 5.14 and the initial condition. If $\psi^\xi(0,\cdot)$ is zero at some points or is not sufficiently smooth, then F_0' may not be well defined or $I(\psi^\xi(0,\cdot) \,|\, \psi^\xi_\infty)$ may be infinite (which is in contradiction with Assumption 5.14). But since we show that ψ^ξ satisfies a simple diffusion equation (see Proposition 5.11), these difficulties disappear as soon as $t > 0$. Therefore, up to considering the problem for $t \geq t_* > 0$, we can suppose that $\psi^\xi(0,\cdot) > 0$.

5.2.2.5 *A convergence result for the adaptive dynamics* (5.62)–(5.61)

In this section, we present a weaker convergence result for another adaptive overdamped Langevin dynamics, namely (5.62)–(5.61):

$$dq_t = -\nabla\Big(V - F_t \circ \xi\Big)(q_t)\,dt + \sqrt{\frac{2}{\beta}}\,dW_t, \tag{5.74}$$

with the same definition as before for F_t, namely

$$F_t'(z) = \mathbb{E}\left(f(q_t)\,\Big|\xi(q_t) = z\right), \tag{5.75}$$

for any $z \in \mathcal{M}$, f being the local mean force. For simplicity, we only consider the case

$$\mathcal{M} = \mathbb{T} \quad \text{and} \quad W = 0,$$

but the results can be extended to the case $\mathcal{M} = \mathbb{R}$ with a suitable confining potential $W \neq 0$, as in Section 5.2.2.2 (see Assumption 5.14 and Corollary 5.16). One interest of the biased dynamics (5.62)–(5.61) and the result of convergence presented here, is that the extension to the case of a multi-dimensional reaction coordinate is straightforward (see also Section 5.2.2.4 above). For the sake of conciseness, we do not provide the details of the result in this case since it follows exactly the same lines as the case $m = 1$ detailed here.

The nonlinear Fokker-Planck equation associated to (5.74) is now:

$$\begin{cases} \partial_t \psi = \mathrm{div}\Big(\nabla(V - F_t \circ \xi)\psi + \beta^{-1}\nabla\psi\Big), \\ F_t'(z) = \dfrac{\displaystyle\int_{\Sigma(z)} f\,|\nabla\xi|^{-1}\psi(t,\cdot)d\sigma_{\Sigma(z)}}{\displaystyle\int_{\Sigma(z)} |\nabla\xi|^{-1}\psi(t,\cdot)d\sigma_{\Sigma(z)}}. \end{cases} \tag{5.76}$$

The main difference with the dynamics (5.60)–(5.61) considered in Theorem 5.15 is that the marginal distribution ψ^ξ does not satisfy a closed partial differential equation in general. Therefore, we do not know *a priori* that the Fisher information $I(\psi^\xi \,|\, \psi^\xi_\infty)$ converges to 0 (compare with item (i) in Corollary 5.16). The strategy here is to directly estimate the derivative of the total entropy E. We obtain a convergence result under two additional assumptions.

Assumption 5.17. The potential V and the function ξ are such that ψ_∞ satisfies LSI($\overline{R}$) for some $\overline{R} > 0$.

Assumption 5.18. The constants m, M, ρ defined in Assumptions 5.12 and 5.13 are such that

$$\frac{mM\beta}{2\sqrt{\rho}} < 1.$$

Theorem 5.19 (Longtime convergence for (5.62)–(5.61)). *Suppose that Assumptions* 5.6, 5.12, 5.13, 5.17 *and* 5.18 *are satisfied, and consider a smooth solution* (ψ, F_t') *to* (5.76). *Then the total entropy E decreases exponentially:*

$$\sqrt{E(t)} \leq \sqrt{E(0)} \exp(-\lambda t),$$

where

$$\lambda = \beta^{-1}\left(1 - \frac{mM\beta}{2\sqrt{\rho}}\right)\overline{R} > 0.$$

In particular, as in Theorem 5.15, *the biasing force F_t' converges exponentially fast to the mean force F'.*

The proof of this result is given in Section 5.2.3.4 below.

Remark 5.20 (On Assumption 5.17). *It is shown in Theorem* 2 *of [Otto and Reznikoff (2007)] that if* (i) $\mu(dx_1\,dx_2) = \exp(-H(x_1, x_2))\,dx_1\,dx_2$ *is a probability measure on a product space $X = X_1 \times X_2$ (where X_i are Euclidean spaces);* (ii) *the conditional probabilities $\mu(dx_2\,|\,x_1)$ satisfy LSI(ρ), with ρ independent of x_1;* (iii) *the marginal $\overline{\mu}(dx_1)$ satisfies LSI(r);* (iv) *the coupling between the two directions is bounded: there exists $\kappa_{1,2} > 0$ such that for all $(x_1, x_2) \in X_1 \times X_2$, $\left|\partial^2_{x_1,x_2} H(x_1, x_2)\right| \leq \kappa_{1,2}$; then there exists $\overline{R} > \min(\rho, r) > 0$ such that μ satisfies LSI($\overline{R}$).*

Thus, in the simple framework of Section 5.2.3.1 *for example, where the configuration space is $\mathbb{T} \times \mathbb{R}$ and the reaction coordinate is $\xi(x, y) = x$, the fact that ψ_∞ satisfies a LSI (Assumption* 5.17*) can be deduced from the fact that the conditioned distributions $\nu^\xi(\infty, \cdot\,|\,z)$ satisfy a LSI (which is Assumption* 5.13*), the marginal ψ^ξ_∞ satisfy a LSI (which is related to Assumption* 5.14*) and the coupling is bounded (which is Assumption* 5.12*). This result can be generalized to the case where X is not a product space [Lelièvre (2009)]. Thus Assumption* 5.17 *is in general satisfied in the framework of Theorem* 5.15, *and does not introduce any further requirement on the potential and the reaction coordinate.*

5.2.3 *Proofs*

One remark to simplify the presentation of the proofs is that we can suppose $\beta = 1$ up to the following change of variable: $\tilde{t} = \beta^{-1}t$, $\widetilde{\psi}(\tilde{t}, x) = \psi(t, x)$, $\widetilde{V}(x) = \beta V(x)$ and $\widetilde{W}(x) = \beta W(x)$. Therefore, we can consider without loss of generality that

$$\beta = 1. \tag{5.77}$$

5.2.3.1 *Proof of Proposition* 5.11 *and Theorem* 5.15 *in a simple case*

In this section, we propose to prove Proposition 5.11 and Theorem 5.15 in the simple case when

$$n = 2, \quad q = (x, y) \in \mathcal{D} = \mathbb{T} \times \mathbb{R}, \quad \xi(x, y) = x.$$

We therefore use in this section the notation x instead of z for the reaction coordinate variable. Since ξ has values in $\mathbb{T}$ ($\mathcal{M} = \mathbb{T}$), the choice $W = 0$ can be made (see Corollary 5.16). Note also that the local mean force f is simply given by $f = \partial_x V$. Our aim is to introduce the main arguments in this simple case before presenting the general proof in Section 5.2.3.2.

In this simple setting, the system (5.64) writes (recall $\beta = 1$):

$$\begin{cases} \partial_t \psi = \mathrm{div}\,(\nabla V \psi + \nabla \psi) - \partial_x (F_t' \, \psi), \\ F_t'(x) = \dfrac{\displaystyle\int_{\mathbb{R}} \partial_x V(x, y) \psi(t, x, y) \, dy}{\psi^\xi(t, x)}, \end{cases} \tag{5.78}$$

where $\psi^\xi(t, x) = \displaystyle\int_{\mathbb{R}} \psi(t, x, y) dy$. Note that in this case $\psi^\xi_\infty \equiv 1$.

Assumptions 5.12 and 5.13 are satisfied in this context for a potential V of the form:

$$V(x, y) = V_0(x, y) + V_1(x, y),$$

where

$$\inf_{\mathbb{T}\times\mathbb{R}} \partial_{y,y} V_0 > 0, \quad \|V_1\|_{L^\infty} < \infty, \quad \|\partial_{x,y}(V_0 + V_1)\|_{L^\infty} < \infty.$$

The potential V is thus a bounded perturbation of an α-convex potential, with a bounded mixed derivative $\partial_{x,y} V$. In this case, Assumptions 5.12 and 5.13 are satisfied with $m = 1$, $M = \|\partial_{x,y} V\|_{L^\infty}$ and $\rho = (\inf_{\mathbb{T}\times\mathbb{R}} \partial_{y,y} V_0) \exp(-\mathrm{osc}\, V_1)$, where $\mathrm{osc}\, V_1 = \sup_{\mathbb{T}\times\mathbb{R}} V_1 - \inf_{\mathbb{T}\times\mathbb{R}} V_1$ (see [Ané *et al.* (2000)]).

Proposition 5.11 is simply obtained by integrating (5.78) with respect to $y \in \mathbb{R}$:

Lemma 5.21. *The density ψ^ξ satisfies the following diffusion equation on $\mathbb{T}$:*

$$\partial_t \psi^\xi = \partial_{x,x} \psi^\xi. \tag{5.79}$$

As stated in Corollary 5.16, this result already yields the exponential convergence to zero of the macroscopic Fisher information $I(\psi^\xi | \psi^\xi_\infty)$ (this is the matter of Lemma 5.29 below), and thus Assumption 5.13 is indeed satisfied with $I_0 = I(\psi^\xi(0,\cdot) \,|\, \psi^\xi_\infty)$ and $r = 4\pi^2$.

A fundamental lemma needed in the sequel is:

Lemma 5.22. *The difference between the biasing force F'_t and the mean force F' can be expressed in terms of the densities as*

$$F'_t - F' = \int_{\mathbb{R}} \partial_x \ln\left(\frac{\psi}{\psi_\infty}\right) \frac{\psi}{\psi^\xi}\, dy - \partial_x \ln\left(\frac{\psi^\xi}{\psi^\xi_\infty}\right).$$

Proof. A simple computation, using the fact that $\psi^\xi_\infty \equiv 1$, gives:

$$\begin{aligned}
&\int_{\mathbb{R}} \partial_x \ln\left(\frac{\psi}{\psi_\infty}\right) \frac{\psi}{\psi^\xi}\, dy - \partial_x \ln\left(\frac{\psi^\xi}{\psi^\xi_\infty}\right) \\
&\quad = \int_{\mathbb{R}} \partial_x \ln \psi \frac{\psi}{\psi^\xi}\, dy - \int_{\mathbb{R}} \partial_x \ln \psi_\infty \frac{\psi}{\psi^\xi}\, dy - \partial_x \ln \psi^\xi \\
&\quad = \int_{\mathbb{R}} \frac{\partial_x \psi}{\psi^\xi}\, dy + \int_{\mathbb{R}} \partial_x (V - F) \frac{\psi}{\psi^\xi}\, dy - \partial_x \ln \psi^\xi \\
&\quad = F'_t - F',
\end{aligned}$$

which conclude the proof. □

We will also use the following two estimates:

Lemma 5.23. *Suppose Assumptions* 5.12 *and* 5.13 *are satisfied. Then, for all $t \geq 0$ and for all $x \in \mathbb{T}$,*

$$|F'_t(x) - F'(x)| \leq \|\partial_{x,y} V\|_{L^\infty} \sqrt{\frac{2}{\rho} e_m(t,x)}.$$

Proof. For any coupling measure $\pi \in \Pi(\mu_{t,x}, \mu_{\infty,x})$, it holds:

$$\begin{aligned} |F_t'(x) - F'(x)| &= \left| \int_{\mathbb{R}\times\mathbb{R}} (\partial_x V(x,y) - \partial_x V(x,y'))\, \pi(dy,dy') \right| \\ &\le \|\partial_{x,y} V\|_{L^\infty} \int_{\mathbb{R}\times\mathbb{R}} |y-y'| \pi(dy,dy') \\ &\le \|\partial_{x,y} V\|_{L^\infty} \sqrt{\int_{\mathbb{R}\times\mathbb{R}} |y-y'|^2 \pi(dy,dy')}. \end{aligned}$$

Taking now the infimum over all $\pi \in \Pi(\mu_{t,x}, \mu_{\infty,x})$ and using Assumption 5.13 together with Lemma 5.10,

$$|F_t'(x) - F'(x)| \le \|\partial_{x,y} V\|_{L^\infty} W(\mu_{t,x}, \mu_{\infty,x}) \le \|\partial_{x,y} V\|_{L^\infty} \sqrt{\frac{2}{\rho} H(\mu_{t,x}|\mu_{\infty,x})},$$

which gives the result. □

Lemma 5.24. *Suppose Assumption* 5.13 *holds. Then for all* $t \ge 0$,

$$E_m(t) \le \frac{1}{2\rho} \int_{\mathbb{T}\times\mathbb{R}} \left| \partial_y \ln\left(\frac{\psi}{\psi_\infty}\right) \right|^2 \psi.$$

Proof. Using Assumption 5.13, it holds:

$$\begin{aligned} E_m &= \int_{\mathbb{T}} e_m \psi^\xi \, dx \\ &\le \int_{\mathbb{T}} \frac{1}{2\rho} \int_{\mathbb{R}} \left| \partial_y \ln\left(\frac{\psi}{\psi^\xi} \Big/ \frac{\psi_\infty}{\psi_\infty^\xi} \right) \right|^2 \frac{\psi}{\psi^\xi} \, dy \, \psi^\xi \, dx, \end{aligned}$$

which yields the result since ψ^ξ/ψ_∞^ξ does not depend on y. □

We are now in position to prove the exponential convergence of $E_m(t)$ to zero, as stated in Theorem 5.15 (see Eq. (5.68)). Equation (5.78) on ψ can be rewritten as:

$$\partial_t \psi = \operatorname{div}\left(\psi_\infty \nabla \left(\frac{\psi}{\psi_\infty} \right) \right) + \partial_x((F' - F_t')\psi).$$

The derivative dE/dt can be obtained by multiplying this equation by $\ln(\psi/\psi_\infty)$ and integrating over $\mathbb{T} \times \mathbb{R}$. After some integration by parts,

$$\begin{aligned} \frac{dE_m}{dt} &= \frac{dE}{dt} - \frac{dE_M}{dt} \\ &= -\int_{\mathbb{T}}\int_{\mathbb{R}} \left| \nabla \ln\left(\frac{\psi}{\psi_\infty}\right) \right|^2 \psi + \int_{\mathbb{T}}\int_{\mathbb{R}} (F_t' - F')\partial_x \ln\left(\frac{\psi}{\psi_\infty}\right) \psi \qquad (5.80) \\ &\quad + \int_{\mathbb{T}} \left| \partial_x \ln\left(\frac{\psi^\xi}{\psi_\infty^\xi}\right) \right|^2 \psi^\xi. \end{aligned}$$

Using Lemma 5.22, it holds:

$$\begin{aligned}\frac{dE_m}{dt} = &- \int_{\mathbb{T}}\int_{\mathbb{R}} \left|\partial_y \ln\left(\frac{\psi}{\psi_\infty}\right)\right|^2 \psi \\ &- \int_{\mathbb{T}}\int_{\mathbb{R}} \left|\partial_x \ln\left(\frac{\psi}{\psi_\infty}\right)\right|^2 \psi + \int_{\mathbb{T}} \left(\int_{\mathbb{R}} \partial_x \ln\left(\frac{\psi}{\psi_\infty}\right) \psi \, dy\right)^2 \frac{1}{\psi^\xi} \, dx \\ &- \int_{\mathbb{T}}\int_{\mathbb{R}} \partial_x \ln\left(\frac{\psi^\xi}{\psi^\xi_\infty}\right) \partial_x \ln\left(\frac{\psi}{\psi_\infty}\right) \psi + \int_{\mathbb{T}} \left|\partial_x \ln\left(\frac{\psi^\xi}{\psi^\xi_\infty}\right)\right|^2 \psi^\xi.\end{aligned}$$

Note that by the Cauchy-Schwarz inequality, the term on the second line above is non-positive. Therefore, using again Lemma 5.22,

$$\frac{dE_m}{dt} \leq - \int_{\mathbb{T}}\int_{\mathbb{R}} \left|\partial_y \ln\left(\frac{\psi}{\psi_\infty}\right)\right|^2 \psi - \int_{\mathbb{T}} \partial_x \ln\left(\frac{\psi^\xi}{\psi^\xi_\infty}\right) \psi^\xi (F'_t - F').$$

With Lemmas 5.23 and 5.24, it follows

$$\begin{aligned}\frac{dE_m}{dt} &\leq -2\rho E_m + \sqrt{\int_{\mathbb{T}} |F'_t - F'|^2 \psi^\xi} \sqrt{\int_{\mathbb{T}} \left|\partial_x \ln\left(\frac{\psi^\xi}{\psi^\xi_\infty}\right)\right|^2 \psi^\xi} \\ &\leq -2\rho E_m + \|\partial_{x,y} V\|_{L^\infty} \sqrt{\frac{2}{\rho} E_m} \sqrt{I(\psi^\xi | \psi^\xi_\infty)}.\end{aligned}$$

Using Assumption 5.13, it holds

$$\frac{d\sqrt{E_m}}{dt} \leq -\rho \sqrt{E_m} + \|\partial_{x,y} V\|_{L^\infty} \sqrt{\frac{I_0}{2\rho}} \exp(-rt),$$

from which (5.68) is deduced. In particular, in the special case when $\rho m^{-2} = r$, the above inequality implies

$$\sqrt{E_m(t)} \leq \left(\sqrt{E_m(0)} + M \sqrt{\frac{I_0}{2\rho}}\, t\right) \exp(-\beta^{-1} r\, t),$$

which indeed implies the statement of the theorem.

Items (ii) and (iii) of Theorem 5.15 are then easily deduced. Note the fact that E and $\|\psi(t,\cdot) - \psi_\infty\|_{L^1(\mathcal{D})}$ converge exponentially fast to zero with rate λ is an immediate consequence of (5.68), Assumption 5.13, Lemma 5.8 and the Csiszár-Kullback inequality (2.84). Equation (5.70) is then easily obtained with Lemma 5.23.

5.2.3.2 *Proof of Proposition* 5.11 *and Theorem* 5.15 *in the general case*

We now present the proof of Proposition 5.11 and Theorem 5.15 in the more general setting of Section 5.2.2.2. The proof follows the same lines as in the simple case presented in Section 5.2.3.1, but with additional difficulties related to the geometry of the submanifold $\Sigma(z)$.

We recall that the derivative of ψ^ξ with respect to the reaction coordinate value reads (see Lemma 3.10 with $m = 1$):

$$\partial_z \psi^\xi(t, z) = \int_{\Sigma(z)} \left(\frac{\nabla\xi \cdot \nabla\psi(t, \cdot)}{|\nabla\xi|^2} + \operatorname{div}\left(\frac{\nabla\xi}{|\nabla\xi|^2}\right) \psi(t, \cdot) \right) |\nabla\xi|^{-1} d\sigma_{\Sigma(z)}. \tag{5.81}$$

Using this expression, it can be shown that ψ^ξ satisfies a simple diffusion equation, which is Proposition 5.11.

Lemma 5.25. *The density ψ^ξ satisfies the following diffusion equation on $\mathcal{M}$:*

$$\partial_t \psi^\xi = \partial_z \left(W' \psi^\xi + \partial_z \psi^\xi \right). \tag{5.82}$$

Proof. For any smooth test function $g : \mathcal{M} \to \mathbb{R}$, the co-area formula (3.12), (5.64) and an integration by parts imply

$$\begin{aligned}
\frac{d}{dt} \int_{\mathcal{M}} \psi^\xi(t, \cdot) g \, dz &= \frac{d}{dt} \int_{\mathcal{D}} \psi(t, \cdot) g \circ \xi \, dx \\
&= \int_{\mathcal{D}} \operatorname{div}\left(|\nabla\xi|^{-2} \left(\nabla(V - F_t \circ \xi + W \circ \xi)\psi + \nabla\psi \right) \right) g \circ \xi \, dx \\
&= -\int_{\mathcal{D}} |\nabla\xi|^{-2} \left(\nabla(V - F_t \circ \xi + W \circ \xi)\psi + \nabla\psi \right) \cdot \nabla\xi \, g' \circ \xi \, dx \\
&= -\int_{\mathcal{D}} |\nabla\xi|^{-2} \left(\nabla V \cdot \nabla\xi \, \psi + \nabla\psi \cdot \nabla\xi \right) g' \circ \xi \, dx \\
&\quad + \int_{\mathcal{D}} F_t' \circ \xi \, g' \circ \xi \, \psi \, dx - \int_{\mathcal{D}} W' \circ \xi \, g' \circ \xi \, \psi \, dx.
\end{aligned}$$

Using again the co-area formula and finally (5.81), it holds

$$
\begin{aligned}
&\frac{d}{dt}\int_{\mathcal{M}}\psi^{\xi}(t,\cdot)g\,dz\\
&=-\int_{\mathcal{M}}\int_{\Sigma(z)}|\nabla\xi|^{-3}\left(\nabla V\cdot\nabla\xi\,\psi+\nabla\psi\cdot\nabla\xi\right)d\sigma_{\Sigma(z)}g'(z)\,dz\\
&\quad+\int_{\mathcal{M}}F_t'(z)g'(z)\psi^{\xi}(z)\,dz-\int_{\mathcal{M}}W'(z)g'(z)\psi^{\xi}(z)\,dz\\
&=-\int_{\mathcal{M}}\int_{\Sigma(z)}\left(|\nabla\xi|^{-3}\nabla\psi\cdot\nabla\xi+\mathrm{div}(\nabla\xi|\nabla\xi|^{-2})|\nabla\xi|^{-1}\psi\right)d\sigma_{\Sigma(z)}g'(z)\,dz\\
&\quad-\int_{\mathcal{M}}W'(z)\psi^{\xi}(z)g'(z)\,dz\\
&=-\int_{\mathcal{M}}\left(\partial_z\psi^{\xi}(t,z)+W'(z)\psi^{\xi}(z)\right)g'(z)\,dz,
\end{aligned}
$$

which is a weak formulation of (5.82). □

As stated in Corollary 5.16, this result already yields the exponential convergence to zero of the macroscopic Fisher information $I(\psi^{\xi}\,|\,\psi^{\xi}_{\infty})$ under adequate assumptions on W (this is the matter of Corollary 5.16 and Lemmas 5.29 and 5.30 below). We suppose in the following that Assumption 5.13 is indeed satisfied.

Lemma 5.22 may be generalized as

Lemma 5.26. *The difference between the biasing force F_t' and the mean force F' can be expressed in terms of the densities as*

$$
F_t'(z)-F'(z)=\int_{\Sigma(z)}\frac{\nabla\xi}{|\nabla\xi|}\cdot\nabla\ln\left(\frac{\psi}{\psi_{\infty}}\right)\frac{\psi}{\psi^{\xi}}|\nabla\xi|^{-2}\,d\sigma_{\Sigma(z)}-\partial_z\ln\left(\frac{\psi^{\xi}}{\psi^{\xi}_{\infty}}\right).
$$

Proof. Using (5.81) and the definition of F_t', it holds:

$$\begin{aligned}
&\int_{\Sigma(z)} \frac{\nabla\xi}{|\nabla\xi|}\cdot\nabla\ln\left(\frac{\psi}{\psi_\infty}\right)\frac{\psi}{\psi^\xi}|\nabla\xi|^{-2}\,d\sigma_{\Sigma(z)} - \partial_z\ln\left(\frac{\psi^\xi}{\psi^\xi_\infty}\right)\\
&\quad= \int_{\Sigma(z)} \frac{\nabla\xi}{|\nabla\xi|}\cdot\nabla\ln\psi\frac{\psi}{\psi^\xi}|\nabla\xi|^{-2}\,d\sigma_{\Sigma(z)}\\
&\qquad- \int_{\Sigma(z)} \frac{\nabla\xi}{|\nabla\xi|}\cdot\nabla\ln\psi_\infty\frac{\psi}{\psi^\xi}|\nabla\xi|^{-2}\,d\sigma_{\Sigma(z)}\\
&\qquad-\partial_z\ln\psi^\xi + \partial_z\ln\psi^\xi_\infty\\
&\quad= \frac{1}{\psi^\xi}\int_{\Sigma(z)} \frac{\nabla\xi\cdot\nabla\psi}{|\nabla\xi|}|\nabla\xi|^{-2}\,d\sigma_{\Sigma(z)}\\
&\qquad+ \int_{\Sigma(z)} \frac{\nabla\xi}{|\nabla\xi|}\cdot\nabla\left(V - F\circ\xi + W\circ\xi\right)\frac{\psi}{\psi^\xi}|\nabla\xi|^{-2}\,d\sigma_{\Sigma(z)}\\
&\qquad-\partial_z\ln\psi^\xi - W'(z)\\
&\quad= \frac{\partial_z\psi^\xi}{\psi^\xi} - \frac{1}{\psi^\xi}\int_{\Sigma(z)}\mathrm{div}\left(\frac{\nabla\xi}{|\nabla\xi|^2}\right)|\nabla\xi|^{-1}\psi\,d\sigma_{\Sigma(z)}\\
&\qquad+ \int_{\Sigma(z)} \frac{\nabla\xi\cdot\nabla V}{|\nabla\xi|^3}\frac{\psi}{\psi^\xi}\,d\sigma_{\Sigma(z)} - F'(z) - \partial_z\ln\psi^\xi\\
&\quad= F_t'(z) - F'(z).
\end{aligned}$$

□

Lemmas 5.23 and 5.24 are extended as follows:

Lemma 5.27. *Suppose Assumptions* 5.12 *and* 5.13 *hold. Then for all* $t \geq 0$ *and for all* $z \in \mathcal{M}$,

$$|F_t'(z) - F'(z)| \leq M\sqrt{\frac{2}{\rho}e_m(t,z)}.$$

Proof. For any coupling measure $\pi \in \Pi(\nu^\xi(t,\cdot|z), \nu^\xi(\infty,\cdot|z))$ defined on $\Sigma(z)\times\Sigma(z)$, it holds:

$$\begin{aligned}
|F_t'(z) - F'(z)| &= \left|\int_{\Sigma(z)\times\Sigma(z)} (f(x) - f(x'))\pi(dx,dx')\right|\\
&\leq \left\|\nabla_{\Sigma(z)}f\right\|_{L^\infty}\sqrt{\int_{\Sigma(z)\times\Sigma(z)} d_{\Sigma(z)}(x,x')^2\,\pi(dx,dx')}.
\end{aligned}$$

Taking now the infimum over all $\pi \in \Pi(\nu^\xi(t,\cdot|z), \nu^\xi(\infty,\cdot|z))$ and using Assumptions 5.12 and 5.13 together with Lemma 5.10, it follows

$$\begin{aligned} |F_t'(z) - F'(z)| &\le MW(\nu^\xi(t,\cdot|z), \nu^\xi(\infty,\cdot|z)) \\ &\le M\sqrt{\frac{2}{\rho} H(\nu^\xi(t,\cdot|z)|\nu^\xi(\infty,\cdot|z))}, \end{aligned} \tag{5.83}$$

which concludes the proof. □

Lemma 5.28. *Suppose Assumption* 5.13 *holds. Then for all* $t \ge 0$,

$$E_m(t) \le \frac{1}{2\rho}\int_{\mathcal{D}} \left|\nabla_{\Sigma(z)} \ln\left(\frac{\psi(t,\cdot)}{\psi_\infty}\right)\right|^2 \psi.$$

Proof. The proof is based on the fact that the conditional probability measures $\nu^\xi(\cdot|z)$ satisfy a LSI(ρ). In the case ν is a probability measure on the (Riemannian) submanifold $\Sigma(z)$, the gradient ∇ in the definition of the Fisher information (2.83) actually denotes the surface gradient on $\Sigma(z)$, namely $\nabla_{\Sigma(z)}$ defined in (5.66). Therefore, the Fisher information of the conditional probability measures $\nu^\xi(t,\cdot\,|\,z)$ and $\nu^\xi(\infty,\cdot\,|\,z)$ writes

$$I\big(\nu^\xi(t,\cdot|z)\,|\,\nu^\xi(\infty,\cdot|z)\big) = \int_{\Sigma(z)} \left|\nabla_{\Sigma(z)} \ln\left(\frac{\psi(t,\cdot)}{\psi_\infty}\right)\right|^2 \frac{\psi(t,\cdot)|\nabla\xi|^{-1} d\sigma_{\Sigma(z)}}{\psi^\xi(t,z)}.$$

With Assumption 5.13, it follows:

$$\begin{aligned} E_m &= \int_{\mathcal{M}} e_m \psi^\xi \, dz \\ &\le \int_{\mathcal{M}} \frac{1}{2\rho}\int_{\Sigma(z)} \left|\nabla_{\Sigma(z)} \ln\left(\frac{\psi(t,\cdot)}{\psi_\infty}\right)\right|^2 \frac{\psi(t,\cdot)|\nabla\xi|^{-1} d\sigma_{\Sigma(z)}}{\psi^\xi(t,z)} \psi^\xi \, dz, \end{aligned}$$

which yields the result, using the co-area formula (3.12). □

We are now in position to prove the exponential convergence of $E_m(t)$ to zero stated in Theorem 5.15 (see Eq. (5.68)). Equation (5.64) on ψ can be rewritten as:

$$\partial_t \psi = \operatorname{div}\left(|\nabla\xi|^{-2}\psi_\infty \nabla\left(\frac{\psi}{\psi_\infty}\right)\right) + \operatorname{div}\Big(|\nabla\xi|^{-2}\nabla\left((A - F_t)\circ\xi\right)\psi\Big).$$

The derivative dE/dt can be obtained by multiplying this equation by $\ln(\psi/\psi_\infty)$ and integrating over $\mathcal{D}$. Thus, after some integration by parts, using the co-area formula (3.12) and Lemma 5.26:

$$\begin{aligned}
\frac{dE_m}{dt} &= \frac{dE}{dt} - \frac{dE_M}{dt} \\
&= -\int_{\mathcal{D}} \left| \nabla \ln\left(\frac{\psi}{\psi_\infty}\right)\right|^2 |\nabla\xi|^{-2}\psi \\
&\quad + \int_{\mathcal{D}} (F_t' - F') \circ \xi \, \nabla\xi \cdot \nabla \ln\left(\frac{\psi}{\psi_\infty}\right) |\nabla\xi|^{-2}\psi \\
&\quad + \int_{\mathcal{M}} \left| \partial_z \ln\left(\frac{\psi^\xi}{\psi^\xi_\infty}\right)\right|^2 \psi^\xi \\
&= -\int_{\mathcal{D}} \left| \nabla_{\Sigma(z)} \ln\left(\frac{\psi}{\psi_\infty}\right)\right|^2 |\nabla\xi|^{-2}\psi \\
&\quad - \int_{\mathcal{D}} \left(\frac{\nabla\xi}{|\nabla\xi|} \cdot \nabla \ln\left(\frac{\psi}{\psi_\infty}\right)\right)^2 |\nabla\xi|^{-2}\psi \\
&\quad + \int_{\mathcal{M}} (F_t' - F')(z) \int_{\Sigma(z)} \frac{\nabla\xi}{|\nabla\xi|} \cdot \nabla \ln\left(\frac{\psi}{\psi_\infty}\right) |\nabla\xi|^{-2}\psi \, d\sigma_{\Sigma(z)} dz \\
&\quad + \int_{\mathcal{M}} \left| \partial_z \ln\left(\frac{\psi^\xi}{\psi^\xi_\infty}\right)\right|^2 \psi^\xi \\
&= -\int_{\mathcal{D}} \left| \nabla_{\Sigma(z)} \ln\left(\frac{\psi}{\psi_\infty}\right)\right|^2 |\nabla\xi|^{-2}\psi \\
&\quad - \int_{\mathcal{D}} \left(\frac{\nabla\xi}{|\nabla\xi|} \cdot \nabla \ln\left(\frac{\psi}{\psi_\infty}\right)\right)^2 |\nabla\xi|^{-2}\psi \\
&\quad + \int_{\mathcal{M}} \left(\int_{\Sigma(z)} \frac{\nabla\xi}{|\nabla\xi|} \cdot \nabla \ln\left(\frac{\psi}{\psi_\infty}\right) |\nabla\xi|^{-2}\psi \, d\sigma_{\Sigma(z)} \right)^2 (\psi^\xi)^{-1} dz \\
&\quad - \int_{\mathcal{M}} \int_{\Sigma(z)} \frac{\nabla\xi}{|\nabla\xi|} \cdot \nabla \ln\left(\frac{\psi}{\psi_\infty}\right) |\nabla\xi|^{-2}\psi \, d\sigma_{\Sigma(z)} \partial_z \ln\left(\frac{\psi^\xi}{\psi^\xi_\infty}\right) dz \\
&\quad + \int_{\mathcal{M}} \left| \partial_z \ln\left(\frac{\psi^\xi}{\psi^\xi_\infty}\right)\right|^2 \psi^\xi.
\end{aligned}$$

Using the Cauchy-Schwarz inequality:

$$\begin{aligned}
&\left(\int_{\Sigma(z)} \frac{\nabla\xi}{|\nabla\xi|} \cdot \nabla \ln\left(\frac{\psi}{\psi_\infty}\right) |\nabla\xi|^{-1} \frac{|\nabla\xi|^{-1}\psi \, d\sigma_{\Sigma(z)}}{\psi^\xi(z)} \right)^2 \\
&\qquad \leq \int_{\Sigma(z)} \left(\frac{\nabla\xi}{|\nabla\xi|} \cdot \nabla \ln\left(\frac{\psi}{\psi_\infty}\right) |\nabla\xi|^{-1} \right)^2 \frac{|\nabla\xi|^{-1}\psi \, d\sigma_{\Sigma(z)}}{\psi^\xi(z)},
\end{aligned}$$

and Lemma 5.26 again, it follows

$$\frac{dE_m}{dt} \leq -\int_{\mathcal{D}} \left| \nabla_{\Sigma(z)} \ln\left(\frac{\psi}{\psi_\infty}\right)\right|^2 |\nabla\xi|^{-2}\psi - \int_{\mathcal{M}} \partial_z \ln\left(\frac{\psi^\xi}{\psi^\xi_\infty}\right)\psi^\xi(F'_t - F').$$

With Assumption 5.12, Lemmas 5.27 and 5.28, it holds

$$\begin{aligned}\frac{dE_m}{dt} &\leq -2\rho m^{-2}E_m + \sqrt{\int_{\mathcal{M}} |F'_t - F'|^2\psi^\xi}\sqrt{\int_{\mathcal{M}}\left|\partial_z \ln\left(\frac{\psi^\xi}{\psi^\xi_\infty}\right)\right|^2\psi^\xi} \\ &\leq -2\rho m^{-2}E_m + M\sqrt{\frac{2}{\rho}E_m}\sqrt{I(\psi^\xi|\psi^\xi_\infty)}.\end{aligned}$$

With Assumption 5.13,

$$\frac{d\sqrt{E_m}}{dt} \leq -\rho m^{-2}\sqrt{E_m} + M\sqrt{\frac{I_0}{2\rho}}\exp(-rt),$$

from which we deduce (5.68). The end of the proof of Theorem 5.15 then follows exactly the same lines as in the simple case considered in Section 5.2.3.1.

5.2.3.3 *Proof of Corollary* 5.16

Convergence of the macroscopic Fisher information. Let us first show that in both cases considered in Corollary 5.16, the exponential convergence of the macroscopic Fisher information (Assumption 5.13) indeed holds.

Consider first the case $\mathcal{M} = \mathbb{T}$ and $W = 0$. Equation (5.67) shows that ψ^ξ satisfies $\partial_t\psi^\xi = \partial_{z,z}\psi^\xi$ on $\mathbb{T}$. This allows us to conclude the exponential convergence of the Fisher information $I(\psi^\xi(t,\cdot)\,|\,\psi^\xi_\infty)$:

Lemma 5.29 (Macroscopic convergence when $\mathcal{M} = \mathbb{T}$ and $W = 0$). *Let ϕ be a function defined for $t \geq 0$ and $x \in \mathbb{T}$ satisfying*

$$\partial_t\phi = \partial_{x,x}\phi \text{ on } \mathbb{T},$$

and such that

$$\int_{\mathbb{T}} \phi(0,\cdot) = 1, \quad \phi(0,\cdot) \geq 0, \quad I(\phi(0,\cdot)|\phi_\infty) < \infty,$$

where $\phi_\infty \equiv 1$ is the longtime limit of ϕ. Then, $\forall t \geq 0$,

$$I(\phi(t,\cdot)\,|\,\phi_\infty) \leq I(\phi(0,\cdot)\,|\,\phi_\infty)\exp(-8\pi^2 t).$$

Proof. Denote $u = \sqrt{\phi}$, and note that

$$I(\phi \,|\, \phi_\infty) = \int_{\mathbb{T}} |\partial_x \ln \phi|^2 \phi = 4 \int_{\mathbb{T}} |\partial_x u|^2.$$

Moreover, from (5.79),

$$\partial_t u = \partial_{x,x} u + \frac{(\partial_x u)^2}{u}.$$

Therefore,

$$\begin{aligned}
\frac{d}{dt} \int_{\mathbb{T}} (\partial_x u)^2 &= 2 \int_{\mathbb{T}} \partial_{x,x,x} u \, \partial_x u + 2 \int_{\mathbb{T}} \partial_x \left(\frac{(\partial_x u)^2}{u} \right) \partial_x u \\
&= -2 \int_{\mathbb{T}} (\partial_{x,x} u)^2 - 2 \int_{\mathbb{T}} \frac{(\partial_x u)^2}{u} \partial_{x,x} u \\
&= -2 \int_{\mathbb{T}} (\partial_{x,x} u)^2 - 2 \int_{\mathbb{T}} \frac{\partial_x ((\partial_x u)^3)}{3u} \\
&= -2 \int_{\mathbb{T}} (\partial_{x,x} u)^2 - \frac{2}{3} \int_{\mathbb{T}} \frac{(\partial_x u)^4}{u^2} \\
&\leq -8\pi^2 \int_{\mathbb{T}} (\partial_x u)^2,
\end{aligned}$$

where we have used the Poincaré-Wirtinger inequality on $\mathbb{T}$, applied to $\partial_x u$: For any function $g \in H^1(\mathbb{T})$,

$$\int_{\mathbb{T}} \left(g - \int_{\mathbb{T}} g \right)^2 \leq \frac{1}{4\pi^2} \int_{\mathbb{T}} (\partial_x g)^2.$$

□

Let us now consider the case when $\mathcal{M} = \mathbb{R}$ and W is a confining potential such that W'' is bounded from below and $Z_W^{-1} \mathrm{e}^{-\beta W}$ satisfies a logarithmic Sobolev inequality (as stated in Corollary 5.16). Equation (5.67) shows that ψ^ξ satisfies

$$\partial_t \psi^\xi = \partial_z \left(W' \psi^\xi + \partial_z \psi^\xi \right)$$

on $\mathbb{R}$, and we would like to conclude to the exponential convergence of the Fisher information $I(\psi^\xi(t,\cdot) \,|\, \psi^\xi_\infty)$.

Lemma 5.30 (Macroscopic onvergence when $\mathcal{M} = \mathbb{R}$ and $W \neq 0$). *Let ϕ be a function defined for $t \geq 0$ and $x \in \mathbb{R}$ satisfying*

$$\partial_t \phi = \partial_x \left(W' \phi + \partial_x \phi \right) \ \text{ on } \mathbb{R},$$

and such that

$$\int_{\mathbb{R}} \phi(0,\cdot) = 1, \quad \phi(0,\cdot) \geq 0, \quad I(\phi(0,\cdot) \,|\, \phi_\infty) < \infty,$$

where $\phi_\infty \equiv Z_W \, \mathrm{e}^{-\beta W}$ *is the longtime limit of* ϕ. *Suppose that* W'' *is bounded from below by a constant* α *and* ϕ_∞ *satisfies LSI(*$\overline{r}$*), with* $\overline{r} > 0$. *Without loss of generality, it can be assumed that*

$$\overline{r} \geq \alpha.$$

Then there exist $I_0 > 0$ *and* $r > 0$ *such that, for all* $t \geq 0$,

$$I(\phi(t,\cdot) \,|\, \phi_\infty) \leq I_0 \exp(-2rt).$$

More precisely, when $\alpha = \overline{r} > 0$, *it is possible to take* $I_0 = I(\phi(0,\cdot) \,|\, \phi_\infty)$ *and* $r = \alpha$. *When* $\alpha < \overline{r}$, *for any* $\varepsilon \in (0, \overline{r})$, *it is possible to consider* $r = \overline{r} - \varepsilon$ *for a well-chosen constant* $I_0 > 0$.

Proof. The fact that $\overline{r} \geq \alpha$ is clear since either $\alpha \leq 0$, or $\alpha > 0$, in which case it is well known that ϕ_∞ satisfies LSI(α) (see for example [Ané *et al.* (2000)]). In the latter case, the choice $\overline{r} = \alpha$ can be made.

Recall that the relative entropy and the Fisher information are respectively

$$H(\phi(t,\cdot) \,|\, \phi_\infty) = \int_{\mathbb{R}} \ln\left(\frac{\phi}{\phi_\infty}\right) \phi, \quad I(\phi(t,\cdot) \,|\, \phi_\infty) = \int_{\mathbb{R}} \left| \partial_x \ln\left(\frac{\phi}{\phi_\infty}\right) \right|^2 \phi.$$

Since ϕ_∞ satisfies LSI($\overline{r}$),

$$H(\phi(t,\cdot)|\phi_\infty) \leq \frac{1}{2\overline{r}} I(\phi(t,\cdot)|\phi_\infty).$$

Moreover, by standard computations (see for example [Arnold *et al.* (2001)]),

$$\frac{d}{dt} H(\phi(t,\cdot) \,|\, \phi_\infty) = -I(\phi(t,\cdot) \,|\, \phi_\infty),$$

and

$$\begin{aligned} \frac{d}{dt} I(\phi(t,\cdot)|\phi_\infty) = -2 & \int_{\mathbb{R}} \frac{\phi}{\phi_\infty} \left| \partial_{x,x} \ln\left(\frac{\phi}{\phi_\infty}\right) \right|^2 \phi_\infty \\ & - 2 \int_{\mathbb{R}} \frac{\phi}{\phi_\infty} \left| \partial_x \ln\left(\frac{\phi}{\phi_\infty}\right) \right|^2 W'' \phi_\infty. \end{aligned} \tag{5.84}$$

If $\alpha = \overline{r}$, we thus obtain from (5.84) that $\frac{d}{dt} I(\phi(t,\cdot)|\phi_\infty) \leq -2\alpha I(\phi(t,\cdot)|\phi_\infty)$ which concludes the proof in this case.

Suppose now $\alpha < \overline{r}$. The technique of proof we propose is taken from [Villani (2009)]. For any $\lambda \in \left(0, \frac{1}{2|\alpha|}\right)$,

$$\begin{aligned}
&\frac{d}{dt}\left(H(\phi(t,\cdot)|\phi_\infty) + \lambda I(\phi(t,\cdot)|\phi_\infty)\right) \\
&\quad = -\int_{\mathbb{R}} \frac{\phi}{\phi_\infty}\left|\partial_x \ln\left(\frac{\phi}{\phi_\infty}\right)\right|^2 \phi_\infty - 2\lambda \int_{\mathbb{R}} \frac{\phi}{\phi_\infty}\left|\partial_{x,x} \ln\left(\frac{\phi}{\phi_\infty}\right)\right|^2 \phi_\infty \\
&\qquad - 2\lambda \int_{\mathbb{R}} \frac{\phi}{\phi_\infty}\left|\partial_x \ln\left(\frac{\phi}{\phi_\infty}\right)\right|^2 W''\phi_\infty \\
&\quad \leq -\int_{\mathbb{R}} (1 + 2\lambda W'')\frac{\phi}{\phi_\infty}\left|\partial_x \ln\left(\frac{\phi}{\phi_\infty}\right)\right|^2 \phi_\infty \\
&\quad \leq -(1 + 2\lambda \inf W'') I(\phi(t,\cdot)|\phi_\infty) \\
&\quad \leq -\frac{1+2\alpha\lambda}{\lambda + 1/(2\overline{r})}\left(H(\phi(t,\cdot)|\phi_\infty) + \lambda I(\phi(t,\cdot)|\phi_\infty)\right).
\end{aligned}$$

Then, for any $\lambda \in \left(0, \frac{1}{2|\alpha|}\right)$,

$$\begin{aligned}
&H(\phi(t,\cdot)\,|\,\phi_\infty) + \lambda I(\phi(t,\cdot)\,|\,\phi_\infty) \\
&\quad \leq \Big(H(\phi(0,\cdot)\,|\,\phi_\infty) + \lambda I(\phi(0,\cdot)\,|\,\phi_\infty)\Big) \exp\left(-\frac{1+2\alpha\lambda}{\lambda + 1/(2\overline{r})}t\right),
\end{aligned}$$

and therefore

$$I(\phi(t,\cdot)\,|\,\phi_\infty) \leq \left(\frac{1}{\lambda}H(\phi(0,\cdot)\,|\,\phi_\infty) + I(\phi(0,\cdot)\,|\,\phi_\infty)\right) \exp\left(-\frac{1+2\alpha\lambda}{\lambda + 1/(2\overline{r})}t\right).$$

Since $\dfrac{1+2\alpha\lambda}{\lambda + 1/(2\overline{r})}$ goes to $2\overline{r}$ when λ goes to 0, for any $\varepsilon \in (0, \overline{r})$, it is possible to find $\lambda > 0$ such that $\dfrac{1+2\alpha\lambda}{\lambda + 1/(2\overline{r})} = 2(\overline{r} - \varepsilon)$, which concludes the proof. □

Convergence of the biasing force. We prove here the convergence result (5.71) for the biasing force. This is a consequence of (5.70).

In the case $\mathcal{M} = \mathbb{T}$ and $W = 0$, we can prove the convergence of $\|F'_t - F'\|_{L^2(\mathbb{T})}$ to zero in the following sense (which implies (5.71), using (5.68)): For any $\varepsilon \in (0,1)$ and $t \geq t_\varepsilon$,

$$\|F'_t - F'\|^2_{L^2(\mathbb{T})} \leq \frac{2}{1-\varepsilon}\frac{M^2}{\rho}E_m(t), \tag{5.85}$$

where

$$t_\varepsilon = \min\left(0, (4\pi^2)^{-1} \ln\left(\varepsilon^{-1}\sqrt{\int_{\mathbb{T}} (\partial_z \psi^\xi(0,\cdot))^2}\right)\right).$$

This is shown using (i) the fact that

$$\int_{\mathbb{T}} (\partial_x \psi^\xi(t,\cdot))^2 \leq \int_{\mathbb{T}} (\partial_x \psi^\xi(0,\cdot))^2 \exp(-8\pi^2 t),$$

the proof of this estimate being similar to the one of Lemma 5.29; and (ii) the fact that for any function $g \in H^1(\mathbb{T})$,

$$\left\| g - \int_{\mathbb{T}} g \right\|_{L^\infty}^2 \leq \int_{\mathbb{T}} (\partial_x g)^2,$$

applied to $g = \psi^\xi$. Thus, $\left\|\psi^\xi - 1\right\|_{L^\infty}^2 \leq \int_{\mathbb{T}} (\partial_x \psi^\xi(0,\cdot))^2 \exp(-8\pi^2 t)$ which implies that for $t \geq t_\varepsilon$, $\psi^\xi(t,\cdot) \geq 1 - \varepsilon$, and thus, (5.85) follows from (5.70).

Let us now prove (5.71) in the case $\mathcal{M} = \mathbb{R}$, under the assumptions on W stated in item (2) of Corollary 5.16. Consider a compact set $K \subset \mathcal{M}$. Since $H^1(K) \subset L^\infty(K)$ (with continuous injection), there exists $c > 0$ such that

$$\begin{aligned}
\left\| \frac{\psi^\xi}{\psi^\xi_\infty} - 1 \right\|_{L^\infty(K)} &\leq c \left(\left\| \frac{\psi^\xi}{\psi^\xi_\infty} - 1 \right\|_{L^2(K)} + \left\| \partial_z \left(\frac{\psi^\xi}{\psi^\xi_\infty} - 1 \right) \right\|_{L^2(K)} \right) \\
&\leq \frac{c}{\inf_K \sqrt{\psi^\xi_\infty}} \left(\sqrt{\int_{\mathbb{R}} \left(\frac{\psi^\xi}{\psi^\xi_\infty} - 1 \right)^2 \psi^\xi_\infty} + \sqrt{\int_{\mathbb{R}} \left(\partial_z \left(\frac{\psi^\xi}{\psi^\xi_\infty} - 1 \right) \right)^2 \psi^\xi_\infty} \right).
\end{aligned}$$

Thus, for any $\varepsilon \in (0, \overline{r})$, there exists $C > 0$ such that

$$\left\| \frac{\psi^\xi}{\psi^\xi_\infty} - 1 \right\|_{L^\infty(K)} \leq C \exp(-rt),$$

with $r = \overline{r} - \varepsilon$. This inequality is obtained from the fact that since ψ^ξ_∞ satisfies LSI($\overline{r}$), then ψ^ξ_∞ also satisfies a Poincaré inequality with the same constant $\overline{r}$ (see for example [Ané *et al.* (2000)]), and a proof similar to that of Lemma 5.30 for the convergence of the Fisher information $\int_{\mathbb{R}} \left(\partial_z \left(\frac{\psi^\xi}{\psi^\xi_\infty} - 1 \right) \right)^2 \psi^\xi_\infty$ associated with the Poincaré inequality. Now,

$$\begin{aligned}
\int_K |F'_t - F'| \psi^\xi_\infty &= \int_K |F'_t - F'| \psi^\xi - \int_K |F'_t - F'| \left(\frac{\psi^\xi}{\psi^\xi_\infty} - 1 \right) \psi^\xi_\infty \\
&\leq \int_{\mathbb{R}} |F'_t - F'|^2 \psi^\xi + C \exp(-rt) \int_K |F'_t - F'| \psi^\xi_\infty.
\end{aligned}$$

Thus, for t sufficiently large, $\int_K |F'_t - F'| \psi^\xi_\infty$ is bounded from above by some constant times $\int_{\mathbb{R}} |F'_t - F'|^2 \psi^\xi$, which yields (5.71) (using (5.70) and (5.68)).

5.2.3.4 *Proof of Theorem* 5.19

Let us finally prove Theorem 5.19. We still assume, up to a change of variable, that $\beta = 1$. It holds:

$$\begin{aligned}\frac{dE}{dt} &= -\int_{\mathcal{D}} \left|\nabla \ln\left(\frac{\psi}{\psi_\infty}\right)\right|^2 \psi + \int_{\mathcal{D}} (F_t' - F') \circ \xi \nabla\xi \cdot \nabla \ln\left(\frac{\psi}{\psi_\infty}\right) \psi \\ &\leq -\int_{\mathcal{D}} \left|\nabla \ln\left(\frac{\psi}{\psi_\infty}\right)\right|^2 \psi + \sqrt{\int_{\mathcal{M}} |F_t' - F'|^2 \psi^\xi} \sqrt{\int_{\mathcal{D}} \left|\nabla\xi \cdot \nabla \ln\left(\frac{\psi}{\psi_\infty}\right)\right|^2 \psi}.\end{aligned}$$

Since, by Lemmas 5.27 and 5.28,

$$\int_{\mathcal{M}} |F_t' - F'|^2 \psi^\xi \leq \frac{M^2}{\rho} \int_{\mathcal{D}} \left|\nabla_{\Sigma(z)} \ln\left(\frac{\psi}{\psi_\infty}\right)\right|^2 \psi,$$

it follows

$$\begin{aligned}\frac{dE}{dt} &\leq -\int_{\mathcal{D}} \left|\nabla \ln\left(\frac{\psi}{\psi_\infty}\right)\right|^2 \psi \\ &\quad + \frac{Mm}{\sqrt{\rho}} \sqrt{\int_{\mathcal{D}} \left|\nabla_{\Sigma(z)} \ln\left(\frac{\psi}{\psi_\infty}\right)\right|^2 \psi} \sqrt{\int_{\mathcal{D}} \left|\frac{\nabla\xi}{|\nabla\xi|} \cdot \nabla \ln\left(\frac{\psi}{\psi_\infty}\right)\right|^2 \psi} \\ &\leq \left(-1 + \frac{Mm}{2\sqrt{\rho}}\right) \int_{\mathcal{D}} \left|\nabla \ln\left(\frac{\psi}{\psi_\infty}\right)\right|^2 \psi,\end{aligned}$$

where we have used the fact that, for any function $g : \mathcal{D} \to \mathbb{R}$, $|\nabla g|^2 = |\nabla_{\Sigma(z)} g|^2 + \left|\frac{\nabla\xi}{|\nabla\xi|} \cdot \nabla g\right|^2$. The logarithmic Sobolev inequality with respect to ψ_∞ (see Assumption 5.17) allows us to conclude the proof.

Chapter 6

Selection

This chapter presents some strategies to improve the numerical efficiency of multiple replica implementations used for free energy computations. A replica is here understood as a configuration of the system at a given time (and not a complete trajectory). An ensemble of replicas is therefore a collection of configurations $(x_t^{k,K})_{k=1,\dots,K}$ of the system, distributed according to some probability measure. The superscript K in $x_t^{k,K}$ indicates the total number of replicas, while the first index k is the replica label. The key point for efficient computations is to increase the number of replicas in areas containing some new information about the global free energy landscape. A *selection mechanism* between replicas is introduced to operate the appropriate elimination of replicas performing relatively unuseful calculations, and the appropriate duplication of replicas performing relatively informative calculations. A special kind of *genetic algorithm* is therefore obtained.

This chapter introduces the mathematical formalism enabling consistent selection procedures. As for all genetic algorithms, some fitness function (allowing to compare the relative importance of replicas) has to be introduced. This function depends on the application at hand. We focus our presentation on two examples:

- alchemical nonequilibrium transitions (using the Jarzynski equality, see Section 4.1.2) are studied in Section 6.1. The selection criterion is then naturally given by the work performed by the external switching exerted on the system, and the aim is to increase the number of replicas with low works in order to reduce the weight degeneracy mentioned in Section 4.1.4;
- adaptive biasing methods (presented in Chapter 5) are considered in Section 6.2. In this case, a selection strategy to improve the

diffusive behavior of the marginal distribution in the reaction coordinate space is sought for. Some examples of possible selection criteria are proposed. For this example, it is even possible to tune the intensity of the selection, which is actually an excellent opportunity to highlight the necessary *compromise between exploration and selection* in selection methods.

Of course, these selection strategies may be used in many other contexts and applications.

Parallel selection strategies are of particular interest for free energy landscapes with a multi-channel shape along the known reaction coordinates, see a typical picture in the very simple case of Figure 1.8. Such situations happen when the reaction coordinate indexing the transition is not rich enough. When applying a nonequilibrium or adaptive method to the associated reaction coordinate, the convergence of free energy calculations is limited by the time for a single simulation to find the side bifurcations between each channel (a local rare event). This is a major limiting factor since exploration of different channels is necessary for instance to determine the dominant channel of the free energy profile. As a consequence, it is highly desirable to enable the duplication of replicas that have found a still unexplored channel, in order to increase the computational time dedicated to these rarely visited areas. There may also be some situations where an initially favorable channel ends abruptly, and the replicas are stuck in a dead-end. If some replicas in an alternative channel are able to explore further values of the reaction coordinate, a selection procedure may be used to eliminate the replicas stuck in the dead-end, which allows to concentrate more rapidly the simulation burden on potentially more interesting regions.

Techniques such as replica exchange and parallel tempering [Geyer (1991); Hukushima and Nemoto (1996)] are well known to practitioners. In this case, several replicas of the system are simulated in parallel in different conditions, the typical case being systems simulated at different temperatures. Exchanges are then attempted between the replicas, and accepted or rejected according to some Metropolis criterion. The bottom line of the method is that the dynamics of the system at higher temperatures is less metastable, so that free-energy barriers may be overcome for systems at the lowest temperatures relying on the exchanges with replicas at higher temperatures.

The multiple replica selection strategies we present in this chapter are somewhat different from replica exchange and parallel tempering since all the replicas evolve under the same dynamics and in the same conditions. The so-obtained empirical distribution approximates the distribution of the configurations of the system. A weight, computed from a *fitness* or *selection* potential, is associated to each replica, depending on the chosen criterion. Resamplings are then performed to avoid the degeneracy of these weights. These resamplings induce exchanges between replicas favoring areas with high selection potential. For practical convenience, we restrict ourselves to selection methods with a *fixed* number of replicas, and such that the weights of the replicas after resampling are all equal.

Selection mechanisms are a fundamental tool in Monte Carlo methods, especially for "Population Monte Carlo" methods. Many algorithms have been developed (see for instance the recent review on population methods [Jasra *et al.* (2007)]), which are widely used in some quantum chemistry computations (Diffusion Monte Carlo methods, see [Assaraf *et al.* (2000)]), or in Bayesian Statistics (Sequential Monte Carlo, see [Doucet *et al.* (2001, 2006)]). An exhaustive mathematical analysis has been developed by Del Moral and his co-workers, see [Del Moral (2004)]. Note that in the probability and statistics fields, each replica of the system is called a *walker* or a *particle*. We use here the name *replica*, which is more appropriate to the context of computational statistical physics.

6.1 Replica selection framework

We present in this section the selection formalism for systems where some selection function is given *a priori*, choosing the alchemical Jarzynski nonequilibrium switching as a running example, as was done in [Rousset and Stoltz (2006)].

6.1.1 *Weighted replica ensembles*

6.1.1.1 *An example: alchemical transitions with nonequilibrium switching dynamics*

We first recall some notation from Section 4.1. Consider a transition indexed by an alchemical parameter with a prescribed time evolution:

$$\Lambda : [0,T] \longrightarrow [0,1],$$

with $\Lambda(0) = 0$ and $\Lambda(T) = 1$. The system evolves according to some dynamics, described by its time-dependent infinitesimal generator $\mathcal{L}_t$. For a given (fixed) value $\lambda \in [0,1]$ of the parameter, the energy of the system is denoted by

$$E_\lambda : \mathcal{S} \to \mathbb{R},$$

with $x = q \in \mathcal{S} = \mathcal{D}$ when the underlying dynamics is the overdamped Langevin dynamics for instance, in which case $E_\lambda(x) = V_\lambda(q)$ is the potential energy of the system; or $x = (q,p) \in \mathcal{S} = T^*\mathcal{D}$ when the underlying dynamics is of Langevin or Hamiltonian type, in which case $E_\lambda(x) = H_\lambda(q,p)$ is the total energy. The canonical equilibrium distribution is therefore

$$Z_\lambda^{-1}\, \mathrm{e}^{-\beta E_\lambda(x)}\, dx,$$

and the free energy difference to be computed reads

$$F(\Lambda(T)) - F(0) = -\frac{1}{\beta} \ln \frac{\displaystyle\int_{\mathcal{S}} \mathrm{e}^{-\beta E_{\Lambda(T)}(x)}\, dx}{\displaystyle\int_{\mathcal{S}} \mathrm{e}^{-\beta E_{\Lambda(0)}(x)}\, dx}.$$

In order to perform some selection of replicas, a fitness function should be chosen. The Jarzynski identity (4.9) states that the canonical equilibrium distribution $\pi_t(dx) = Z_t^{-1}\, \mathrm{e}^{-\beta E_{\Lambda(t)}(x)}\, dx$ can be recovered by a proper reweighting of the nonequilibrium distribution:

$$\int_{\mathcal{S}} \varphi(x)\, \pi_t(dx) = \frac{\mathbb{E}\left(\varphi(x_t)\, \mathrm{e}^{-\beta \mathcal{W}_t}\right)}{\mathbb{E}\left(\mathrm{e}^{-\beta \mathcal{W}_t}\right)}, \tag{6.1}$$

where the work exerted on the system during the switching process reads

$$\mathcal{W}_t = \int_0^t \frac{\partial E_{\Lambda(s)}}{\partial \lambda}(x_s)\, \Lambda'(s)\, ds. \tag{6.2}$$

In practice, the expectation in the right-hand side of (6.1) can be approximated for instance as

$$\frac{\mathbb{E}\left(\varphi(x_t)\, \mathrm{e}^{-\beta \mathcal{W}_t}\right)}{\mathbb{E}\left(\mathrm{e}^{-\beta \mathcal{W}_t}\right)} \simeq \frac{\displaystyle\sum_{k=1}^{K} \varphi(x_t^{k,K})\, \mathrm{e}^{-\beta \mathcal{W}_t^{k,K}}}{\displaystyle\sum_{k=1}^{K} \mathrm{e}^{-\beta \mathcal{W}_t^{k,K}}},$$

where the configurations $(x_t^{k,K})_{1 \le k \le K}$ are obtained by independent realizations of the nonequilibrium dynamics.

The selection paradigm relies on a reinterpretation of the statistical sample of replicas $(x_t^{k,K})_{1 \le k \le K}$ as a weighted sample, each replica having a weight proportional to $\mathrm{e}^{-\beta \mathcal{W}_t^{k,K}}$. The corresponding weighted empirical probability measure (see (6.6) below) approximates π_t, see (6.1). In order to enhance the number of relevant configurations, replicas with large weights should be favored, while replicas with low weights should be eliminated. This amounts to favoring lower work values, the work $\mathcal{W}_t$ quantifying how close to equilibrium the system has remained and how informative the corresponding configuration is for the free energy computation. It also suggests considering the selection potential given by the (opposite of the) instantaneous work exerted on the system:

$$S_t(x) := -\beta \frac{\partial E_{\Lambda(t)}}{\partial \lambda}(x)\, \Lambda'(t). \tag{6.3}$$

The selection mechanism should then be designed in such a way that the fundamental fluctuation identity (6.1) is preserved.

6.1.1.2 *General presentation of weighted ensembles of replicas*

We present a general description of an ensemble of K replicas of a given system, described by variables $x_t^{k,K} \in \mathcal{S}$ $(1 \leq k \leq K)$, and evolving independently according to some Markov process with generator $\mathcal{L}_t$. We assume that the law of the process x_t with associated generator $\mathcal{L}_t$ has a density with respect to the Lebesgue measure, which is indeed the case when $\mathcal{L}_t^*$ is (hypo)elliptic. Recall that the equation governing the evolution of this law is the Fokker-Planck equation (2.23):

$$\partial_t \psi = \mathcal{L}_t^* \psi.$$

The processes $x_t^{k,K}$ share the same law, and they are exchangeable, in the sense that the law of the vector $(x_t^{\sigma(k),K})_{1 \leq k \leq K}$ is the same as the law of $(x_t^{k,K})_{1 \leq k \leq K}$ for any permutation σ of $\{1, \ldots, K\}$. Exchangeability is of course a consequence of independence. When selection is considered, the evolutions are no longer independent, but, as we will see below, exchangeability is a key property which is preserved.

Assume that a family of fitness functions

$$S_t : \mathcal{S} \to \mathbb{R}$$

is given (such as (6.3) above), and that each replica has a non-normalized *importance weight*[1]

$$W_t^{k,K} = \exp\left(\int_0^t S_s\left(x_s^{k,K}\right) ds \right). \tag{6.4}$$

[1]The notation $W_t^{k,K}$ for the weights should not be confused with the notation $\mathcal{W}_t^{k,K}$ for the works (6.2).

Replicas which have visited areas with higher fitnesses have exponentially increasing weights. The weights $W_t^{k,K} \geq 0$ can be normalized upon defining

$$w_t^{k,K} = \frac{W_t^{k,K}}{\sum_{l=1}^{K} W_t^{l,K}}. \tag{6.5}$$

The (weighted) empirical probability distribution of replicas is denoted by

$$\pi_t^K(dx) := \sum_{k=1}^{K} w_t^{k,K} \delta_{x_t^{k,K}}(dx). \tag{6.6}$$

The expectation of an observable $\phi(x)$ with respect to π_t^K is then a stochastic process:

$$\langle \phi \rangle_{t,K} = \int_{\mathcal{S}} \phi(x)\, \pi_t^K(dx) = \sum_{k=1}^{K} w_t^{k,K} \phi\left(x_t^{k,K}\right). \tag{6.7}$$

The process $t \mapsto \langle \phi \rangle_{t,K}$ may have a large variance when the weight distribution $(w_t^{k,K})_{k=1,\dots,K}$ is degenerate (a notion which will be made precise below, see (6.14)), leading to large statistical errors in the estimate. We will see in Section 6.1.2 how the degeneracy of weights can be avoided, or at least limited (see Algorithms 6.3 and 6.4).

An important quantity to compute unbiased averages from weighted empirical probability distributions is the weight normalization (the denominator in (6.5), up to a factor $1/K$):

$$Z_{0,t}^K = \frac{1}{K} \sum_{k=1}^{K} W_t^{k,K} = \frac{1}{K} \sum_{k=1}^{K} \exp\left(\int_0^t S_s\left(x_s^{k,K}\right) ds \right), \tag{6.8}$$

which is actually the average of the non-normalized weights $W_t^{k,K}$. There is an alternative expression of the normalization, which is interesting since it will still hold when the selection between replicas is switched on.

Lemma 6.1. *The normalization of weights* (6.8) *is related to the average fitness as*

$$Z_{0,t}^K = \exp\left(\int_0^t \langle S_s \rangle_{s,K}\, ds \right), \tag{6.9}$$

where the weighted average $\langle \cdot \rangle_{t,K}$ *is defined by* (6.7).

Proof. A differentiation with respect to time of the logarithm of the weight normalization (6.8) gives

$$\begin{aligned} \partial_t \left(\ln Z_{0,t}^K \right) &= \partial_t \left[\ln \left(\sum_{k=1}^{K} \exp\left(\int_0^t S_s\left(x_s^{k,K}\right) ds \right) \right) \right] \\ &= \frac{\sum_{k=1}^{K} S_t\left(x_t^{k,K}\right) W_t^{k,K}}{\sum_{k=1}^{K} W_t^{k,K}} = \langle S_t \rangle_{t,K}. \end{aligned}$$

A time integration on the interval $[0, t]$ then leads to (6.9). □

The exchangeability of the replicas allows to recover weighted expectations as expectations over the realizations of the replica ensemble:

$$\mathbb{E}\Big[\langle\phi\rangle_{t,K} Z_{0,t}^K\Big] = \mathbb{E}\left[\phi(x_t)\exp\left(\int_0^t S_s(x_s)\,ds\right)\right], \tag{6.10}$$

where $t \mapsto x_t$ is distributed at time $t = 0$ according to the initial distribution of the replica system, and evolves according to the dynamics of each single replica. The interest of replica ensembles is that the variance of the estimator decreases as the number K of replicas increases, thanks to the Central Limit Theorem.

The unbiasedness property (6.10) is an important feature that selection mechanisms will have to preserve. On the other hand, the independence property will be lost during the selection process. This is actually not a concern as long as the exchangeability of the replicas is preserved.

The weighted empirical probability measure (6.6) is fully characterized by the initial condition and its time evolution equation, which, in the large population limit ($K \to +\infty$), is a diffusion equation with source terms.

Proposition 6.2. *Expectations with respect to the empirical probability distribution of the replicas converge when $K \to +\infty$ almost surely towards expectations with respect to a limiting distribution $\psi(t,x)\,dx$:*

$$\langle\phi\rangle_{t,K} = \int_{\mathcal{S}} \phi(x)\,\pi_t^K(dx) \xrightarrow[K\to+\infty]{} \int_{\mathcal{S}} \phi(x)\,\psi(t,x)\,dx \quad \text{a.s.}, \tag{6.11}$$

the limiting probability density function ψ satisfying the nonlinear evolution equation:

$$\partial_t\psi = \mathcal{L}_t^*\psi + \left(S_t - \int_{\mathcal{S}} S_t(x)\psi(t,x)\,dx\right)\psi. \tag{6.12}$$

Proof. The proof of this proposition follows from a standard application of the Feynman-Kac equality. The evolution of the law of the replicas in the absence of fitness function is described by the operator $\mathcal{L}_t^*$. When fitness functions are introduced, (6.10) can be rewritten as

$$\mathbb{E}\left[\phi(x_t)\exp\left(\int_0^t S_s(x_s)\,ds\right)\right] = \int_{\mathcal{S}} \phi(x)\,\Psi(t,x)\,dx$$

where Ψ satisfies the Fokker-Planck equation

$$\partial_t\Psi = \mathcal{L}_t^*\Psi + S_t\Psi.$$

The term $S_t \Psi$ can be interpreted as a source term in the above equation, so that the normalization condition $\int_{\mathcal{S}} \Psi(t,x)\,dx = 1$ is not preserved. In order to restore it, the probability measure

$$\psi(t,x)\,dx = \frac{\Psi(t,x)\,dx}{\displaystyle\int_{\mathcal{S}} \Psi(t,y)\,dy}$$

should be considered instead. A straightforward computation shows that the corresponding density ψ still satisfies a Fokker-Planck equation, but with a modified source term $S_t - \overline{S}_t$:

$$\partial_t \psi = \mathcal{L}_t^* \psi + \left(S_t - \int_{\mathcal{S}} S_t(y)\psi(t,y)\,dy \right) \psi.$$

Notice that the modified source term is such that

$$\int_{\mathcal{S}} \left[S_t(x) - \left(\int_{\mathcal{S}} S_t(y)\psi(t,y)\,dy \right) \right] \psi(t,x)\,dx = 0,$$

so that the normalization of the probability density function ψ is indeed preserved. □

Application to nonequilibrium switchings. We now come back to the Jarzynski equality (4.10). With the notation of Section 4.1, $\psi(t,x)\,dx = \pi_t(dx)$ is the canonical measure associated with the value $\Lambda(t)$ of the alchemical parameter.

An estimator of the free energy difference $\Delta F_t = F(\Lambda(t)) - F(0)$ can be obtained from the weight normalization at time t as

$$\Delta \widehat{F}_t^K = -\frac{1}{\beta} \ln Z_{0,t}^K = -\frac{1}{\beta} \ln \left(\frac{1}{K} \sum_{k=1}^{K} W_t^{k,K} \right), \tag{6.13}$$

where the non-normalized weights are defined in (6.4), and the fitness function is given by (6.3). The Jarzynski equality shows that the weight normalization $Z_{0,t}^K = \mathrm{e}^{-\beta \Delta \widehat{F}_t^K}$ is an unbiased estimator of $\mathrm{e}^{-\beta \Delta F_t}$. By independence of the replicas, $\Delta \widehat{F}_t^K$ is an asymptotically normal estimator of the free energy difference, with bias and variance of order K^{-1} (see Section 4.1.5 for further precision).

However, $\Delta \widehat{F}_t^K$ suffers from the fact that only a few work values

$$\mathcal{W}_t^{k,K} = -\frac{1}{\beta} \ln W_t^{k,K} = -\frac{1}{\beta} \int_0^t S_s\left(x_s^{k,K}\right)\,ds$$

are really important in (6.13), so that the estimator is plagued by large statistical errors. It should therefore be interesting to select replicas in order to avoid the weight degeneracy and to hopefully reduce the variance of the estimator.

6.1.2 *Resampling strategies*

To avoid weight degeneracies, a *resampling* or *selection strategy* has to be designed. A quantitative criterion for weight degeneracy has to be chosen. When the associated threshold is attained (at some random time τ), some resampling is triggered. The idea is that, on average, the kth replica should be replicated $Kw_\tau^{k,K}$ times at the resampling times τ. This is the key property which ensures the preservation of the unbiasedness property (6.10).

6.1.2.1 *Measuring the degeneracy*

Let us first describe the degeneracy criterion. The most standard choice is given by the relative entropy between the current normalized weights $w_t^K = (w_t^{1,K}, \dots, w_t^{K,K})$ and the initial uniform weights $w_0^K = (1/K, \dots, 1/K)$:

$$\mathrm{Ent}(w_t^K) = \sum_{k=1}^{K} w_t^{k,K} \ln\left(\frac{w_t^{k,K}}{w_0^{k,K}}\right) = \sum_{k=1}^{K} w_t^{k,K} \ln\left(K\, w_t^{k,K}\right). \tag{6.14}$$

It is easily shown that the relative entropy has values in $[0, \ln K]$. The nonnegativity of the relative entropy is a consequence of the nonnegativity of the function $x \mapsto x \ln x - x + 1$:

$$\mathrm{Ent}(w_t^K) = \frac{1}{K}\sum_{k=1}^{K} K\, w_t^{k,K} \ln\left(K\, w_t^{k,K}\right) \geq \frac{1}{K}\sum_{k=1}^{K}\left(K\, w_t^{k,K} - 1\right) = 0.$$

The relative entropy (6.14) is minimal when the weights are uniform (minimal information):

$$\mathrm{Ent}\left(\left\{\frac{1}{K}, \dots, \frac{1}{K}\right\}\right) = 0,$$

and maximal when the distribution is concentrated on one single replica (maximal information):

$$\mathrm{Ent}\Big(\{1, 0, \dots, 0\}\Big) = \ln K.$$

6.1.2.2 *Resampling algorithm*

A replica system with resamplings is evolved according to the underlying dynamics of the replicas, until the weight distribution is too degenerate, *i.e.* until the entropy of the weights is too large. At these times, some resampling is performed, and a new ensemble of replicas is obtained, each one having a weight $1/K$. The simulation is then continued.

The precise algorithm is given below. An important step in the resampling procedure is the generation of the so-called *branching numbers* at a resampling time τ, which are random integers $\{n^{k,K} \in \mathbb{N},\ k = 1,\ldots,K\}$ such that

$$\sum_{k=1}^{K} n^{k,K} = K, \qquad \mathbb{E}\left(n^{k,K} \,\middle|\, \{w_\tau^{1,K},\ldots,w_\tau^{K,K}\}\right) = K w_\tau^{k,K}. \tag{6.15}$$

The integer $n^{k,K}$ is the number of new replicas starting from the configuration $x_{\tau^-}^{k,K}$. More details about the methods used in practice to obtain these branching numbers are given in Section 6.1.2.3.

The update of the weighted replica sample $\left\{x_{\tau^-}^{1,K},\ldots,x_{\tau^-}^{K,K}\right\}$ at a resampling time τ, according to the generated set of branching numbers $\{n^{k,K} \in \mathbb{N},\ k = 1,\ldots,K\}$, consists in creating a new replica sample

$$\{x_\tau^{1,K},\ldots,x_\tau^{K,K}\}$$

with uniform weights $1/K$, in which there are exactly $n^{k,K}$ replicas whose configuration is equal to $x_{\tau^-}^{k,K}$. This defines a stochastic process with jumps which is right continuous with left limits.

Algorithm 6.3 (Ensemble resampling).
Choose a simulation time $T_{\rm simu}$ and a degeneracy threshold $\alpha > 0$. Generate an initial distribution of replicas $(x_0^{1,K},\ldots,x_0^{K,K})$, independently sampled from a probability distribution π_0, each replica starting with the uniform weight $w_0^{k,K} = 1/K$. Initialize the sequence of resampling times $\{T_p\}_{p\geq 0}$ as $T_0 = 0$. For $0 \leq t \leq T_{\rm simu}$,

(1) for $t \in [T_p, T_{p+1}]$, let the replicas $\{x_t^{k,K}\}_{1\leq k\leq K}$ evolve according to their original dynamics, and compute their weights

$$w_t^{k,K} = \frac{W_t^{k,K}}{\displaystyle\sum_{1\leq l\leq K} W_t^{l,K}},$$

where $W_t^{k,K} = \exp\left(\displaystyle\int_{T_p}^{t} S_s\left(x_s^{k,K}\right)\,ds\right)$;

(2) at random times $\{T_{p+1}, p \geq 0\}$ defined by

$$T_{p+1} = \inf\left\{t \geq T_p \,\middle|\, \mathrm{Ent}(w_t^K) \geq \alpha\right\},$$

resample replicas according to their respective weights by generating random integers $\{n^{k,K} \in \mathbb{N},\ k = 1,\ldots,K\}$ under the constraints (6.15) (with $\tau = T_{p+1}$). A new replica ensemble is generated according to the branching numbers $\{n^{k,K} \in \mathbb{N},\ k = 1,\ldots,K\}$, and the weights are reset to unity: $W_{T_{p+1}}^{k,K} = 1$ for $k = 1,\ldots,K$;

(3) *the total weight normalization* $Z^K_{T_p,T_{p+1}}$ *between each resampling is updated according to* (6.8) *or* (6.9), *and the total weight normalization on* $[0,t]$ *is*

$$Z^K_{0,t} := Z^K_{0,T_1} \cdots Z^K_{T_p,t} = \exp\left(\int_0^t \langle S_s \rangle_{s,K} ds\right). \tag{6.16}$$

6.1.2.3 *Computation of the branching numbers*

Several resampling methods can be used to compute the branching numbers required in Algorithm 6.3. A classical method is the so-called *residual multinomial resampling*, see [Del Moral (2004); Doucet *et al.* (2001)]. This method has the advantage of satisfying usual limit theorems, and still remaining stable in terms of variance when the number of resamplings per unit time grows to infinity (*i.e.* in the limit $\alpha \to 0$).

For simplicity, we drop the time index in the subscripts, and denote by $\lfloor x \rfloor$ the integer part of x. We also use the convention $\sum_{l=1}^{0} \cdot = 0$. The multinomial residual resampling consists in setting $n^{k,K} = \lfloor K\, w^{k,K} \rfloor + m^{k,K}$, with $m^{k,K} \in \mathbb{N}$ such that $\mathbb{E}(m^{k,K} \mid \{w^{1,K}, \dots, w^{K,K}\}) = K\, w^{k,K} - \lfloor K\, w^{k,K} \rfloor$. Therefore, the procedure focuses on determining the so-called *residual branching numbers* $m^{k,K}$, the final branching numbers being the sum of the integer parts and the residual branching numbers (see (6.17)).

Considering a set of weights $(w^{k,K})_{k=1,\dots,K}$, the method consists in the following steps:

(1) for each replica, compute the residual branching weight

$$r^{k,K} = K w^{k,K} - \lfloor K\, w^{k,K} \rfloor,$$

the corresponding sum

$$\sum_{k=1}^{K} r^{k,K} = K_{\mathrm{res}} \leq K,$$

and decompose the interval $[0, K_{\mathrm{res}}]$ in K sub-intervals:

$$[0, r^{1,K}] \bigcup [r^{1,K}, r^{1,K} + r^{2,K}] \bigcup \cdots \bigcup [r^{1,K} + \cdots + r^{k-1,K}, K_{\mathrm{res}}];$$

(2) draw K_{res} independent variables $\{U^{i,K},\ i = 1, \dots, K_{\mathrm{res}}\}$ uniformly distributed in $[0, K_{\mathrm{res}}]$, and denote by $m^{k,K}$ the number of such variables that fall in the k-th interval $\left[\sum_{l=1}^{k-1} r^{l,K}, \sum_{l=1}^{k} r^{l,K}\right]$;

(3) for $k = 1, \dots, K$, set the branching numbers to:

$$n^{k,K} = \lfloor K w^{k,K} \rfloor + m^{k,K}. \tag{6.17}$$

Note that the random vector $\{m^{k,K},\ k = 1, \dots, K\}$ is distributed according to a multinomial law with parameters $\left(\frac{r^{1,K}}{K_{\text{res}}}, \dots, \frac{r^{K,K}}{K_{\text{res}}}\right)$ (distribution) and K_{res} (trials). For this method, the new replica configurations are independent conditionally on the former configurations. This enables a mathematical study of the method.

An alternative resampling scheme which performs well in many cases, but for which no convergence result holds (since the replica independence conditionally on the former configurations is lost), is the so-called *systematic resampling*. It consists in setting

$$n^{k,K} = \left\lfloor K \sum_{l=1}^{k} w_{\mathcal{T}}^{l,K} + U \right\rfloor - \left\lfloor K \sum_{l=1}^{k-1} w_{\mathcal{T}}^{l,K} + U \right\rfloor,$$

where the random number U used in this procedure is distributed according to a uniform law on $[0, 1]$. It can heuristically be argued (and is numerically observed) that this method behaves better in terms of variance since fewer random numbers are required.

6.1.2.4 *Continuous-in-time selection*

There is an equivalent description of Algorithm 6.3 when the resampling threshold α tends to 0. In this case, continuous-in-time resamplings are performed by assigning to each replica birth and death times, and all replicas have a weight $1/K$ at all times. When a replica explores a region of lower fitness, its death time is decreased and it is therefore more likely that it will be suppressed in a near future. This is consistent with the fact that replicas of lower fitnesses should count less in the weighted average. In a similar manner, the birth time of replicas exploring regions of larger fitnesses is decreased, so that at some point a replica is killed at random, and restarted from this region of higher fitness, therefore enhancing the average in this zone. The following notation is used in the following:

$$x_+ = \max(0, x), \qquad x_- = \max(0, -x) \geq 0.$$

The branching process proposed in [Rousset (2006a, b)] reads:

Algorithm 6.4 (Individual resampling). *Generate an initial distribution of replicas* $(x_0^{1,K}, \dots, x_0^{K,K})$*, sampled independently from a probability distribution* π_0*, each replica starting with the uniform weight* $w_0^{k,K} = 1/K$*. Generate independent times* $\{\mathcal{T}_1^{k,\mathrm{b}}, \mathcal{T}_1^{k,\mathrm{d}},\ k = 1, \dots, K\}$ *from an exponential law of mean* 1 *(the superscripts* b *and* d *refer to "birth" and "death"* *respectively), and initialize the jump times as* $T_0^{k,\mathrm{d}} = 0$, $T_0^{k,\mathrm{b}} = 0$. *For* $t \geq 0$ *and* $n \geq 0$,

- (mutation) *between two jump times, the replicas* $x_t^{k,K}$ *evolve according to their original dynamics, the weights being* $1/K$ *at all times;*
- (death) *at random times* $T_{p+1}^{k,\mathrm{d}}$ *defined by*

$$\int_{T_p^{k,\mathrm{d}}}^{T_{p+1}^{k,\mathrm{d}}} \left[S_t\left(x_t^{k,K}\right) - \langle S_t \rangle_{t,K} \right]_- dt = \mathscr{T}_{p+1}^{k,\mathrm{d}},$$

an index $l \in \{1, \ldots, K\}$ *is picked at random, and the configuration of the* k*th replica is replaced by the configuration of the* l*th replica:* $x_{T_{p+1}^{k,\mathrm{d}}}^{k,K} \leftarrow x_{T_{p+1}^{k,\mathrm{d}}}^{l,K}$. *A new death time* $\mathscr{T}_{p+2}^{k,\mathrm{d}}$ *is generated from an exponential law of mean* 1*;*
- (birth) *at random times* $T_{p+1}^{k,\mathrm{b}}$ *defined by*

$$\int_{T_p^{k,\mathrm{b}}}^{T_{p+1}^{k,\mathrm{b}}} \left[S_t\left(x_t^{k,K}\right) - \langle S_t \rangle_{t,K} \right]_+ dt = \mathscr{T}_{p+1}^{k,\mathrm{b}},$$

an index $l \in \{1, \ldots, K\}$ *is picked at random, and the configuration of the* l*th replica is replaced by the configuration of the* k*th replica:* $x_{T_{p+1}^{k,\mathrm{d}}}^{l,K} \leftarrow x_{T_{p+1}^{k,\mathrm{d}}}^{k,K}$. *A time* $\mathscr{T}_{p+2}^{k,\mathrm{b}}$ *is generated from an exponential law of mean* 1*;*
- *update the weight normalization according to* (6.9)*:*

$$Z_{0,t}^K = \exp\left(\int_0^t \langle S_s \rangle_{s,K}\, ds \right). \tag{6.18}$$

This procedure, called "Interacting Replica Ensemble" in the sequel, can therefore be seen as a self-adjusted time-continuous resampling (formally, Algorithm 6.4 is a limit of Algorithm 6.3 in the case $\alpha \to 0$). All replicas have the same weight at any time of the simulation.

6.1.2.5 *Consistency and convergence*

The distributions of the replica ensemble obtained with Algorithm 6.3 or 6.4 is consistent with the weighted replica distribution (6.6), in the sense that it satisfies (6.10). This might be referred to as the "unbiasedness condition" of resampling schemes.

Proposition 6.5. *Consider the weighted empirical distribution π_t^K given by (6.6):*

$$\pi_t^K(dx) := \sum_{k=1}^{K} w_t^{k,K} \delta_{x_t^{k,K}}(dx),$$

where the replica distribution $(x_t^{k,K})_{1\le k\le K}$ and the weights $(w_t^{k,K})_{1\le k\le K}$ are obtained either from Algorithm 6.3 or 6.4 (in the latter case they are all equal to $1/K$), or from a standard evolution without resampling (as presented in Section 6.1.1.2).

Then, for any observable ϕ, the following equilibrium expectation does not depend on the chosen algorithm:

$$\mathbb{E}\left[\langle\phi\rangle_{t,K} Z_{0,t}^K\right] = \mathbb{E}\left[\phi(x_t)\exp\left(\int_0^t S_s(x_s)\,ds\right)\right], \tag{6.19}$$

where x_t evolves according to the underlying dynamics characterized by the generator $\mathcal{L}_t$, and $\langle\phi\rangle_{t,K}$ is defined in (6.7).

We already showed the result when no resampling is performed (see Section 6.1.1.2). For replicas evolving according to Algorithm 6.3, the result holds by construction of the resampling scheme since the expectations in (6.19) are the same right before and right after a resampling step. For continuous-in-time resampling, the proof can be read in [Rousset (2006a, b)].

Moreover, at least formally, the random empirical distribution π_t^K of replicas converges in the sample size limit $K \to +\infty$ towards the probability density $\psi(t,x)\,dx$ solution of the nonlinear forward evolution equation (6.12).

Several convergence results and statistical properties of the replicas distribution can be proven, see for instance [Del Moral (2004); Del Moral and Miclo (2000)]. They can be summarized by the following sentence: The estimator obtained from the empirical distribution of replicas π_t^K in the left-hand side of (6.11) is a consistent and asymptotically normal estimator of the right-hand side of (6.11), with bias and variance of order K^{-1}. See Lemma 3.20, Proposition 3.25 and Theorem 3.28 in [Del Moral and Miclo (2000)] for a proof of the latter assertion.

6.1.2.6 *Application to the computation of free energy differences*

In the context of nonequilibrium switching dynamics, the free energy difference (6.13) using replicas evolving according to a resampling strategy

("RS"), such as Algorithm 6.3 or 6.4, is obtained from (6.16) or (6.18). Therefore, using Proposition 6.5 with $\phi = 1$, an estimator of the free energy difference is

$$\Delta \widehat{F}_t^{\mathrm{RS},K} = \int_0^t \langle S_s \rangle_{s,K} \, ds,$$

where $\langle S_s \rangle_{s,K}$ is defined in (6.7). In particular, for continuous-in-time resampling as decribed in Algorithm 6.4 (or right after a resampling step in Algorithm 6.3), all the weights are equal and

$$\Delta \widehat{F}_t^{\mathrm{RS},K} = \frac{1}{K} \sum_{k=1}^{K} \mathcal{W}_t^{k,K} = \int_0^t \int_{\mathcal{S}} \frac{\partial E_{\Lambda(s)}}{\partial \lambda}(x) \, \pi_s^K(dx) \, \Lambda'(s) \, ds, \qquad (6.20)$$

where π_s^K is given by (6.6).

The estimator $\Delta \widehat{F}_t^{\mathrm{RS},K}$ shares, as mentioned above, the same asymptotic statistical properties as the estimator $\Delta \widehat{F}_t^K$ defined by (6.13) (bias and variance of order K^{-1}). The hope is that, for M fixed, the variance of $\Delta \widehat{F}_t^{\mathrm{RS},K}$ is lower than the variance of $\Delta \widehat{F}_t^K$.

6.1.3 *Discrete-time version*

The numerical implementation is done by discretizing the time-continuous processes considered in the previous sections. Denote by Δt the fixed time-step, and by $n \in \mathbb{N}$ the iteration index. With a slight abuse of notation, we replace the subscripts t in the objects introduced in the previous sections by subscripts n.

The kernel $P_n(x, dy)$ of the Markov chain at step n is assumed to be consistent with the continuous dynamics. For instance, in the context of nonequilibrium switching dynamics, the kernel P_n depends on the current value $\lambda^n = \Lambda(n\Delta t)$ of the alchemical parameter, see Section 4.1.3 for further precision. The discrete-time dynamics therefore evolves a set of replicas $(x_n^{k,K})_{k=1,\ldots,K}$ at step n into a set $(x_{n+1}^{k,K})_{k=1,\ldots,K}$ at step $n+1$.

The fitness or selection function is also defined at discrete times, and will be denoted by $S_n(x)$ in the sequel. An expression of the discrete-time fitness function can be obtained from an expression of a time-continuous fitness function using some quadrature rule. In the context of nonequilibrium switching dynamics as described in Section 4.1, the selection function given by (6.3) may be approximated as

$$S_n(x) = -\frac{\beta}{2} \left(\frac{\partial E_{\lambda^n}}{\partial \lambda}(x) + \frac{\partial E_{\lambda^{n+1}}}{\partial \lambda}(x) \right) \frac{\lambda^{n+1} - \lambda^n}{\Delta t}.$$

The (weighted) empirical probability distribution of replicas at iteration n is still defined as

$$\pi_n^K(dx) = \sum_{k=1}^{K} w_n^{k,K} \delta_{x_n^{k,K}}(dx), \tag{6.21}$$

the integrals in the definition of the discrete weights (6.4) being replaced by discrete sums for indices lower than or equal to $n-1$. When no resampling occurs in the n first steps $\{1, \dots, n\}$,

$$w_n^{k,K} = \frac{\exp\left(\sum_{m=0}^{n-1} S_m\left(x_m^{k,K}\right)\Delta t\right)}{\sum_{l=1}^{K} \exp\left(\sum_{m=0}^{n-1} S_m\left(x_m^{l,K}\right)\Delta t\right)}.$$

The empirical average value of a given function ϕ at step n is denoted by:

$$\langle \phi \rangle_{n,K} = \int_S \phi(x)\, \pi_n^K(dx) = \sum_{k=1}^{K} w_n^{k,K}\, \phi(x_n^{k,K}).$$

The discrete-time normalization of weights on any iteration interval $\{n, \dots, n'\}$ is the counterpart of (6.9):

$$Z_{n,n'}^K = \prod_{m=n}^{n'-1} \left\langle e^{S_m \Delta t} \right\rangle_{m,K}. \tag{6.22}$$

An alternative expression can be obtained as in Section 6.1.1.2. Define

$$\widetilde{Z}_{0,n}^K = \frac{1}{K} \sum_{k=1}^{K} \exp\left(\sum_{m=0}^{n-1} S_m\left(x_m^{k,K}\right)\right).$$

When no resampling occurs for the time indices $\{1, \dots, n\}$,

$$\frac{\widetilde{Z}_{0,n+1}^K}{\widetilde{Z}_{0,n}^K} = \frac{\sum_{l=1}^{K} w_n^{k,K}\, e^{S_n(x_m^{k,K})}}{\sum_{l=1}^{K} w_n^{k,K}} = \left\langle e^{S_n \Delta t} \right\rangle_{n,K},$$

so that, using $\widetilde{Z}_{0,0}^K = 1$,

$$\widetilde{Z}_{0,n}^K = Z_{0,n}^K.$$

This equation is the time-discrete version of (6.8) and Lemma 6.1. This gives a Feynman-Kac representation for $n \geq 0$, similar to (6.10):

$$\mathbb{E}\left[\langle \phi \rangle_{n,K} Z_{0,n}^K\right] = \mathbb{E}\left[\phi(x_n) \exp\left(\sum_{m=0}^{n-1} S_m(x_m)\Delta t\right)\right],$$

where x_0 is initially distributed according to the same probability distribution as the replica ensemble at step 0, and $x_n \mapsto x_{n+1}$ evolves according to the dynamics described by the kernel P_n.

The time discretizations of Algorithms 6.3 and 6.4 are now straightforwardly obtained, leading respectively to Algorithms 6.6 and 6.8.

Algorithm 6.6 (Discrete-time version of ensemble resampling). *Choose a degeneracy threshold $\alpha > 0$. Generate an initial distribution of replicas $(x_0^{1,K}, \dots, x_0^{K,K})$, sampled independently from a probability distribution π_0, each replica starting with the uniform weight $w_0^{k,K} = 1/K$. Initialize the sequence of resampling times $\{T_p\}_{p\geq 0}$ at $T_0 = 0$, and the weight normalization $Z_{0,0}^K = 1$. Iterate on $n \geq 0$,*

- *evolve each replica according to the underlying time-discrete dynamics: $x_{n+1}^{k,K} \sim P_n(x_n^{k,K}, \cdot)$;*
- *update the corresponding weights as*

$$W_{n+1}^{k,K} = W_n^{k,K} \exp\Big(S_n\left(x_n^{k,K}\right)\Delta t\Big), \qquad w_{n+1}^{k,K} := \frac{W_{n+1}^{k,K}}{\sum_{l=0}^{K} W_{n+1}^{k,K}};$$

- *update the weight normalization according to* (6.22), i.e.*:*

$$Z_{0,n+1}^K = Z_{0,n}^K \left\langle \mathrm{e}^{S_n \Delta t} \right\rangle_{n,K};$$

- *if* $\mathrm{Ent}\left(w_{n+1}^{1,K}, \dots, w_{n+1}^{K,K}\right) \geq \alpha$*, resample replicas according to their respective weights, as done in step* (2) *of Algorithm* 6.3.

Remark 6.7 (Resampling at each time-step). *If $\alpha = 0$, resampling occurs at each iteration step $n \geq 0$, and $w_n^{k,K} = 1/K$ for all replicas and at any time index n.*

Algorithm 6.8 (Discrete-time version of individual resampling). *Generate an initial distribution of replicas $(x_0^{1,K}, \dots, x_0^{K,K})$, sampled independently from a probability distribution π_0, each replica having at all times the uniform weight $w_n^{k,K} = 1/K$. Generate independent jump times $\tau_0^{k,\mathrm{d}}$ and $\tau_0^{k,\mathrm{b}}$ $(1 \leq k \leq K)$ from an exponential law of mean* 1*, and initialize the weight normalization as $Z_{0,0}^K = 1$. Iterate on $n \geq 0$,*

- *evolve each replica according to the underlying time-discrete dynamics: $x_{n+1}^{k,K} \sim P_n(x_n^{k,K}, \cdot)$;*

- *update the birth and death times as*

$$\tau_{n+1}^{k,\mathrm{d}} = \tau_n^{k,\mathrm{d}} - \left[S_n\left(x_n^{k,K}\right) - \langle S_n \rangle_{n,K}\right]_- \Delta t,$$

$$\tau_{n+1}^{k,\mathrm{b}} = \tau_n^{k,\mathrm{b}} - \left[S_n\left(x_n^{k,K}\right) - \langle S_n \rangle_{n,K}\right]_+ \Delta t;$$

- *if* $\tau_{n+1}^{k,\mathrm{d}} \leq 0$, *generate a new death time* $\tau_{n+1}^{k,\mathrm{d}}$ *from an exponential law of mean* 1, *choose an index* $l \in \{1, \dots, K\}$ *at random, and replace the configuration of the* k*th replica by the configuration of the* l*th replica;*
- *if* $\tau_{n+1}^{k,\mathrm{b}} \leq 0$, *generate a new birth time* $\tau_{n+1}^{k,\mathrm{b}}$ *from an exponential law of mean* 1, *choose an index* $l \in \{1, \dots, K\}$ *at random, and replace the configuration of the* l*th replica by the configuration of the* k*th replica;*
- *update the weight normalization according to* (6.22), i.e.*:*

$$Z_{0,n+1}^K = Z_{0,n}^K \left\langle \mathrm{e}^{S_n \Delta t} \right\rangle_{n,K}.$$

The selection procedure given by Algorithm 6.6 verifies the following "unbiasedness condition" at the discrete time level.

Proposition 6.9. *Consider the weighted empirical distribution of replicas* π_n^K *given by* (6.21), *and constructed either from Algorithm 6.6 or without resampling. Then, for any observable* ϕ,

$$\mathbb{E}\left[\langle \phi \rangle_{n,K} Z_{0,n}^K\right] = \mathbb{E}\left[\phi(x_n) \exp\left(\sum_{m=0}^{n-1} S_m(x_m) ds\right)\right],$$

where the Markov chain (x_n) *evolves according to the kernel family* $(P_n)_{n \geq 0}$, *and* x_0 *is distributed according to the initial distribution of the replicas.*

The selection performed by Algorithm 6.8 verifies this property only up to time-step errors in the computation of the birth and death times.

6.1.4 *Numerical application*

We consider the Widom insertion problem, with the parameters of Section 2.4.1.1. A nonequilibrium switching with the schedule $\Lambda(t) = (t/T)^2$ is used as in Section 4.1.5.4, the initial conditions being obtained by subsampling a Langevin trajectory at $\lambda = 0$, with a time spacing be-

Table 6.1 Comparison of free energy differences obtained with the standard estimator (6.13) and the estimator (6.20) when selection is turned on. The standard error over 100 independent realizations of the switching process are reported in brackets. The reference value is $\mu^{\text{ex}} = 1.317$.

Method N_{replicas}	$T = 1$ $N = 100$	$T = 1$ $N = 1000$	$T = 4$ $N = 250$	$T = 4$ $N = 2500$	$T = 10$ $N = 100$	$T = 10$ $N = 1000$
Standard	1.56 (0.62)	1.34 (0.19)	1.34 (0.19)	1.32 (0.06)	1.33 (0.17)	1.31 (0.05)
Selection	1.45 (0.72)	1.43 (0.23)	1.31 (0.19)	1.32 (0.05)	1.32 (0.16)	1.31 (0.05)

tween two configurations being $T_{\text{sample}} = 0.5$. The selection is implemented with Algorithm 6.8, using the fitness function (6.3), and a (small) time-step $\Delta t = 5 \times 10^{-4}$ for the Langevin dynamics (in order to avoid time-step errors in the unbiasedness property, see the discussion after Proposition 6.9).

The selection procedure changes dramatically the work distributions, see Figure 6.1. When selection is used, the work distributions are approximately Gaussian for all switching rates considered, whereas the work distributions obtained through standard nonequilibrium methods are much wider, so that the relevant part of the work distribution (the lower tail) is only of small relative importance. Of course, the widths of both work distributions decrease as the transition is made slower, and the shapes of the plain nonequilibrium process work distribution come closer to a Gaussian shape.

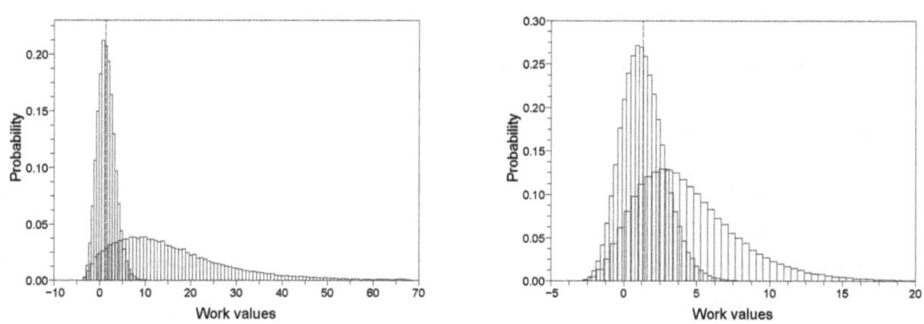

Fig. 6.1 Comparison of the work distribution for $T = 1$ (left) and $T = 4$ (right). In each case, the distribution with the smallest width is obtained with the selection procedure turned on. The vertical dashed line is the reference value of the free energy difference given in Section 2.4.1.1.

The free-energy estimates presented in Table 6.1 show that resorting to selection does not change much the quality of the free-energy estimator.

This may seem surprising since the work distributions are much better behaved when selection is turned on. This may be understood by noticing that resampling steps increase the variance of the estimators since they reduce the number of different configurations, and are therefore at the origin of some degeneracy in the distribution of replicas. There are therefore two competing effects, the variance increase when resampling, and the variance decrease arising from the lower weight degeneracy. This suggests that there is some optimal resampling criterion α to be chosen. See [El Makrini *et al.* (2007)] for a careful study of these competing effects.

Let us however end this section with a positive tone, by quoting also applications where the selection procedure enhances the quality of the estimators, such as the toy example in [Rousset and Stoltz (2006)].

6.2 Selection in adaptive methods

In this section, we present an application of selection methods to adaptive methods. In this case, there is no natural fitness or selection function at hand. We nonetheless propose to superimpose a jump process on the diffusive dynamics in order to enhance the exploration of the reaction coordinate values. As will be seen below, such a superimposed selection process is particularly interesting at the early stages of the simulation, when the reaction coordinate space has hardly been explored. It can also be switched off once it is no longer necessary.

6.2.1 *Motivation for the selection term*

6.2.1.1 *Description of the dynamics without selection*

We consider for simplicity the case of overdamped Langevin dynamics, and a one-dimensional reaction coordinate with values in $\mathbb{T}$. To obtain rigorous convergence results on the ABF dynamics, it was argued in Section 5.2 that some additional terms depending on $|\nabla\xi|$ should be added in the typical ABF dynamics (5.19), in order to obtain a simple diffusive behavior for the law of $\xi(q_t)$. This leads to (5.60)-(5.61):

$$\begin{cases} dq_t = -\nabla\Big(V - F_t \circ \xi - \beta^{-1}\ln(|\nabla\xi|^{-2})\Big)(q_t)\, |\nabla\xi|^{-2}(q_t)\, dt \\ \qquad\qquad + \sqrt{2\beta^{-1}}|\nabla\xi|^{-1}(q_t)\, dW_t, \\ F_t'(z) = \mathbb{E}\left(f(q_t)\,\Big|\, \xi(q_t) = z\right), \end{cases}$$

where the local mean force f is

$$f = \frac{\nabla V \cdot \nabla \xi}{|\nabla \xi|^2} - \beta^{-1} \mathrm{div}\left(\frac{\nabla \xi}{|\nabla \xi|^2}\right).$$

The adaptive dynamics is characterized by a nonlinear operator acting on the law $\psi(t,q)\,dq$ of the process, so that the evolution of the law of the process is governed by the nonlinear equation (5.64):

$$\begin{cases} \partial_t \psi = A(\psi) = \mathrm{div}\left(\dfrac{\nabla(V - F_t \circ \xi)\psi + \beta^{-1}\nabla\psi}{|\nabla\xi|^{-2}}\right), \\ F_t'(z) = \dfrac{\displaystyle\int_{\Sigma(z)} f\,|\nabla\xi|^{-1}\psi(t,\cdot)\,d\sigma_{\Sigma(z)}}{\displaystyle\int_{\Sigma(z)} |\nabla\xi|^{-1}\psi(t,\cdot)\,d\sigma_{\Sigma(z)}}. \end{cases} \tag{6.23}$$

It can then be shown (see Proposition 5.11) that the marginal distribution

$$\psi^\xi(t,z) = \int_{\Sigma(z)} |\nabla\xi(q)|^{-1}\,\psi(t,q)\,\sigma_{\Sigma(z)}(dq),$$

satisfies the diffusion equation

$$\partial_t \psi^\xi = \beta^{-1}\partial_{z,z}\psi^\xi. \tag{6.24}$$

The law of the process may be approximated by the empirical probability distribution generated by a multiple replica implementation (see for instance (5.35)):

$$\psi_K(t,q)\,dq = \frac{1}{K}\sum_{k=1}^{K} \delta_{q_t^{k,K}}(dq),$$

where the replicas $q_t^{k,K}$ evolve according to the underlying dynamics and interact only through their common biasing function. We refer to Section 5.1.3.1 for precise convergence results, which hold when some regularization of the empirical marginal is performed. Formally however, ψ_K is expected to satisfy (6.23) in the large sample size limit $K \to +\infty$.

6.2.1.2 *Motivation for the selection term*

It is shown in Theorem 5.15 that the rate of convergence of the dynamics is limited by the rate of convergence of the diffusion equation on the marginal when the convergence of the sampling at fixed values of the reaction coordinate is fast enough. A possible way to enhance the convergence rate is therefore to add some selection procedure to improve the exploration of the values of the reaction coordinate, for instance by duplicating replicas visiting underexplored regions. The selection process is completely characterized by the fitness function S_t.

The formalism of resampling methods as described in Section 6.1 still holds in the present context, except that no unbiasedness property such as (6.10) can be satisfied because of the nonlinear interaction introduced among replicas. Thus, the selection criterion may be motivated, from a mathematical perspective, only through the (formal) nonlinear evolution equation satisfied by the weighted or resampled replica ensemble density in the large sample size limit $K \to +\infty$ (see Proposition 6.2):

$$\partial_t \psi = A(\psi) + \left(S_t - \int_{\mathcal{D}} S_t(q)\, \psi(t,q)\, dq \right) \psi. \tag{6.25}$$

When the fitness function depends on $\xi(q)$ only:

$$S_t(q) = \mathcal{S}_t(\xi(q)),$$

it is easily shown that the evolution of the marginal law of ξ inferred from (6.25) is the diffusion equation (6.24) with some additional term:

$$\partial_t \psi^\xi = \beta^{-1} \partial_{z,z} \psi^\xi + \left(\mathcal{S}_t - \int_{\mathbb{T}} \mathcal{S}_t(z)\, \psi^\xi(t,z)\, dz \right) \psi^\xi.$$

A possible selection criterion is then

$$\mathcal{S}_t(z) = c(t)\, \frac{\partial_{z,z} \psi^\xi(t,z)}{\psi^\xi(t,z)}, \tag{6.26}$$

where $c(t) \geq 0$ has to be adjusted. With this choice, the evolution of the marginal density is

$$\partial_t \psi^\xi = \left(\beta^{-1} + c(t) \right) \partial_{z,z} \psi^\xi. \tag{6.27}$$

This comes from the fact that

$$\int_{\mathbb{T}} \mathcal{S}_t(z)\, \psi^\xi(t,z)\, dz = c(t) \int_{\mathbb{T}} \partial_{z,z} \psi^\xi(t,z)\, dz = 0.$$

The comparison with (6.24) shows that the diffusion is indeed enhanced in the reaction coordinate direction.

The fitness function favors replicas in the convex areas of the marginal density ψ_t^ξ. These areas correspond to the free energy barriers associated with ξ that the biasing algorithm still needs to overcome.

In practice, the number of replicas is finite, and a trade-off between exploration and selection has to be found. A way to determine a maximal selection intensity $c(t)$ is given by the requirement that the local exploration time of the underlying stochastic dynamics (related to the decorrelation time of some observables) should be small enough compared to the typical jump time in the selection procedure (for instance, when using the continuous-in-time resampling method in Algorithm 6.8, the average birth or death times). If this is not the case, the replica sample has no time to explore or change ("mutate", in the genetic optimization language), and the selection procedure reduces the total amount of information by enhancing a few replicas, therefore increasing the statistical error in the computation. This is best seen in numerical examples, see Section 6.2.2 below.

It may be desirable to decrease $c(t)$ as the exploration of the reaction coordinate values advances, in order to reduce the extra variance arising from resampling steps. A possible automatic procedure to this end is for instance to monitor the difference between the largest and the lowest values of the marginal distribution, and to decrease c each time the difference is smaller that some given threshold (similarly to how the parameter is set to zero in the original implementation of the Wang-Landau algorithm [Wang and Landau (2001a)], for which some elements are given in Section 5.1.4.2). The relative entropy of the marginal distribution with respect to the uniform distribution may also be a relevant indicator.

Remark 6.10 (Other selection criteria). *Since the aim is to increase the exploration of underexplored values of the reaction coordinate, many other selection criteria based on the (empirical) marginal distribution could be considered. For instance, the fitness function*

$$S_t(q) = c(t)\left(1 - \frac{\psi^\xi(t,\xi(q))}{\int_{\mathbb{T}} \psi^\xi(t,z)\,dz}\right)$$

compares the current value of the marginal density to the average value. If the current value is lower than the average, the fitness is positive, so that replicas visiting this region will be duplicated; while replicas staying in regions whose average visit time is already larger than the average visit time will be killed.

This method sounds reasonable, but, contrary to the choice (6.26), *there is no mathematical insight on whether it will really increase the convergence rate. Numerical tests would therefore be in order to find the best strategies.*

6.2.2 *Numerical application*

We present some numerical results for the dimer in a WCA solvent. The physical parameters of the system are the same as in Section 2.5.2.3. The ABF method is used with $N_{\text{replicas}} = 10^4$ replicas of the system evolved with an overdamped Langevin dynamics using $\Delta t = 0.00025$. The truncated reaction coordinate space $[\xi_{\min}, \xi_{\max}] = [-0.2, 1.2]$ is separated into $N_\xi = 50$ bins of equal size. The mean force is estimated in each bin by a combination of plain pathwise averages and averages over replicas, see Sections 5.1.3.1 and 5.1.3.4. All replicas are started in the compact state.

The marginal ψ^ξ in the selection term is approximated by the histogram of the current distribution of replicas, which is possible since N_{replicas} is much larger than N_ξ. We denote by $\{\psi_k^{\xi,n}\}_{1 \leq k \leq N_\xi}$ the current discrete distribution of replicas, obtained by representing the marginal distribution by a histogram. The numbers $\psi_k^{\xi,n}$ are then equal to the number of replicas whose values of ξ fall into the nth bin at the kth simulation step, divided by the total number of replicas. The fitness function (6.26) for a replica whose reaction coordinate value is in the kth bin is approximated with finite differences as

$$S_k^n = c\,\frac{\psi_{k+1}^{\xi,n} - 2\psi_k^{\xi,n} + \psi_{k-1}^{\xi,n}}{\psi_k^{\xi,n}},$$

where we have considered for simplicity a constant selection factor c. Note that

$$S_k^n = \frac{3c}{\psi_k^{\xi,n}} \left(\frac{\psi_{k+1}^{\xi,n} + \psi_k^{\xi,n} + \psi_{k-1}^{\xi,n}}{3} - \psi_k^{\xi,n} \right),$$

so that the selection term is proportional to the difference between the value of the marginal distribution in the considered bin, and its local average. The selection therefore favors a local averaging of the marginal distribution.

In Figure 6.2, we compare the proportions of stretched states as a function of time in the early stages of the dynamics, depending on the selection intensity (compare with Figure 5.4, which was however obtained with more replicas so that the proportion of stretched states increased faster than for the results obtained here in the case $c = 0$). The results show that the selection process improves the diffusion in the reaction coordinate space at the early stages of the algorithm, hence the convergence of the algorithm. Once the proportion of stretched states is roughly equal to 0.5, the selection procedure is not as interesting as at the beginning, and some increased variance can be read off from the wiggles in the curves. This suggests that selection should be turned off once the proportion is close to 0.5.

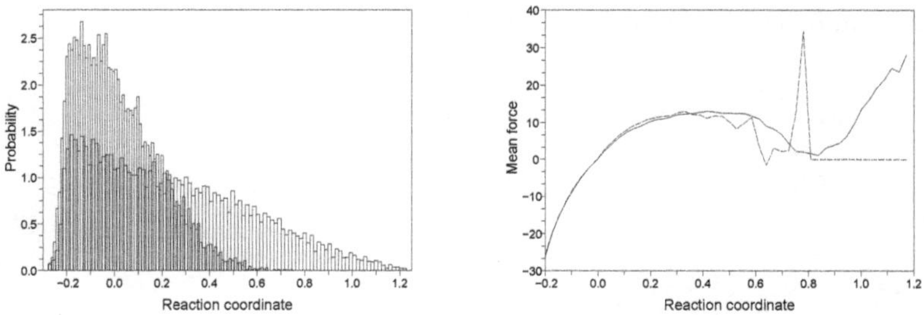

Fig. 6.2 Comparison of marginal distributions (Left) and mean forces (Right) at time $t = 0.15$ for dynamics with selection ($c = 1$) or without selection ($c = 0$). When selection is turned on, the marginal distribution is more spread out. The mean force is also smoother (solid line on the right picture), and estimates are available everywhere whereas the bias is still zero in some unvisited parts of the reaction coordinate space for dynamics without selection.

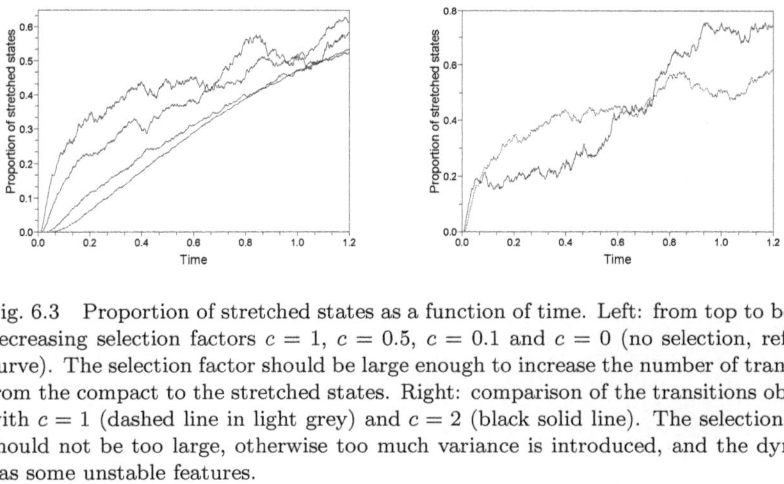

Fig. 6.3 Proportion of stretched states as a function of time. Left: from top to bottom, decreasing selection factors $c = 1$, $c = 0.5$, $c = 0.1$ and $c = 0$ (no selection, reference curve). The selection factor should be large enough to increase the number of transitions from the compact to the stretched states. Right: comparison of the transitions obtained with $c = 1$ (dashed line in light grey) and $c = 2$ (black solid line). The selection factor should not be too large, otherwise too much variance is introduced, and the dynamics has some unstable features.

The magnitude of the selection term has to be chosen carefully, see Figure 6.3. To be efficient, the selection coefficient c must be large enough, but not too large. This can be understood as a trade-off between exploration and selection. If too much selection is considered, the sample becomes degenerate, and the replicas concentrate in certain regions of the reaction coordinate space. In the specific example considered here, the optimal value of the selection factor is around $c = 1$.

When values $c \geq 5$ are considered, the selection is too high and the variance explodes. For those higher values of c, a much larger number of replicas should be simulated in order to obtain meaningful results.

Appendix A

Most important notation used throughout this book

A.1 General notation

- *Trace and determinant*: For a matrix A in $\mathbb{R}^{n\times n}$, $\boxed{\mathrm{tr}(A)}$ denotes the trace of A and $\boxed{\det(A)}$ the determinant of A.
- *Composition*: For two matrices $A \in \mathbb{R}^{n\times m}$ and $B \in \mathbb{R}^{m\times p}$, $AB \in \mathbb{R}^{n\times p}$ is the matrix with components $(\sum_{k=1}^{m} A_{i,k}B_{k,j})_{i,j}$. For two functions $\phi_1 : \mathbb{R}^m \to \mathbb{R}^n$ and $\phi_2 : \mathbb{R}^p \to \mathbb{R}^m$, $\boxed{\phi_1 \circ \phi_2}$ is the composition of ϕ_1 and ϕ_2: $\phi_1 \circ \phi_2(x) = \phi_1(\phi_2(x))$.
- *Frobenius product*: For two matrices A and B in $\mathbb{R}^{n\times m}$, $\boxed{A : B} = \mathrm{tr}(A^T B) = \sum_{i=1}^{n}\sum_{j=1}^{m} A_{i,j}B_{i,j}$.
- *Inverse*: For an invertible matrix $A \in \mathbb{R}^{n\times n}$, $\boxed{A^{-1}_{i,j}}$ denotes the (i,j)th component of the inverse matrix A^{-1}.
- *Transposition*: For a matrix $A \in \mathbb{R}^{n\times m}$, $\boxed{A^T,}$ denotes the transpose of A.
- *Identity matrix*: The identity matrix is the matrix $\boxed{\mathrm{Id}}$, with components $\boxed{\delta_{i,j}}$, where $\delta_{i,j}$ is the Kronecker symbol, equals to 1 if $i = j$ and 0 otherwise.
- *Tensor product*: For two vectors u and v in $\mathbb{R}^n$, $\boxed{u \otimes v}$ is the $n \times n$ matrix with components $(u_i\, v_j)$.
- *Partial differentiation*: For a function $\phi : \mathbb{R}^n \to \mathbb{R}$, $\boxed{\nabla_i \phi}$ or $\boxed{\partial_{x_i}\phi}$ denote the partial derivative $\dfrac{\partial \phi}{\partial x_i}$.
- *Gradient*: For a function $\phi : \mathbb{R}^n \to \mathbb{R}^m$, $\boxed{\nabla \phi}$ denotes the $n \times m$ matrix with components $\left(\dfrac{\partial \phi_j}{\partial x_i}\right)_{i,j}$, where $i \in \{1, \ldots, n\}$ and $j \in$

$\{1, \ldots, m\}$ (where $(\phi_j)_{1 \le j \le m}$ denotes the components of ϕ). [2] If ϕ is a function of two variables $(q,p) \in \mathbb{R}^n \times \mathbb{R}^m$, $\boxed{\nabla_q \phi} \in \mathbb{R}^n$ denotes the gradient in q, $\boxed{\nabla_p \phi} \in \mathbb{R}^m$ denotes the gradient in p, and $\boxed{\nabla \phi} \in \mathbb{R}^{n+m}$ the total gradient.

- *Divergence*: For a function $\phi : \mathbb{R}^n \to \mathbb{R}^n$, $\boxed{\mathrm{div}(\phi)} = \sum_{i=1}^n \partial_{x_i} \phi_i$.
- *Hessian*: For a function $\phi : \mathbb{R}^n \to \mathbb{R}$, $\boxed{\mathrm{Hess}_x \phi}$ denotes the bilinear form; for $u, v \in \mathbb{R}^n$, $\mathrm{Hess}_x \phi(u,v) = u^T \nabla^2 \phi(x) v$, where $\boxed{\nabla^2 \phi(x)}$ denotes the symmetric matrix with components $\left(\dfrac{\partial^2 \phi(x)}{\partial x_i \partial x_j} \right)_{i,j}$. If $\phi : \mathbb{R}^n \to \mathbb{R}^m$, this is generalized as: for $u, v \in \mathbb{R}^n$, $\mathrm{Hess}_x \phi(u,v) = \left(u^T \nabla^2 \phi_1(x) v, \ldots, u^T \nabla^2 \phi_m(x) v\right) \in \mathbb{R}^m$.
- *Expectations*: For a probability measure μ, a random variable X and a function ϕ, $\boxed{\mathbb{E}_\mu(\phi(X))}$ denotes the expectation of the random variable $\phi(X)$, when the law of X is μ. For a probability measure μ, a Markov chain $(X_n)_{n \ge 0}$ (or a Markov process $(X_t)_{t \ge 0}$) and a functional Φ, $\boxed{\mathbb{E}_\mu(\Phi((X_n)_{n \ge 0}))}$ (resp. $\boxed{\mathbb{E}_\mu(\Phi((X_t)_{t \ge 0}))}$) denotes the expectation of the random variable $\Phi((X_n)_{n \ge 0})$ (resp. $\Phi((X_t)_{t \ge 0})$), when the law of the initial condition X_0 is μ.
- *Functional spaces and norms*: $\boxed{\mathrm{C}^1_c(\mathbb{R}^n)}$ (resp. $\boxed{\mathrm{C}^\infty_c(\mathbb{R}^n)}$) denotes continuously differentiable (resp. infinitely continuously differentiable) scalar functions defined on $\mathbb{R}^n$, with compact support. For a real $p > 1$, the functional space $\boxed{L^p(\mathbb{R}^n)}$ contains scalar functions ϕ defined on $\mathbb{R}^n$ such that $\int_{\mathbb{R}^n} |\phi|^p < \infty$. It is equipped with the norm $\|\phi\|_{L^p} = \left(\int_{\mathbb{R}^n} |\phi|^p\right)^{1/p}$. The functional space $\boxed{L^\infty(\mathbb{R}^n)}$ contains scalar functions ϕ defined on $\mathbb{R}^n$ such that $\sup_{x \in \mathbb{R}^n} |\phi(x)| < \infty$. It is equipped with the norm $\|\phi\|_{L^\infty} = \sup_{x \in \mathbb{R}^n} |f(x)|$.
- *Total variation*: For two probability measures π_1 and π_2, $\boxed{\|\pi_1 - \pi_2\|_{\mathrm{TV}}}$ denotes the total variation of the signed measure $\pi_1 - \pi_2$:

$$\|\pi_1 - \pi_2\|_{\mathrm{TV}} = \sup \left\{ \left| \int \phi \, d\pi_1 - \int \phi \, d\pi_2 \right|, \ \phi \text{ measurable}, \ |\phi| \le 1 \right\}.$$

[2] This convention for the definition of the gradient implies that for a scalar function ϕ, $\nabla \phi$ is a column vector. It is consistent with the formula: for any $x, \delta x \in \mathbb{R}^n$, $(\nabla \phi(q))^T \delta q = \lim_{\varepsilon \to 0} \frac{\phi(x + \varepsilon \delta x) - \phi(x)}{\varepsilon}$. This gradient is also sometimes denoted by $\nabla \otimes \phi$.

In cases π_1 (resp. π_2) admits a density ψ_1 (resp. ψ_2) with respect to the Lebesgue measure, $\|\pi_1 - \pi_2\|_{\rm TV} = \|\psi_1 - \psi_2\|_{L^1}$.

- *Entropy and Ficher information*: Let π_1 and π_2 be two probability measures. The notation $\boxed{\pi_1 \ll \pi_2}$ indicates that π_1 is absolutely continuous (*i.e.* admits a density) with respect to π_2. The relative entropy of π_1 with respect to π_2 is $\boxed{H(\pi_1|\pi_2)} = \int \ln\left(\frac{d\pi_1}{d\pi_2}\right) d\pi_1$.
The Fisher information of π_1 with respect to π_2 is $\boxed{I(\pi_1|\pi_2)} = \int \left|\nabla \ln\left(\frac{d\pi_1}{d\pi_2}\right)\right|^2 d\pi_1$.

A.2 Physical spaces and energies

- *Positions and momenta*: The positions of the $\boxed{N}$ particles considered are denoted by $\boxed{q}$ and vary in a domain $\boxed{\mathcal{D}} \subset \mathbb{R}^{3N}$. For periodic boundary conditions, $\mathcal{D} \subset \mathbb{T}^{3N}$, where $\mathbb{T} = \mathbb{R}/\mathbb{Z}$ is the one-dimensional torus. The dimension of $\mathcal{D}$ is also denoted by $n = 3N$. The momenta of the N particles are denoted by $\boxed{p} \in \mathbb{R}^{3N}$. The cotangent space to $\mathcal{D}$, where couples (q,p) live, is $\boxed{T^*\mathcal{D}} = \mathcal{D} \times \mathbb{R}^{3N}$. The velocity vector is $\boxed{v} = M^{-1}p$.
- *Hamiltonian*: The Hamiltonian, assumed to be separable, is denoted $\boxed{H(q,p)} = \frac{1}{2}p^T M^{-1} p + V(q)$, where $\boxed{M} \in \mathbb{R}^{3N \times 3N}$ is the (symmetric, positive) mass matrix (supposed to be constant in this monograph) and $\boxed{V} : \mathcal{D} \to \mathbb{R}$ is the potential energy. The kinetic part is denoted $\boxed{E_{\rm kin}(p)} = \frac{1}{2}p^T M^{-1}p$. In the so-called alchemical case, the potential energy (and thus the Hamiltonian) depends on an external parameter λ: $H_\lambda(q,p) = \frac{1}{2}p^T M^{-1} p + V_\lambda(q)$.
- *Reaction coordinate*: The reaction coordinate is a smooth function $\boxed{\xi}$ of the positions $q \in \mathcal{D}$, with values in $\boxed{\mathcal{M}}$, where $\mathcal{M} \subset \mathbb{R}^m$ or $\mathcal{M} \subset \mathbb{T}^m$, and $m < 3N$. A free variable in $\mathcal{M}$ is typically denoted by z.
- *Summation convention on repeated indices*: Latin indices $i, j, k, \ldots$ refer to Cartesian coordinates and vary between 1 and $n = 3N$. Greek Indices $\alpha, \beta, \gamma, \ldots$ refer to position constraints and vary between 1 and m. Latin indices $a, b, c, \ldots$ refer to position and momenta constraints, and vary between 1 and $2m$. For the ease of notation, in some proofs, the summation convention on repeated indices is used.

- *Momentum reversal, backward and time-reversed flow*: The momentum reversal function is denoted $\boxed{S(q,p)} = (q,-p)$ and the momentum reversal operator is denoted $\boxed{\mathcal{R}(\phi)} = \phi \circ S$, where ϕ is any function defined on $T^*\mathcal{D}$. For a given time-homogeneous flow $\phi_t : T^*\mathcal{D} \to T^*\mathcal{D}$ associated to a phase-space dynamics, the backward flow is $S \circ \phi_t \circ S$, and the time-reversed flow is ϕ_{-t}. The flow is symmetric if $\phi_{-t} = \phi_t^{-1}$ and time-reversible if $S \circ \phi_t \circ S = \phi_{-t}$.

A.3 Spaces with constraints, projection operators

- *Submanifolds*: The space of positions at a fixed value of the reaction coordinate is $\boxed{\Sigma(z)} = \{q \in \mathcal{D},\, \xi(q) = z\}$, where $z \in \mathcal{M}$. The tangent (resp. cotangent) space to $\Sigma(z)$ is $\boxed{T\Sigma(z)}$ (resp. $\boxed{T^*\Sigma(z)}$) and the tangent (resp. cotangent space) to $\Sigma(z)$ at a fixed position $q \in \Sigma(z)$ is denoted $\boxed{T_q\Sigma(z)}$ (resp. $\boxed{T_q^*\Sigma(z)}$). For a Hamiltonian dynamics constrained to a fixed value z of the reaction coordinate, the position-velocity pairs (q,v) are in the tangent space $T\Sigma(z)$, and the position-momentum pairs (q,p) are in the cotangent space $T^*\Sigma(z)$.
- *Projection operators*: The (symmetric) $m \times m$ Gram matrix associated to ξ is denoted $\boxed{G = (\nabla\xi)^T\nabla\xi}$. Componentwise, it reads $G_{\alpha,\gamma} = \nabla\xi_\alpha \cdot \nabla\xi_\gamma$, for $\alpha, \gamma \in \{1, \ldots, m\}$. At a given position $q \in \mathcal{D}$, the orthogonal projection operator onto the tangent space $T_q\Sigma(\xi(q))$ is $\boxed{P(q)} = \mathrm{Id} - \nabla\xi(q)G^{-1}(q)\nabla\xi(q)^T$. More explicitly, $\nabla\xi(q)G^{-1}(q)\nabla\xi(q)^T = \sum_{\alpha,\gamma=1}^m G_{\alpha,\gamma}^{-1}(q)\nabla\xi_\alpha(q) \otimes \nabla\xi_\gamma(q)$.
- *More on projection operators*: In Sections 3.3 and 4.3.2, we will need projection operators orthogonal for the scalar product (in the cotangent space) induced by M^{-1}, namely $\langle p, \tilde{p}\rangle_{M^{-1}} = p^T M^{-1}\tilde{p}$. Let us first introduce the mass dependent Gram matrix: $\boxed{G_M} = \nabla\xi^T M^{-1}\nabla\xi$. Then, the projection operator onto the cotangent space $T_q^*\Sigma(\xi(q))$ which is orthogonal for the scalar product $\langle\cdot,\cdot\rangle_{M^{-1}}$ is $\boxed{P^M(q)} = \mathrm{Id} - \nabla\xi(q)G_M^{-1}(q)\nabla\xi(q)^T M^{-1}$.
- *More on submanifolds*: In Sections 3.3 and 4.3.2, more notation for spaces with constraints is needed. The effective momentum (resp. velocity) associated with ξ is $\boxed{p_\xi(q,p)} = G_M^{-1}(q)\nabla\xi(q)^T M^{-1}p$ (resp. $\boxed{v_\xi(q,p)} = \nabla\xi(q)^T M^{-1}p$). The

submanifold associated with constraints on (ξ, p_ξ) is denoted $\boxed{\Sigma_{\xi,p_\xi}(z,p_z)} = \{(q,p) \in T^*\mathcal{D},\ (\xi(q), p_\xi(q,p)) = (z,p_z)\}$. Likewise, the submanifold associated with constraints on (ξ, v_ξ) is denoted $\boxed{\Sigma_{\xi,v_\xi}(z,v_z)} = \{(q,p) \in T^*\mathcal{D},\ (\xi(q), v_\xi(q,p)) = (z,v_z)\}$. Thus, $T^*\Sigma(z) = \Sigma_{\xi,p_\xi}(z,0) = \Sigma_{\xi,v_\xi}(z,0)$.

- *Symplecticity and Poisson bracket*: The symplectic matrix is $\boxed{J} = \begin{bmatrix} 0 & \mathrm{Id}_{3N} \\ -\mathrm{Id}_{3N} & 0 \end{bmatrix}$. It defines the symplectic 2-form $\omega(u,v) = u^T J v$, where $u, v \in \mathbb{R}^{6N}$. More generally, $\boxed{J_{2d\times 2d}} = \begin{bmatrix} 0 & \mathrm{Id}_d \\ -\mathrm{Id}_d & 0 \end{bmatrix}$ denotes the symplectic matrix in dimension $2d$. For two scalar smooth functions ϕ_1 and ϕ_2 defined on $T^*\mathcal{D}$, the Poisson bracket is denoted: $\boxed{\{\phi_1, \phi_2\}} = (\nabla\phi_1)^T J \nabla\phi_2 = (\nabla_q\phi_1)^T \nabla_p\phi_2 - (\nabla_p\phi_1)^T \nabla_q\phi_2$. If $\phi_1 : T^*\mathcal{D} \to \mathbb{R}^n$ and $\phi_2 : T^*\mathcal{D} \to \mathbb{R}^m$, this is generalized as $\mathbb{R}^{n\times m} \ni \{\phi_1, \phi_2\} = (\nabla\phi_1)^T J \nabla\phi_2$.
- *More on symplecticity and Poisson bracket*: In Sections 3.3 and 4.3.2, we will need a generalization of the Poisson bracket to the submanifolds $\Sigma_{\xi,p_\xi}(z,p_z)$ and $\Sigma_{\xi,v_\xi}(z,v_z)$. Let us denote $\Xi : \mathcal{D} \to \mathbb{R}^{2m}$ one of the two generalized constraints: $\boxed{\Xi = (\xi, p_\xi) \text{ or } \Xi = (\xi, v_\xi)}$. The free variable in $\mathbb{R}^{2m}$ is denoted by ζ ($\zeta = (z, p_z)$ or $\zeta = (z, v_z)$). The (skew-symmetric) Gram matrix associated with Ξ is $\boxed{\Gamma} = \nabla\Xi^T J \nabla\Xi$. The constrained (skew-symmetric) symplectic matrix is $\boxed{J_\Xi} = J - J\,\nabla\Xi\,\Gamma^{-1}\,\nabla\Xi^T\,J$. The Poisson bracket associated with constraints on Ξ is: For two scalar smooth functions ϕ_1 and ϕ_2 defined on $\mathbb{R}^{6N}$, $\boxed{\{\phi_1, \phi_2\}_\Xi} = \nabla\phi_1^T J_\Xi \nabla\phi_2 = \{\phi_1, \phi_2\} - \{\phi_1, \Xi\}\Gamma^{-1}\{\Xi, \phi_2\}$. The Poisson bracket $\{\phi_1, \phi_2\}_\Xi$ on $\Sigma_\Xi(\zeta)$ actually depends only on the values of ϕ_1 and ϕ_2 on $\Sigma_\Xi(\zeta)$.
- *Surface gradient and divergence*: The surface gradient on $\Sigma(z)$ is denoted $\boxed{\nabla_\Sigma\phi} = P\nabla\phi$ (where ϕ is a given smooth scalar function defined on $\mathcal{D}$). The surface divergence on $\Sigma(z)$ is denoted $\boxed{\mathrm{div}_\Sigma\Phi} = \mathrm{tr}(P\nabla\Phi) = \sum_{i,j=1}^{3N} P_{i,j}\nabla_j\Phi_i$ (where Φ is a given smooth function defined on $\mathcal{D}$ with values in $\mathbb{R}^{3N}$). Notice that ∇_Σ and div_Σ are defined independently of z. Moreover, $\nabla_\Sigma\phi(q)$ (resp. $\mathrm{div}_\Sigma\Phi(q)$) depends only on the values of ϕ (resp. Φ) on $\Sigma(\xi(q))$.

- *Mean curvature vector*: The mean curvature vector of $\Sigma(z)$ is denoted $\boxed{\mathcal{H}}$. The mean curvature vector is defined on $\Sigma(z)$, with values in $\mathbb{R}^n$.

A.4 Measures

- *Canonical measure*: The canonical probability measure is denoted $\boxed{\mu(dq\,dp)} = Z_\mu^{-1} \exp(-\beta H(q,p))\,dq\,dp$, where $\boxed{\beta} = \frac{1}{k_B T}$ is the inverse temperature. It writes $\mu(dq\,dp) = \nu(dq)\kappa(dp)$ where $\boxed{\nu(dq)} = Z_\nu^{-1} \exp(-\beta V(q))\,dq$ (resp. $\boxed{\kappa(dp)} = Z_\kappa^{-1} \exp(-\beta p^T M^{-1} p/2)\,dp$) is the marginal of μ in q (resp. in p). For a parametrized potential V_λ (and thus Hamiltonian H_λ), the associated canonical probability measures are denoted by μ_λ and ν_λ.
- *Conditional measure*: The notation $\boxed{\delta_{\xi(q)-z}(dq)}$ indicates a conditional measure in q, with respect to $\xi(q) = z$. It is defined by $dq = \delta_{\xi(q)-z}(dq)\,dz$ (where dq in the left-hand side denotes the Lebesgue measure in $\mathbb{R}^n$). We also use a similar notation to denote a Dirac mass: $\boxed{\delta_{q_0}(dq)}$ is a Dirac mass in q at point $q_0 \in \mathcal{D}$.[3]
- *More on conditional measures*: In Sections 3.3 and 4.3.2, the notation $\boxed{\delta_{\xi(q)-z,p_\xi(q,p)-p_z}(dq\,dp)}$ (resp. $\boxed{\delta_{\xi(q)-z,v_\xi(q,p)-v_z}(dq\,dp)}$) indicates a conditional measure in (q,p) associated to the constraints $(\xi(q), p_\xi(q,p)) = (z, p_z)$ (resp. $(\xi(q), v_\xi(q,p)) = (z, v_z)$). It is defined by $dq\,dp = \delta_{\xi(q)-z,p_\xi(q,p)-p_z}(dq\,dp)\,dz\,dp_z$ (resp. $dq\,dp = \delta_{\xi(q)-z,v_\xi(q,p)-v_z}(dq\,dp)\,dz\,dv_z$) (where $dq\,dp$ on the left-hand side denotes the Lebesgue measure in $\mathbb{R}^n \times \mathbb{R}^n$). Notice that $\delta_{\xi(q)-z,p_\xi(q,p)-p_z}(dq\,dp) = \delta_{p_\xi(q,p)-p_z}(dp)\,\delta_{\xi(q)-z}(dq)$ (and $\delta_{\xi(q)-z,v_\xi(q,p)-v_z}(dq\,dp) = \delta_{v_\xi(q,p)-v_z}(dp)\,\delta_{\xi(q)-z}(dq)$).
- *Marginal and conditional canonical measures*: The image of μ (resp. ν) by ξ is denoted $\boxed{\mu^\xi(dz)} = \left(Z_\mu^{-1} \int_{\Sigma(z)\times\mathbb{R}^{3N}} \exp(-\beta H(q,p))\delta_{\xi(q)-z}(dq)dp\right) dz$ (resp. $\boxed{\nu^\xi(dz)} = \left(Z_\nu^{-1} \int_{\Sigma(z)} \exp(-\beta V(q))\delta_{\xi(q)-z}(dq)\right) dz$). Notice that, since the mass matrix does not depend on the position, $\mu^\xi = \nu^\xi$. This measure is also called the marginal of μ (resp. ν) in ξ. The probability measure μ (resp. ν) conditional to $\xi(q) = z$ is

[3]Notice that $\delta_{q_0}(dq) = \delta q - q_0(dq)$.

denoted $\boxed{\mu^{\xi}(dq\,dp|z)} = \dfrac{\exp(-\beta H(q,p))\,\delta_{\xi(q)-z}(dq)\,dp}{\displaystyle\int_{\Sigma(z)\times\mathbb{R}^{3N}} \exp(-\beta H(q,p))\,\delta_{\xi(q)-z}(dq)\,dp}$

(resp. $\boxed{\nu^{\xi}(dq|z)} = \dfrac{\exp(-\beta V(q))\,\delta_{\xi(q)-z}(dq)}{\displaystyle\int_{\Sigma(z)} \exp(-\beta V(q))\,\delta_{\xi(q)-z}(dq)}$).

- *Surface measures*: The surface measure $\boxed{\sigma_{\Sigma(z)}(dq)}$ on the submanifold $\Sigma(z)$ is the measure on $\Sigma(z)$ induced by the Lebesgue measure in the ambient space $\mathbb{R}^n$ and the standard scalar product: $\sigma_{\Sigma(z)}(dq) = (\det G)^{1/2}\delta_{\xi(q)-z}(dq)$. If $\mathbb{R}^n$ is equipped with the scalar product $\langle q, \tilde{q}\rangle_M = q^T M \tilde{q}$, the surface measure is denoted $\boxed{\sigma^M_{\Sigma(z)}(dq)} = (\det M)^{1/2}(\det G_M)^{1/2}\delta_{\xi(q)-z}(dq)$.
- *More on surface measures*: In Sections 3.3 and 4.3.2, other surface measures are needed. The surface measure $\boxed{\sigma_{T^*\Sigma(z)}(dq\,dp)}$ on $T^*\Sigma(z)$ is the phase-space measure (also called the symplectic volume measure or the Liouville measure). More generally, the phase-space measure on $\Sigma_{\xi,p_\xi}(z,p_z)$ (resp. on $\Sigma_{\xi,v_\xi}(z,v_z)$) is denoted by $\boxed{\sigma_{\Sigma_{\xi,p_\xi}(z,p_z)}(dq\,dp)}$ (resp. $\boxed{\sigma_{\Sigma_{\xi,v_\xi}(z,v_z)}(dq\,dp)}$). Notice that $\sigma_{\Sigma_{\xi,p_\xi}(z,p_z)}(dq\,dp) = \delta_{\xi(q)-z,p_\xi(q,p)-p_z}(dq\,dp)$ and $\sigma_{\Sigma_{\xi,v_\xi}(z,v_z)}(dq\,dp) = \det(G_M)\delta_{\xi(q)-z,v_\xi(q,p)-v_z}(dq\,dp)$. Therefore, $\sigma_{T^*\Sigma(z)}(dq\,dp) = \sigma_{\Sigma_{\xi,p_\xi}(z,0)}(dq\,dp) = \sigma_{\Sigma_{\xi,v_\xi}(z,0)}(dq\,dp) = \delta_{\xi(q)-z,p_\xi(q,p)}(dq\,dp) = \det(G_M)\delta_{\xi(q)-z,v_\xi(q,p)}(dq\,dp)$. In terms of marginals, $\sigma_{\Sigma_{\xi,p_\xi}(z,p_z)}(dq\,dp) = \sigma^{M^{-1}}_{\Sigma_{p_\xi(q,\cdot)}(p_z)}(dp)\sigma^M_{\Sigma(z)}(dq)$ and $\sigma_{\Sigma_{\xi,v_\xi}(z,v_z)}(dq\,dp) = \sigma^{M^{-1}}_{\Sigma_{v_\xi(q,\cdot)}(v_z)}(dp)\sigma^M_{\Sigma(z)}(dq)$. In the special case $p_z = v_z = 0$, this writes $\sigma_{T^*\Sigma(z)}(dq\,dp) = \sigma^{M^{-1}}_{T^*_q\Sigma(z)}(dp)\sigma^M_{\Sigma(z)}(dq)$, where $\boxed{\sigma^{M^{-1}}_{T^*_q\Sigma(z)}(dp)} = \sigma^{M^{-1}}_{\Sigma_{p_\xi(q,\cdot)}(0)}(dp) = \sigma^{M^{-1}}_{\Sigma_{v_\xi(q,\cdot)}(0)}(dp)$.
- *Canonical measures on submanifolds*: The canonical measure on $T^*\Sigma(z)$ is $\boxed{\mu_{T^*\Sigma(z)}(dq\,dp)} = \dfrac{\exp(-\beta H(q,p))\sigma_{T^*\Sigma(z)}(dq\,dp)}{\displaystyle\int_{T^*\Sigma(z)} \exp(-\beta H(q,p))\sigma_{T^*\Sigma(z)}(dq\,dp)}$. It is also the conditional probability measure $\mu_{T^*\Sigma(z)}(dq\,dp) = \mu^{\xi,p_\xi}(dq\,dp|(z,0))$. In terms of marginals, $\mu_{T^*\Sigma(z)}(dq\,dp) = \kappa^{M^{-1}}_{T^*_q\Sigma(z)}(dp)\,\nu^M_{\Sigma(z)}(dq)$. The marginal in q is $\boxed{\nu^M_{\Sigma(z)}(dq)} =$

$\dfrac{\exp(-\beta V(q))\,\sigma^M_{\Sigma(z)}(dq)}{\displaystyle\int_{\Sigma(z)} \exp(-\beta V(q))\,\sigma^M_{\Sigma(z)}(dq)}$. The marginal in p is supported by $T^*q\Sigma(z)$ and writes $\boxed{\kappa^{M^{-1}}_{T^*_q\Sigma(z)}(dp)} = \left(\dfrac{\beta}{2\pi}\right)^{(3N-m)/2} \times \exp\left(-\beta\frac{p^T M^{-1} p}{2}\right)\sigma^{M^{-1}}_{T^*_q\Sigma(z)}(dp)$. In case $M = \mathrm{Id}$, we denote $\boxed{\nu_{\Sigma(z)}(dq)} = \dfrac{\exp(-\beta V(q))\,\sigma_{\Sigma(z)}(dq)}{\displaystyle\int_{\Sigma(z)} \exp(-\beta V(q))\,\sigma_{\Sigma(z)}(dq)}$.

A.5 Free energy

- *Free energy*: The free energy in the alchemical case (*i.e.* for a parametrized Hamiltonian H_λ) is denoted $\boxed{F(\lambda)} = -\beta^{-1}\ln\left(\displaystyle\int_{T^*\mathcal{D}} \exp(-\beta H_\lambda(q,p))\,dq\,dp\right)$. The free energy associated to the reaction coordinate ξ is denoted $\boxed{F(z)} = -\beta^{-1}\ln\left(\displaystyle\int_{\Sigma(z)\times\mathbb{R}^{3N}} \exp(-\beta H(q,p))\,\delta_{\xi(q)-z}(dq)\,dp\right)$. The free energy is defined up to an additive constant. The free energy differences are denoted $\boxed{\Delta F(\lambda)} = F(\lambda) - F(0)$ or $\boxed{\Delta F(z)} = F(z) - F(0)$. Sometimes, in order to simplify the notation, we simply write $\boxed{\Delta F}$ instead of $\Delta F(1) = F(1) - F(0)$.
- *Rigid free energy*: In the reaction coordinate case, the free energy rewrites (up to an additive constant) $F(z) = -\beta^{-1}\ln\left(\displaystyle\int_{\Sigma(z)} \exp(-\beta V(q))\,\delta_{\xi(q)-z}(dq)\right) = -\beta^{-1}\ln\left(\displaystyle\int_{\Sigma(z)} \exp(-\beta V^\xi(q))\,\sigma_{\Sigma(z)}(dq)\right)$, where $\boxed{V^\xi} = V + \dfrac{\ln(\det G)}{2\beta}$. The rigid free energy is defined as $\boxed{F_{\mathrm{rgd}}(z)} = -\beta^{-1}\ln\left(\displaystyle\int_{\Sigma(z)} \exp(-\beta V(q))\,\sigma_{\Sigma(z)}(dq)\right)$.
- *Local mean force*: In the reaction coordinate case, the mean force is the derivative of F with respect to z, and it writes $\nabla F(z) =$

$\int_{\Sigma(z)} f(q)\,\nu^\xi(dq|z)$ where f is the local mean force with components $\boxed{f_\alpha} = \sum_{\gamma=1}^m G^{-1}_{\alpha,\gamma} \nabla\xi_\gamma \cdot \nabla V - \beta^{-1}\mathrm{div}\left(\sum_{\gamma=1}^m G^{-1}_{\alpha,\gamma}\nabla\xi_\gamma\right) = \sum_{\gamma=1}^m G^{-1}_{\alpha,\gamma}\nabla\xi_\gamma \cdot (\nabla v_\xi + \beta^{-1}\mathcal{H})$, for $\alpha \in \{1,\dots,m\}$. Likewise, $\nabla F_{\mathrm{rgd}}(z) = \int_{\Sigma(z)} f_{\mathrm{rgd}}\, d\nu_{\Sigma(z)}$ where $\boxed{f_{\mathrm{rgd},\alpha}} = \sum_{\gamma=1}^m G^{-1}_{\alpha,\gamma}\nabla\xi_\gamma \cdot (\nabla V + \beta^{-1}\mathcal{H})$, for $\alpha \in \{1,\dots,m\}$.

- *More on free energies and local mean forces*: In Sections 3.3 and 4.3.2, we will need additional free energies. First, the rigid free energy is generalized as $\boxed{F^M_{\mathrm{rgd}}(z)} = -\beta^{-1}\ln\left(\int_{\Sigma(z)} \exp(-\beta V(q))\sigma^M_{\Sigma(z)}(dq)\right)$. Moreover, free energies associated with the generalized constraints Ξ ($\Xi = (\xi, p_\xi)$ or $\Xi = (\xi, v_\xi)$) are defined as $\boxed{F^\Xi_{\mathrm{rgd}}(\zeta)} = -\beta^{-1}\ln\left(\int_{\Sigma_\Xi(\zeta)} \exp(-\beta H(q,p))\,\sigma_{\Sigma_\Xi(\zeta)}(dq\,dp)\right)$. Notice that $F^M_{\mathrm{rgd}}(z) = F^{\xi,p_\xi}_{\mathrm{rgd}}(z,0) = F^{\xi,v_\xi}_{\mathrm{rgd}}(z,0)$. The rigid mean force can be expressed as $\nabla F^M_{\mathrm{rgd}}(z) = \int_{T^*\Sigma(z)} f^M_{\mathrm{rgd}} d\mu_{T^*\Sigma(z)}$, where, for $(q,p) \in T^*\Sigma(z)$, $\boxed{f^M_{\mathrm{rgd}}(q,p)} = G^{-1}_M(q)\nabla\xi(q)^T M^{-1}\nabla V(q) - G^{-1}_M(q)\mathrm{Hess}_q(\xi)(M^{-1}p, M^{-1}p) \in \mathbb{R}^m$. For free energies associated with the generalized constraints Ξ, the mean force is $\nabla F^\Xi_{\mathrm{rgd}}(\zeta) = \dfrac{\int \begin{pmatrix} f^\Xi_{\mathrm{rgd}} \\ g^\Xi_{\mathrm{rgd}} \end{pmatrix} \exp(-\beta H) d\sigma_{\Sigma_\Xi(\zeta)}}{\int \exp(-\beta H) d\sigma_{\Sigma_\Xi(\zeta)}}$ where $\boxed{\begin{pmatrix} f^\Xi_{\mathrm{rgd}} \\ g^\Xi_{\mathrm{rgd}} \end{pmatrix}} = \Gamma^{-1}\{\Xi, H\}$. When (q,p) are such that $p_\xi(q,p) = v_\xi(q,p) = 0$, then $g^\Xi_{\mathrm{rgd}} = 0$ and $f^\Xi_{\mathrm{rgd}} = f^M_{\mathrm{rgd}}$.

Bibliography

Abraham, R. and Marsden, J. E. (1978). *Foundations of mechanics* (Benjamin/Cummings Publishing Co. Inc. Advanced Book Program).

Adjanor, G., Athènes, M. and Calvo, F. (2006). Free energy landscape from path-sampling: Application to the structural transition in LJ_{38}, *Eur. Phys. J. B* **53**, 1, pp. 47–60.

Akhmatskaya, E., Bou-Rabee, N. and Reich, S. (2009). A comparison of generalized Hybrid Monte-Carlo methods with and without momentum flip, *J. Comput. Phys.* **228**, 6, pp. 2256–2265.

Alder, B. J. and Wainwright, W. T. (1956). Molecular dynamics by electronic computers, in I. Prigogine (ed.), *Proc. of the Int. Symp. on Statistical Mechanical Theory of Transport Processes (Brussels, 1956)* (Interscience, Wiley), pp. 97–131.

Ambrosio, L., Fusco, N. and Pallara, D. (2000). *Functions of Bounded Variation and Free Discontinuity Problems* (Oxford Science Publications).

Ambrosio, L. and Soner, H. (1996a). Flow by mean curvature of surfaces of any codimension, in *Variational methods for discontinuous structures (Como, 1994), Progr. Nonlinear Differential Equations Appl.*, Vol. 25 (Birkhäuser), pp. 123–134.

Ambrosio, L. and Soner, H. (1996b). Level set approach to mean curvature flow in arbitrary codimension, *J. Differential Geom.* **43**, 4, pp. 693–737.

Ané, C., Blachère, S., Chafaï, D., Fougères, P., Gentil, I., Malrieu, F., Roberto, C. and Scheffer, G. (2000). *Sur les ingalités de Sobolev logarithmiques* (Société Mathématique de France), in French.

Arnold, A., Markowich, P., Toscani, G. and Unterreiter, A. (2001). On convex Sobolev inequalities and the rate of convergence to equilibrium for Fokker-Planck type equations, *Comm. Part. Diff. Eq.* **26**, pp. 43–100.

Arnol'd, V. I. (1989). *Mathematical methods of classical mechanics, Graduate Texts in Mathematics*, Vol. 60 (Springer-Verlag).

Assaraf, R., Caffarel, M. and Khelif, A. (2000). Diffusion Monte Carlo with a fixed number of walkers, *Phys. Rev. E* **61**, 4, pp. 4566–4575.

Atchade, Y. F. and Liu, J. S. (2010). The Wang-Landau algorithm for Monte-Carlo computation in general state spaces, *Stat. Sinica* **20**, 1, pp. 209–233.

Athènes, M. (2004). A path-sampling scheme for computing thermodynamic properties of a many-body system in a generalized ensemble, *Eur. Phys. J. B* **38**, pp. 651–663.

Atilgan, E. and Sun, S. X. (2004). Equilibrium free energy estimates based on nonequilibrium work relations and extended dynamics, *J. Chem. Phys.* **121**, pp. 10392–10400.

Babin, V., Roland, C. and Sagui, C. (2008). Adaptively biased molecular dynamics for free-energy calculations, *J. Chem. Phys.* **128**, p. 134101.

Bakry, D. (1997). On Sobolev and logarithmic Sobolev inequalities for Markov semigroups, in *New trends in stochastic analysis* (World Scientific), pp. 43–75.

Bakry, D. and Emery, M. (1985). *Séminaire de Probabilités XIX*, *Lect. Notes Math.*, Vol. 1123 (Springer-Verlag).

Balian, R. (2007). *From Microphysics to Macrophysics. Methods and Applications of Statistical Physics*, Vol. I - II (Springer).

Bally, V. and Talay, D. (1995). The law of the Euler scheme for stochastic differential equations: I. Convergence rate of the distribution function, *Probab. Theory Rel.* **104**, pp. 43–160.

Bally, V. and Talay, D. (1996). The law of the Euler scheme for stochastic differential equations: II. Convergence rate of the density, *Monte-Carlo Methods and Applications* **2**, pp. 93–128.

Barducci, A., Bussi, G. and Parrinello, M. (2008). Well-tempered metadynamics: A smoothly converging and tunable free-energy method, *Phys. Rev. Lett.* **100**, p. 020603.

Baskes, M. I. (1992). Modified embedded-atom potentials for cubic materials and impurities, *Phys. Rev. B* **46**, 5, pp. 2727–2742.

Bennett, C. H. (1976). Efficient estimation of free energy differences from Monte-Carlo data, *J. Comput. Phys.* **22**, pp. 245–268.

Blondel, A. (2004). Ensemble variance in free energy calculations by thermodynamic integration: theory, optimal "alchemical" path, and practical solutions, *J. Comput. Chem.* **25**, pp. 985–993.

Bobkov, S. and Götze, F. (1999). Exponential integrability and transportation cost related to logarithmic Sobolev inequalities, *J. Funct. Anal.* **163**, 1, pp. 1–28.

Bond, S. D., Leimkuhler, B. J. and Laird, B. B. (1999). The Nosé-Poincaré method for constant temperature molecular dynamics, *J. Comput. Phys.* **151**, 1, pp. 114–134.

Bonnans, J. F., Gilbert, J. C., Lemaréchal, C. and Sagastizábal, C. A. (2002). *Numerical optimization* (Springer).

Bornemann, F. and Schütte, C. (1992). Homogenization of Hamiltonian system with a strong constraining potential, *Physica D* **102**, pp. 57–77.

Bou-Rabee, N. and Vanden-Eijnden, E. (2009). Pathwise accuracy and ergodicity of metropolized integrators for SDEs, *Commun. Pure Appl. Math.* **63**, 5, pp. 655–696.

Brünger, A., Brooks, C. B. and Karplus, M. (1984). Stochastic boundary conditions for molecular dynamics simulations of ST2 water, *Chem. Phys. Lett.*

105, 5, pp. 495–500.

Bussi, G., Laio, A. and Parinello, M. (2006). Equilibrium free energies from nonequilibrium metadynamics, *Phys. Rev. Lett.* **96**, p. 090601.

Cancès, E., Castella, F., Chartier, P., Faou, E., Le Bris, C., Legoll, F. and Turinici, G. (2004). High-order averaging schemes with error bounds for thermodynamical properties calculations by molecular dynamics simulations, *J. Chem. Phys.* **121**, 21, pp. 10346–10355.

Cancès, E., Castella, F., Chartier, P., Faou, E., Le Bris, C., Legoll, F. and Turinici, G. (2005). Long-time averaging for integrable Hamiltonian dynamics, *Numer. Math.* **100**, 2, pp. 211–232.

Cancès, E., Defranceschi, M., Kutzelnigg, W., Le Bris, C. and Maday, Y. (2003). Computational quantum chemistry: A primer, in P. G. Ciarlet and C. Le Bris (eds.), *Handbook of Numerical Analysis (Special volume on computational chemistry)*, Vol. X (Elsevier), pp. 3–270.

Cancès, E., Legoll, F. and Stoltz, G. (2007). Theoretical and numerical comparison of sampling methods for molecular dynamics, *Math. Model. Numer. Anal.* **41**, 2, pp. 351–390.

Canuto, C., Hussaini, Y., Quarteroni, A. and Zang, T. (2006). *Spectral Methods: Fundamentals in Single Domains* (Springer-Verlag).

Car, R. and Parrinello, M. (1985). Unified approach for molecular dynamics and density-functional theory, *Phys. Rev. Lett.* **55**, 22, pp. 2471–2474.

Carter, E. A., Ciccotti, G., Hynes, J. T. and Kapral, R. (1989). Constrained reaction coordinate dynamics for the simulation of rare events, *Chem. Phys. Lett.* **156**, 5, pp. 472–477.

Chen, M.-H. and Shao, Q.-M. (1997). On Monte-Carlo methods for estimating ratios of normalizing constants, *Ann. Stat.* **25**, 4, pp. 1563–1594.

Chipot, C., Hénin, J. and Lelièvre, T. (2010). Generalized Adaptive Biasing Force methods, In preparation.

Chipot, C. and Pohorille, A. (2007a). Calculating free energy differences using perturbation theory, in C. Chipot and A. Pohorille (eds.), *Free energy calculations* (Springer), pp. 33–75.

Chipot, C. and Pohorille, A. (eds.) (2007b). *Free Energy Calculations*, *Springer Series in Chemical Physics*, Vol. 86 (Springer).

Ciccotti, G., Kapral, R. and Vanden-Eijnden, E. (2005). Blue Moon sampling, vectorial reaction coordinates, and unbiased constrained dynamics, *Chem. Phys. Chem.* **6**, 9, pp. 1809–1814.

Ciccotti, G., Lelièvre, T. and Vanden-Eijnden, E. (2008). Projection of diffusions on submanifolds: Application to mean force computation, *Commun. Pure Appl. Math.* **61**, 3, pp. 371–408.

Cohen, D., Jahnke, T., Lorenz, K. and Lubich, C. (2006). Numerical integrators for highly oscillatory Hamiltonian systems: A review, in A. Mielke (ed.), *Analysis, Modeling and Simulation of Multiscale Problems* (Springer), pp. 553–576.

Crooks, G. E. (1998). Nonequilibrium measurements of free energy-differences for microscopically reversible markovian systems, *J. Stat. Phys.* **90**, 5, pp. 1481–1487.

Crooks, G. E. (1999). Entropy production fluctuation theorem and the nonequilibrium work relation for free-energy differences, *Phys. Rev. E* **60**, 3, pp. 2721–2726.

Darve, E. (2007). Thermodynamic integration using constrained and unconstrained dynamics, in C. Chipot and A. Pohorille (eds.), *Free Energy Calculations* (Springer), pp. 119–170.

Darve, E. and Porohille, A. (2001). Calculating free energy using average forces, *J. Chem. Phys.* **115**, pp. 9169–9183.

Darve, E., Wilson, M. and Pohorille, A. (2002). Calculating free energies using a scaled-force molecular dynamics algorithm, *Mol. Sim.* **28**, 1-2, pp. 113–144.

Davies, E. (1982a). Dynamical stability of metastable states, *J. Funct. Anal.* **46**, 3, pp. 373–386.

Davies, E. (1982b). Metastable states of symmetric Markov semigroups I, *Proc. London Math. Soc.* **45**, 3, pp. 133–150.

Davies, E. (1982c). Metastable states of symmetric Markov semigroups II, *J. London Math. Soc.* **26**, 3, pp. 541–556.

Davies, E. (1983). Spectral properties of metastable Markov semigroups, *J. Funct. Anal.* **52**, pp. 315–329.

Del Moral, P. (2004). *Feynman-Kac Formulae, Genealogical and Interacting Particle Systems with Applications*, Probability and its Applications (Springer).

Del Moral, P. and Miclo, L. (2000). Branching and interacting particle systems approximations of Feynman-Kac formulae with applications to nonlinear filtering, *Lect. Notes Math.* **1729**, pp. 1–145.

Dellago, C., Bolhuis, P. G. and Chandler, D. (1999). On the calculation of reaction rate constants in the transition path ensemble, *J. Chem. Phys.* **110**, 14, pp. 6617–6625.

Dellago, C., Bolhuis, P. G. and Geissler, P. L. (2002). Transition path sampling, *Adv. Chem. Phys.* **123**, pp. 1–78.

Dembo, A. and Zeitouni, O. (1998). *Large Deviations Techniques and Applications* (Springer).

Demmel, J. (1997). *Applied Numerical Linear Algebra* (SIAM).

den Otter, W. and Briels, W. J. (1998). The calculation of free-energy differences by constrained molecular-dynamics simulations, *J. Chem. Phys.* **109**, 11, pp. 4139–4146.

den Otter, W. K. (2000). Thermodynamic integration of the free energy along a reaction coordinate in Cartesian coordinates, *J. Chem. Phys.* **112**, 17, pp. 7283–7292.

Dickson, B., Legoll, F., Lelivre, T., Stoltz, G. and Fleura-Lessard, P. (2010). Free energy calculations: An efficient adaptive biasing potential method, *J. Phys. Chem. B* **114**, 17, pp. 5823–5830.

Doucet, A., Del Moral, P. and Jasra, A. (2006). Sequential Monte-Carlo samplers, *J. Roy. Stat. Soc. B* **68**, 3, pp. 411–436.

Doucet, A., Freitas, N. D. and Gordon, N. J. (eds.) (2001). *Sequential Monte Carlo Methods in Practice*, Statistics for Engineering and Information Science (Springer).

Duane, S., Kennedy, A. D., Pendleton, B. J. and Roweth, D. (1987). Hybrid

Monte-Carlo, *Phys. Lett. B* **195**, 2, pp. 216–222.

Duflo, M. (1997). *Random Iterative Models* (Springer).

E, W. and Vanden-Eijnden, E. (2004). Metastability, conformation dynamics, and transition pathways in complex systems, in S. Attinger and P. Koumoutsakos (eds.), *Multiscale Modelling and Simulation, Lect. Notes Comput. Sci. Eng.*, Vol. 39 (Springer, Berlin), pp. 35–68.

Efron, B. (1979). Bootstrap methods: Another look at the jackknife, *Ann. Stat.* **7**, pp. 1–26.

El Makrini, M., Jourdain, B. and Lelièvre, T. (2007). Diffusion Monte-Carlo method: numerical analysis in a simple case, *Math. Model. Numer. Anal.* **41**, 2, pp. 189–213.

Ellis, R. S. (1985). *Entropy, Large Deviations and Statistical Mechanics* (Springer).

Ern, A. and Guermond, J.-L. (2004). *Theory and Practice of Finite Elements* (Springer-Verlag).

Ethier, S. N. and Kurtz, T. G. (1986). *Markov Processes: Characterization and Convergence*, Wiley Series in Probability and Statistics (John Wiley & Sons).

Evans, L. C. and Gariepy, R. F. (1992). *Measure Theory and Fine Properties of Functions*, Studies in Advanced Mathematics (CRC Press).

Faou, E. (2006). Nosé-Hoover dynamics in a shaker, *J. Chem. Phys.* **124**, p. 184104.

Faou, E. and Lelièvre, T. (2009). Conservative stochastic differential equations: Mathematical and numerical analysis, *Math. Comput.* **78**, pp. 2047–2074.

Ferrenberg, A. M. and Swendsen, R. H. (1989). Optimized Monte-Carlo data analysis, *Phys. Rev. Lett.* **63**, 12, pp. 1195–1198.

Fishman, G. S. (1996). *Monte Carlo: Concepts, Algorithms and Applications* (Springer).

Fixman, M. (1974). Classical statistical mechanics of constraints: A theorem and application to polymers, *Proc. Nat. Acad. Sci. USA* **71**, 8, pp. 3050–3053.

Fixman, M. (1978). Simulation of polymer dynamics. I. General theory, *J. Chem. Phys.* **69**, pp. 1527–1537.

Flyvbjerg, H. and Petersen, H. G. (1989). Error estimates on averages of correlated fata, *J. Chem. Phys.* **91**, pp. 461–466.

Freddolino, P. L., Arkhipov, A. S., Larson, S. B., McPherson, A. and Schulten, K. (2006). Molecular dynamics simulations of the complete satellite tobacco mosaic virus, *Structure* **14**, pp. 437–449.

Freidlin, M. (2004). Some remarks on the Smoluchowski-Kramers approximation, *J. Stat. Phys.* **117**, 3-4, pp. 617–634.

Freidlin, M. I. and Wentzell, A. D. (1998). *Random Perturbations of Dynamical Systems* (Springer).

Frenkel, D. and Smit, B. (2002). *Understanding Molecular Simulation, From Algorithms to Applications (2nd ed.)* (Academic Press).

Gallavotti, G. (1999). *Statistical Mechanics*, Texts and Monographs in Physics (Springer).

Gelman, A. and Meng, X. L. (1998). Simulating normalizing constants: from importance sampling to bridge sampling to path sampling, *Stat. Sci.* **13**, 2, pp. 163–185.

Geyer, C. J. (1991). Markov chain Monte-Carlo maximum likelihood, in E. Karamigas (ed.), *Computing Science and Statistics: The 23rd symposium on the interface* (Interface foundation), pp. 156–163.

Geyer, C. J. (1992). Practical Markov chain Monte Carlo (with discussion), *Stat. Sci.* **7**, 4, pp. 473–511.

Geyer, C. J. (1994). Estimating normalizing constants and reweighting mixtures in Markov chain Monte Carlo, Tech. Rep. 565, School of Statistics, University of Minnesota.

Gibbs, J. W. (1902). *Elementary Principles in Statistical Mechanics* (Yale University Press).

Gilks, W. R., Richardson, S. and Spiegelhalter, D. J. (1996). *Markov Chain Monte Carlo in Practice* (Chapman and Hall).

Glowinski, R. and Tallec, P. L. (1989). *Augmented Lagrangian and operator-splitting methods in nonlinear mechanics*, Studies in Applied Mathematics (SIAM).

Gore, J., Ritort, F. and Bustamante, C. (2003). Bias and error in estimates of equilibrium free-energy differences from nonequilibrium measurements, *Proc. Nat. Acad. Sci. USA* **100**, 22, pp. 12564–12569.

Gross, L. (1975). Logarithmic Sobolev inequalities, *Amer. J. Math.* **97**, 4, pp. 1061–1083.

Grunewald, N., Otto, F., Villani, C. and Westdickenberg, M. G. (2009). A two-scale approach to logarithmic Sobolev inequalities and the hydrodynamic limit, *Ann. Inst. H. Poincar Probab. Statist.* **45**, 2, pp. 302–351.

Hahn, A. M. and Then, H. (2009). Characteristic of Bennett's acceptance ratio method, *Phys. Rev. E* **80**, 3, p. 031111.

Hairer, E., Lubich, C. and Wanner, G. (2003). Geometric numerical integration illustrated by the Störmer-Verlet method, *Acta Numerica* **12**, pp. 399–450.

Hairer, E., Lubich, C. and Wanner, G. (2006). *Geometric Numerical Integration: Structure-Preserving Algorithms for Ordinary Differential Equations*, *Springer Series in Computational Mathematics*, Vol. 31 (Springer-Verlag).

Hartmann, C. (2008). An ergodic sampling scheme for constrained Hamiltonian systems with applications to molecular dynamics, *J. Stat. Phys.* **130**, 4, pp. 687–711.

Hartmann, C. and Schütte, C. (2005a). A constrained Hybrid Monte-Carlo algorithm and the problem of calculating the free energy in several variables, *Z. Angew. Math. Mech.* **85**, 10, pp. 700–710.

Hartmann, C. and Schütte, C. (2005b). A geometric approach to constrained molecular dynamics and free-energy, *Commun. Math. Sci.* **3**, 1, pp. 1–20.

Has'minskii, R. Z. (1980). *Stochastic Stability of Differential Equations* (Sijthoff and Noordhoff).

Hastings, W. K. (1970). Monte Carlo sampling methods using Markov chains and their applications, *Biometrika* **57**, pp. 97–109.

Hendrix, D. A. and Jarzynski, C. (2001). A "fast growth" method of computing free-energy differences, *J. Chem. Phys.* **114**, pp. 5974–5981.

Hénin, J. and Chipot, C. (2004). Overcoming free-energy barriers using unconstrained molecular dynamics simulations, *J. Chem. Phys.* **121**, pp. 2904–2914.

Hinch, E. J. (1994). Brownian motion with stiff bonds and rigid constraints, *J. Non-Newtonian Fluid Mech.* **271**, pp. 219–234.

Holley, R. and Stroock, D. (1987). Logarithmic Sobolev inequalities and stochastic Ising models, *J. Stat. Phys.* **46**, pp. 1159–1194.

Hoover, W. G. (1985). Canonical dynamics - Equilibrium phase-space distributions, *Phys. Rev. A* **31**, 3, pp. 1695–1697.

Hörmander, L. (1967). Hypoelliptic second order differential equations, *Acta Math.* **119**, pp. 147–171.

Horowitz, A. M. (1991). A generalized guided Monte Carlo algorithm, *Phys. Lett. B* **268**, pp. 247–252.

Huisinga, W. (2001). *Metastability of Markovian systems: A transfer operator approach in application to molecular dynamics*, Ph.D. thesis, Freie Universität Berlin.

Huisinga, W. and Schmidt, B. (2006). Metastability and dominant eigenvalues of transfer operators, in B. Leimkuhler, C. Chipot, R. Elber, A. Laaksonen, A. Mark, T. Schlick, C. Schütte and R. Skeel (eds.), *New Algorithms for Macromolecular Simulation* (Springer), pp. 167–182.

Hukushima, K. and Nemoto, K. (1996). Exchange Monte-Carlo method and application to spin glass simulations, *J. Phys. Soc. Jpn.* **65**, 6, pp. 1604–1608.

Hummer, G. (2007). Nonequilibrium methods for equilibrium free-energy calculations, in C. Chipot and A. Pohorille (eds.), *Free Energy Calculations* (Springer), pp. 171–198.

Hummer, G. and Szabo, A. (2001). Free-energy reconstruction from nonequilibrium single-molecule pulling experiments, *Proc. Nat. Acad. Sci. USA* **98**, 7, pp. 3658–3661.

Iannuzzi, M., Laio, A. and Parrinello, M. (2003). Efficient exploration of reactive potential energy surfaces using Car-Parrinello molecular dynamics, *Phys. Rev. Lett.* **90**, 23, p. 238302.

Ikeda, N. and Watanabe, S. (1989). *Stochastic Differential Equations and Diffusion Processes* (North-Holland).

Jarzynski, C. (1997a). Equilibrium free-energy differences from nonequilibrium measurements: A master-equation approach, *Phys. Rev. E* **56**, 5, pp. 5018–5035.

Jarzynski, C. (1997b). Nonequilibrium equality for free energy differences, *Phys. Rev. Lett.* **78**, 14, pp. 2690–2693.

Jarzynski, C. (2006). Rare events and the convergence of exponentially averaged work values, *Phys. Rev. E* **73**, p. 046105.

Jasra, A., Stephens, D. A. and Holmes, C. C. (2007). On population-based simulation for static inference, *Stat. Comput.* **17**, 3, pp. 263–279.

Jourdain, B., Lelièvre, T. and Roux, R. (2009). Existence, uniqueness and convergence of a particle approximation for the Adaptive Biasing Force process, Accepted for publication in *Math. Model. Numer. Anal.*

Karatzas, I. and Shreve, S. E. (1988). *Brownian Motion and Stochastic Calculus*

(Springer).

Kirkwood, J. G. (1935). Statistical mechanics of fluid mixtures, *J. Chem. Phys.* **3**, 5, pp. 300–313.

Kliemann, W. (1987). Recurrence and invariant measures for degenerate diffusions, *Ann. Probab.* **15**, 2, pp. 690–707.

Kong, A., McCullagh, P., Meng, X. L., Nicolae, D. and Tan, Z. (2003). A theory of statistical models for Monte-Carlo integration, *J. Roy. Stat. Soc. B* **65**, 3, pp. 585–618.

Kumar, S., Bouzida, D., Swendsen, R. H., Kollman, P. A. and Rosenberg, J. M. (1992). The weighted histogram analysis method for free-energy calculations on biomolecules. I. The method, *J. Comput. Chem.* **13**, 8, pp. 1011–1021.

Kupferman, R., Stuart, A. M., Terry, J. R. and Tupper, P. F. (2002). Long-term behaviour of large mechanical systems with random initial data, *Stoch. Dynam.* **2**, 4, pp. 1–30.

Kushner, H. J. S. (1984). *Approximation and Weak Convergence Methods for Random Processes, with Applications to Stochastic Systems Theory* (MIT Press).

Laio, A. and Parrinello, M. (2002). Escaping free-energy minima, *Proc. Natl. Acad. Sci. U.S.A* **99**, pp. 12562–12566.

Lapeyre, B., Pardoux, E. and Sentis, R. (2003). *Introduction to Monte-Carlo Methods for Transport and Diffusion Equations* (Oxford).

Laudenbach, F. (2001). *Calcul différentiel et intégral* (Ecole Polytechnique), in French.

Le Bris, C. and Legoll, F. (2007). Derivation of symplectic numerical schemes for highly oscillatory Hamiltonian systems, *C. R. Acad. Sci. Paris Série I* **344**, 4, pp. 277–282.

Lechner, W. and Dellago, C. (2007). On the efficiency of path sampling methods for the calculation of free energies from non-equilibrium simulations, *J. Stat. Mech.-Theory E.* , p. P04001.

Legoll, F., Luskin, M. and Moeckel, R. (2007). Non-ergodicity of the Nosé-Hoover thermostatted harmonic oscillator, *Arch. Ration. Mech. Anal.* **184**, pp. 449–463.

Legoll, F., Luskin, M. and Moeckel, R. (2009). Non-ergodicity of Nosé-Hoover dynamics, *Nonlinearity* **22**, pp. 1673–1694.

Leimkuhler, B. J. and Reich, S. (2005). *Simulating Hamiltonian Dynamics, Cambridge Monographs on Applied and Computational Mathematics*, Vol. 14 (Cambridge University Press).

Leimkuhler, B. J. and Skeel, R. D. (1994). Symplectic numerical integrators in constrained Hamiltonian systems, *J. Comput. Phys.* **112**, 1, pp. 117–125.

Lelièvre, T. (2009). A general two-scale criteria for logarithmic Sobolev inequalities, *J. Funct. Anal.* **256**, 7, pp. 2211–2221.

Lelièvre, T. and Minoukadeh, K. (2010). Longtime convergence of an adaptive biasing force method: The bi-channel case, *Hal preprint* **00477302**.

Lelièvre, T., Rousset, M. and Stoltz, G. (2007a). Computation of free energy differences through nonequilibrium stochastic dynamics: The reaction coordinate case, *J. Comput. Phys.* **222**, 2, pp. 624–643.

Lelièvre, T., Rousset, M. and Stoltz, G. (2007b). Computation of free energy profiles with adaptive parallel dynamics, *J. Chem. Phys.* **126**, p. 134111.

Lelièvre, T., Rousset, M. and Stoltz, G. (2008). Long-time convergence of an Adaptive Biasing Force method, *Nonlinearity* **21**, pp. 1155–1181.

Lelièvre, T., Rousset, M. and Stoltz, G. (2010). Langevin dynamics with constraints and computation of free energy differences, In preparation.

Liu, J. S. (2001). *Monte Carlo strategies in scientific computing* (Springer).

Lu, N. and Kofke, D. A. (2001). Accuracy of free-energy perturbation calculations in molecular simulation. I. Modeling, *J. Chem. Phys.* **114**, 17, pp. 7303–7311.

Mackenzie, P. B. (1989). An improved hybrid Monte Carlo method, *Phys. Lett. B* **226**, 3-4, pp. 369–371.

Maragliano, L. and Vanden-Eijnden, E. (2006). A temperature accelerated method for sampling free energy and determining reaction pathways in rare events simulations, *Chem. Phys. Lett.* **426**, 1-3, pp. 168–175.

Marinari, E. and Parisi, G. (1992). Simulated tempering - a new Monte-Carlo scheme, *Europhys. Lett.* **19**, 6, pp. 451–458.

Markowich, P. A. and Villani, C. (2000). On the trend to equilibrium for the Fokker-Planck equation: An interplay between physics and functional analysis, *Mat. Contemp.* **19**, pp. 1–29.

Marsili, S., Barducci, A., Chelli, R., Procacci, P. and Schettino, V. (2006). Self-healing Umbrella Sampling: A non-equilibrium approach for quantitative free energy calculations, *J. Phys. Chem. B* **110**, 29, pp. 14011–14013.

Mattingly, J. C., Stuart, A. M. and Higham, D. J. (2002). Ergodicity for SDEs and approximations: locally Lipschitz vector fields and degenerate noise, *Stoch. Process. Appl.* **101**, 2, pp. 185–232.

Mazonka, O. and Jarzynski, C. (1999). Exactly solvable model illustrating far-from-equilibrium predictions, arXiv:condmat **9912121**.

Meng, X.-L. and Wong, W. H. (1996). Simulating ratios of normalizing constants via a simple identity: A theoretical exploration, *Stat. Sinica* **6**, pp. 831–860.

Metropolis, N., Rosenbluth, A. W., Rosenbluth, M. N., Teller, A. H. and Teller, E. (1953). Equations of state calculations by fast computing machines, *J. Chem. Phys.* **21**, 6, pp. 1087–1091.

Meyn, S. P. and Tweedie, R. L. (1993). *Markov Chains and Stochastic Stability*, Communications and Control Engineering Series (Springer).

Minh, D. D. L. and Adib, A. B. (2008). Optimized free energies from bidirectional single-molecule force spectroscopy, *Phys. Rev. Lett.* **100**, p. 180602.

Minlos, R. A. (2000). *Introduction to Mathematical Statistical Physics*, *University Lecture Series*, Vol. 19 (American Mathematical Society).

Minoukadeh, K., Chipot, C. and Lelièvre, T. (2010). Potential of mean force calculations: A multiple-walker adaptive biasing force approach, *J. Chem. Th. Comput.* **6**, 4, pp. 1008–1017.

Morse, D. (2004). Theory of constrained Brownian motion, *Adv. Chem. Phys.* **128**, pp. 65–189.

Niklasson, A. M. N., Tymczak, C. J. and Challacombe, M. (2006). Time-reversible Born-Oppenheimer molecular dynamics, *Phys. Rev. Lett.* **97**, p. 123001.

Nosé, S. (1984). A unified formulation of the constant temperature molecular-dynamics methods, *J. Chem. Phys.* **81**, 1, pp. 511–519.

Oberhofer, H. and Dellago, C. (2008). Optimum bias for fast-switching free energy calculations, *Comput. Phys. Commun.* **179**, pp. 41–45.

Oberhofer, H., Dellago, C. and Geissler, P. L. (2005). Biased sampling of nonequilibrium trajectories: Can fast switching simulations outperform conventional free-energy calculation methods? *J. Phys. Chem. B* **109**, 14, pp. 6902–6915.

Oksendal, B. (1992). *Stochastic Differential Equations: An Introduction with Applications (3rd ed.)* (Springer).

Öttinger, H. C. (1995). *Stochastic Processes in Polymeric Fluids* (Springer).

Otto, F. and Reznikoff, M. G. (2007). A new criterion for the logarithmic Sobolev inequality and two applications, *J. Funct. Anal.* **243**, pp. 121–157.

Otto, F. and Villani, C. (2000). Generalization of an inequality by Talagrand and links with the logarithmic Sobolev inequality, *J. Funct. Anal.* **173**, 2, pp. 361–400.

Peskun, P. (1973). Optimum Monte Carlo sampling using Markov chains, *Biometrika* **60**, pp. 607–612.

Peters, F. (2000). *Polymers in flow, modelling and simulation*, Ph.D. thesis, TU Delft.

Plechac, P. and Rousset, M. (2010). Exact and non-stiff sampling of highly oscillatory systems: An implicit mass-matrix penalization approach, *Multiscale Model. Simul.* **8**, pp. 498–539.

Rapaport, D. C. (1995). *The Art of Molecular Dynamics Simulations* (Cambridge University Press).

Reed, M. and Simon, B. (1975). *Methods of Modern Mathematical Physics. II. Fourier Analysis and Self-Adjointness* (Academic Press).

Reich, S. (1995). Smoothed dynamics of highly oscillatory Hamiltonian systems, *Physica D* **89**, pp. 28–42.

Reich, S. (2000). Smoothed Langevin dynamics of highly oscillatory systems, *Physica D* **138**, pp. 210–224.

Rey-Bellet, L. (2006). Open classical systems, *Lect. Notes Math.* **1881**, pp. 41–78.

Rickman, J. M. and LeSar, R. (2002). Free-energy calculations in materials research, *Ann. Rev. Mater. Res.* **32**, pp. 195–217.

Roberts, G. O., Gelman, A. and Gilks, W. R. (1997). Weak convergence and optimal scaling of random walk Metropolis algorithms, *Ann. Appl. Probab.* **7**, pp. 110–120.

Roberts, G. O. and Rosenthal, J. S. (1998). Optimal scaling of discrete approximations to Langevin diffusions, *J. Roy. Stat. Soc. B* **60**, pp. 255–268.

Rousset, M. (2006a). *Continuous time "Population Monte Carlo" and Computational Physics*, Ph.D. thesis, Université Paul Sabatier.

Rousset, M. (2006b). On the control of an interacting particle approximation of Schrödinger groundstates, *SIAM J. Math. Anal.* **38**, 3, pp. 824–844.

Rousset, M. and Stoltz, G. (2006). Equilibrium sampling from nonequilibrium dynamics, *J. Stat. Phys.* **123**, 6, pp. 1251–1272.

Royer, G. (2007). *An Initiation to Logarithmic Sobolev Inequalities* (American

Mathematical Society).

Rubin, H. and Ungar, P. (1957). Motion under a strong constraining force, *Commun. Pure Appl. Math.* **10**, pp. 65–87.

Rubinstein, R. (1981). *Simulation and the Monte Carlo Method* (Wiley Series in Probability and Statistics).

Ryckaert, J. P. and Bellemans, A. (1978). Molecular-dynamics of liquid alkanes, *Faraday Discuss.* **66**, pp. 95–106.

Scemama, A., Lelièvre, T., Stoltz, G., Cancès, E. and Caffarel, M. (2006). An efficient sampling algorithm for Variational Monte-Carlo, *J. Chem. Phys.* **125**, p. 114105.

Schlick, T. (2002). *Molecular Modeling and Simulation* (Springer).

Schlitter, J. (1991). Methods for minimizing errors in linear thermodynamic integration, *Molecular Simulation* **7**, 1, pp. 105–112.

Schmiedl, T. and Seifert, U. (2007). Optimal finite-time processes in stochastic thermodynamics, *Phys. Rev. Lett.* **98**, p. 108301.

Schütte, C. (1999). *Habilitation Thesis* (Freie Universitt Berlin).

Shirts, M. R., Bair, E., Hooker, G. and Pande, V. S. (2003). Equilibrium free energies from nonequilibrium measurements using maximum-likelihood methods, *Phys. Rev. Lett.* **91**, p. 140601.

Shirts, M. R. and Chodera, J. D. (2008). Statistically optimal analysis of samples from multiple equilibrium states, *J. Chem. Phys.* **124**, 12, p. 124105.

Sprik, M. and Ciccoti, G. (1998). Free energy from constrained molecular dynamics, *J. Chem. Phys.* **109**, 18, pp. 7737–7744.

Stoltz, G. (2007). Path sampling with stochastic dynamics: Some new algorithms, *J. Comput. Phys.* **225**, pp. 491–508.

Straub, J. E., Borkovec, M. and Berne, B. J. (1988). Molecular-dynamics study of an isomerizing diatomic in a Lennard-Jones fluid, *J. Chem. Phys.* **89**, 8, pp. 4833–4847.

Stroock, D. W. and Varadhan, S. R. S. (1979). *Multidimensional Diffusion Processes, Grundlehren der Mathematischen Wissenschaften,* Vol. 233 (Springer).

Sun, S. X. (2003). Equilibrium free energies from path sampling of nonequilibrium trajectories, *J. Chem. Phys.* **118**, 13, pp. 5769–5775.

Takens, F. (1980). Motion under the influence of a strong constraining force, in *Global theory of dynamical systems (Proc. Internat. Conf., Northwestern Univ., Evanston, Ill., 1979), Lect. Notes Math.*, Vol. 819 (Springer), pp. 425–445.

Talay, D. (1991). Approximation of upper Lyapunov exponents of bilinear stochastic differential systems, *SIAM J. Numer. Anal.* **28**, 4, pp. 1141–1164.

Talay, D. and Tubaro, L. (1990). Expansion of the global error for numerical schemes solving stochastic differential equations, *Stoch. Anal. Appl.* **8**, 4, pp. 483–509 (1991).

Tan, Z. (2004). On a likelihood approach for Monte-Carlo integration, *J. Am. Stat. Assoc.* **99**, 468, pp. 1027–1036.

Tersoff, J. (1989). Modeling solid-state chemistry: Interatomic potentials for multicomponent systems, *Phys. Rev. B* **39**, pp. 5566–5568.

Torrie, G. M. and Valleau, J. P. (1974). Monte-Carlo free energy estimates using non-Boltzmann sampling: Application to the sub-critical Lennard-Jones fluid, *Chem. Phys. Lett.* **28**, 4, pp. 578–581.

Torrie, G. M. and Valleau, J. P. (1977). Non-physical sampling distributions in Monte-Carlo free-energy estimation - Umbrella sampling, *J. Comput. Phys.* **23**, 2, pp. 187–199.

Trefethen, L. N. and Bau, D. (1997). *Numerical Linear Algebra* (SIAM).

Vaikuntanathan, S. and Jarzynski, C. (2008). Escorted free energy simulations: Improving convergence by reducing dissipation, *Phys. Rev. Lett.* **100**, p. 190601.

van der Vaart, A. W. (1998). *Asymptotic Statistics* (Cambridge University Press).

van Duin, A. C. T., Dasgupta, S., Lorant, F. and Goddard III, W. A. (2001). ReaxFF: A reactive force field for hydrocarbons, *J. Phys. Chem. A* **105**, pp. 9396–9409.

van Kampen, N. G. (1981). Statistical mechanics for trimers, *App. Sci. Res.* **37**, pp. 67–75.

van Kampen, N. G. (1985). Elimination of fast variables, *Phys. Rep.* **124**, 2, pp. 9–160.

Verlet, L. (1967). Computer "experiments" on classical fluids. I. Thermodynamical properties of Lennard-Jones molecules, *Phys. Rev.* **159**, pp. 98–103.

Villani, C. (2002). A review of mathematical topics in collisional kinetic theory, in *Handbook of Mathematical Fluid Dynamics, Vol. I* (North-Holland, Amsterdam), pp. 71–305.

Villani, C. (2003). *Topics in Optimal Transportation, Graduate Studies in Mathematics*, Vol. 58 (American Mathematical Society).

Villani, C. (2009). Hypocoercivity, *Mem. Am. Math. Soc.* **202**, 950.

Wang, F. and Landau, D. (2001a). Determining the density of states for classical statistical models: A random walk algorithm to produce a flat histogram, *Phys. Rev. E* **64**, p. 056101.

Wang, F. G. and Landau, D. P. (2001b). Efficient, multiple-range random walk algorithm to calculate the density of states, *Phys. Rev. Lett.* **86**, 10, pp. 2050–2053.

Williams, D. (1991). *Probability with Martingales* (Cambridge University Press).

Ytreberg, F. M., Swendsen, R. H. and Zuckerman, D. M. (2006). Comparison of free energy methods for molecular systems, *J. Chem. Phys.* **125**, p. 184114.

Ytreberg, F. M. and Zuckerman, D. M. (2004). Single-ensemble nonequilibrium path-sampling estimates of free energy differences, *J. Chem. Phys.* **120**, 23, pp. 10876–10879.

Zheng, H. and Zhang, Y. (2008). Determination of free energy profiles by repository based adaptive Umbrella sampling: Bridging nonequilibrium and quasiequilibrium simulations, *J. Chem. Phys.* **128**, p. 204106.

Zuckerman, D. M. and Woolf, T. B. (2002). Theory of a systematic computational error in free energy differences, *Phys. Rev. Lett.* **89**, p. 180602.

Zuckerman, D. M. and Woolf, T. B. (2004). Systematic finite-sampling inaccuracy in free energy differences and other nonlinear quantities, *J. Stat. Phys.* **114**, 5-6, pp. 1303–1323.

Zwanzig, R. (1973). Nonlinear generalized Langevin equations, *J. Stat. Phys.* **9**, pp. 215–220.

Zwanzig, R. W. (1954). High-temperature equation of state by a perturbation method I. Nonpolar gases, *J. Chem. Phys.* **22**, 8, pp. 1420–1426.

Index

Lightning Source UK Ltd.
Milton Keynes UK
UKHW022246040321
379802UK00004B/160

9 781848 162471